Cahiers de Logique et d'Épistémologie
Volume 3

Hugh MacColl et la Naissance du Pluralisme Logique
suivi d'extraits majeurs de son oeuvre

Volume 1
Prolog, tout de suite!
Patrick Blackburn, Johan Bos et Kristina Striegnitz

Volume 2
Gottlob Frege. Une Introduction
Markus Stepanians
Traduit de l'allemand par Alexandre Thiercelin

Volume 3
Hugh MacColl et la Naissance du Puralisme Logique: suivi d'extraits majeurs de son oeuvre
Shahid Rahman et Juan Redmond
Traduit par Sébastien Magnier

Volume 4
Lecture de Quine
François Rivenc

Cahiers de Logique et d'Épistémologie Series Editors
Dov Gabbay   dov.gabbay@kcl.ac.uk
Shahid Rahman   shahid.rahman@univ-lille3.fr

Assistance Technique
Juan Redmond   juanredmond@yahoo.fr

Comité Scientifique: Daniel Andler (Paris – ENS); Diderik Baetens (Gent); Jean Paul van Bendegem (Vrije Universiteit Brussel); Johan van Benthem (Amsterdam/Stanford); Walter Carnielli (Campinas-Brésil); Pierre Cassou-Nogues (Lille 3 – UMR 8163-CNRS); Jacque Dubucs (Paris 1); Jean Gayon (Paris 1); François De Gandt (Lille 3 – UMR 8163-CNRS); Paul Gochet (Liège); Gerhard Heinzmann (Nancy 2); Andreas Herzig (Université de Toulouse – IRIT: UMR 5505-NRS); Bernard Joly (Lille 3 – UMR 8163-CNRS); Claudio Majolino (Lille 3 – UMR 8163-CNRS); David Makinson (London School of Economics); Gabriel Sandu (Paris 1); Hassan Tahiri (Lille 3 – UMR 8163-CNRS)

Hugh MacColl et la Naissance du Pluralisme Logique

suivi d'extraits majeurs de son oeuvre

Shahid Rahman et Juan Redmond

Traduit par Sébastien Magnier

© Individual author and College Publications 2008. All rights reserved.

ISBN 978-1-904987-82-6

College Publications
Scientific Director: Dov Gabbay
Managing Director: Jane Spurr
Department of Computer Science
King's College London, Strand, London WC2R 2LS, UK

http://www.collegepublications.co.uk

Original cover design by orchid creative   www.orchidcreative.co.uk
Printed by Lightning Source, Milton Keynes, UK

All rights reserved. No part of this publication may be reproduced, stored in a retrieval system or transmitted in any form, or by any means, electronic, mechanical, photocopying, recording or otherwise without prior permission, in writing, from the publisher.

A Nargis et Ashraf
Shahid Rahman

A Delia et Aldo
Juan Redmond

Mot du traducteur

Hugh MacColl est surtout connu aujourd'hui pour le versant logique de son œuvre. Pourtant celle-ci comporte une partie littéraire, certes moins conséquente, mais non moins intéressante. Par le biais de M. Stranger, personnage de l'un de ses romans, MacColl a voulu, dans une visée pédagogique, rendre la science et les découvertes de son temps accessibles à un large public. C'est dans ce même esprit que s'insère cette collection : permettre une plus large diffusion des idées et concepts à l'œuvre dans la logique et la philosophie des sciences contemporaines, aujourd'hui majoritairement publiées en langue anglo-saxonne. A l'occasion, de la publication de *Hugh MacColl : An Overview of his Logical Work with Anthology*, il m'a été proposé par le Professeur Shahid Rahman de prendre en charge la présente traduction. Je tiens ici à lui témoigner toute ma gratitude pour m'avoir fait confiance en me proposant de traduire pour la première fois en langue française un texte sur le système formel de Hugh MacColl. Mes remerciements s'adressent aussi à Juan Redmond, qui, tout au long de ce travail, a toujours répondu présent à mes nombreuses sollicitations. Je le remercie pour sa patience ainsi que pour ses conseils avisés. Je remercie également les fils de Paul pour la relecture attentive et minutieuse dont ils firent preuve.

Sébastien Magnier

TABLE DES MATIERES

INTRODUCTION.. xiii
    HUGH MACCOLL ET LA NAISSANCE DU PLURLISME
    LOGIQUE.. xiii
    NOTES BIOGRAPHIQUES xvii

1. LES ELEMENTS DE LA PHILOSOPHIE DE LA
    LOGIQUE ET DU LANGAGE DE MACCOLL 1
    1.1 DEUX NOTIONS ELEMENTAIRES DANS LA
        GRAMMAIRE FORMELLE DE MACCOLL : *ENONCE*
        ET *PROPOSITION*. 3
    1.2 LA LOGIQUE DE LA NON-EXISTENCE DE
        MACCOLL.. 12
    1.2.1 Le domaine symbolique et sa dynamique 12
    1.2.2. Propositions avec et sans engagement ontologique 17

2. LA PHILOSOPHIE DU CONDITIONNEL 20
    2.1. L'INVENTION DE LA LOGIQUE MODALE T : LA
        CRITIQUE DE L'IMPLICATION PAR MACCOLL 20
    2.2. VERS UNE LOGIQUE CONNEXE ET PERTINENTE 29
    2.2.1 La logique connexe 29
    2.2.2. La logique connexe de MacColl et ses recherches
        sur la pertinence 33
    2.2.2.1. Connexivité et subalternation 33
    2.2.2.2. Vers la pertinence suivie de remarques conclusives 35

3. TABLEAU POUR LA LOGIQUE CONNEXE DE
    MACCOLL.. 37
    3.1. INTRODUCTION... 37
    3.2. TABLEAU POUR LA LOGIQUE CONNEXE 38
        DEFINITION 40
        Le conditionnel connexe... 44
        Règles de tableau pour le conditionnel connexe 44
    3.3. LA SEMANTIQUE DE KRIPKE POUR LA LOGIQUE
        CONNEXE 51
    3.4. PERTINENCE ET CONNEXIVITE... 55

Remerciements 57

APPENDICE 58
A) MacColl et la fiction littéraire 58
B) MacColl et la probabilité conditionnelle 62

BIBLIOGRAPHIE DE MACCOLL 67
Références 78
Index des noms 82

EXTRAITS MAJEURS DE L'OEUVRE DE HUGH
MACCOLL 85

[1906a]: Symbolic Logic and its Applications,
Longmans, Green and Co., London. 87
[1877o]: The Calculus of Equivalent Statements and
Integration Limits. Proceedings of the London
Mathematical Society, (1877-1878), vol. 9, pp. 9-20. 237
[1877p]: The Calculus of Equivalent Statements (II).
Proceedings of the London Mathematical Society,
vol. 9, pp. 177-186. 249
[1878d]: The Calculus of Equivalent Statements (III).
Proceedings of the London Mathematical Society,
(1878-1879), vol. 10, pp. 16-28. 259
[1878e]: Symbolical or Abbreviated Language, with an
Application to Mathematical Probability. *The
Educational Times (Reprint)*, vol. 28, pp. 20-23. 272
[1878f]: Symbolical Language:--No. 2. *The
Educational Times (Reprint)*, vol. 28, p.100. 276
[1879e]: The Calculus of Equivalent Statements (IV).
Proceedings of the London Mathematical Society,
vol. 11, pp. 113-21. 277
[1880n]: On the Diagrammatic and Mechanical
Representation of Propositions and Reasoning. The
London, Edinburgh and Dublin philosophical
Magazine and Journal of Science, vol. 10, pp. 168-
171. 286

[1880o]: Implication and Equational Logic. The London, Edinburgh and Dublin philosophical Magazine and Journal of Science, vol. 11, pp. 40-43. 290
[1880p]: Symbolical Reasoning (I). *Mind,* vol. 5, pp. 45-60. 295
[1897a] Symbolic Reasoning(II). *Mind,* Vol. 6, pp. 493-510. 311
[1901f] La Logique Symbolique et ses Applications. Bibliothèque du 1° Congrès International de Philosophie. Logique et Histoire des Sciences, pp. 135-183. 329
[1902i]: Symbolic[al] Reasoning (IV). *Mind,* Vol. 11, pp. 352-368. 380
[1903i]: Symbolic[al] Reasoning (V). *Mind,* Vol. 12, pp. 355-364. 397
[1905o]: Symbolic[al] Reasoning (VI). *Mind,* Vol. 14, pp. 74-81. 407
[1905p]: Existential Import. *Mind,* Vol. 14, pp. 295-296. 415
[1905q]: Symbolic[al] Reasoning (VII). *Mind,* Vol. 14, pp. 390-397 416
[1905r]: The Existential Import of Propositions [A Reply to Bertrand Russell]. *Mind,* Vol. 14, pp. 401-402. 424
[1905s]: The Existential Import of Propositions. *Mind,* Vol. 14, No. 56. pp. 578-580. 429
[1906k]: Symbolic Reasoning (VIII). *Mind,* Vol. 15, pp. 504-518. 432
[1908b]: 'If' and 'Imply'. *Mind,* Vol. 17, pp. 151-152. 447
[1908c]: 'If' and 'Imply'. *Mind,* Vol. 17, pp. 453-455. 449
[1910c]: Linguistic Misunderstandings (I). *Mind,* Vol. 19, pp. 186-199. 452
[1910d]: Linguistic Misunderstandings (II). *Mind,* Vol. 19, pp. 337-355. 467

Introduction

Hugh MacColl et la naissance du pluralisme logique

Hugh MacColl (1837-1909) fut un mathématicien et logicien qui passa les premières années de sa vie en Écosse. Après quelques années de travail en différents lieux de Grande-Bretagne, il s'installa à Boulogne-sur-Mer (France), où il développa la majeure partie de son œuvre et devint citoyen français. Hugh MacColl fut connu en son temps pour ses contributions novatrices dans le monde de la logique. En dépit du fait que l'on pourrait difficilement dire que son travail satisfasse la rigueur de la philosophie des mathématiques et de la logique de Frege, MacColl représente la première approche du pluralisme logique. Sa première contribution à l'algèbre logique du 19ième siècle fut son calcul qui autorise non seulement une classe d'interprétation (comme dans l'algèbre de Boole) mais aussi une interprétation propositionnelle. Qui plus est, MacColl donna sa préférence à l'interprétation propositionnelle en raison de sa généralité et l'appela logique pure. Le connecteur principal de sa logique pure est le conditionnel et par conséquent, son algèbre contient un opérateur spécifique pour ce connecteur. Dans *Symbolic Logic and its Applications* (1906) (réimprimé dans le présent volume), MacColl publia la version achevée de sa(ses) logique(s) où des propositions sont qualifiées soit de certaines, soit de impossibles, soit de contingentes, ou encore de vraies ou de fausses. Après sa mort, son travail subit un triste destin. Contrairement aux autres logiciens qui lui étaient contemporains tels que L. Couturat, G. Frege, W.S. Jevons, J. Venn, G. Peano, C.S. Pierce, B. Russell et E. Schröder, qui connaissaient le travail de MacColl, ses contributions au monde logique ne semblent pas avoir reçues ni les remerciements ni les études systématiques qu'elles auraient méritées. Plus encore, nombre de ses idées furent attribuées à ses successeurs ; les exemples les plus connus sont : la notion d'implication stricte, la première approche formelle de la logique modale et la discussion des paradoxes de l'implication matérielle, habituellement attribuée à C.I. Lewis. Il en va de même pour ce qui est de ses contributions à la logique probabiliste (probabilité conditionnelle), logique plurivalente (relationnelle), logique de la pertinence et logique connexe. Le fait qu'il ait aussi exploré la possibilité de construire un système formel capable de raisonner avec des fictions est moins connu. Ce dernier point semble être lié à sa reconstruction formelle du syllogisme aristotélicien par le biais de la logique connexe.

Deux raisons majeures semblent avoir été déterminantes dans le fait que son travail soit tombé dans l'oubli. L'une s'apparente à des problèmes techniques et l'autre à la position philosophique qu'il adopte.

La forte influence de la méthodologie de la logique, amorcée par le travail de Frege immédiatement après la mort de MacColl, justifie le premier facteur. De fait, la méthode logique de présentation d'un système logique comme un ensemble d'axiomes clos sous une relation de conséquence, instiguée par Frege puis développée plus en profondeur par Peano, Russell et d'autres, remplaça rapidement la méthode algébrique de calcul du 19ième siècle, employée par MacColl.

La seconde raison est liée à la philosophie de sa logique. Les idées philosophiques de MacColl étaient basées sur une sorte d'instrumentalisme étendu au delà des deux paradigmes principaux de la logique formelle du 19ième, respectivement la mathématique comme logique (logicisme) et la logique comme algèbre (approche algébrique de Boole). Il est intéressant de noter que sa position philosophique se rapproche plus du conventionnalisme et de l'instrumentalisme français tels qu'ils sont développés par ses jeunes contemporains Henri Poincaré et Pierre Duhem ainsi que du pragmatisme américain de Charles Saunders Peirce que de l'empirisme ou encore du logicisme.

Au delà de ses contributions scientifiques, MacColl s'intéressa à la littérature. En suivant l'esprit de son siècle, il publia deux romans, *Mr Stranger's Sealed Packet* (1889) et *Ednor Whitlock* (1891) ainsi qu'un essai *Man's Origin, Duty and Destiny* (1909). Le premier est un roman de science fiction, un voyage sur Mars. Il s'agit là du troisième roman sur Mars publié en anglais. Dans ces deux derniers travaux, MacColl discute du conflit entre science et religion et des problèmes liés à la foi, au doute et à l'incroyance.

Il est impossible de résister à la tentation de comparer la contribution scientifique de MacColl avec ses incursions dans le domaine littéraire. Son œuvre renferme à la fois le plus conservateur des livres de science fiction de l'époque victorienne et un des propos les plus novateurs pour la logique du 19ième siècle.

Permettez-nous de retracer, brièvement, le regain d'intérêt pour le travail de MacColl. Durant les années soixante, Storrs McCall (1963-1967) concentre son attention sur les travaux de MacColl, particulièrement sur les paradoxes de l'implication matérielle, la logique connexe ainsi que sa reconstruction du syllogisme de Aristote[1]. Entre les années quatre-vingts et quatre-vingt-dix, à l'université Erlangen-Nuremberg, Christian Thiel et ses collaborateurs redécouvrent le travail de MacColl à l'occasion d'un projet de recherche historico-social sur la logique du 19ième siècle. Le projet déclenche la première exploration systématisée du travail de Hugh MacColl par Christian Thiel (1966), Volker Peckhaus (1986) et Anthony Christie (1986, 1990). Le groupe de recherche de l'université Erlangen-Nuremberg incite Michael Astroh à approfondir les recherches sur la logique et la philosophie du langage de Hugh MacColl (1993, 1995, 1996, 1999), comme le fit également Shahid Rahman (1997a, 1997b, 1999, 2000). Ensemble Michael Astroh et Shahid Rahman conçurent un atelier, qui finalement prit forme dans Greifswald, « Hugh MacColl and the Tradition of Logic » avec la participation de Stephen Read, Peter Simons, Volker Peckhaus, Göran Sundholm, Christian Thiel et Ian Wolenski. L'atelier a abouti à un numéro spécial du *Nordic Journal of Philosophical Logic*, édité par Michael Astroh et Stephen Read (1999). Ce volume constitue, après les travaux de MacColl, la source principale de notre propos.

---

[1] Cf. également Spencer [1973].

Notes biographiques

Malheureusement il n'y a pas de documents sur la vie de MacColl. Nous nous fierons donc aux informations contenues dans le second contrat de son mariage, signé de sa propre main. Nous suivrons ici la méthode utilisée dans la biographie de MacColl publiée en 2001 par M. Astroh, I. Grattan-Guinness et S. Read qui avec l'article de Rahman (1997b) contiennent les contours principaux de la présente biographie.

Hugh MacColl naquit à Strontian, Argyllshire, le 11 ou 12 Janvier 1837. Il fut le fils de John MacColl et de Martha Marc Rae. Hugh fut leur plus jeune enfant. Il avait trois frères et deux sœurs. Son père était un berger et agriculteur du Glencamgarry dans le Kilmalie (entre Glenfinnan et Fort William), il se maria avec Martha Mac Rae dans la paroisse de Kilmalie le 6 Février 1823.

Le décès prématuré du père de Hugh eut un impact considérable sur toute la famille. Le père était âgé de 45 ans et le petit Hugh seulement de trois ans. Après sa mort, la mère et les enfants

MacColl en 1886
(avec la bienveillance de Michael Astroh)

partirent pour Letterfearn puis pour Ballachulish. Ce n'est qu'à partir de ce moment que les enfants se mirent à apprendre l'anglais, avant leur mère ne leur parlait qu'en gaélique. En 1841, Malcolm, le frère aîné de Hugh, alors âgé de neuf ans vivait seul avec sa tante et sa grand-mère. Malcolm étudiait tellement bien à l'école qu'une richissime femme paya pour lui afin qu'il puisse suivre un séminaire à Dalkeith, près d'Edinburgh, là où les professeurs d'écoles étaient formés. Il enseigna en différents endroits et fut ordonné prêtre à St Mary de Glasgow par l'évêque de Glasgow en Août 1857.

En ce temps, Malcolm soutenait son jeune frère Hugh pendant ses études. Mais malheureusement cette aide cessa avant que MacColl n'eût le temps de les finir : son frère fut impliqué dans une discorde qui divisa l'église épiscopale en 1857-1858 ; refusant de soutenir la position de l'évêque, il fut excommunié.

À partir 1858, Hugh occupa différentes fonctions telles que celle d'enseignant en Grande-Bretagne, et ce, jusqu'à ce qu'il quitta le pays quelques années plus tard. En 1865 il partit pour Boulogne-sur-Mer (France) et s'y établit jusqu'à la fin de ses jours. Les raisons pour lesquelles il quitta son pays sont inconnues mais il n'est pas difficile d'imaginer des motivations économiques. Si nous prenons en considération l'immense flux migratoire qu'a subi la Grande-Bretagne au $19^{\text{ième}}$ siècle, il n'est pas difficile de concevoir que changer de pays soit chose courante. À cette époque, Boulogne-sur-Mer était une ville prospère avec des liens économiques et culturels très étroits avec les britanniques ; par conséquent, un endroit agréable pour des personnes désireuses de quitter la Grande-Bretagne.

Avant de partir, MacColl se maria avec Mary Elisabeth Johnson de Loughborough dans le Leicestershire. Elle vint en France avec lui, en Avril 1866 leur première fille, Mary Janet, vit le jour. En tout cinq enfants, quatre filles et un garçon, virent le jour à Boulogne-sur-Mer.

Pendant ces années, la situation financière était difficile voire souvent précaire. Hugh travaillait essentiellement en tant que professeur particulier, enseignant les mathématiques, l'anglais et la logique.

En contraste avec cette description, une publication nécrologique dans *La France du Nord* du 30 Décembre 1909 dépeint MacColl comme un professeur du *Collège Communal*. Toutefois il n'est pas possible d'identifier MacColl comme étant l'un des membres de l'équipe enseignante puisque son nom n'apparaît nullement dans les documents ou brochures du collège.

Néanmoins, ni les restrictions budgétaires ni l'agrandissement de la famille ne l'ont empêché de continuer à étudier. Il s'est préparé et a obtenu le BA en qualité d'étudiant extérieur à l'université de Londres en 1876[2].

Grâce à ses fréquentes publications dans différents journaux scientifiques et à ses échanges avec les plus grands hommes de sciences de son temps, nous pouvons présumer que MacColl espérait une reconnaissance envers ses travaux logique. On peut tout aussi bien imaginer qu'il aspirait à un poste d'universitaire. Ce point est étayé par une lettre adressée à Bertrand Russell

---

[2] Université de Londres. 1877. *Calendrier de l'année 1877*. London: Taylor and Francis.

en 1901 dans laquelle MacColl, âgé de 64 ans, se recommande lui-même comme maître de conférence en logique.[3]

Sa première femme, Mary Elisabeth mourut le 2 Février 1884 après une longue maladie. Trois années plus tard, le 17 Août 1887, MacColl se remaria avec Mlle Hortense Lina Marchal, native de Thann (Alsace). Les conditions de vie de MacColl s'améliorèrent sur le plan économique et social, notamment grâce à la position financière stable de la famille de sa nouvelle épouse. Hortense et sa sœur, Mme Busch-Marchal, étaient gestionnaires du très connu pensionnat de jeunes filles dans le milieu chic de Boulogne. Les parents de Hortense vivaient de leurs rentes privées, pendant que son frère, Jules Marchal et son beau frère Gustave Busch tenaient une boutique à Boulogne. Le couple constitua une vie de famille harmonieuse avec de solides bases économiques.

Dans les années qui suivirent la mort de sa première femme, MacColl abandonna ses publications habituelles au profit d'intérêts littéraires. En 1888 et 1891, il publia deux romans *Mr Stranger's Sealed Packet* et *Ednor Whitlock*. Le premier est un écrit de science fiction.

Bien que dans les livres classiques de l'histoire de la logique on puisse difficilement trouver une étude systématique ou même une description de son travail, à son époque ses contributions scientifiques étaient vivement discutées, en atteste ses publications ainsi que les échanges avec d'autres scientifiques dont, entre autres, Bertrand Russell et Charles Sanders Pierce. Pour preuve, au moins à partir de 1865, MacColl contribua à de nombreux et prestigieux journaux tels que : *The Educational Times and Journal of the College of Preceptors, Proceedings of the London Mathematical Society, Mind, The London, Edinburgh and Dublin philosophical Magazine and Journal of Science* et *L'Enseignement Mathématique*.
Ces idées furent discutées et également très souvent franchement critiquées. Une discussion riche prit place entre MacColl, Russell et T. Sherman sur l'engagement existentiel des propositions. Malheureusement, nombre de logiciens influents venant de la tradition booléenne tels que W.S. Jevons,, furent très hostiles aux innovations de MacColl. Dans un article de 1881, Jevons, critique la formulation propositionnelle du conditionnel « si..., alors... »

---

[3]   MacColl [1901c].

présentée par MacColl dans l'article « *Implication and equational logic* » de la même année[4]. MacColl y écrit :

> *Friendly contests are at present waged in the 'Educational Times' among the supporters of rival logical methods. I hope Prof. Jevons will not take it amiss if I venture to invite him to enter the lists with me, and there make good the charge of "ante-Boolian confusion" which he brings against my method.*

La réponse de Jevons, se fit sans délai :

> *It is difficult to believe that there is any advantage in these innovations [...]. His proposals seem to me to tend towards throwing Formal Logic back into its Ante-Boolian confusion [...] I certainly do not feel bound to sacrifice my peace of mind for the next few years by engaging to solve any problems which the ingenuity and leisure of Mr. MacColl or his friends may enable them to devise.*[5]

En réalité, le point est que MacColl cherchait une formulation logique du conditionnel conciliable avec la notion de *jugement hypothétique* de la tradition philosophique (Rahman 2000). Une notion de conditionnel dont des personnes telles que Jevons et Venn, comme nous venons de la voir, voulaient se débarrasser.

Gottlob Frege avait lui aussi connaissance des travaux de MacColl (ce qui n'était malheureusement pas réciproque). Il compare sa *Begriffsschrift* avec le travail de Boole et avec celui de MacColl dans l'article *Über den Zweck der Begriffsschrift*, où il critique le talon d'Achille du projet de MacColl, à savoir le défaut d'une notion précise reliant le niveau propositionnel avec celui du premier ordre.

Ernst Schröder, qui dans son célèbre *Vorlesungen über die Algebra der Logik* cite et discute énormément les contributions de MacColl, a de prime abord une impression quelque peu négative sur les innovations de MacColl même si plus tard il semble changer d'avis, concédant que l'algèbre de MacColl possède un haut degré de généralisation et de simplicité, particulièrement dans les contextes de la logique appliquée. Par contre, Schröder rejette définitivement l'interprétation propositionnelle que donne MacColl de la syllogistique aristotélicienne.

---

[4] MacColl [1880o], paragraphe 43.
[5] Jevons [1881], p. 486.
[6] Frege [1882], p. 4 ou p. 100 dans l'édition de Angelleli de 1964.

C. Ladd-Franklin, alors reconnu pour être un disciple de Pierce, accuse la communauté scientifique britannique d'avoir ignoré la contribution de MacColl pour la formalisation des propositions universelles à l'aide d'un conditionnel :

> *The logic of the non-symmetrical affirmative copula, "all a is b", was first worked out by Mr. Maccoll. Nothing is stranger in the recent history of Logic in England, than the non-recognition which has befallen the writings of this author. [...], it seems incredible that English logicians should not have seen that the entire task accomplished by Boole has been accomplished by Maccoll with far greater conciseness, simplicity and elegance...*[7]

En effet, les commentaires les plus positifs ne sont pas venus de Grande-Bretagne : dans son *Formulaire de Mathématique* de 1895 G. Peano reconnaît la dette qu'il a envers la logique propositionnelle de MacColl. Il en va de même pour ce que G. Vailati et L. Couturat écrivent de MacColl en 1899 dans le volume VII de la *Revue de métaphysique et morale*. Couturat fait ressortir le travail de MacColl sur la logique propositionnelle, même s'il exclut sa logique modale et sa logique probabiliste :

> *Ceci n'est vrai que pour les propositions à sens constant, qui sont toujours vraies ou toujours fausses, mais non pour les propositions* **à sens variable***, qui sont tantôt fausses, en d'autres termes, qui sont* **probables***. C'est ce qui explique la divergence entre le Calcul logique que nous exposons ici et le* **Calcul des jugements équivalents** *de M. MacColl, fondé sur la considération des probabilités.*[8]

La plus grande influence du travail de MacColl est certainement celle qui s'enracine dans les développements de l'implication stricte et de la logique modale formelle que l'on retrouve chez C. I Lewis.

Mais à la fin des années 1890, la *London Mathematical Society* refusa de publier davantage de contributions de MacColl. Cherchant un autre moyen de présenter la forme achevée de sa logique, il participa au *Premier Congrès International de Philosophie*[9] à Paris (1901) et aux publications de *L'enseignement Mathématique*[10]. Quelques années plus tard il publia une version anglaise augmentée : *Symbolic logic and its Applications* (1906). Trois années après cela, il publia

---

7 Ladd-Franklin [1889].
8 Couturat [1899], p. 621.
9 MacColl [1901f].
10 MacColl [1903e], MacColl [1904j].

Origin, Destiny and Duty[11], un essai avec sa conception de la science et de la religion. Comme nous le verrons dans notre appendice, la posture que MacColl adopte en tant qu'écrivain contraste avec l'image que nous pouvons nous faire de lui au regard des ses travaux scientifiques, à savoir celle d'un logicien novateur et tolérant. Malheureusement, comme c'est si souvent le cas dans l'histoire des sciences, ses conceptions politiques, sociales et éthiques ne se situèrent pas à un niveau aussi élevé que l'ouverture d'esprit dont il sut faire preuve en logique.

MacColl mourut le 27 Décembre 1909 à Boulogne en citoyen français. Hortense, sa femme, mourut, le 13 Octobre 1918.

---

[11] MacColl [1909a].

1. LES ELEMENTS DE LA PHILOSOPHIE DE LA LOGIQUE ET DU LANGAGE DE MACCOLL

> *There are two leading principles which separate my symbolic system from all others. The first is the principle that there is nothing sacred or eternal about symbols; that all symbolic conventions may be altered when convenience requires it, in order to adapt them to new conditions, or to new classes of problems [...]. The second principle which separates my symbolic system from others is the principle that the complete statement or proposition is the real unit of all reasoning.*
> (Symbolic Logic and Its Applications, 1906, pp. 1-2).

> SYMBOLICAL *reasoning may be said to have pretty much the same relation to ordinary reasoning that machine-labour has to manual labour [...]. In the case of symbolical reasoning we find in an analogous manner some regular system of rules and formulae, easy to retain [...], and enabling any ordinary mind to obtain by simple mechanical processes results which would be beyond the reach of the strongest intellect if left entirely to its own resources.*
> (Symbolic Logic I [1880p], p. 45).

> *Mais la logique symbolique fait pour la raison ce que fait le télescope ou le microscope pour l'œil nu.*
> (La logique symbolique, 1903d, p. 420)[12]

Comme cela est déjà annoncé dans l'introduction, la philosophie de MacColl est un genre d'instrumentalisme logique qui le conduit à poser les bases de ce qui pourrait être considéré comme le premier pluralisme logique.[13] Les extraits cités ci-dessus mettent en évidence que certains contextes de raisonnement peuvent nécessiter une logique particulière qui n'est pas applicable à d'autres contextes. En conséquence, en construisant un système symbolique pour un type particulier de logique, les expressions utilisées doivent être considérées avec la plus grande attention. La position que MacColl adopte est basée sur une notion pragmatique de l'énoncé et de la proposition, où l'aspect communicatif du signe (signes utilisés pour véhiculer l'information) est au centre de sa philosophie. Cet aspect de la philosophie du langage de MacColl le mène à explorer les possibilités d'un système formel suffisamment fin pour capturer les nuances et les caractéristiques du langage naturel. Les résultats de ses explorations sont impressionnants : l'utilisation de do-

---

12   Cf. La remarque similaire de Frege dans sa *Begriffsschrift*, xi.
13   Cf. Grattan-Guinness [1999].

maines restreints pour rendre compte de la contrepartie formelle de la notion grammaticale de sujet, l'utilisation de termes pour les individus et les concepts individuels, la distinction entre la négation d'une formule propositionnelle et la négation de prédicats, l'utilisation à la fois du prédicat d'existence et du prédicat de non-existence, la critique des paradoxes de l'implication matérielle dans le cadre de la logique modale et l'implication stricte, la logique plurivalente, la logique probabiliste, la logique de la pertinence, la logique connexe. Malheureusement, nombre de ces idées ne furent pas complètement développées et beaucoup d'autres restèrent dans un état tel qu'il nous est difficile de les comprendre aujourd'hui. L'une des raisons de cet inachèvement caractéristique de son travail est liée au défaut d'une structure technique appropriée capable de mettre en pratique ses nombreuses idées. Non seulement MacColl ne connaissait pas l'approche axiomatique de la logique, mais il n'a pas su réaliser la puissance fournie par les quantificateurs (et les variables liées), bien que son langage formel contienne des opérateurs très proches de la notion de quantificateur (restreint). A cela s'ajoute une tension fondamentale dans sa notion d'énoncé et de proposition, le nœud de cette tension se joue entre une approche sémiotique et pragmatique et une approche sémantique. Cette dernière n'étant pas résolue, elle confronte, dans certains passages, le lecteur à une difficile tâche interprétative. En général, il est assez difficile de réaliser une description systématique de son travail, sujet à de divers changements inattendus. Cependant, certaines critiques auraient pu juger moins hâtivement son travail si nous avions eut une meilleure compréhension de sa conception instrumentaliste du langage formel. Le langage formel de MacColl est essentiellement délimité par le contexte, et par conséquent, sa notation logique ne peut pas être lue indépendamment du contexte informel à l'intérieur duquel œuvrent les instruments formels. Le langage de MacColl n'est pas une notation rigide, universellement applicable, mais plutôt un système flexible, conçu pour s'adapter à différents contextes, s'enracinant sur fond de connaissances tacites (métalogique) de l'environnement en question. Il s'agit en fait, plus d'une structure que d'un système. Nous continuerons d'utiliser le terme *système* en suivant l'utilisation qu'en fait MacColl, mais il est important de toujours se rappeler que son langage logique est une structure formelle. Lorsqu'il étudia les langues naturelles, cette structure prit la forme d'une grammaire formelle – bien que son angle d'approche principal soit celui d'un logicien plus que celui d'un linguiste.

À travers les paragraphes qui vont suivre, nous essaierons de dessiner les contours du chemin qui mène MacColl de sa notion d'énoncé à ses propositions variées pour de nouvelles logiques.

1.1 Deux notions élémentaires dans la grammaire formelle de MacColl : *Enoncé* et *Proposition*.

MacColl présente la version finale de son système logique dans son livre *Symbolic logic and its applications* de 1909 qui sera la principale source de notre propos.

Dans la conception que MacColl se fait de l'énoncé, le son, les signes ou les symboles sont employés pour véhiculer les informations, et une proposition est une phrase qui, au regard de sa forme, peut être divisée en deux parties respectivement nommées sujet et prédicat. Il semble que pour lui, un symbole soit un signe dans un langage artificiel et que les expressions des propositions soient les symboles principaux du langage artificiel. Ici le terme artificiel inclut le langage naturel, ce qui est artificiel est compris comme un produit culturel et le symbole comme un code. Le langage artificiel où les expressions des propositions sont emboîtées est alors conçu comme étant fourni par une structure grammaticale. Une proposition exprime d'une manière plus précise et spécifique ce qu'exprime un énoncé de façon plus vague et générale. Chaque proposition est par conséquent un énoncé alors que le contraire ne vaut pas. Il est clair que dans tous les cas (selon MacColl) nous pouvons transmettre la même information par ces deux moyens. Par exemple, si quelqu'un nous demande si nous aimerions fumer un cigare, nous pouvons répondre soit en secouant la tête : un énoncé, soit avec une proposition : « Je ne fume pas de cigare ».

> *I define a statement as any sound, sign, or symbol (or any arrangement of sounds, signs, or symbols) employed to give information; and I define a proposition as a statement which, in regard to form, may be divided into two parts respectively called subject and predicate. [...] A nod, a shake of the head, the sound of a signal gun, the national flag of a passing ship, and the warning "Caw" of a sentinel rook, are, by this definition, statements, but not propositions. The nod may mean "I see him"; the shake of the head, "I do not see him"; the warning "Caw" of the rook, "A man is coming with a gun", or "Danger approaches"; and so on. These propositions express more specially and precisely what the simpler statements express more vaguely and generally.*[14]

La théorie de l'énoncé de MacColl est assez compliquée et demeure en relation avec sa théorie de l'évolution du langage dans la culture humaine. Ce point fut étudié en profondeur par Michael Astroh, qui a discuté le lien entre la notion d'énoncé de MacColl et la tradition linguistique de son épo-

---

[14] MacColl [1906a], pp. 1-4.

que.¹⁵ En outre, MacColl revisite la théorie traditionnelle des hypothétiques dans les termes de sa définition des énoncés et des *propositions de logique pure* (*grosso modo* : les propositions valides).¹⁶ Pour notre propos, nous retiendrons ici l'idée qu'un énoncé est une chaîne de signe utilisée pour transmettre des informations et pouvant devenir ce que MacColl appelle une *proposition*. Cette dernière assume une structure sujet-prédicat. Cette structure sujet-prédicat représente le lien entre un domaine restreint (sujet de MacColl) et un prédicat (prédicat de MacColl) défini sur le domaine.¹⁷ L'idée que la notion grammaticale de sujet corresponde au concept de domaine restreint d'une grammaire formelle est, selon notre point de vue, l'une des contributions majeures de MacColl. Ce domaine restreint (sujet) peut être soit *déterminé*, *déterminé* par l'introduction d'un terme, soit *indéterminé*. Ce dernier se résume simplement à un domaine et un prédicat défini sur tous les éléments du domaine. Les domaines *indéterminés* (sujets) sont par conséquent la version existentielle et universelle de la quantification chez MacColl. En réalité ce type de quantification ne suppose pas nécessairement un engagement ontologique.¹⁸ Afin d'éviter toute confusion, nous nommerons la notion de *sujet indéterminé* de MacColl : *propositions quantifiées*. De plus à l'intérieur des propositions quantifiées (*sujets indéterminés*), nous distinguons les *A-propositions* (propositions universelles) des *I-propositions* (propositions existentielles). Les expressions de base de son langage formel sont de la forme :

$$H^B$$

Où *H* est le domaine (sujet) et *B* le prédicat. Il donne l'exemple suivant :

H: Le cheval
B: marron
H^*B*: Le cheval est marron

---

15 Astroh [1999b] et Astroh [1995].
16 Cf. Sundholm [1999], Rahman [1998], [2000].
17 Comme mentionné plus haut, MacColl semble penser que le passage – à l'intérieur d'une communauté d'hommes donnée – du niveau où l'information a était d'une certaine manière transmise par la signification d'un énoncé, à l'articulation d'une proposition adéquate – qui assume une structure sujet-prédicat – est le signe d'un degré plus élevé dans l'évolution de cette communauté d'hommes. De plus, MacColl croit, que ce degré supérieur d'élévation peut être introduit par convention dans le stock linguistique du langage d'origine. (cf. MacColl [1906a], pp. 3-4).
18 Cf. Rahman/Redmond [2005].

MacColl observe que le mot « cheval » est une classe. Pour différencier les éléments de la classe dénotée par H, nous pouvons soit :

i) introduire des suffixes numériques : H_1, H_2, etc.
ou
ii.1) assigner à chaque individu un attribut différent, dans ce cas H_B représente « le cheval marron » ou H_W « le cheval particulier qui a gagné la course ».
ii.2) introduire une sous-classe (propre ou impropre), dans ce cas H_W représente « tous les chevaux qui ont gagné la course ».[19]

Le terme de classe semble correspondre à notre notion moderne de domaine restreint dont la portée des quantificateurs est telle que nous pouvons la trouver de nos jours dans la grammaire formelle. Ici, MacColl suit strictement le langage naturel. Il suppose aussi un domaine universel (tacite), qui, comme nous le discuterons ultérieurement, inclut la non-existence, mais à ce niveau de langage objet, seulement quelques restrictions (*portions*) de cet univers peuvent être exprimées. Ces *portions* (restrictions du domaine universel) constituent le domaine sur lequel portent les propositions.

> *Let S denote our Symbolic Universe or "Universe of Discourse" consisting of all the things S_1, S_2, &c, real, unreal, existent, or non-existent, expressly mentioned or tacitly understood in our argument or discourse. Let X denote any class of individuals X_1, X_2, &c., forming a portion of the Symbolic Universe S.*[20]

L'utilisation de suffixes par MacColl – comme décrit en i et ii.1 – semble correspondre à notre notion contemporaine de *terme individuel* (incluant les termes pour les individus et les concepts individuels).

> *On the other hand, S_B and H_k [...]. These are not complete propositions; they are merely qualified subjects waiting for their predicates.*[21]

Les symboles tels que B, H, W,... peuvent être utilisés à la fois comme classes et comme suffixes. Tout en explorant leurs différentes utilisations, Mac-

---

[19] L'utilisation que fait MacColl des articles du langage naturel « le » et « un » est difficile à suivre. Parfois MacColl les utilise pour faire la différence entre « tous » et « quelques uns » et parfois pour faire la différence entre l'expression quantifiée et une instance particulière de ces expressions.
[20] MacColl [1906a], pp. 1-4.
[21] MacColl [1906a], p. 5.

Coll discute de la différence entre une utilisation en qualité d'adjectif et un usage prédicatif des ces symboles.

> *Thus the suffix W is adjectival; the exponent S is predicative* [...].
>
> *The symbol H^W, without an adjectival suffix, merely asserts that a horse, or the horse, won the race without specifying which horse of the series H_1, H_2, &c.*[22]

L'expression H^W est de fait ambiguë : elle peut signifier à la fois « *le cheval de l'ensemble* » (tel que « le cheval est un animal ») qui doit être lu comme une proposition universelle « *tous les chevaux de l'ensemble* » :

> *"The horse has been caught".* [...] *asserts that every horse of the series H_1, H_2, &c., has been caught.*[23]

Il se peut toutefois qu'elle ne désigne qu'une *portion* du domaine restreint et en conséquence, elle doit être lue comme une proposition particulière « *le cheval a été attrapé* », autrement dit, « certains chevaux [de l'ensemble] ont été attrapés ».

Dans le but de lever toute ambiguïté issue de cette expression, MacColl utilise deux exposants spéciaux, à savoir ε pour l'universel, et θ pour le particulier.[24] Ces outils mènent respectivement aux *A*- et *I-propositions*.

$$(H^W)^\varepsilon \text{ et } (H^W)^\theta$$

---

[22] MacColl [1906a], p. 5. L'interprétation de ce passage n'est pas si directe. Ici, nous choisissons de comprendre H^W comme une proposition quantifiée, qui semble être compatible avec l'utilisation qu'en fait MacColl dans la suite de son livre. Un autre choix pourrait être l'interprétation (et critique) de Russell ; il semble comprendre l'expression de MacColl comme une fonction propositionnelle (cf. la réponse de MacColl à Russell dans MacColl [1910c,d]). La lecture de Russell peut être soutenue par le fait que MacColl s'efforce de montrer que de telles expressions n'ont pas d'engagement ontologique. Le problème avec la lecture de Russell est qu'elle ne s'accorde pas avec la plupart des explications et commentaires donnés par MacColl lui-même.
[23] MacColl [1906a], p. 41.
[24] Cf. MacColl [1906a], pp. 40-41.

Malheureusement, MacColl utilise le même exposant comme opérateur modale. Vraisemblablement il les envisageait comme un genre général d'expression quantifiée (cf. chapitre 2.1 et suivants). L'expression $(H^W)^{\varepsilon}$ stipule que le prédicat W s'applique à tous les éléments du domaine H. Réciproquement, θ stipule que le prédicat s'applique à une sous-classe de chevaux.

C'est ici que MacColl s'approche au plus près de la notion de quantificateur (restreint). Avec un peu de recul, les expressions ε et θ peuvent être perçues de façon similaire à l'interprétation des quantificateurs de second ordre développée par Frege. Tristement, MacColl n'a pas utilisé explicitement de variables liées pour les individus, ce qui l'empêcha de mesurer toute la profondeur et pertinence de la notion de quantificateur qu'il fournit.

Une autre interprétation à donner de l'utilisation par MacColl des termes en tant qu'adjectifs porte sur les sous-classes (possiblement avec un seul élément) du domaine. Ainsi, dire, l'utilisation de H comme un adjectif (pour cheval) dans A_H représente que la sous-classe du domaine A (animaux) contient soit tous les animaux qui sont des chevaux, soit quelques animaux qui sont des chevaux. Par le biais de cette interprétation, nous nous confrontons à la notion non-restreinte (et restreinte) de l'utilisation de prédicats chez MacColl : si la sous-classe contient tous les animaux qui sont des chevaux, alors l'utilisation du terme comme adjectif est dite non-restreinte (nous pouvons écrire $A_{(H)_n}$) et réciproquement pour la sous-classe restreinte contenant quelques animaux qui sont des chevaux (nous pouvons écrire $A_{(H)_r}$).

Nous pouvons combiner ceci avec les exposants pour les quantificateurs de la façon suivante : dans *Tous les chevaux ont gagné la course* $(H^W)^{\varepsilon}$ le prédicat *gagné* est dit être non-restreint, en opposition à *Tous les chevaux marron ont gagné la course* $(H_B{}^W)^{\varepsilon}$, qui est un dispositif différent que MacColl utilise pour exprimer la *I-proposition* : *Au moins quelques uns des chevaux (à savoir, tous ceux qui sont marron) ont gagné la course*. MacColl propose de remplacer toutes les indications spécifiques d'un terme utilisé en qualité d'adjectif par *r*, qui se comporte ici comme une variable portant sur ces adjectifs (ou sous-classe)[25] pour obtenir : $(H_r{}^W)^{\varepsilon}$ *Au moins quelques chevaux (à savoir, tous ceux qui sont des éléments*

---

[25] A noter que ce point est très proche de la théorie médiévale de la restriction *suppositio*.

de la sous-classe donnée) ont gagné la course, c'est-à-dire : *Quelques chevaux ont gagné la course.*

Assez souvent, si le prédicat est non-restreint et fait partie d'une *A-proposition*, MacColl omet l'exposant ε. S'il y a un terme utilisé en tant qu'adjectif non-restreint, il omet aussi l'indication explicite de cette supposition – c'est-à-dire, que dans ce cas MacColl n'introduit pas *u*.

MacColl montre que l'utilisation de *B* (pour marron) comme adjectif dans $H_B{}^W$ suppose H^B. Manifestement, si c'est possible de choisir un individu qui soit marron ou une sous-classe d'individus marron de la classe *H* des chevaux, cela suppose que la classe des chevaux soit une sous-classe (propre ou impropre) de la classe des objets marron.

Il est important de noter que dans le système de MacColl, les classes impliquées dans les expressions telles $H_B{}^W$ ne sont jamais vides et qu'elles n'ont aucun engagement ontologique. Notez encore que dès lors où le domaine universel de MacColl inclut des objets non-existants, ni H^B ni $H_1{}^B$ n'ont nécessairement d'engagement ontologique. Cette condition non négligeable nous permet de dire que l'utilisation de termes en qualité d'adjectif de prédicats logiques (ou prédicats) *quand ils sont utilisés comme des expressions pour des termes individuels* (c'est-à-dire utilisé dans le sens décrit en ii.1) semblent être proches de l'utilisation contemporaine des descriptions définies – rappelons que de nos jours, dans le cas de descriptions définies, nous introduisons ces termes à l'aide de l'opérateur iota. Une étude approfondie et poussée sur l'évolution de la théorie des descriptions définies de Russell par rapport aux discussions qu'il eut pu avoir MacColl fait encore défaut à ce jour. Peut-être qu'en raison du manque d'engagement ontologique des termes utilisés en tant qu'adjectifs par MacColl, ceux-ci sont plutôt compris comme désignateurs non-rigides plutôt que comme des descriptions définies russeliennes.

Il est important de souligner que l'utilisation par MacColl de ces termes en tant qu'adjectifs comportent certains avantages par rapport à la logique de premier ordre de Frege-Russell. En effet, suivant cet exemple :

La tortue rapide est marron (la tortue qui est rapide est marron)

En logique standard de premier ordre nous traduisons ce type d'expression à l'aide d'une conjonction : un individu du domaine est rapide, cet individu est une tortue et c'est individu est marron. Cette transcription exprime quelque chose qui est généralement faux. Il se peut très bien que la tortue en question soit rapide (pour une tortue), mais qu'au regard du royaume animal, les tortues même rapides soient lentes. Le langage logique de MacColl permet une traduction qui soit plus juste

$(T_f)^B$

où l'expression T_f signale qu'à partir de l'univers des tortues, nous sélectionnons celle qui est rapide et pour qui le prédicat B (être marron) peut être asserté avec vérité.

Les prédicats n-aires et les prédicats de second ordre peuvent être emboités dans le langage formel de MacColl (vraisemblablement) de la manière suivante. Un prédicat L (aimer) tel que dans *Hugh aime Hortense* peut être traité dans son système notationnel comme une expression qui, lorsqu'elle est appliquée à une constante individuelle (suffixes de la forme 1, 2...) qui désigne un élément du domaine H des êtres humains, donne lieu à un prédicat unaire. Ce prédicat unaire exprime la propriété d'aimer l'individu désigné par le suffixe 1 (Hortense).

$(H_1)^L$ (où H est le domaine des êtres humains, L représente *aimer* et 1 désigne *Hortense*)

Écrivons maintenant le résultat de cette opération, à savoir le prédicat *aimer Hortense* comme un nouveau prédicat, c'est-à-dire *L1*. Ce prédicat peut être appliqué à une constante individuelle qui désigne Hugh, ce qui donne lieu à une formule stipulant que l'individu désigné par le suffixe 2 (Hugh) a la propriété de « *aimer l'individu 1 (Hortense)* » :

$(H_2)^{L(1)}$ (où H est le domaine des êtres humains, *L(1)* représente *aimer Hortense*, et 2 désigne *Hugh*)

Un prédicat de second ordre peut également être exprimé dans la structure de MacColl. Prenez un prédicat unaire comme *Rouge est une couleur*. Le prédicat *C(ouleur)* est traité comme une expression qui lorsqu'elle s'applique à un élément du domaine produit le prédicat unaire R(ouge) qui est le résultat de la proposition exprimée dans *Rouge est une couleur*. C'est-à-dire que nous construisons premièrement le prédicat de premier ordre R (rouge)

$(O_1)^R$ (où O est le domaine des objets, R représente *rouge*, et 1 désigne un objet donné rouge) ensuite nous prédiquons sur R que c'est une couleur

$((O_1)^R)^C$ (où O est le domaine des objets, R représente *rouge*, 1 désigne un objet donné rouge et C représente *couleur*)

Une autre possibilité, plus simple, serait de capturer le résultat de la construction de second ordre à l'aide des termes utilisés comme des adjectifs.

$((P)_R)^C$ (où P est le domaine des prédicats unaires, R représente un élément spécifique du domaine de P, à savoir le prédicat unaire *rouge*, et C représente *couleur*).

Cette possibilité est très simple mais elle ne montre pas formellement la structure de second ordre de l'expression. Il est plus probable que MacColl aurait préféré cette dernière expression formelle en lui ajoutant informellement une interprétation adéquate.

Comme mentionné précédemment son approche formelle du langage naturelle ressemble fortement à une version des sémantiques formelles modernes telles que la DRT et la DPL. MacColl a essayé de faire transparaître les distinctions du langage naturel dans son système formel plus que n'importe lequel de ses contemporains, comme par exemple Frege. Ceci est directement lié à son instrumentalisme, où l'importance contextuelle joue un rôle primordial.

Le rapprochement du système notationnel de MacColl avec la manière de penser dans les grammaires formelles modernes devient encore plus prégnant lorsque nous considérons la façon dont il conçoit la négation. Le langage logique de MacColl contient deux symboles pour la négation. La première est une négation externe qui porte sur l'ensemble de l'expression (*de dicto*) tel que dans $(A^B)'$, $(A_B)'$ et la seconde porte uniquement sur le prédicat (*de re*) tel que dans A^{-B}, A_{-B}.[26] Le second symbole nécessite une certaine structure de prédicat. Testons une fois de plus la notation de MacColl par rapport à la logique standard de premier ordre en nous aidant de l'exemple suivant :

Fumer est imprudent

---

[26] Cf. Rahman [1999] et Rahman [2001].

Dans cet exemple, le problème ne réside pas seulement dans le prédicat de second ordre ici en jeu, mais aussi dans la traduction de *imprudent*. Effectivement dans la logique standard de premier et second ordre la négation s'applique exclusivement sur les formules exprimées dans une proposition et non pas aux prédicats eux-mêmes. Frege et Russell ont tous deux explicitement rejeté la négation des prédicats. Néanmoins, dans de nombreuses langues naturelles réside un processus permettant de produire de la négation avec des expressions de genres divers, dont certaines incluent des prédicats. Les grammaires formelles modernes fournissent des structures de prédicats (notamment via l'opérateur λ). MacColl ne possédait pas de tels outils et fut sévèrement critiqué pour avoir fait porter la négation sur des prédicats. Pourtant, sa notation peut faire transparaître la structure de la négation du prédicat de notre exemple. Plus précisément, la négation d'un terme utilisé comme adjectif lui permet de capturer des expressions telles que *imprudent, impatient* etc., dans son langage formel avec la structure négation+prédicat des langues naturelles. Par conséquent, dans la notation de MacColl l'analyse formelle de la phrase ci-dessus se transpose dans la formule :

$((P)s)^{-W}$ (où P est le domaine des prédicats unaires, S représente un élément spécifique de P, à savoir le prédicat unaire *fumer* et -W représente la négation de *prudent*).

Évidemment un défaut de cette transcription formelle est qu'elle ne montre pas la différence entre une prédication de premier et de second ordre. Mais celle-ci peut être introduite de la manière que nous avons décrite précédemment lorsque nous discutions de la formulation des prédicats de second ordre.

Notez que la négation d'une prédication comme opposée à la négation de toute la formule propositionnelle où cette prédication a une occurrence, possède de fortes similarités avec la théorie de Russell de la double portée de la négation lorsqu'elle est appliquée aux prédicats d'une description définie. La distinction de Russell entre :

1) Ce n'est pas le cas que (il y a un et seulement un individu qui est actuellement le Roi de France et cet individu est chauve)

2) Il y a un et seulement un individu qui est actuellement le Roi de France et cet individu n'est pas chauve.

peut être transcrite dans la notation de MacColl de la manière suivante :

1*) $(H_K{}^B)'$ (où H est le domaine des individus, K : l'actuel Roi de France et B : être chauve)
2*) $(H_K)^{-B}$

Malheureusement, MacColl n'a pas réalisé toute la puissance d'expressivité de la distinction de son propre système de notation et considérait les formules 1* et 2* comme équivalentes.[27]

Une difficulté particulière du système de MacColl émerge lorsqu'il introduit les propositions quantifiées avec un engagement ontologique. Dans ce contexte, il introduit le prédicat **0** qui doit être la négation *de re*. La manière dont il le combine avec les propositions quantifiées est malheureusement embarrassante (cf. 2.2 et suivant).

Concluons cette section par la remarque suivante : l'utilisation, dans un langage formel, de domaines de quantifications restreints, de la distinction entre une utilisation des termes en qualité d'adjectif et une fonction prédicative des prédicats, de la négation de cette dernière, de l'introduction d'un prédicat de non-existence, de la distinction entre les expressions quantifiées avec et sans engagement ontologique sont autant d'idées audacieuses, mais malgré celles-ci, MacColl n'accorde pas à son système notationnelle l'attention qu'il requiert et ne réalise pas lui-même la pleine puissance de ce dernier.

Permettez nous maintenant de vous présenter la logique de la non-existence de MacColl dans le détail.

1.2 La logique de la non-existence de MacColl

1.2.1 Le domaine symbolique et sa dynamique

L'approche la plus influente de la logique de la non-existence est certainement celle provenant de la tradition Frege-Russell. L'idée principale est relativement simple mais aussi quelque peu décevante : raisonner avec des fictions, c'est raisonner avec des propositions soit (trivialement) vraies, parce que nous en nions l'existence, ou fausses trivialement de la même manière. Le problème naît du deuxième membre de cette alternative : toute proposi-

---

[27] MacColl [1909a], p. 5.

tion (sauf si c'est une proposition existentielle niée) qui contient des termes fictionnels que l'on peut asserter est fausse. Par exemple, si, relativement à un contexte donné, *Pegasus* est un nom vide, dans ce cas les phrases « *Pegasus* a deux ailes » et « *Pegasus* a trois ailes », expriment toutes deux des propositions fausses par rapport au domaine donné, bien que la phrase « *Pegasus* n'existe pas » exprime une proposition vraie.

La justification de cette manière de se débarrasser du problème est aussi directe, peut-être trop : en science, nous ne parlons avec intérêt qu'à propos des choses qui comptent comme réelles dans notre domaine. Voulez-vous pouvoir raisonner à l'aide d'expériences mentales où des propositions contrefactuelles autres que des propositions existentielles négatives sont assertées ? Considérez alors que les objets de votre expérience mentale sont des éléments de votre domaine et appliquez leur notre bonne vieille logique de premier ordre. C'est-à-dire raisonnez comme si le monde décrit par votre fiction était réel, pour cela rien de plus que la logique classique standard n'est requis. On pourrait dès lors commencer à suspecter que quelque chose soit faux ici. Assez souvent, lorsque nous introduisons des fictions nous voulons établir le lien entre deux domaines classés en royaumes ontologiques distincts. En d'autres termes, l'intérêt du raisonnement avec des propositions contrefactuelles est de pouvoir être en mesure de raisonner avec une structure qui met en lumière les liens entre ce qui est considéré comme réel et non-réel dans notre domaine. Le défi est en de raisonner dans un mode parallèle. De plus, de tels raisonnements nécessitent de comprendre comment le flux d'informations entre ces mondes parallèles s'exerce. Contrairement à ce qu'en a retenu la tradition et l'histoire des développements modernes de la logique, la manière qu'a Frege de traiter le problème n'est pas exactement la même que Russell, et nous pouvons trouver des dissidents à la solution mentionnée précédemment. L'un des plus importants est bien sûr MacColl. C'est au regard de sa notion d'existence et des arguments impliquant des fictions que son travail montre une profonde différence avec celui de ses contemporains. En réalité, il est le premier à essayer d'introduire dans un système formel l'idée que l'on puisse intégrer des fictions dans des contextes logiques avec pour objectif, non seulement d'introduire la distinction entre ce qui est réel et ce qui est fictif avec diverses (ontologiquement parlant) sortes de langages, mais aussi avec les outils techniques afin de mettre en évidence les connections entre ces différents royaumes ontologiques. Qui plus est, son approche dynamique de la logique des fictions a incité à de nouvelles recherches approfondies de son travail, notamment dans les contextes intentionnels.

Pour achever son projet de logique de la non-existence, MacColl introduit en premier lieu deux classes *mutuellement complémentaires et contextuellement déterminées* :

- la classe des existants et la classe des non-existants. Il nomme la classe de ce qui existe réellement « **e** » contenant les éléments : e_1, e_2, ... Chaque individu, dans des circonstances données, peut être qualifié avec vérité d'existant au regard de cette classe.

- la classe des non-existants, « l'ensemble vide **0** ». Il appelle cette classe l'ensemble vide, bien qu'elle soit en fait pleine, ce qui est malencontreux. Cette classe contient les objets 0_1, 0_2,... qui ne correspondent à rien de notre univers que nous admettons être réel. Les objets tels que les centaures et les cercles-carrés, se trouvent dans cette classe. L'erreur de notation que commet MacColl en appelant la classe des non-existants « ensemble vide » ouvrit la porte à la critique telle que celle de Bertrand Russell et amusa la communauté scientifique avec laquelle il publiait nombre de ses contributions. Le problème est que Frege utilise le terme d'ensemble vide aussi dans le contexte des entités fictives. De plus Frege l'utilise même comme un objet.[28]

---

[28] Rappelons que pour Frege une expression sensée peut n'avoir aucune dénotation. Donc, si nous supposons la compositionnalité de la dénotation, ces phrases qui contiennent des expressions sans dénotation ne doivent pas avoir de dénotation non plus. Maintenant, pour Frege la dénotation d'une assertion est sa valeur de vérité. Donc, si une assertion donnée contient un terme fictif (ou vide), cette assertion manquera de valeur de vérité. Mais ceci serait, selon le point de vue de Frege, exclu en tant que type d'assertion interdit dans le paradis de la science. De plus nous ne pouvons même pas asserter avec vérité que les termes fictifs ou vides n'existent pas. Si, au contraire, nous voulons pouvoir asserter la négation de propositions existentielles à propos des entités fictives, la solution de Frege est de supposer que chaque terme fictionnel différent dénote le même ensemble vide (*Grundlagen der Arithmethik*, paragraphe 53). Cet artifice a comme conséquence que n'importe quelle proposition contenant des entités fictives est fausse, à moins qu'il ne s'agisse de la négation d'une proposition existentielle fictive. MacColl pense à l'ensemble vide aussi, mais il est selon lui ni vide ni ne doit nécessairement produire des propositions fausses. Plus précisément, comme l'ensemble vide de MacColl n'est pas vide, cela autorise le fait qu'il contienne différentes entités non-existantes comme ses éléments.

et ensuite une troisième :

Un univers symbolique qui inclut les deux distinctions précédentes.

En outre, si un ensemble est contenu dans l'ensemble des non-existants, alors son pendant est lui-même inclut dans l'ensemble des existants et vice et versa. Nous avons donc une structure du domaine comme suit :

> **e** : *Existence réel*. Les éléments sont aussi des éléments de la classe **S** mais non de la classe **0**.
>
> **0** : *Non existant ou seulement existence symbolique*. Les éléments sont aussi des éléments de la classe S mais non de la classe **e**.
>
> **S** : *Existence symbolique*. Les éléments de cette classe sont à la fois des éléments de **0** et de **e**.

MacColl introduit cette classe dans le langage objet pour représenter les prédications correspondantes, *e. g.* $(H_3)^\mathbf{S}$, $((H_1)^B)^\mathbf{S}$, $((H_1)^B)^\mathbf{0}$, $(H_2)^\mathbf{e}$ (le cheval 3 est un élément de la classe symbolique, le cheval 1 est marron et est un élément à la fois de la classe symbolique et de la classe des non-existants et le cheval 2 est un élément de la classe des existants). Ce système de notation permet donc l'introduction des différents modes d'existence comme domaines de discours, *e.g.* : $(\mathbf{S}_3)^H$, $((\mathbf{0}_1)^H)^B$, $((\mathbf{S}_1)^H)^B$, $(\mathbf{e}_2)^H$ (l'élément 3 de la classe symbolique est un cheval, etc.).[29]

Si les suffixes se comportent comme des adjectifs représentant des classes, et qu'ils sont des éléments de la classe **0** ou **e**, l'hypothèse de MacColl, selon laquelle les pendants de ses adjectifs sont des éléments des classes ontologiques complémentaires autorise ce qui suit : si la classe **e** renferme en elle l'unique élément cheval, nommé *Rocinante*, alors le pendant de n'importe quelle classe des chevaux de **0** est une sous-classe de **e**. La signification des autres sens des classes d'inclusion semble difficile de compréhension. Si le singleton *Rocinante* est une sous-classe de la classe des chevaux H, alors l'hypothèse de MacColl nécessite que le pendant du singleton soit une sous-classe de **e**. Cette même hypothèse rend difficile la compréhension des suffixes en tant qu'individus, une théorie des pendants des individus serait nécessaire.

---

[29] MacColl [1905o], pp. 74-76 et [1906a], pp. 76-77.

Ce qui est intéressant c'est que MacColl suppose que ces domaines interagissent, ou plus précisément qu'il y a une interaction entre la classe symbolique et les deux autres classes ontologiques. En fait, son point de vue semble plus être guidé par des considérations épistémiques et dynamiques qu'ontologiques. Par exemple, supposez qu'à un moment donné d'une argumentation, la proposition suivante soit assertée :

$(H_3)^S$ (*le cheval 3 a une existence symbolique*)

Cette proposition peut être assertée parce qu'au moment de l'assertion, le contexte ne contient pas assez d'informations précises concernant le statut ontologique du cheval en question. Mais dans un contexte ultérieur, de nouvelles informations au sujet de cet objet peuvent apparaître. Ce qui nous autorise à préciser l'assertion que nous faisions précédemment sur ce cheval en concédant que ce cheval n'existe pas réellement. MacColl fournit quelques exemples de cette dynamique dans des cas de déceptions qui établissent une connexion entre le dynamisme de son univers symbolique et des contextes intentionnels.[30] Exemples tels que :

L'homme que tu as vu dans le jardin est vraiment un ours.
L'homme que tu as vu dans le jardin n'est pas un ours.

Cet exemple est intéressant et encourageant. MacColl adopte le point de vue d'un observateur qui asserte les propositions ci-dessus et étudie ce qui se passe avec les suppositions ontologiques qu'elles impliquent. Il en conclut que le statut ontologique de l'homme que décrit la proposition est de ne pas exister dans le premier exemple et d'exister dans le second. MacColl n'a pas analysé la dynamique produite dans la phrase :

L'objet que tu vois dans le jardin et dont tu penses que c'est un homme qui existe est véritablement un ours qui existe.

Toujours est-il que les exemples de MacColl sont passionnants et mériteraient une exploration détaillée.

---

30   MacColl [1905o], p. 78.

1.2.2. Propositions avec et sans engagement ontologique

MacColl essaie de capturer dans son langage formel des propositions quantifiées avec et sans engagement ontologique.

Permettez-nous de revenir une fois de plus à l'expression d'une *I-proposition* concernant le paradigmatique cheval marron de MacColl :

$((H_r)^B)^\varepsilon$ (*Au moins quelques chevaux sont marron*)

selon le propre point de vue de MacColl, cette proposition n'engage en rien quant à l'existence de quelconques chevaux. Ainsi l'expression universelle :

$(H_r^B)'$ (*Il n'est pas le cas qu'il y ait un cheval marron = aucun cheval n'est marron*)

Dans le but d'obtenir des expressions du type *I-proposition* avec un engagement ontologique, MacColl introduit, comme nous avons pu le voir précédemment, un univers non-vide regroupant à la fois les objets existants et les objets non-existants. À la page 5 de son livre, MacColl introduit le prédicat de non-existence **0** dans la formulation des *I-propositions* qui peuvent faire l'objet d'une négation *de re*. À l'aide du prédicat de non-existence, nous obtenons des expressions telles que :

$(H_c)^{-\mathbf{0}}$

qui se lit conformément à la lecture de MacColl : *Chacun des chevaux attrapés existe*. Ou alors *Au moins certains des chevaux (c'est-à-dire, tous ceux qui ont été attrapé) existent.*[31] MacColl utilise ici l'adjectif non-restreint, C, en tant que sous-classe de tous les chevaux attrapés et la négation du prédicat de non-existence comme assertion qu'une telle classe est inclue dans l'ensemble des objets (non-non-) existants. Par conséquent la formule $(H_c)^{-\mathbf{0}}$ exprime une *I-proposition* avec engagement ontologique. Une formulation plus explicite pourrait être :

$((H_r)^\mathbf{e})^\varepsilon$ (*Certains chevaux existent*)

---

[31] MacColl [1906a], p. 5.

> *Au moins certains des chevaux (à savoir, tous ceux qui sont des éléments d'une sous-classe donnée) existent.*

MacColl utilise un cas particulier de la dernière notation dans sa reconstruction du syllogisme traditionnel où il omet l'exposant ε[32]:

$$(X_Y)^e$$

> *Au moins certains éléments de X (c'est-à-dire, tous ceux qui sont des éléments de Y) existent.*

C'est-à-dire,

> *Certains éléments de X sont Y (et existent).*

Pareillement pour les *O-propositions* :

$$(X_{-Y})^e$$

> *Certains éléments de X ne sont pas des Y (et existent).*

MacColl introduit également dans son système la formulation pour les *A-propositions* avec engagement ontologique, telle que :

$$(H_{-c})^o$$

qui par conséquent s'exprime ainsi :

> *Chacun des chevaux non-attrapés n'existe pas.*

Mais aux dires de MacColl, cela implique (souvenez-vous de son hypothèse de complémentarité ontologique) que

> *Chaque cheval qui a été attrapé existe*, ou *Chaque cheval a été attrapé (et est existant).*[33]

---

[32] MacColl [1906a], p. 44.
[33] MacColl [1906a], p. 5.

Son système de notation est plus direct dans sa manière de tourner les *I-propositions* avec engagement ontologique, à savoir :

$((H_r)^{e})^{\varepsilon}$ (*Certains chevaux sont existants*).

La notion de proposition en tant que véhicule informatif semble fournir le fond de motivation de toute sa conception de la logique, où les données d'un contexte peuvent avoir des conséquences logiques. Dans la section suivante, nous mettrons en évidence la manière dont les données du statut ontologique du sujet sur lequel porte la proposition sont construites et comment elles peuvent déterminer l'ensemble de conséquences qui peuvent en être issues. C'est un fait intéressant que MacColl ait essayé de formuler une logique où des données en tout genre peuvent avoir des conséquences logiques. Le fruit de ses recherches a ouvert la voie aux bases de la logique modale moderne et au commencement de la philosophie du conditionnel.

2. LA PHILOSOPHIE DU CONDITIONNEL

La notion centrale de MacColl est celle de conditionnel. Il ne reconnaît pas seulement que ce connecteur tient une place primordiale dans sa logique, il fait du conditionnel le cœur de sa philosophie. Cela entraîna toute une série de réflexions sur le conditionnel dont les répercussions sont toujours aussi vives que depuis les premières recherches sur la signification logique de l'implication matérielle soulevées par MacColl. En outre, nombre de discussions contemporaines supposant l'aspect pragmatique de la signification peuvent être perçues comme partageant un fond philosophique commun avec MacColl, à savoir : les éléments constitutifs de la signification en logique sont liés à la transmission d'informations. MacColl n'élabora pas complètement sa théorie de la signification, mais il l'approcha par divers aspects, tous semblent être liés à l'idée que la transmission d'informations doit aboutir à une logique où l'intégration de nouvelles de données peut faire une différence.

2.1. L'invention de la logique modale T : La critique de l'implication par MacColl

Permettez nous de commencer par deux citations de Stephen Read qui présentent le point historique de manière concise[34] :

> *The received wisdom is that strict implication was invented and developed by the American logician C. I. Lewis.*[35]
>
> *C. I Lewis repeatedly exempts MacColl from criticisms of his predecessors in their accounts of implication. They had all taken a true implication, or conditional, to be one with false antecedent or true consequent. MacColl uniquely, and correctly in Lewis' view, rejected this account, identifying a true implication with the impossibility of true antecedent and false consequent. Lewis's development of the calculus of strict implication arises directly and explicitly out of MacColl's work.*[36]

Effectivement, *les idées reçues* sur les origines de la logique modale formelle sont fausses - et Lewis fut loin d'être innocent dans la propagation de ces

---

34	Cf. également Rahman [1997b].
35	Read [1999], p. 59.
36	Read [1999], p. 59.

fausses idées.[37] La contribution de MacColl dans ce domaine est arrivée au moment où les traditions historiques grecque, arabe et médiévale furent mises à l'écart, suspectées de psychologisme. En réalité, MacColl entreprit son travail comme la construction d'un pont entre l'approche philosophique de la logique et l'approche mathématique de cette dernière[38] et discuta les principales idées de sa logique modale dans des échanges épistolaires, principalement avec Bertrand Russell. Le premier article contenant la version définitive de sa logique modale fut présenté à Paris (1901) au *Premier Congrès International de Philosophie. Logique et Histoire des Sciences*, où entre autres, Couturat, Frege, Peano et Russell étaient membres du comité scientifique. D'autres publications sur le même sujet furent publiées en France entre 1903 et 1904. Finalement, la formulation la plus aboutie de sa logique fut présentée dans son livre de 1906 *Symbolic logic and its applications*. Mais revenons à présent sur ses réflexions sur l'implication matérielle :

> *For nearly thirty years I have been vainly trying to convince them that this assumed invariable equivalence between a conditional (or implication) and a disjunctive is an error, and now Mr. Shearman's quotation supplies me with a welcome test case which ought, I think, to decide the question finally in my favour. Take the two statements "He is a doctor" and "He is red-haired", each of which, is a variable, because it may be true or false. Is it really the fact that one of these statements implies the other? Speaking of any Englishman taken at random out of those now living, can we truly say of him "If he is a doctor he is red-haired", or "if he is red-haired he is a doctor?" Is it really a certainty that either "all English doctors are red-haired", or else "all red-haired Englishman are doctors?"*
> *[...]*
> *Thus, Mr. Russell, arguing correctly from the customary convention of logicians, arrives at the strange conclusion that (among Englishmen) we may conclude from a man's red hair that he is a doctor, or from his being a doctor that (whatever appearances may say to the contrary) his hair is red.*[39]

L'insatisfaction de MacColl vis-à-vis de l'implication matérielle illustrée dans les extraits précédents est si grande qu'elle le mène à sa notion d'implication stricte. Sa notion de conditionnel est la définition connue de l'implication

---

37   Cf. Rahman [1997b] et Read [1999], p. 59.
38   *The writer of this paper would like to contribute his humble share as a peacemaker between the two sciences, both of which he profoundly respects and admires. He would deprecate all idea of aggression or conquest [...]. Do not Englishmen and Scotchmen alike now both "glory" as George III said he did, "in the name of Briton"? Why should not logicians and mathematicians unite in like manner under some common appellation ?* (MacColl [1880p], p. 46-47).
39   MacColl [1908b], p. 152.

stricte. « A implique B » est entendu chez MacColl par « il est impossible que A et non-B » ($(AA')^n$ selon son formalisme) :

> Let W denote the first proposition and E the second. It is surely an awkward assumption (or convention) that leads here to the conclusion that "either W implies E or else E implies W". War in Europe does not necessarily imply a disastrous earthquake the same year in Europe; nor does a disastrous earthquake in Europe necessarily imply a great war the same year in Europe.
> [...] with Mr Russell the proposition "A implies B" means $(AB')^\iota$, whereas with me it means $(AB')^\eta$.[40]

Le langage modal de MacColl est conçu comme une extension des valuations booléennes « vrai » et « faux » qu'il pense ne pas être suffisamment générales, voire même une description non fidèle vis-à-vis des capacités intellectuelles humaines.[41] Il a introduit les modalités suivantes : « certain » (ou « nécessaire »), « impossible » et « contingent » (ou « ni impossible, ni nécessaire »). Une motivation à cela était une interprétation probabiliste. À la page 7 de son livre de 1906, nous trouvons la présentation suivante :

---

[40] MacColl [1908c], p. 453.
[41] MacColl semble pouvoir désormais aboutir à des conclusions via la signification de l'implication stricte comme le signe d'un degré d'évolution supérieur à celui des animaux. Le texte suivant est probablement pointé contre Boole et ses successeurs qui sont les tenants du courant dominant dans la logique, mais qui n'ont pas de connecteur pour exprimer le conditionnel :
Brute and man alike are capable of concrete reasoning; man alone is capable of abstract reasoning [...]. The brute as well as is capable of the concrete inductive reasoning [...] that is to say, from experience -often painful experience- the brute as well as man can learn that the combination of events A and B is invariably followed by the event C [...]. We have seen that from two elementary prémisses A and B, brutes as well as men can [...] draw a conclusion C. But no brute can, from the two implicational premisses [...] draw the implicational conclusion A:C [it is impossible that A not C]. *It is evident that the latter is not only more difficult, but also that it is on a higher and totally different plane. In the former the two premisses and the conclusion are all three elementary statements [...] while the whole reasoning is a simple implication. In the latter, the two premisses and the conclusion are all three implications, while the whole reasoning is an implication of the second order.* MacColl [1902i], pp. 367-368.

Vrai	τ	A^{τ} : A est vrai dans le cas considéré.
Faux	ι	A^{ι} : A est faux dans le cas considéré.
Certain	ε	A^{ε} : A est toujours vrai (vrai dans tous les cas), compatible avec nos données et nos définitions, sa probabilité est de 1.
Impossible	η	A^{η} : A contredit quelques données ou définitions, sa probabilité est de 0.
Variable	θ	A^{θ} : A est possible mais incertain (A est contingent). Ce qui est équivalent à $A^{-\eta}A^{-\varepsilon}$: A est ni impossible ni certain; A est possible mais incertain. La probabilité n'est ni 0 ni 1, mais est comprise entre les deux.

À la page 14 MacColl ajoute la modalité « possible » (π) définie comme non-impossible, et explique que cette modalité indique que la probabilité n'est pas 0 mais peut être de 1 ou moins que 1.

Sur la même page, nous trouvons la solution de la négation interne (*de re*) de cet opérateur. MacColl ne fait pas ici usage de deux signes différents, l'un pour la négation *de re* l'autre pour la négation *de dicto* (comme il a pu le faire auparavant pour la négation des prédicats). En introduisant cette distinction, il fait usage de la différence de position des termes comme adjectif ou comme prédicat. Nous ne suivrons pas la procédure de MacColl, nous changerons simplement les positions du précédent tableau :

Négation interne de la Vérité	$(A')^{\tau}$	A^{ι}
Négation de la Fausseté	$(A')^{\iota}$	A^{τ}
Négation interne du Certain	$(A')^{\varepsilon}$	A^{η} Afin d'être distingué de la négation externe $(A^{\varepsilon})'$: A n'est pas nécessaire
Négation interne de l'Impossibilité	$(A')^{\eta}$	A^{ε} Afin d'être distingué de la négation externe $(A^{\eta})'$: A n'est pas impossible
Négation interne du Variable	$(A')^{\theta}$	A^{θ} Afin d'être distingué de la négation externe $(A^{\theta})'$: A n'est pas contingent

| Négation interne du Possible | $(A')^\pi$ | Il est possible que non A Afin d'être distingué de la négation externe (A^π) : A est impossible |

Ces modalités se combinent avec la logique propositionnelle de la façon suivante :

$A+B$	disjonction
A'	négation
AB	conjonction
$A:B$	implication stricte
$(AB)^\eta$	définition de l'implication stricte

A de nombreuses reprises, MacColl a soutenu que son système logique était supérieur à celui utilisé par les « logiciens booléens » (Jevons, Schröder, Venn et al.) en raison de sa plus grande facilité à résoudre des problèmes philosophique et mathématique, particulièrement ceux liés à la probabilité. MacColl expose explicitement que sa logique puise son origine dans des problèmes de probabilité (Cf. Appendice **B**). Par la suite, il donna une interprétation plus générale de sa logique modale. Malheureusement l'interprétation qu'il donne de ses modalités a toujours oscillé entre valeurs, prédicats, opérateurs et variables pour des propositions de genre correspondant. Peter Simons soutient que les modalités de MacColl se comprennent mieux sous le joug d'une logique probabiliste qu'en tant que logique fonctionnelle plurivalente.[42] Werner Stelzner et Ian Woleński optent pour une interprétation opérationnelle.[43] Read reconstruit la logique modale de MacColl avec le style de son temps, c'est-à-dire avec une algèbre appropriée pour diverses interprétations.[44] Nous souhaiterions rappeler que, comme nous l'exposions en 1.1, les modalités de MacColl sont aussi utilisées en tant qu'opérateurs pour produire des expressions quantifiées. C'est exactement comme si parfois, dans le but d'aboutir à la généralité, il considérait les données sur lesquelles les modalités sont définies soit comme des individus, soit en tant que contextes, ou encore comme probabilités et ainsi de suite. Toutefois, dans ce qui suit, nous traiterons ces modalités comme s'il s'agissait d'exposants

---

[42] Cf. Simons [1999].
[43] Cf Stelzner [1999], Woleński [1999].
[44] Cf. Read [1999].

tenant lieu d'opérateurs portant sur les formules et l'usage non-exponentiel de ces modalités comme représentant des formules. Malgré tout, Read montre que le théorème de la logique modale de MacColl qui suit apparaît de nos jours sous l'axiome **T**.

MacColl suppose la validité des modalités suivantes[45] :

1. $(A+A')^\varepsilon$
2. $(A^\tau + A^\iota)^\varepsilon$
3. $(AA')^\eta$
4. $(A^\varepsilon + A^\eta + A^\theta)^\varepsilon$
5. $A^\varepsilon : A^\tau$
6. $A^\eta : A^\iota$
7. $A^\varepsilon = (A')^\eta$
8. $A^\eta = (A')^\varepsilon$
9. $A^\theta = (A')^\theta$

Il est intéressant de comparer la formule 2 avec la formule 1. Dans la seconde formule, la vérité de la proposition en question est introduite dans le langage objet. Stelzner interprète ce dispositif de MacColl comme étant l'introduction dans le langage objet de noms propres pour les contextes.[46]
La formule 4 (ainsi que sa duale 5) correspond(ent) à l'axiome qui caractérise de nos jours la très connue logique modale standard **T**. Qui plus est, comme Read le fait remarquer, MacColl rejette explicitement l'axiome **S4**.

$A^\varepsilon : A^{\varepsilon\varepsilon}$ [47]

et accepte l'axiome de nécessitation

$\eta^\eta = \varepsilon$ [48]

où la notation exponentielle signale l'utilisation de la modalité en tant qu'opérateur.

45	MacColl [1906a], p. 8.
46	Cf. Stelzner [1999].
47	Cf. MacColl [1896c], p. 13 et Read [1999], p. 579.
48	Cf. MacColl [1906a], p. 13 et Read [1999], p. 74.

Tout ceci confirme la reconstruction que Read fait du système de MacColl, et atteste même que le système **T** était contenu dans sa logique bien avant qu'il reçoive le nom de **T**. La revendication de Read s'oppose à la lecture de Storrs MacCall qui pense que la logique de MacColl correspond à une logique non-normale. Jusqu'à ce point de la discussion, l'argument de Read semble tout de même être juste, comme nous le verrons plus loin sur ce sujet, il peut y avoir un certain genre de compromis. Le point est le suivant : MacColl n'a pas un seul système, ceci est encore plus vrai pour ce qui est des problèmes touchant à l'implication. Permettez-nous de parcourir le cheminement de MacColl sur les paradoxes de l'implication matérielle et de présenter sa solution dans la structure de sa logique modale.

MacColl réalise que son implication stricte échappe aux paradoxes de l'implication matérielle que nous avons pu voir précédemment. C'est-à-dire que les formules :

   1) $(A:B)+(B:A)$
   2) $B:(A:B)$
   3) $A':(A:B)$

ne sont bien évidemment pas valides.

L'idée de MacColl est que si les formules qui précèdent étaient valides, comme c'est le cas avec l'implication matérielle (c'est-à-dire lorsque « : » est remplacé par le symbole de l'implication matérielle « → »), alors nous pourrions avoir des exemples dans lesquels ce type de disjonctions est valides : « si un homme est roux, c'est un docteur ou si un homme est docteur, il est roux ».

Remarquez que si A et B sont des formules avec une structure : opérateur de nécessité+variable propositionnelle, alors la disjonction (1) n'est pas valide non plus. Nous savons aujourd'hui que ce cas particulier de la disjonction caractérise la logique **S.4.3**, qui, formulée sous la structure d'une sémantique de Kripke, équivaut à supposer la réflexivité, la transitivité et la linéarité de la structure. Pourtant si A représente une contradiction et/ou B une tautologie, alors ces trois formules se révèlent être valides. MacColl montre en fait que les cas suivants sont valides dans son système :

   4) $A:\varepsilon$
   5) $\eta:A$

où ε représente une formule exprimant une proposition arbitrairement nécessaire, et pareillement pour η. Manifestement les formules suivantes le sont aussi :

6) $B^\varepsilon{:}(A{:}B)$
7) $A^\eta{:}(A{:}B)$

Idem pour celle-ci :

8) $\eta{:}\varepsilon$

instanciée telle que

9) $(BB'){:}(A+A')$

MacColl concède que ces formules peuvent donner l'impression de *formules apparemment paradoxales*, mais elles ne le sont pas. Il déclare simplement, comme le font nombre de logiciens modaux contemporains, qu'elles doivent être acceptées parce que la signification qui découle de la définition de l'implication stricte n'est pas paradoxale.[49] Sur la même page où il concède la validité des formules que nous venons de présenter, il rejette explicitement :

10) $A^\eta A^{\varepsilon}$ [50]

L'argumentation de MacColl à propos de ce rejet est délicate à suivre, mais si nous regardons ensemble toutes ses explications sur le sujet[51] cela apparaît comme s'il rejetait les substitutions telles que :

11) $(AA')^\eta{:}(AA')^\varepsilon$

La construction de cette formule semble découler de la distinction entre la position prédicative de la modalité définie sur une classe et leurs interprétations en tant que formules. Sa principale explication est basée sur l'observation que des formules telles que 10 asserte que « *every impossibility of the class η*

---

[49] MacColl [1906a], p. 13.
[50] MacColl [1906a], p. 13.
[51] MacColl [1906a], pp. 13, 78-79.

is also an individual of the class of certainties ε which is absurd ».[52] Ce qui rend difficile la compréhension de la formule qui suit :

12) $(A+A')^\eta:(A+A')^\varepsilon$

D'une part, elle devrait être acceptée parce que le conséquent (ou l'antécédent) est valide (une contradiction), de plus la formule-sujet de l'antécédent $(A+A')$ est la même que la formule-sujet du conséquent (qui est apparemment contenue dans la classe ε). D'autre part, la lecture de cette formule en termes de classe, proposée par MacColl, rend la lecture de l'antécédent difficile : $A+A'$ n'est pas de la classe des formules impossibles. On pourrait encore soutenir que si tel est le cas pour cette dernière, alors 12 devrait être valide aussi.

Une autre possibilité est de conclure que la construction de règles pour les formules implicites dans 8 et 10 n'autorise pas de formules telles que 12. Les deux formules suivantes peuvent être construites à partir de 8 et 10 :

13) $(AA'):(AA')$

14) $(A+A'):(A+A')$

MacColl accepte-t-il les formules 4, 5 et 8 en supposant que la portée des modalités de l'antécédent et du conséquent ne soit pas de la même formule (Cf. 2.2.1.2 et suivant) ? Bien entendu, ceci ne signifie pas nécessairement qu'il rejette 13 et 14 ; simplement qu'elles ne sont pas dérivées de 8.
L'usage prédicatif de la modalité n'est après tout pas un opérateur, mais un genre de prédication métalogique sur les propositions. Effectivement cela semble être lié à l'utilisation que fait MacColl des thèses de la logique connexe – qui ne peuvent pas être prouvées dans la version standard de la logique modale **T** – avec sa méthode d'élimination des formules redondantes dans une implication. Dans la logique connexe de MacColl, selon notre reconstruction, ni 13, ni 14 ne sont valides. De plus l'utilisation de prédicats métalogiques « vrai » et « faux » dans $(A\tau + A\iota)^\varepsilon$ semble être liée à des restrictions métalogiques de l'implication connexe.[53] Nous nous trouvons ici au commencement d'une nouvelle piste de réflexion.

---

52   MacColl [1906a], p. 78.
53   Voir l'opérateur **V** et **F** dans la section 3. À l'aide de cet opérateur la for-

2.2. Vers une logique connexe et pertinente

MacColl refond la tradition philosophique de la logique connexe dans son système formel. La logique connexe est intimement liée à la critique qu'il fait de l'implication matérielle. Il est vrai que dans la logique connexe la substitution de la disjonction $(A:B)+(B:A)$ n'est pas seulement fausse, mais contradictoire.

$$(A':A)+(A:A')$$

en effet, la logique connexe contient comme thèses les négations de disjonctions. C'est-à-dire :

$$(A':A)'$$
$$(A:A')'$$

Plus encore, la logique connexe est au coeur de la reconstruction que MacColl opère du syllogisme qui lui permet certaines généralisations sur la notion de subalternation par le biais des thèses qui en sont dérivées.

$$(A:B):(A':B)'$$
$$(A:B):(A:B')'^{54}$$

Implanter ces thèses dans un système logique n'est pas une tâche aisée. MacColl y est parvenu en introduisant quelques restrictions métalogiques sur les modalités de l'antécédent et du conséquent qui sont en lien avec les observations que nous faisions à propos de l'identification qu'il opère entre nécessité et propositions valides. Commençons par présenter brièvement une description plus générale de la logique connexe.

2.2.1 La logique connexe

Une grande partie de la discussion à propos du conditionnel peut être posée en ces termes : ces conditionnels qui induisent une relation d'implication peuvent-ils être formulés dans un système formel ? La raison de l'échec de l'approche classique de l'implication est, d'une manière générale, d'avoir

---

mule $(A^{\tau}+A^{\iota})^{\varepsilon}$ peut être lue comme exprimant qu'il est valide que la proposition A est logiquement soit vraie soit fausse de façon contingente (cf. MacColl [1906a], p. 16).

[54] MacColl [1906a], pp. 49-65, 92-93.

ignoré le *sens de la signification* entre l'antécédent et le conséquent dans une implication valide. L'une des premières théories dans l'histoire de la logique traitant du sens de cette connexion provient de discussions stoïciennes visant à renforcer la relation entre antécédent et conséquent d'un conditionnel, qui dans ce contexte fut nommé συναρτησις (connexion) et joua un rôle majeur dans l'histoire de la philosophie.

Cette théorie donna une justification pour la validité de ce qu'aujourd'hui nous exprimons dans la logique classique standard à travers des formules telles que $\neg(a \rightarrow \neg a)$ et $\neg(\neg a \rightarrow a)$ (à partir de maintenant nous adopterons la terminologie standard pour l'implication matérielle et la noterons « $\rightarrow$ »).

Commençons premièrement par discuter deux exemples qui permettent de faire ressortir l'idée sous-jacente de la logique connexe. Le premier exemple est une modification d'une idée de Stephan Read (1994), qu'il utilisa contre la défense de l'implication matérielle par Grice. Le second est basé sur une idée de Lewis Carroll.

L'exemple de Read :
Cette exemple montre comment une disjonction donnée de propositions conditionnelles, aucune des deux n'étant vraies, est, d'un point de vue classique, néanmoins valide. Imaginez la situation suivante :
Stephen Read asserte que la logique de la pertinence sous sa forme dialogique ne fait pas vraiment partie de la logique. Supposez alors que Jacques Dubucs rejette l'assertion de Read.[55] Désormais, considérez les formules suivantes où a et b représentent des variables propositionnelles :

(1)   Si Read avait raison, Dubucs aurait raison aussi : $(a \rightarrow b)$

Maintenant (1) est manifestement faux. La proposition suivante l'est aussi :

(2)   Si Dubucs avait raison, Read aurait raison aussi : $(b \rightarrow a)$

Les disjonctions de (1) et (2) doivent donc être fausses :

(3) $(a \rightarrow b) \vee (b \rightarrow a)$

---

[55]   Le contenu de ces exemples est fictif.

Pourtant du point de vue classique, cette disjonction est valide bien que cela semble tout à fait contre-intuitif. Souvenez-vous de la critique de MacColl au sujet de l'implication matérielle basée précisément sur cette forme de disjonction.

Si nous reformulons (3) de la manière suivante :

(4) $(a \rightarrow \neg a) \vee (\neg a \rightarrow a)$

l'analyse vérifonctionnelle de cette disjonction, qui la conçoit comme valide, montre à quel point une telle théorie peut être embarrassante. Bien sûr cette disjonction n'est pas valide du point de vue intuitionniste, mais la double négation de cette dernière, valide selon le même point de vue, n'est pas non plus plausible.

Le point de la logique connexe est précisément qu'elle rend cette disjonction non-valide. La logique connexe vérifie les formules qui suivent :

(5) $\neg((a \rightarrow \neg a) \vee (\neg a \rightarrow a))$

ou

(6) $\neg(a \rightarrow \neg a)$

et

(7) $\neg(\neg a \rightarrow a)$.

La proposition (6) est connue sous le nom de *première thèse connexe boécienne*, la (7), *première thèse connexe aristotélicienne*.

À partir de maintenant, nous utiliserons un autre symbole pour le conditionnel connexe :

(8) $\neg(a \Rightarrow \neg a)$ (*première thèse connexe boécienne*)

(9) $\neg(\neg a \Rightarrow a)$ (*première thèse connexe aristotélicienne*)[56]

---

[56]   Il nous faut préciser que la thèse de la connexivité a reçu divers noms dans la littérature. Ce que nous appelons *première thèse boécienne* est quelques fois confondu avec la thèse aristotélicienne et ce que nous avons appelé *seconde thèse boécienne* est parfois simplement appelé thèse de Boèce.

L'exemple de Lewis Carroll :

Au 19ième siècle, Lewis Carroll a présenté un conditionnel que John Venn appela *le problème d'Alice* et qui donna lieu à plusieurs articles et discussions.

(10) $((a\rightarrow b)\wedge(c\rightarrow(a\rightarrow\neg b)))\rightarrow\neg c$

Si nous considérons $a\rightarrow\neg b$ et $a\rightarrow b$ comme étant incompatibles, le conditionnel doit être valide. Considérons par exemple les propositions suivantes :

(11)   Si Read avait raison, Dubucs aurait raison aussi : $(a\rightarrow b)$

(12)   Si Read avait raison, Dubucs aurait tort : $(a\rightarrow\neg b)$

Ces formules sont à première vue incompatibles, mais une fois encore, l'analyse vérifconctionnelle de ces propositions ne confirme pas cette intuition : si *a* est faux, les deux implications sont vraies. Boèce a présupposé cette incompatibilité dans nombre d'occasions. Ceci motiva Storrs MacCall à formuler la *seconde thèse connexe boécienne* :

(13) $(a\Rightarrow b)\Rightarrow\neg(a\Rightarrow\neg b)$ (*seconde thèse connexe boécienne*)

Aristote a plutôt utilisé des preuves relatives à cette formule :

(14) $(a\Rightarrow b)\Rightarrow\neg(\neg a\Rightarrow b)$ (*seconde thèse connexe aristotélicienne*)

qui est aujourd'hui appelée *seconde thèse connexe aristotélicienne*, Aristote montra même comment la première ainsi que la seconde de ses thèses connexes sont liées dans les *Premières analytiques* (57a36-b18). Aristote argumente contre $(a\Rightarrow b)\Rightarrow(\neg a\Rightarrow b)$ de la manière suivante : à partir de $a\Rightarrow b$ nous obtenons par contra-position $\neg b\Rightarrow\neg a$, et à partir de $\neg b\Rightarrow\neg a$ et $\neg a\Rightarrow b$, nous obtenons par transitivité $\neg b\Rightarrow b$, qui contredit la thèse $\neg(\neg b\Rightarrow b)$.

Le problème avec cette logique est que si nous essayons d'introduire la première thèse connexe aristotélicienne dans une logique telle que la logique classique standard, alors, en raison de la validité de :

$\neg(\neg a\rightarrow a)\rightarrow\neg a$

par modus ponens, pour n'importe quel *a* nous obtenons comme théorème ¬*a*.

Ce qui donne quelque chose de pire encore si nous essayons de faire de même avec la thèse boécienne, parce que nous obtenons *a*.

Il doit désormais être clair que les thèses connexes ne peuvent pas être introduites aussi facilement.[57] Comme trop souvent avec son travail, MacColl a touché du doigt la bonne idée, mais ne l'a pas explorée de façon systématique.

2.2.2. La logique connexe de MacColl et ses recherches sur la pertinence

2.2.2.1. Connexivité et subalternation

MacColl fut le premier à essayer d'introduire les thèses de la connexivité dans un système formel. Dans son article *The Calculus of Equivalent Statements*, il fournit les conditions suivantes pour la seconde thèse de Boèce :

> RULE 18- *If A (assuming it to be a consistent statement) implies B, then A does not imply B'* [i.e. not-B].
>
> Note.- *The implication $\alpha:\beta'$ asserts that α and β are inconsistent with each other; the non implication $\alpha \dot{=} \beta'$ asserts that α and β are consistent with each other.*
>
> [...] *α is a consistent statement – i.e., one which may be true.*[58]

MacColl introduit un pré requis métalogique de cohérence dans le langage objet tout en maintenant la seconde thèse boécienne de la connexivité. MacColl explique dans la note de bas de page citée ci-dessus que pour que la proposition *A* soit dite cohérente, elle doit pouvoir être vraie, c'est-à-dire qu'aucune contradiction logique ne doit pouvoir être dérivée de l'hypothèse de la vérité de *A*. Dans la version achevée de sa logique, il fait usage des modalités de contingence. De façon plus précise, il exige que l'antécédent et le conséquent d'un conditionnel connexe soient des propositions contingen-

---

[57] Routley et Montgomery [1968] ont étudiés les incidences de l'ajout de la thèse connexe dans la logique classique.
[58] MacColl [1877p], p. 184.

tes.⁵⁹ Désormais, contingence signifie ici une formule qui ne soit ni une contradiction ni une formule valide. L'interprétation de la modalité que nous présentons ici est l'une de celles mentionnées ci-dessus en relation avec les différentes interprétations que MacColl donne à la modalité suivant la position prédicative. Cette modalité introduit des propriétés métalogiques dans le langage objet, ceci a certainement dû concourir à l'idée qu'il se soit approché de considérations touchant aux logiques non-normales, non loin de la logique **T**. MacColl avertit explicitement que sa modalité peut être aussi bien *formelle* (liée à aucune donnée ou information) que *matérielle* (liée à des données ou des informations).⁶⁰

Comme mentionné précédemment, MacColl fait usage de la thèse de la connexivité pour formuler une version générale de la subalternation avec et sans engagement ontologique ; tandis que Aristote l'utilise pour prouver certaines propriétés de sa syllogistique. Il est intéressant de remarquer que la reconstruction qu'opère MacColl de la logique aristotélicienne aboutit à une syllogistique sans engagement ontologique. Pour mettre ce point plus en évidence, permettez nous de combiner l'écriture quantificationnelle standard contemporaine avec celle propositionnelle de MacColl. Avec une telle association la subalternation doit alors être formulée comme suit :

$$\forall x(Ax{:}Bx){:}\forall x(Ax{:}B\;'x)\;'$$

Il apparaît très clairement que la version de la subalternation de MacColl n'est valide que si vous assumez que $(A{:}B){:}(A{:}B\;)\;'$ est valide. La validité de cette forme de subalternation ne suppose aucun engagement ontologique.

Une fois de plus, l'idée principale contenue ici, aussi loin que nous avons pu pousser la reconstruction de la logique de MacColl, est que le conditionnel connexe est un genre d'implication stricte à laquelle s'ajoute deux restrictions ; qu'il n'y ait ni contradiction dans l'antécédent et ni tautologie dans le conséquent. Cette façon de produire divers systèmes logiques semble être typique au style de MacColl. En réalité, il ne fait qu'explorer l'idée qui consiste à éliminer les *redondances* dans une implication dans le but d'introduire un lien de pertinence entre antécédent et conséquent.

---

59   MacColl [1906a], pp. 45, 49-65, 92-93.
60   MacColl [1906a], p. 97.

2.2.2.2. Vers la pertinence suivie de remarques conclusives

Déjà en 1878 dans le troisième article de sa série *The Calculus of Equivalent Statements*, MacColl semble être intéressé par le développement de méthodes d'élimination des redondances pour une formule donnée. Il déclare à propos de ce type d'élimination dans une disjonction (qu'il nomme énoncé *indéterminé*) :

> *Any term of an indeterminate statement may be omitted as redundant when this term, multiplied by the denial of the sum of all its co-terms, gives the product 0.*[61]

Comme MacColl le met en évidence, plus loin dans le passage, la méthode est basée sur la transformation d'une disjonction en une implication matérielle, où les présumés disjoints redondants constituent l'antécédent. Si l'implication matérielle en résultant est valide, alors la disjonction en jeu est redondante.

Étant donné A^0+B+C, l'implication matérielle $(A^0(B+C)')'$ est valide. Il s'agit bien d'un cas du paradoxe de l'implication matérielle que nous avons discuté précédemment. De même, nous obtenons l'autre cas du paradoxe de l'implication matérielle si l'un des éléments est matériellement impliqué par un ou plusieurs autres. Donc en prenant A^1BC, l'implication matérielle $(BC(A^1)')'$ en résultant est effectivement valide.

Partant de là, on pourrait certainement utiliser la même méthode pour obtenir une notion plus stricte d'implication, qui soit une notion d'implication sans redondance. Cela semble être le lieu commun de ses diverses méthodes visant à obtenir une *prémisse plus forte* ou/et une conclusion affaiblie.[62] Rahman (1997a) suggère que la méthode de MacColl pour éliminer les redondances contient le concept suivant de pertinence. Souvenez-vous que dans cette section et dans la précédente, nous sommes revenus à la notation et utilisation classiques standards « → ».

Nommons un ensemble (d'occurrences indexées) de variables propositionnelles ayant un poids dans la *détermination de la vérité* pour une formule A si et seulement si la valeur de vérité de A peut être déterminée comme vraie ou fausse quelque soit l'assignation de vérité, vrai ou faux, de l'ensemble. Permettez nous de dire que (l'occurrence indexée i d') une variable proposi-

---

[61] MacColl [1878d], p. 17.
[62] MacColl [1906a], pp. 27-33.

tionnelle dans A est *redondante* si et seulement si il est connu qu'il y ait un ensemble dont la vérité est déterminée pour A qui ne contienne pas (d'occurrence indexée i de) cette variable propositionnelle. Donc, l'ensemble $\{a\}$ n'a manifestement pas de vérité déterminée pour $a \rightarrow b$, mais l'ensemble $\{a_1, a_2\}$ est déterminant pour la vérité de $a \rightarrow a$.[63]

En conformité à cela, toutes les formules suivantes qui contiennent des redondances doivent être exclues du contexte de l'inférence :

$((a \rightarrow b) \rightarrow a) \rightarrow a$ $a \rightarrow (b \rightarrow a)$
$((a \rightarrow a) \rightarrow a) \rightarrow a$ $a \rightarrow (a \rightarrow a)$
$a \rightarrow (b \vee \neg b)$ $a \rightarrow (a \vee b)$
$a \rightarrow (a \vee \neg a)$ $a \rightarrow (a \vee a)$
$(a \wedge \neg a) \rightarrow b$ $(a \wedge b) \rightarrow a$
$(a \wedge \neg a) \rightarrow a$ $(a \wedge a) \rightarrow a$

Dans son dernier article ainsi que dans son livre, MacColl combine ce concept de redondance avec son implication stricte. Notez que dans ce contexte l'exclusion de la formule 11 (en 2.1 ci dessus) peut être expliquée par le fait qu'elle contient une *occurrence* redondante de l'antécédent.

Un défi consisterait à de rassembler toutes ces idées et de produire une logique connexe pertinente recoupant les différentes suggestions de MacColl. Un propos hautement plus philosophique chercherait à lier tout ceci avec la notion d'information que contient sa notion d'énoncé. La notion de redondance comprise comme une perte d'information semble être un bon point de départ pour une telle enquête.

Permettez nous de conclure notre article par un travail en l'honneur de MacColl : une reconstruction de sa notion de logique connexe dans le style de preuve théorétique qui commença à se développer quelques années après sa mort.

---

[63] Cf. Rahman [1998], p. 37.

3. TABLEAU POUR LA LOGIQUE CONNEXE DE MACCOLL[64]

3.1. Introduction

MacColl, et plus récemment R. Angell (1962), S. McCall (de 1963 à 1975), et C. Pizzi (1977, 1993, 1996) ont cherché un système formel dans lequel la validité des formules connexes pouvait être exprimée. De nouveaux résultats ont été mis à jour par R. Angell (2002), M. Astroh (1999a), Cl. Pizzi/T. Williamson (1997), G. Priest (1999), S. Rahman/H. Rückert (2001) et H. Wansing (2005). Wansing a aussi écrit un aperçu très consciencieux sur la logique connexe dans l'encyclopédie philosophique de Stanford.

Notre article présente un nouveau système de tableaux pour la logique connexe ayant pour but d'approfondir les idées originales de Hugh MacColl, il est en même temps une version non-dialogique et une réécriture de Rahman/Rückert (2001). Dans ce contexte rappelons tout de même que Stephan Read a montré que la logique modale de MacColl, sous une reconstruction algébrique donne naissance au système modal **T** et qu'il conteste les propos de Storrs MacCall sur la logique de MacColl faisant de cette dernière ce qui est connu depuis Lewis comme la logique **LS3**. Notre propos sera de vous montrer qu'il peut y avoir entre ces deux conceptions de la logique de MacColl une certaine forme de compromis. Effectivement MacColl eut le système **T**, mais lorsqu'il traita du conditionnel connexe, qui est essentiel pour sa reconstruction du syllogisme, il préfigura une logique qui implique des propriétés métalogiques du système **T** et ce qui semble être en lien direct avec ce que nous appelons contemporainement logique modale non-normale (contextes dans lesquels des tautologies peuvent être fausses).

En outre nous fournissons l'idée principale pour une sémantique de Kripke pour ces tableaux. Nous suivons l'idée de MacColl selon laquelle le conditionnel connexe est un genre d'implication stricte avec deux conditions de restrictions supplémentaires ; à savoir, qu'il n'y ait ni contradiction dans l'antécédent ni tautologie dans le conséquent. C'est-à-dire que les mondes (du modèle correspondant) où l'antécédent et le conséquent sont évalués, ne doivent contenir ni contradictions ni tautologies.

---

[64] Cette section a était principalement écrite par Shahid Rahman.

Dans la présente reconstruction de la logique connexe de MacColl, dans le cadre des tableaux, nous ferons usage de deux opérateurs avec l'interprétation : « α n'est pas une vérité logique » et « il y a au moins un modèle dans lequel α est vrai ». Ces opérateurs donnent lieu à un genre de logique modale implicite. Dans la troisième partie de cette réflexion, nous introduirons la sémantique de Kripke dans laquelle deux sortes distinctes d'ensembles de mondes traitent respectivement de la vérité logiquement contingente et de la fausseté logiquement contingente d'une formule. Pour finir, nous esquisserons la façon de combiner la logique connexe de MacColl avec sa notion de pertinence logique de la section précédente (2.2.1.2).

Contrairement au courant dominant, nous partirons de la sémantique de la preuve théorétique pour aboutir à la sémantique du modèle théorétique. La philosophie est controversée, pour reprendre les mots de Graham Priest, mais c'est ainsi que les logiques de la pertinence contemporaines furent développées. Qui plus est, nous pensons que l'interprétation dialogique est une structure suffisante pour introduire les intuitions sémantiques en fond tel le difficile sujet de la pertinence ainsi que celui de la « connexion logique ».

3.2. Tableau pour la logique connexe

Le système de tableaux présenté est basé sur la première formulation du conditionnel connexe dialogique introduit par Rahman (1997a) dans *Habilitationsschrift*, et davantage développée dans Rahman/Rückert (2001).

Il nécessite deux paires de signes, à savoir {**O**, **P**} et {●, ○}. L'interprétation choisie pour {**O**, **P**} est *Opposant, Proposant* et pour ce qui est de {●, ○} : *en charge de la validité de la preuve, n'est pas en charge de la validité de la preuve*. Si cette interprétation dialogique ne vous donne pas entière satisfaction, vous pouvez la changer pour *vrai, faux, négation, affirmation* ou autre. En fait c'est la seconde paire {●, ○} qui corresp aux symboles *f, t* (règles de droite et de gauche dans le calcul standard des séquents).

Les règles du tableau doivent alors inclurent la combinaison des deux paires de signes. Pour des raisons de simplicité, les règles doivent être formulées pour **X** et **Y**, qui sont utilisées pour **O** et **P**. Pour les constantes de la logique standard nous avons donc l'ensemble de règles suivantes :

Cas de (Y°)	*Cas de* (X•)
$\Sigma, (Y°)\alpha \vee \beta$	$\Sigma, (X•)\alpha \vee \beta$
---------------	---------------
$\Sigma, (Y°)\alpha \mid \Sigma, (Y°)\beta$	$\Sigma, (X•)\alpha$ $\Sigma, (X•)\beta$
$\Sigma, (Y°)\alpha \wedge \beta$	$\Sigma, (X•)\alpha \wedge \beta$
---------------	---------------
$\Sigma, (Y°)\alpha$ $\Sigma, (Y°)\beta$	$\Sigma, (X•)\alpha \mid \Sigma, (X•)\beta$
$\Sigma, (Y°)\alpha \to \beta$	$\Sigma, (X•)\alpha \to \beta$
---------------	---------------
$\Sigma, (X•)\alpha \mid \Sigma, (Y°)\beta$	$\Sigma, (Y°)\alpha$ $\Sigma, (X•)\beta$
$\Sigma, (Y°)\neg\alpha$	$\Sigma, (X•)\neg\alpha$
---------------	---------------
$\Sigma, (X•)\alpha$	$\Sigma, (Y°)\alpha$

Les règles de clôture sont les suivantes :

> Un tableau pour $(X•)\alpha$ (i.e. commençant avec $(X•)\alpha$) est clos si et seulement si chaque branche (incluant chacun des *sous-tableaux* possibles) est clôturée soit par l'occurrence d'une paire de formules atomiques de la forme $((Y°)a, (X•)a)$ ou *par une règle de clôture spéciale*. Autrement il est dit ouvert.

Les raisons qui nous poussent à inclure des clauses sur les *sous-tableaux* et des *règles spéciales de clôture* seront données dans la prochaine section.

Les opérateurs **V** et **F**

Comme mentionné précédemment, la présente reconstruction du conditionnel connexe de MacColl fait usage des opérateurs de *satisfiabilité* **V** et **F**, son opérateur dual. L'opérateur **F** est en lien avec le très connu *opérateur d'échec* de Prolog[65]. Introduisons les règles de leurs tableaux respectifs.

L'opérateur **V**
L'interprétation choisie pour cet opérateur est : « il y a au moins un modèle où α est vrai ». Dans le cadre des tableaux, l'interprétation que nous choisissons pour les occurrences d'une formule telle que **V**α dans une branche est : « il y a un **(sous)tableau ouvert** pour α ».

En réalité, on pourrait concevoir un tableau pour α comme une séquence finie de sous-tableaux telle que le premier tableau ait α pour origine, et les autres membres de la séquence sont obtenus par application des règles de tableaux. Pourtant, nous indexerons explicitement uniquement les sous-tableaux ouverts par **V** et **F**. Pour garder une trace entre un tableau et un sous-tableau, nous ferons usage d'index. Si la branche dans laquelle l'un des deux opérateurs apparaît contient l'index **i**, alors le sous-tableau sera indexé **i.1**.

D'une manière plus générale, l'intuition est qu'un index **i** nomme un sous-tableau et **i**A nous indique que A est évalué dans le sous-tableau que **i** nomme. Nos index sont des séquences finies de nombres entiers positifs tels que 1.1.1 et 1.1.2.

DEFINITION

> ➤ Un *index* est une séquence finie d'entiers positifs. Une *formule indexée* est une expression de la forme **i**φ, où **i** est l'index de la formule φ.

> ➤ Si l'index **i** est une séquence de longueur >1, l'entier positif de la séquence sera séparé par périodes. Donc, si **i** est un index et **n** est un entier positif, alors **i.n** est un nouvel index, appelé *extension de* **i**. L'index est alors un *segment initial* de **i.n**.

---

[65] Gabbay [1987] a utilisé cet opérateur pour la logique modale. Hoepelmann et Van Hoof [1988] ont appliqué cette idée de Gabbay à la logique non-monotone. Pour finir Rahman ([1997a], chapitre II(A).4.2), introduit l'opérateur **F** dans la formulation de la sémantique des tableaux et des stratégies dialogiques pour la logique connexe.

Supposez que l'expression (**P•**)**V**α apparaisse dans une branche avec l'interprétation suivante :

> « le **P**roposant, qui dans cette branche a en charge la preuve de la validité, déclare qu'il y a un tableau ouvert pour α ».

Cette formule générera un sous-tableau pour **P°**α avec pour interprétation :

> « le **P**roposant, qui dans ce sous-tableau, n'est pas en charge de la preuve de la validité, déclare qu'il y a un (sous)tableau ouvert pour α ».

Les règles de tableaux pour l'opérateur **V** doivent inclure la combinaison de deux paires de signes {**O**, **P**} et {•, ○} :

Cas de (**Y**°)	*Cas de* (**X•**)
n, Σ, (**Y**°)**V**α i.	n, Σ, (**X•**)**V**α i.
n+1, Σ, (**Y**°)α i.1	n+1Σ, (**X**°)α i.1

« n » est le nombre de pas dans le (sous)tableau **i** où **V**α apparaît.

Les conditions de clôture de **toutes** les branches (la branche qui commence avec la formule principale du tableau et qui se termine avec un sous-tableau avec l'index **i.1**) doivent être désormais claires :

> ➢ La branche d'un (sous)tableau **i** dans laquelle **V**α apparaît est **ouverte au coup n** si le sous-tableau **i.j** est clos et réciproquement si le sous-tableau est ouvert alors il **clôturera en n**.

Exemples :

Dans les exemples suivants « n » indique le nœud de la branche du tableau **i** où **V** apparaît et **i.1** est le sous-tableau généré par une application de la règle **V**.

Exemple 1

$$\begin{array}{c}
\vdots\\
n\ (\mathbf{P^\bullet})\ \mathbf{V}a\wedge\neg a\ \mathbf{i}\\
(\mathbf{P^\circ})\ a\wedge\neg a\ \mathbf{i.1}\\
(\mathbf{P^\circ})\ a\ \mathbf{i.1}\\
(\mathbf{P^\circ})\ \neg a\ \mathbf{i.1}\\
(\mathbf{O^\bullet})\ a\ \mathbf{i.1}
\end{array}$$

La branche de **i** *est* **ouverte en** **n** *parce que le sous-tableau* **i.1 ferme** *avec*
$\{(\mathbf{P^\circ})\ a, (\mathbf{O^\bullet})\ a\}$

Notez que le sous-tableau correspond au tableau standard pour $t\ a\wedge\neg a$.

Exemple 2

$$\begin{array}{c}
\vdots\\
n\ (\mathbf{P^\circ})\ \mathbf{V}\neg a\ \mathbf{i}\\
(\mathbf{P^\circ})\ \neg a\ \mathbf{i.1}\\
(\mathbf{O^\bullet})\ a\ \mathbf{i.1}
\end{array}$$

La branche se **ferme en** **n** *parce que le sous-tableau est*
ouvert

Notez que le sous-tableau correspond au tableau standard pour $t\ \neg a$.

L'opérateur **F**

L'opérateur **F** est le dual de **V**. L'interprétation choisie pour cet opérateur est : « la formule α n'est pas une vérité logique ». Dans le cadre des tableaux, l'interprétation choisie pour l'occurrence de la formule **F**α dans une branche est : « il y a un **(sous)tableau non-clôturé** pour α ».

Plus précisément, permettez nous de supposer que l'expression $(\mathbf{P^\bullet F})\alpha$ apparaisse dans une branche avec l'interprétation suivante :

« le **P**roposant qui, dans cette branche, a la charge de la preuve de la validité, déclare qu'il y a un tableau non clôturé pour α ».

Cette formule génèrera un sous-tableau pour **O**•α avec l'interprétation choisie :

« L'**O**pposant déclare qu'il y a un (sous)tableau clos pour α pour lequel il est en charge de la preuve de la validité ».

Les règles sont les suivantes :

Cas de (Y°)-	*Cas de* (X•)
n, Σ, (Y°)**F**α i	n, Σ, (X•)**F**α i
n+1 Σ, (X•)α i.1	n+1 Σ, (Y•)α i.1

Notez que cet opérateur produit le changement de (X•) en (Y•) et vice versa.

➢ La branche du (sous)tableau **i** où **F**α apparaît est **ouverte au coup n** si le sous-tableau **i.j** est clos et réciproquement si le sous-tableau est ouvert alors **i** est **clos en n**.

Exemples :

Exemple 3

.n (**P•**)**F**$a \vee \neg a$ i
(**O•**) $a \vee \neg a$ i.1
(**O•**) a i.1
(**O•**) $\neg a$ i.1
(**P°**)a i.1

La branche est **ouverte en n** *parce que le sous-tableau*
se ferme *avec*
{(**P°**) a, (**O•**) a}

Exemple 4

.

n (**P°**) **F**$\neg a$ i
(**O•**) $\neg a$ i.1
(**P°**) a i.1

La branche est **clôturée en n** *parce que le sous-tableau*
est **ouvert**

Le conditionnel connexe

Règles de tableaux pour le conditionnel connexe

Comme vous avez pu le comprendre, la reformulation du conditionnel connexe comprend les restrictions qui suivent sur l'implication stricte :

1. L'antécédent doit être contingent logiquement (pas incohérent).

2. Le conséquent ne doit pas être tautologique.

3. L'implication stricte est pensée dans une structure réflexive. C'est-à-dire que l'implication matérielle est valable dans un monde où l'implication stricte l'est aussi.

Ces trois conditions peuvent être exprimées dans la structure d'une logique modale implicite et dans le système de tableaux tel que nous venons de le décrire via les opérateurs **V** et **F** :

> $\alpha \Rightarrow \beta$ est connexivement valide si l'antécédent (de l'implication stricte) donne lieu à un sous-tableau ouvert, *i.e.* il y a au moins un modèle dans lequel l'antécédent est vrai (c'est-à-dire, si **V**α est valable). Ici, l'idée est que ***ex contradictione nihil sequitur*** (rien n'est dérivable d'une contradiction).

Pareillement :

> $\alpha \Rightarrow \beta$ est *non-connexe* si le conséquent (de l'implication stricte) donne lieu à un sous-tableau clôturé, *i.e.* il y a au moins un modèle dans lequel le conséquent est faux (c'est-à-dire, si **F**β est valable). Ici, l'idée est que ***ex quodlibet verum nequitur*** (il n'y a pas de proposition à partir de laquelle découle une tautologie ou une vérité supposée).

Ce qui donne les règles suivantes :

Cas de (Y°)-	*Cas de* (X•)
Σ, (Y°) $\alpha \Rightarrow \beta$ i	Σ, (X•) $\alpha \Rightarrow \beta$ i
-----	Σ, (X•)$\alpha \rightarrow \beta$ i | (X•)**V**α i | (X•)**F**β i
Σ, (Y°)$\alpha \rightarrow \beta$ i Σ, (Y°)**V**α i Σ, (Y°)**F**β i	

L'application des règles de tableaux du conditionnel connexe pour les cas de (**X•**) produit une conjonction avec trois membres, à savoir les formules avec les deux opérateurs **V** et **F** ainsi que *l'implication matérielle* standard. Nous allons dès à présent introduire les règles qui ont pour effet de produire une logique modale implicite.

> Les sous-tableaux découlant de **V**α et **F**α peuvent contenir : les sous-formules de **V**α et **F**α, des implications matérielles et aucune autre formule que les sous-formules de ces implications. Pensez les sous-tableaux comme des mondes où seules les formules qui sont

issues de la branche supérieure sont précisément les implications matérielles correspondantes. C'est exactement comme si chaque implication du tableau supérieur fonctionnait comme une implication stricte.

> ➤ Les sous-tableaux produits par **V** et **F** sont différents. Bien que les sous-tableaux aient des liens de filiation (en l'occurrence l'implication matérielle), il n'y a pas d'accessibilité entre eux.

Pour introduire ceci, nous utilisons une idée proche du dispositif d'élimination de la logique intuitionniste : les formules **V** (ou **F**) génèrent des sous-tableaux qui ne contiennent aucune autre formule que les sous-formules dérivées des formules **V** (ou des formules **F**), et le conditionnel standard (implication matérielle) du tableau supérieur. Les sous-tableaux générés par les opérateurs **V** et **F** s'ouvrent avec l'ensemble $\Sigma_{[\rightarrow]}$ où « $\Sigma_{[\rightarrow]}$ » se lit comme suit :

> ➤ **règle $\Sigma_{[\rightarrow]}$:**
> Si α est une sous-formule d'une formule de **V** (ou formule de **F**) et $\Sigma \cup \alpha$ le commencement du tableau correspondant, alors il faut remplacer dans $\Sigma \cup \alpha$ **chaque** conditionnel connexe ayant une occurrence dans le tableau supérieur comme connecteur principal par l'implication matérielle correspondante, changer la **charge de la preuve** de(s) l'implication(s) matérielle(s) si nécessaire, en fonction de la règle pour l'opérateur en jeu. Aucune autre formule n'est un élément de $\Sigma_{[\rightarrow]}$.

En d'autres termes, chaque sous-tableau doit au départ contenir soit l'antécédent, soit le conséquent du conditionnel connexe et l'implication matérielle lui correspondant. L'importance mise sur *chaque conditionnel* est rendue claire dans l'exemple 3.

Cas de (Y°)	*Cas de* $(X^\bullet)$
$\Sigma, (Y^\circ)\mathbf{V}\alpha\,i.$	$\Sigma, (X^\bullet)\mathbf{V}\alpha\,i.$
$\Sigma_{[\to]}, (Y^\circ)\alpha\,i.n$	$\Sigma_{[\to]}, (X^\circ)\alpha\,i.n$
$\Sigma, (Y^\circ)\mathbf{F}\alpha\,i.$	$\Sigma, (X^\bullet)\mathbf{F}\alpha\,i.$
$\Sigma_{[\to]}, (Y^\circ)\alpha\,i.m$	$\Sigma_{[\to]}, (X^\circ)\alpha\,i.m$

Le tableau principal comme par α avec $(\mathbf{P}^\bullet)\alpha$.

> **Règle de commencement stratégique pour les logiques connexes :**
> Nous supposons qu'un tableau pour α commence avec $(\mathbf{P}^\bullet)\alpha$. Donc, un tableau clos (un tableau dont toutes les branches sont fermées) pour α prouve que α est valide.

Bien que la répétition des clôtures des branches soit sans détour, la formulation précise de la règle de clôture adéquate est un peu plus gênante parce qu'elle doit inclure des cas où la branche se termine avec un sous-tableau et des branches où le sous-tableau n'apparaît pas.

> **R-C** : Règles de clôture pour Ω (Ω est soit **V**, soit **F**) : Supposez un tableau ayant pour point de départ $(\mathbf{P}^\bullet)\alpha$. Si une branche de ce tableau se termine part un sous-tableau généré soit par une formule $(\mathrm{P}^\bullet)\Omega$, soit par une formule $(\mathrm{P}^\circ)\Omega$, alors l'ensemble des branches se clôture si et seulement si le sous-tableau ne se ferme pas. Une branche du même tableau ne se terminant pas par un sous-tableau de la forme décrite ci-dessus[66] est close si et seulement si elle se termine par une paire de la forme $(\mathrm{O}^\circ)\alpha$, $(\mathbf{P}^\bullet)\alpha$. Autrement il est dit ouvert.

> Notez que nous considérons qu'une branche comprend la séquence entière de sous-tableaux générés.

---

[66] C'est-à-dire, si le jeu se termine sans aucun sous-tableau ou s'il se termine par un sous-tableau généré par une formule de type $(\mathbf{O}^\circ)\Omega$, ou de type $(\mathbf{O}^\bullet)\Omega$.

Exemples :

Dans les exemples suivants, la numérotation à gauche des formules indique le pas dans la preuve alors que celui de droite est l'index du tableau et du sous-tableau :

Exemple 5

1   (**P•**)¬(a⇒¬a) **1**
2   (**O°**)a⇒¬a **1** (règle de négation sur 1)
3   (**O°**)a→¬a **1** (règle de gauche pour le conditionnel connexe sur 2)
4   (**O°**)**V**a **1** (règle de gauche pour le conditionnel connexe sur 2)
5   (**O°**)**F**¬a **1** (règle de gauche pour le conditionnel connexe sur 2)
6   (**O°**)a **1.1** (règle de **V** sur 4)
7   (**O°**)a→¬a **1.1** (règle de $\Sigma_{[\to]}$ sur 4)

Appliquons maintenant la règle standard sur l'implication matérielle (**O°**)a→¬a au sous-tableau **1.1** :

    8   (**P•**)a **1.1** | 9   (**O°**)¬a **1.1**
    10 (**P•**)a **1.1** |

La branche (unique) du tableau se termine avec un sous-tableau généré par une formule (**O°**)**V**. Les deux branches du sous-tableau (**1.1**) se ferment avec (**O°**)a et (**P•**)a. Donc, selon **R-C**, l'ensemble des branches du tableau sont fermées, par conséquent le tableau lui-même **est clos**.

La preuve pour ¬(¬a⇒¬a) est sensiblement la même, mais nécessite l'utilisation de l'autre opérateur :

Exemple 6

1   (**P•**)¬(¬a⇒a) **1**
2   (**O°**)¬a⇒a **1** (règle de négation sur 1)
3   (**O°**)¬a→a **1** (règle de gauche pour le conditionnel connexe sur 2)
4   (**O°**)**V**¬a **1** (règle de gauche pour le conditionnel connexe sur 2)
5   (**O°**)**F**a **1** (règle de gauche pour le conditionnel connexe sur 2)
6   (**P•**)a **1.1** (règle de **F** sur 5)
7   (**O°**)¬a→a **1.1** (règle de $\Sigma_{[\to]}$ sur 5)

Appliquons désormais la règle standard sur l'implication matérielle (**O°**)¬a→a au sous-tableau **1.1** :

    8   (**P•**)¬a **1.1** | 9   (**O°**)a **1.1**
    10   (**O°**)a **1.1** |

Les deux branches du sous-tableau **1.1** générées par $(\mathbf{O}°)\mathbf{F}a$ se ferment avec $(\mathbf{O}°)a$ et $(\mathbf{P}^\bullet)a$. Donc, selon **R-C**, la branche (unique) du tableau est close, par conséquent le tableau lui-même **est clos**.

Exemple 7

Dans les exemples suivants, toute la force de la règle $\Sigma_{[\to]}$-entre en action. Cette règle permet de « charrier » **tous les conditionnels connexes** comme des implications matérielles dans les sous-tableaux générés.

$$1\ (\mathbf{P}^\bullet)(a{\Rightarrow}b) \Rightarrow \neg(a{\Rightarrow}\neg b)\ \mathbf{1}$$

2 $(\mathbf{P}^\bullet)(a{\Rightarrow}b){\to}\neg(a{\Rightarrow}\neg b)$ **1** | 3 $(\mathbf{P}^\bullet)\mathbf{V}(a{\Rightarrow}b)$ **1** | 4 $(\mathbf{P}^\bullet)\mathbf{F}\neg(a{\Rightarrow}\neg b)$ **1**

A ce stade, il doit être clair que le développement d'une formule **V** et **F** (3, 4) doit produire des branches closes, parce que l'antécédent ainsi que le conséquent du conditionnel connexe sont logiquement contingents.

Regardons ce qui se passe pour la branche (2) située à l'extrémité gauche du tableau si nous déroulons les règles de l'implication matérielle.

BRANCHE DE GAUCHE (GAUCHE)

        5 $(\mathbf{O}°)$ $(a{\Rightarrow}b)$ **1** (implication matérielle sur 2)
        6 $(\mathbf{P}^\bullet)\neg(a{\Rightarrow}\neg b)$ **1** (implication matérielle sur 2)
        7 $(\mathbf{O}°)a{\Rightarrow}\neg b$ **1** (négation sur 6)
        8 $(\mathbf{O}°)a{\to}b$ **1** (règle de gauche pour le conditionnel connexe sur 5)
        9 $(\mathbf{O}°)\mathbf{V}a$ **1** (règle de gauche pour le conditionnel connexe sur 5)
        10 $(\mathbf{O}°)\mathbf{F}b$ (règle de gauche pour le conditionnel connexe sur 5)
        11 $(\mathbf{O}°)$ a **1.1** (règle de **V** sur 9)
        12 $(\mathbf{O}°)a{\to}b$ **1.1** (règle de $\Sigma_{[\to]}$ sur 5)
        13 $(\mathbf{O}°)a{\to}{\neg}b$ **1.1** (règle de $\Sigma_{[\to]}$-sur 7)
14 $(\mathbf{P}^\bullet)a$ ...**1.1** | 15 $(\mathbf{O}°)b$ **1.1**
       | 16 $(\mathbf{P}^\bullet)a$ **1.1** | 17 $(\mathbf{O}°)\neg b$ **1.1**
       | | 18 $(\mathbf{P}^\bullet)b$ **1.1**

Le sous-tableau se clôture en raison des paires $((\mathbf{O}°)a, (\mathbf{P}^\bullet)a)$ et $((\mathbf{O}°)a, (\mathbf{P}^\bullet)a)$. D'après **R-C**, l'ensemble des branches **sont closes**. Comme nous avons déjà montré que les branches (3, 4) sont closes elles aussi, c'est donc le tableau entier qui est clôturé.

L'exemple qui suit expose le cas d'une formule qui est dangereuse pour la logique connexe, c'est une formule qui montre que si nous ajoutons les deux thèses de la logique connexe, elle explose dans la trivialité. En effet si on ajoute $\neg(\alpha\rightarrow\neg\alpha)$ et $\neg(\neg\alpha\rightarrow\alpha)$ à la logique classique, alors les formules suivantes se tiennent :

$$\neg(\alpha\rightarrow\neg\alpha)\rightarrow\alpha$$
$$\neg(\neg\alpha\rightarrow\alpha)\rightarrow\neg\alpha$$

Laissez nous vous montrer que ce type de trivialité ne se produit pas dans notre système. Nous allons prouver qu'un tableau pour $\neg(a\Rightarrow\neg a)\Rightarrow a$ ne se clôture pas. La preuve de la formule duale $\neg(\neg\alpha\Rightarrow\alpha)\Rightarrow\neg\alpha$ est exactement la même.

Exemple 8

$$1\ (\mathbf{P}^\bullet)\neg(a\Rightarrow\neg a)\Rightarrow a\ \mathbf{1}$$

2 $(\mathbf{P}^\bullet)\neg(a\Rightarrow\neg a)\rightarrow a\ \mathbf{1}$ | 3 $(\mathbf{P}^\bullet)\mathbf{V}\neg(a\Rightarrow\neg a)\ \mathbf{1}$ | 4 $(\mathbf{P}^\bullet)\mathbf{F}a\ \mathbf{1}$
(règle de droite pour « $\Rightarrow$ » sur 1)

Les branches 3 et 4 se ferment. La branche de l'extrémité de droite (4) se ferme parce que $(\mathbf{P}^\bullet)\mathbf{F}\ a$ génère un sous-tableau qui restera ouvert.

Montrer que la branche 3 se ferme est un peu plus délicat. Le sous-tableau qui est généré s'ouvre avec $(\mathbf{P}^\circ)\neg(a\Rightarrow\neg a)$, et continue avec $(\mathbf{O}^\bullet)a\Rightarrow\neg a$. Les trois branches de ce sous-tableau se ferment et causent la fermeture de l'ensemble des branches. Explication : la première branche du sous-tableau contenant l'implication matérielle $((\mathbf{O}^\bullet)a\rightarrow\neg a)$ sera ouverte selon les règles standards et cela produira la fermeture de la branche. La seconde branche du sous-tableau contenant $(\mathbf{O}^\bullet)\mathbf{V}a$, génère un sous-tableau dans lequel les formules $(\mathbf{O}^\circ)a$, $(\mathbf{O}^\bullet)a\rightarrow\neg a$ donnent lieu à la clôture de la branche par $(\mathbf{O}^\circ)a$, $(\mathbf{P}^\bullet)a$. Pour la dernière branche du tableau, celle contenant $(\mathbf{O}^\bullet)\mathbf{F}\neg a$, le cas est identique que pour la seconde.

Malheureusement nous n'en avons pas encore fini avec cet arbre. Il y a une branche qui fera que le tableau entier restera ouvert. Voyons cela dans la branche 2 contenant $(\mathbf{P}^\bullet)\neg(a\Rightarrow\neg a)\rightarrow a$:

BRANCHE DE GAUCHE (GAUCHE)

    5 ($\mathbf{O}°$)a⇒¬a) **1** (implication matérielle sur 2)
    6 ($\mathbf{P}^\bullet$)a **1** (implication matérielle sur 2)
    7 ($\mathbf{P}^\bullet$)a⇒¬a **1** (négation sur 6)
8 ($\mathbf{P}^\bullet$)a→¬a **1** | 9 ($\mathbf{P}^\bullet$)$\mathbf{V}a$ **1** | 10 ($\mathbf{P}^\bullet$)$\mathbf{F}$¬a **1**
    (règle de droite pour le conditionnel connexe sur 7)

L'application des règles standards de tableaux sur 8 donne ($\mathbf{O}°$)a. ($\mathbf{P}^\bullet$)¬a qui se ferme avec 6 ($\mathbf{P}^\bullet$)a. Développons la branche du milieu (9) et ouvrons le sous-tableau **1.1**.

    11 ($\mathbf{P}°$) a **1.1** (règle de $\mathbf{V}$ sur 9)
    12 ($\mathbf{P}°$) a→¬a **1.1**(règle $\Sigma_{[\to]}$ sur 6)
  13 ($\mathbf{O}^\bullet$)a ... **1.1** | 14 ($\mathbf{P}°$)¬a **1.1**
        | 15 ($\mathbf{O}^\bullet$)a **1.1**

Le sous-tableau généré par la formule ($\mathbf{P}^\bullet$)$\mathbf{V}a$ ferme en raison de la paire (($\mathbf{P}°$)a, ($\mathbf{O}^\bullet$)a). Donc, d'après **R-C**, l'ensemble des branches du tableau, et par conséquent, le tableau lui-même reste **ouvert**.

3.3. La sémantique de Kripke pour la logique connexe

Maintenant, nous pouvons finalement expliquer clairement la sémantique des modèles théorétiques. L'idée est très proche de celle développée dans la sémantique de Routley/Meyer pour la logique de la pertinence dans laquelle une relation d'accessibilité ternaire est introduite pour le conditionnel $Rw_iw_jw_k$, tel que si le conditionnel est vrai en w_i et l'antécédent vrai en w_j, alors le conséquent du conditionnel est vrai en w_k. Qui plus est, comme il en a déjà été fait mention, l'intégralité de cette idée semble être très proche de celle de Priest dans « *Negation as Cancellation and Connexive Logic* » (1999), bien que le présent travail fut développé indépendamment et inspiré d'un résultat de l'approche dialogique de la logique connexe déjà suggéré dans Rahman (1997a).

Le point de la logique connexe de MacColl, comprise telle que nous avons pu le faire, est que son utilisation des opérateurs de contingence est ambiguë : parfois il en fait une utilisation modale standard (une proposition qui est possible mais non nécessaire dans un modèle de structure de logique modale), parfois ils sont plutôt pensés comme contingence logique (une formule qui n'est pas valide et non-contradictoire). Nous introduisons deux ensembles de mondes $\mathbf{W}^{\thetaت}$ et $\mathbf{W}^{\theta f}$ dans le but d'introduire cette dernière idée.

Étant donné un modèle $<W,R,v>$ de logique modale standard augmenté comme suit $<\mathbf{W}, \mathbf{W}^{\theta t}, \mathbf{W}^{\theta f}, \mathbf{R}, \mathbf{R^*}, v>$ avec :

1) deux ensembles de mondes $\mathbf{W}^{\theta t}$ et $\mathbf{W}^{\theta f}$, $\mathbf{W}^{\theta t} \cap \mathbf{W}^{\theta f} = \emptyset$, tel que la formule φ est dite être contenue en $\mathbf{w}^{\theta t}$, $\mathbf{w}^{\theta t} \in \mathbf{W}^{\theta t}$, si et seulement si la valuation v du modèle rend φ vrai en $\mathbf{w}^{\theta t}$. Réciproquement, la formule φ est dite être contenue en $\mathbf{w}^{\theta f}$, $\mathbf{w}^{\theta f} \in \mathbf{W}^{\theta f}$, si et seulement si la valuation v du modèle rend ¬φ vrai en $\mathbf{w}^{\theta f}$.

Des tautologies ne peuvent pas être contenues dans $\mathbf{W}^{\theta f}$ et réciproquement $\mathbf{w}^{\theta t} \in \mathbf{W}^{\theta t}$ ne peut pas contenir de contradiction. Si l'on suppose que φ soit $a \wedge \neg a$, alors φ ne peut pas être contenue en $\mathbf{w}^{\theta f}$, $\mathbf{w}^{\theta f} \in \mathbf{W}^{\theta f}$, parce qu'il n'y a aucun modèle avec une valuation rendant ¬φ vrai en $\mathbf{w}^{\theta f}$.

2) une relation quaternaire $\mathbf{R^*}$ entre les mondes tels que $\mathbf{w} \in \mathbf{W}$, $\mathbf{w}^{\theta t} \in \mathbf{W}^{\theta t}$, $\mathbf{w}^{\theta f} \in \mathbf{W}^{\theta f}$

$$\mathbf{R^*www}^{\theta t}\mathbf{w}^{\theta f}$$

(c'est-à-dire que la relation $\mathbf{R^*}$ impose que $\mathbf{w}$ soit accessible à partir de $\mathbf{w}$, et qu'à partir de $\mathbf{w}$ il y ait un accès à $\mathbf{w}^{\theta t}$ et à $\mathbf{w}^{\theta f}$)

Intuitivement, l'idée est que l'implication connexe $\alpha \Rightarrow \beta$ est vraie dans un monde donné si et seulement si l'implication matérielle est vraie dans ce même monde ; et qu'il y a deux mondes accessibles différents, l'un dans lequel l'antécédent et le conséquent sont vrais, l'autre dans lequel l'antécédent et le conséquent sont vrais ainsi que l'antécédent et la négation du conséquent.

$\alpha \Rightarrow \beta$ est vrai en $\mathbf{w} \in \mathbf{W}$ si et seulement si :

si et seulement si $\mathbf{w} \in \mathbf{W}$, $\mathbf{w}^{\theta t}$, $\mathbf{w}^{\theta f}$ tels que $\mathbf{R^*www}^{\theta t}\mathbf{w}^{\theta f}$,

et que toutes les conditions suivantes sont réunies

1. $\mathbf{V}_w(\alpha)=0$ ou $\mathbf{V}_w(\beta)=1$
2. $\mathbf{V}_{w\theta t}(\alpha)=1$
3. $\mathbf{V}_{w\theta t}(\beta)=1$
4. $\mathbf{V}_{w\theta f}(\neg\beta)=1$
5. $\mathbf{V}_{w\theta f}(\neg\alpha)=1$

La validité est définie sur les mondes normaux. Il est manifeste qu'aucune formule valide (autre que celles contingentes) ne peut être assumée dans un monde non-normal.

Exemples :

Ce qui se passe si dans l'antécédent nous avons une contradiction ou si dans le conséquent nous avons une tautologie doit être clair : le conditionnel connexe est mis en échec parce que les conditions 2 et 4 ne sont pas respectées.

Exemple 9

Montrons que $\neg(a\Rightarrow\neg a)$ est valide.

Supposons que cette formule soit fausse en **w**
alors $(a\Rightarrow\neg a)$ doit être vrai en **w**.

Les conditions pertinentes sont 3 et 4 – nous laisserons les autres au lecteur. Selon ces conditions, ces formules doivent toujours être vraies en $\mathbf{w}^{\theta t}$

a
$\neg a$

ce qui est contradictoire : elles ne peuvent pas être toutes les deux vraies en $\mathbf{w}^{\theta t}$. Il n'y a donc aucun contre-modèle à $\neg(a\Rightarrow\neg a)$.

Exemple 10

Montrons que $\neg(a\Rightarrow\neg a)\Rightarrow a$ n'est pas valide et ne peut pas rendre notre système trivial.

Nous traiterons uniquement la partie principale de la preuve.

Supposons que $\neg(a{\Rightarrow}\neg a){\Rightarrow}a$ soit faux en **w**.

En raison de la condition 1, $\neg(a{\Rightarrow}\neg a)$ doit être vrai et a doit être faux en **w** par conséquent $a{\Rightarrow}\neg a$ doit être faux en **w**.
Si $a{\Rightarrow}\neg a$ est faux, alors il est également faux que certaines des conditions de vérité données ci-dessus soient respectées par rapport à cette formule. Et en effet c'est le cas. Revenons une fois de plus aux conditions 2 et 3 : selon ces conditions a et $\neg a$ doivent être vrais en $\mathbf{w}^{\theta}{}_t$. Mais ce n'est pas le cas. Par conséquent, il est faux que les conditions 2 et 3 soient respectées, la formule n'est donc pas valide.

Preuves de correction et de complétude

Les preuves peuvent être développées à partir des considérations suivantes :

1. Les sous-tableaux pour **V** et **F** correspondent aux mondes des ensembles $\mathbf{W}^{\theta}{}_t$ et $\mathbf{W}^{\theta}{}_f$.
2. La condition $\mathbf{W}^{\theta}{}_t \cap \mathbf{W}^{\theta}{}_f = \emptyset$ correspond au fait que les sous-tableaux pour **V** et **F** sont toujours dans différentes branches.
3. Étant donné une branche ouverte du tableau qui ne contient pas de sous-tableau et une lecture l'interprétant de manière tout à fait standard : si $(\mathbf{O}°)a$ apparaît à un noeud de la branche, la valuation booléenne 1 est assignée à a; si $(\mathbf{P}^{\bullet})a$ apparaît, c'est la valuation 0 qui lui est assignée.
4. S'il y a une branche ouverte qui commence avec la thèse et se termine avec un sous-tableau et que $(\mathbf{X}°)a$ apparaît à un noeud de la branche du sous-tableau, la valeur 1 est assignée à a ; si $(\mathbf{Y}^{\bullet})a$ apparaît, c'est la valeur 0 qui lui est assignée.
5. Si une branche donnée commence avec la thèse et se termine par un sous-tableau généré soit par une formule $(\mathbf{O}°)\Omega$ ou par une formule $(\mathbf{O}^{\bullet})\Omega$ est ouverte ; alors il y au moins un monde qui en fonction de 1 est un élément de l'un des ensembles $\mathbf{W}^{\theta}{}_t$ et $\mathbf{W}^{\theta}{}_f$ tel qu'il satisfait la valuation décrite en 4.
6. Si une branche donnée commence avec la thèse et se termine par un sous-tableau généré soit par une formule $(\mathbf{P}°)\Omega$ ou une formule $(\mathbf{P}^{\bullet})\Omega$ est ouverte ; alors il n'y aucun monde qui en fonction de 1 soit un élément de l'un des ensembles $\mathbf{W}^{\theta}{}_t$ et $\mathbf{W}^{\theta}{}_f$ tel qu'il satisfait la valuation décrite en 4.

3.4. Pertinence et connexivité

Nombre de partisans de la logique connexe peuvent penser que le système précédent pour la logique connexe n'est pas satisfaisant, certains prétextant qu'il y a des principes intuitivement corrects concernant le conditionnel qui ne sont pas corroborés et qu'il y en d'autres qui devraient l'être. Pour ce qui est du premier groupe, nous avons $(a \lor \neg a) \rightarrow (a \lor \neg a)$ qui n'est pas valide dans le système présenté ci-dessus, et pour le second nous avons les paradoxes de l'implication matérielle. En fait, l'une des motivations de MacColl fut de solutionner les paradoxes de l'implication matérielle. Ainsi que nous l'avons détaillé précédemment, il essaie de développer une méthode via laquelle aucune formule ne contient (d'occurrence) de variables propositionnelles qui soient vérifonctionnellement redondantes pour la vérité du conditionnel impliqué (cf. 2.2.1.2).

Donc, *e.g.*, $a \rightarrow (a \lor a)$ et $a \rightarrow (a \lor b)$ doivent être « éliminés » de l'ensemble des formules valides parce que dans le premier conditionnel, la première occurrence de a peut être substituée par n'importe quelle variable propositionnelle et parce que cette substitution est redondante pour établir ses conditions de vérité. Un argument identique peut être mis en avant pour l'occurrence de b dans le second conditionnel. Il est simple d'introduire ceci dans un système de tableaux, *grosso modo* : chaque occurrence d'une variable propositionnelle est utilisée pour clôturer une branche. En fait, deux systèmes sont possibles, l'un dans lequel *chaque variable propositionnelle* de la formule principale est utilisée pour clôturer une branche (ce qui permet à $(a \lor \neg a) \rightarrow (a \lor \neg a)$ et $a \rightarrow (a \lor a)$ d'être valides) et l'autre nécessitant que *chaque occurrence d'une variable propositionnelle* soit utilisée dans le sens décrit ci-dessus. Dans le second système, les formules $(a \lor \neg a) \rightarrow (a \lor \neg a)$ et $a \rightarrow (a \lor a)$ n'aboutissent pas à des tableaux clos.

La question soulevée ici est de savoir comment combiner cette logique de la pertinence avec une logique connexe d'une manière telle que nous ayons un conditionnel qui soit à la fois connexe et pertinent.

Du point de vue des tableaux, l'idée est assez simple : si la formule en jeu est valide de façon pertinente, alors le conditionnel connexe exprime que cette formule se tient. Si le conditionnel connexe exprime une formule qui soit l'un des axiomes aristotéliciens ou boéciens, alors la formule en jeu est valide.

Écrivons $\alpha \rightarrowtail \beta$ pour un conditionnel pertinent quel que soit le système de pertinence parmi ceux mentionnés précédemment. Le conditionnel connexe pertinent ($\Rightarrow$) surgit grâce à l'ajout de conditions métalogiques sur ce conditionnel par le biais des opérateurs **V** et **F**.

Pourtant, ici, nous supposons le système le plus faible (dans lequel différentes occurrences de variables propositionnels peuvent faire une différence – cf. 2.2.1.2) pour la simple raison que la logique connexe alors surgit comme extension conservatrice de ce type de logique de la pertinence, dans le sens où chaque formule valide dans cette logique de la pertinence sera aussi valide dans la partie où les formules connexes ont été ajouté. Il est particulièrement intéressant de noter que les infâmes banalisations des formules pour la logique connexe, à savoir $\neg(\alpha \rightarrow \neg\alpha) \rightarrow \alpha$, et $\neg(\neg\alpha \rightarrow \alpha) \rightarrow \neg\alpha$ ne tiennent pas, même du point de vue de la pertinence. Ici, nous supposons des tableaux avec des index pour les mondes comme usuellement dans ce type de système de preuve pour la logique modale.

Cas de (**Y**°)	*Cas de* (**X**•)
Σ, (**Y**°)($\alpha \Rightarrow \beta$) wi	Σ, (**X**•)($\alpha \Rightarrow \beta$) wi
Σ, (**Y**°)$\alpha \rightarrowtail \beta$ Σ, (**Y**°) **V**α wi Σ, (**Y**°) **F**β wi	Σ,(**X**•)$\alpha \rightarrowtail \beta$ wi | Σ,(**X**•)**V**α wi | Σ,(**X**•)**F**β wi

Ici, nous avons supposé une sémantique vérifonctionnelle pour la partie pertinente de l'implication. Une sémantique complète pour un modèle théorétique pour ce type de logique connexe reste toujours un problème ouvert, mais les considérations qui suivent peuvent aider :

- $\alpha \Rightarrow \beta$ est vrai en w$\in$ **W** si et seulement si

  w$\in$ **W**, w$^{\theta_t}$, w$^{\theta_f}$ tels que R*www$^{\theta_t}$w$^{\theta_f}$ est défini comme précédemment

  et que toutes les conditions qui suivent sont respectées :

$\mathbf{V}_w(\alpha \rightarrowtail \beta)= 1$. C'est-à-dire, $\mathbf{V}_w(\alpha)=0$ ou $\mathbf{V}_w(\beta)=1$, et il n'y a aucune variable propositionnelle qui ne soit pas un élément de l'ensemble des vérités déterminées pour $\neg \alpha \vee \beta$.

$\mathbf{V}_{w\theta t}\,(\alpha)=1$
$\mathbf{V}_{w\theta t}\,(\beta)=1$
$\mathbf{V}_{w\theta f}\,(\neg \beta)=1$
$\mathbf{V}_{w\theta f}\,(\neg \alpha)=1$

- Etant donné une branche ouverte d'un tableau qui ne contient pas de sous-tableau et une lecture interprétant la branche comme suit : si $(\mathbf{O}°)a$ apparaît à un nœud de la branche, la valuation booléenne 1 est assignée à a, si $(\mathbf{P}^\bullet)a$ apparaît, c'est la valuation 0 qui lui est assignée. Si il y a une paire $(\mathbf{O}°)a$ $(\mathbf{P}^\bullet)a$ et que la branche est ouverte, alors il y a un ensemble de vérités déterminées pour la thèse tel que au moins une occurrence de la variable propositionnelle n'est pas un élément de cet ensemble.

Remerciements

Shahid Rahman remercie Heinrich Wansing (Dresden) qui a lu la première version de la section 3 et suggéra de nombreuses et importantes corrections, ainsi que Cheryl Lobb-Rahman qui, entre les activités avec leurs enfants et ses multiples obligations a su trouver le temps d'améliorer de beaucoup l'anglais de la version non-traduite (cette langue n'étant pas la langue maternelle des auteurs).

Shahid Rahman et Juan Redmond souhaitent aussi remercier Tero Tulenheimo et Ahti Pietarinen (Helsinki) pour les échanges sur le système de notation de MacColl, l'équipe « pragmatisme dialogique » de l'Université de Lille 3 pour ses discussions enrichissantes, particulièrement Nicolas Clerbout, Cédric Dégremont, Laurent Keiff et Gildas Nzokou, ainsi que les « nouveaux venus » du groupe : Mathieu Fontaine, Coline Pauwels et Sébastien Magnier.

APPENDICE

A) MacColl et la fiction littéraire

La planète Mars a toujours eu une présence inquiétante. Déjà dans les temps anciens en raison de ces mouvements dérangeant à travers le ciel, ou plus récemment à cause des similarités physiques avec notre planète. En réalité, il y a un nombre troublant de similarités : sa magnitude, sa surface solide et son atmosphère, la probable présence d'eau. Toutes ces caractéristiques ont mené à penser une possible forme de vie éloignée sur Mars. Qui plus est, en 1877 l'astronome Asaph Hall (1829-1907) a découvert deux lunes à Mars, *Deimos* et *Phobos*. Il y a un autre fait encore plus intéressant si l'on considère comment une fiction peut avoir des répercussions sur notre monde réel. L'astronome italien Giovanni Virginio Schiaparelli (1835-1910) rapporta qu'avec l'aide d'un télescope, il observa des lignes droites sur Mars. Il appela ces lignes « *channels* » (sillons) en italien mais ce nom fut (peut-être intentionnellement) maladroitement traduit en « *canals* » (canaux). Cette traduction encouragea la spéculation à propos de la présence d'êtres intelligents qui auraient construit ces canaux. Sans surprise, la traduction erronée de son propos augmenta les espérances populaires et fantaisistes.

Comme Stein Haugom Oslem l'a déjà mis en évidence dans son article exhaustif sur le travail littéraire de MacColl (1999), tous ces éléments rendirent cette planète très populaire lorsque Hugh MacColl décida d'écrire de la science fiction. En plus de *Man's Origin, Destiny, and Duty* ainsi que son travail logique, MacColl a aussi publié deux romans, *Mr Stranger's Sealed Packet* (Chatto & Windus, London, 1889) et *Ednor Whitlock* (Chatto & Windus, London, 1891). Dans *Mr Stranger's Sealed Packet*, MacColl fut aussi un pionnier dans le choix de son sujet : un voyage sur la planète Mars. Quelques années avant lui, un autre écrivain explora la possibilité de voyager dans l'espace : Jules Vernes. L'écrivain français choisit la Lune pour destination de son roman *De la Terre à la Lune* de 1865 et *Autour de la Lune* de 1870. Le succès qu'eurent les romans de Jules Vernes en France, ont sûrement inspiré MacColl pour étendre l'idée à Mars. MacColl ne fut pas le premier mais le troisième à proposer un roman en anglais impliquant un tel genre d'aventures. En 1880, Percy Gregg publia *Across the Zodiac* (dans lequel un habitant de la Terre utilise une gravité négative pour voyager à travers l'espace, découvrant sur Mars une société utopique disposant d'une supériorité technologique et pratiquant la télépathie) ; en 1887, Hudor Genone publia *Bellona's Bridegroom : A Romance* (dans lequel, encore un habitant de la Terre découvre sur Mars une société idéale anglophone qui rajeunit au lieu de vieillir). Bien sûr MacColl ne fut pas le dernier. Quelques années plus tard le livre le plus populaire sur le sujet fut publié : *The War of the Worlds* (1898) par H.G.Wells.

Ce qu'il y a de frappant, c'est que ce n'est pas un travail fantaisiste, comme *Advntures of Hans Pfaal* (1835) de Edgar Allan Poe, mais belle et bien de la science fiction dans le style de Jules Verne, bien que, tristement, dans le travail de MacColl la visée pédagogique de la vulgarisation de la science se fasse au détriment de la qualité littéraire.
La première approche de ses romans est décevante pour un lecteur de science fiction moderne. En réalité, MacColl ne fut pas réellement inspiré pour imaginer un « monde différent ». Dans *Mr Stranger's Sealed Packet*, il projeta simplement sur Mars le monde autour de lui. À cet égard, le roman ne diffère pas de beaucoup des deux premiers travaux de science fiction publiés à propos de Mars. Le point intéressant est que MacColl n'a pas eu une approche réellement littéraire de la science fiction, mais il comprit ce genre de littérature plutôt comme illustration de la science ou moyen de la rendre populaire. Malheureusement, le lecteur le remarque dès le début. Ce qui est surprenant pour un lecteur connaissant les écrits logiques de MacColl est sa manière on ne peut plus conservatrice de penser les mondes alternatifs au notre. Dans sa logique, MacColl conçoit des mondes où sont présents toutes sortes de fictions, y incluant même des objets contradictoires, alors que ce n'est pas le cas dans le monde de sa science fiction.
Le personnage principal de son écrit de science fiction, Mr Stranger, réalise le souhait de son père décédé, qui se dédia lui-même exclusivement à la science. Mr Stranger continue de développer les théories et découvertes de son père, construisant un véhicule produisant une force gravitationnelle artificielle. Avec ce vaisseau spatial anti-gravitationnel, il visite les lunes de Mars – déjà découvertes dans le monde réel – et ensuite la planète rouge elle-même. Bien que la planète soit rouge, ses habitants sont bleuâtres. Mars, elle-même, est comme la Terre, et les martiens sont des humains, transférés lorsque la proximité de la planète rouge permit le transfert gravitationnel d'un grand nombre de personnes sur Mars lors d'un désastre préhistorique. Les martiens se comportent similairement aux humains, mais il y a des éléments utopiques dans leur description. Leur société est très rationnelle, avec une morale supérieure et rigide, avec une structure uniforme et harmonieuse. Il n'y a pas de maladie, pas de conflit social, et aucune technologie de guerre.

Dans les deux romans de MacColl, nous reconnaissons des thèmes typiquement représentatifs de l'époque victorienne, à savoir le conflit entre science et religion. C'est une période durant laquelle la réexamination des suppositions commença en raison des nouvelles découvertes en science, telles que celles de Charles Darwin et Charles Lydell. A cette époque, se tinrent de nombreuses discussions sur l'Homme et le monde, sur la science et l'histoire, et pour finir sur la religion et la philosophie. Cet inéluctable sens de la nouveauté donna lieu à un intérêt profond pour la relation entre modernité et continuité culturelle. C'est cet intérêt que reflète MacColl dans son roman. Bien que les deux romans soient différents, ils partagent nombre de thèmes et intérêts, et anticipent bon nombre des arguments que MacColl présenta plus tardivement dans *Man's Origin, Destiny, and Duty*.[67]

MacColl fait preuve d'attitudes et d'opinions conservatrices lorsqu'il en vient à ces questions fondamentales, particulièrement lorsqu'il écrit sur le rôle des hommes et des femmes, de la famille, du mariage, ainsi de suite. A cette époque, en Grande-Bretagne, l'importance de la pureté féminine alliée au stress du rôle de la femme dans la gestion de la maison, aida à créer un espace libre de pollution et de la corruption de la société.[68]

Sur Mars, Mr Stranger rencontre une famille dans laquelle il est le bienvenu. Il tombe amoureux de la fille de la famille et se marie avec elle. Sa nouvelle femme possède toutes les caractéristiques d'une chrétienne de l'ère victorienne typique d'Angleterre. C'est une femme docile avec son mari, compatissante, une femme aimante avec suffisamment de force émotionnelle et de sagesse pour devenir la valeur centrale de la famille, de la société (victorienne) en général. Elle meurt lors d'une visite sur Terre à cause de son intolérance aux bactéries terriennes. Un sujet assez commun dans la fiction : l'alternative fictionnelle au monde est pure alors que la Terre est impure. Les personnages fictifs meurent lorsqu'ils se retrouvent sur la Terre. La Terre, lieu où fiction et réalité se rencontrent, fait mourir les personnages fictifs. Comme dans *Don Quijote*, le personnage fictif meurt à cause de la « réalité ». Malheureusement MacColl ne se soucie pas beaucoup de ces excitantes caractéristiques de la fiction. Une fois encore, le lecteur qui connaît MacColl à travers ses écrits sur la logique des fictions peut s'attendre à ce que ce même auteur explore les possibilités logiques et littéraires résultant de la rencontre entre des habitants de différents mondes, puisque MacColl suggère quelques

---

[67] Voir Cuypers [1999] et la remarque déjà mentionnée de l'article de Olsen [1999].

[68] Cf. Cuypers [1999] et Olsen [1999].

idées sur cette possibilité dans sa logique de la ficiton, où certains domaines contiennent des objets du monde réel et d'autres purement fictifs. Pourtant si un même lecteur lit la littérature fictionnelle de MacColl, il sera absolument déçu. MacColl ne s'applique, ici, pas à lui-même.

Comme nous en avons fait mention auparavant, le roman a une visée éducative : instruisant son public aux possibilités ouvertes par la science. MacColl fait un usage abondant de théories et de faits pour rendre l'histoire de Mr Stranger aussi plausible que possible. Sans aucune surprise, ceci contrecarre la valeur littéraire de l'ensemble du projet.

Ednor Whitlock n'est pas un roman de science fiction, mais présente tout de même des caractéristiques similaires. Les premiers moments sont suffisants pour se faire une idée de ce dont il s'agit : Ednor, un jeune homme des années quatre-vingt-dix, cherchant à s'abriter de la pluie dans une librairie, par chance ramasse une édition de *Westminster Review*. Là, il trouve un article et se retrouve absorbé par les arguments contre les croyances religieuses. Sa foi est ébranlée par ce qu'il apprend des nouvelles idées scientifiques et des critiques historiques de la Bible. Les propres croyances de MacColl sont concernées dans ce récit. Ce roman est techniquement plus simple que l'autre mais développe plus attentivement la thèse selon laquelle l'incroyance cause l'immoralité.

Une autre caractéristique que nous voulons souligner ici concerne les rapports que MacColl a avec l'Allemagne et avec la culture germanique. Premièrement, il ne connaissait pas l'allemand – donc il ne pouvait pas lire les travaux des penseurs germaniques. Il le souligna lui-même dans une lettre à bertrand Russell :

> ...*unfortunately all German works are debarred to me because I do not know the language, so that I know nothing of Cantor's and Dedekind's views on infinity.*[69]

Pourtant il semble difficile de croire que MacColl n'a pas lu les penseurs allemands simplement parce qu'il ne connaissait pas leur langue, particulièrement depuis que MacColl, qui habita en France de nombreuses années, fit expérience de situations où différentes langues sont utilisées.

---

[69] MacColl [1909c].

Certains préjugés sociaux et politiques, partagés à la fois par la société britannique et la société française, ont pu jouer un rôle contre les allemands. Comme Cuypers et Olsen l'ont déjà remarqué, l'écrit *Ednor Whitlock* nous fournit des indices appuyant cette thèse. Dans le roman, l'Allemagne prend corps dans le Révèrent Milford et Mademoiselle Hartman. Cette dernière est, dans le roman, un personnage antipathique. MacColl la dote de qualités peu attrayantes, de toute évidence en lien avec ses ancêtres allemands. Le premier développe, quant à lui, des arguments en faveur du théisme.

Ces préjugés politico-sociaux furent assez souvent basés sur ce qui était considéré comme une attaque de la foi chrétienne. L'Allemagne et les universités allemandes furent la source du « Criticisme allemand radical », dû plus particulièrement au rationalisme scientifique qui devint populaire en Allemagne grâce aux efforts de Ernst Haeckel (1834-1919), qui rendit le darwinisme accessible. Il créa les conditions pour que le darwinisme touche un large public. En réalité, Haeckel apparaît comme la principale cible des attaques de *Man's Origin, Destiny, and Duty*, le dernier livre de MacColl qui résume l'ensemble de son point de vue à propos du conflit entre science et religion.

Il est triste que ces préjugés l'empêchèrent de lire les mathématiciens et logiciens allemands de son temps. Cela aurait pu lui fournir les instruments dont il avait besoin pour mener à terme ses diverses propositions novatrices en logique.

B) MacColl et la probabilité conditionnelle

MacColl attache énormément d'importance au fait que sa logique puisse s'appliquer avec succès dans la résolution de problèmes spécifiques en rapport avec la probabilité. MacColl précise explicitement que sa logique tire son origine dans des problèmes de probabilités :

> *It may interest some of the readers to be told that this method owes its origin to a question in probability (No. 3440), proposed by Mr. Stephen Watson in the* EDUCATIONAL TIMES. *My solution of this question, with an introductory article, entitled "Probability Notation", was published in the* EDUCATIONAL TIMES *for August, 1871; and in this introductory article may be seen the germs of the more present method. Shortly after this I gave up all mathematical investigations, and my thoughts did not again revert to the subject till two or three months before the appearance of my article on "Symbolical Language" for July, 1877. [...] I noticed that this "Symbolic Language", as I called it, might also be employed, without change or modification, and with unerring certainty, in tracing to their last hiding-place the limits which often escape so mysteriously from the mathematician's*

> *grasp when he ventures to change the order of integration or the variables in a multiple integral. I still looked upon the method, however, as an essentially mathematical one - grafted on a logical stem, but destined to yield mathematical fruit, and mathematical fruit only.* (MacColl [1878d], p. 27)

Theodor Hailperin, qui étudia les applications de la logique de MacColl dans la probabilité, la relie aux premiers résultats de la probabilité conditionnelle. D'après Hailperin, contrairement à Boole ou Pierce, la notation de MacColl fait clairement la distinction entre un argument (fait, proposition, classe) et la probabilité qui lui est attribuée (cf. Hailperin [1996], pp. 132-134). De fait, la notation de la probabilité conditionnelle de MacColl fait usage de termes utilisés en tant qu'adjectifs que nous avons discutés précédemment et ressemble, tout au moins en apparence, à la notation de Pierce. En effet, la notation de Peirce de la probabilité conditionnelle est la suivante :

> Let b_a denote the frequency of b's among the a's. Then considered as a class, if a and b are events, b_a denotes the fact that if a happens b happens (Peirce [1867], paragraphe 14).

Mais Peirce n'explique pas ce qu'est la classe b_a, ou comment elle dénote un fait.

La formulation de MacColl semble lui être similaire :

> *In applying symbolic logic to probability; the following notation, will I think; be found useful.*
> *Def. 1.* — *The symbol x_a denotes the chance that the statement x is true on the assumption that the statement a is true.* (MacColl [1879e], p. 113).

Cependant, la compréhension de la probabilité conditionnelle telle qu'elle figure chez MacColl diffère de celle de Peirce dans la mesure où les deux composants sont des énoncés plus que des classes. La probabilité absolue ou inconditionnée d'un énoncé x s'écrit x_ε. Nous présentons ici un résumé des principales idées de la probabilité conditionnelle dans MacColl [1879e], basé sur l'analyse de Hailperin (Hailperin [1996], pp. 132-134).

MacColl ne produit pas de développement formel – les démonstrations sont basées sur une compréhension informelle des significations et sur une utilisation d'exemples dans lesquels des points aléatoires sont sélectionnés à partir de régions géométriques avec une probabilité supposée être proportionnelle à l'espace des régions à partir desquelles les sélections sont opérées.

1) MacColl relie la probabilité conditionnelle avec l'implication stricte en stipulant que a implique x, est équivalent à la probabilité de l'énoncé $x_a=1$.

2) Il définit a et x comme étant indépendants dans n'importe laquelle des quatre conditions à respecter :

$$x_a = x_\varepsilon$$
$$a_x = a_\varepsilon$$
$$a_x = a_{x'}$$
$$x_a = x_{a'}$$

Notez que si nous envisageons logiquement ces interprétations, elles expriment les conditions ou redondances qui mènent aux paradoxes de l'implication (voir chapitre 2.2.1.2). Prenez la première condition, si a implique b et b est nécessairement vrai ; alors la valeur de vérité de l'antécédent est redondante, dans le sens où nous ne la connaissons pas. Ceci s'applique aux autres conditions de manière identique.

3) MacColl produit des résultats tels que :

$$x_a = \frac{(ax)_\varepsilon}{a_\varepsilon}$$

c'est en fait la définition usuelle que nous acceptons pour la probabilité conditionnelle :

$$(a+b)_x = a_x + b_x - (ab)_x$$

4) Certains résultats sont obtenus en induisant des logiques tels que :

$$x_{x+y} = \frac{x_\varepsilon}{(x+y)_\varepsilon}$$

$$(xy)_x = y_x$$

5) Sa méthode par inversion est facilement dérivée à partir de ses règles sous la forme :

$$a_x = \frac{a_\varepsilon x_a}{a_\varepsilon x_a + b_\varepsilon x_b + c_\varepsilon x_c + \ldots}$$

en supposant que x implique $a+b+c+\ldots$

Tous ces résultats montrent le gain en clarté, précision et généralité ; amorcé par l'introduction de MacColl d'une notation appropriée pour la probabilité conditionnelle.

BIBLIOGRAPHIE DE MACCOLL

La bibliographie suivante est basée sur les publications ci-dessous :
Shahid Rahman, « Hugh MacColl: Eine bibliographische Erschliessung seine Hauptwerke und Notizen zu ihrer Rezeptionsgeschichte », *History and philosophy of logic*, vol. 18, 1997, pp. 165-83; ainsi que Michael Astro et Johan W. Klüwer, Bibliography : Hugh MacColl, *Nordic Journal of Philosophical Logic*, vol. 3, n°1, November 1999 (version en ligne).
Le « *The Educational Times* » correspond à « *The Educational Times and Journal of the College of Preceptors* ».

[1865a]   [Question No.] 1709. *The Educational Times*, vol. 18, p.20.
[1865b]   [Question No.] 1739. *The Educational Times*, vol. 18, p.69.
[1865c]   [Question No.] 1761. *The Educational Times*, vol. 18, p.94.
[1865d]   [Question No.] 1788. *The Educational Times*, vol. 18, p.114.
[1865e]   [Question No.] 1830. *The Educational Times*, vol. 18, p.166.
[1865f]   Solution to [Question No.] 1739. *The Educational Times*, vol. 18, p.211.
[1865g]   [Question No.] 1709. *The Educational Times (Reprint)*, vol. 4, p.57.
[1865h]   Solution to [Question No.] 1709. *The Educational Times (Reprint)*, vol. 4, pp. 57-58.
[1865i]   [Question No.] 1739. *The Educational Times (Reprint)*, vol. 4, p.102.
[1865j]   [Question No.] 1761. *The Educational Times (Reprint)*, vol. 4, pp. 102-103.
[1865k]   Solution to [Question No.] 1739. *The Educational Times (Reprint)*, vol. 4, p.102.
[1865l]   [Question No.] 1761. *The Educational Times (Reprint)*, vol. 4, pp. 103-105.
[1865m]   On the Numerical Solution of Equations. (Continued from the Solutions of Questions 1739, 1761.), *The Educational Times (Reprint)*, vol. 4, pp. 105-107.

[1866a]   Angular and Linear Notation. A Common Basis for the *Bilinear* (a Transformation of the *Cartesian*), the *Trilinear*, the *Quadrilinear*, &C., Systems of Geometry *The Educational Times*, vol. 19, p.20.
[1866b]   [Question No.] 2008. *The Educational Times*, vol. 19, p.89.
[1866c]   [Question No.] 2245. *The Educational Times*, vol. 19, p.134.
[1866d]   [Question No.] 1898. *The Educational Times*, vol. 19, p.258.
[1866e]   Angular and Linear Notation. A Common Basis for the *Bilinear* (a transformation of the *Cartesian*), the *Trilinear*, the *Quadrilinear*, &c., Systems of Geometry, *The Educational Times (Reprint)*, vol. 5, pp. 74-77.
[1866f]   Solution to [Question No.] 1950. *The Educational Times (Reprint)*, vol. 5, pp. 105-107.

[1870a]   ***Algebraical Exercises and Problems with Elliptical Solutions***, Longmans, Green and Co., London.
[1870b]   Formulae of Reduction, *The Educational Times*, vol. 23.
[1870c]   [Question No.] 3209. *The Educational Times*, vol. 23, p.135.

[1870d]   Solution to [Question No.] 3207. ***The Educational Times***, vol. 23, p.159.
[1870e]   [Question No.] 3264. ***The Educational Times***, vol. 23, p.185.
[1870f]   [Question No.] 3292. ***The Educational Times***, vol. 23, p.209.
[1870g]   An easy way of remembering the *Formulæ of Reduction* in the Integral Calculus, both as regards their Investigation and their Application, ***The Educational Times (Reprint)***, vol. 14, pp. 43-45.
[1870h]   Solution to [Question No.] 3210. ***The Educational Times (Reprint)***, vol. 14, p.73.

[1871a]   Probability Notation. ***The Educational Times***, vol. 23, pp. 230-231.
[1871b]   Solution to [Question No.] 3279. ***The Educational Times***, vol. 23, p.231.
[1871c]   [Question No.] 3322. ***The Educational Times***, vol. 23, p.232.
[1871d]   Solution to [Question No.] 1100. ***The Educational Times***, vol. 23, pp. 259-260.
[1871e]   [Question No.] 3342. ***The Educational Times***, vol. 23, p.261.
[1871f]   Solution to [Question No.] 3322. ***The Educational Times***, vol. 23, p.288.
[1871g]   [Question No.] 3370. ***The Educational Times***, vol. 23, p.290.
[1871h]   [Question No.] 3342. ***The Educational Times***, Vol. 24 (1871-1872), (April 1871), p.19.
[1871i]   Solution to [Question No. 3385]. ***The Educational Times***, Vol. 24 (1871-1872), (Mai 1871), p. 40.
[1871j]   [Question No.] 3292. ***The Educational Times***, vol. 24 (1871-1872), (June 1871), p. 64.
[1871k]   Probability Notation No. 2. ***The Educational Times***, vol. 24 (1871-1872), (August 1871), pp. 111-112.
[1871l]   Solution to [Question No.] 3440. ***The Educational Times***, vol. 24 (1871-1872), (Aug 1871), p. 112.
[1871m]   [Question No.] 3342. ***The Educational Times***, vol. 24 (1871-1872), (October 1871), p. 163.
[1871n]   [Question No.] 3515. ***The Educational Times***, vol. 24 (1871-1872), (November 1871), p. 188.
[1871o]   [Question No.] 3433. ***The Educational Times***, vol. 24 (1871-1872), (November 1871), p. 189.
[1871p]   Solution to [Question No.] 3408. ***The Educational Times***, vol. 24 (1871-1872), (December 1871), p.216.
[1871q]   Probability Notation. ***The Educational Times (Reprint)***, vol. 15, pp. 20-22.
[1871r]   Solution to [Question No.] 3279. ***The Educational Times (Reprint)***, vol. 15, p.23.
[1871s]   Solution to [Question No.] 1100. ***The Educational Times (Reprint)***, vol. 15, pp. 35-36.
[1871t]   [Question No.] 3209. ***The Educational Times (Reprint)***, vol. 15, p.36.
[1871u]   [Question No.] 3264. ***The Educational Times (Reprint)***, vol. 15, p.40.
[1871v]   [Question No.] 3322. ***The Educational Times (Reprint)***, vol. 15, p.55.
[1871w]   Solution to [Question No.] 3322. ***The Educational Times (Reprint)***, vol. 15, pp. 55-56.
[1871x]   [Question No.] 3342. ***The Educational Times (Reprint)***, vol. 15, p. 58.
[1871y]   Solution to [Question No.] 3385. ***The Educational Times (Reprint)***, vol. 15, pp. 83-85.
[1871z]   Solution to [Question No.] 3411. ***The Educational Times (Reprint)***, vol. 15,

pp. 102-103.
[1871aa]   [Question No.] 3292. *The Educational Times (Reprint)*, vol. 15, p. 104.
[1871ab]   [Question No.] 3388. *The Educational Times (Reprint)*, vol. 15, p. 106.
[1871ac]   Probability Notation. No. 2. *The Educational Times (Reprint)*, vol. 16, pp. 29-31.
[1871ad]   Solution to [Question No.] 3440. *The Educational Times (Reprint)*, vol. 16, pp. 31-32.
[1871ae]   [Question No.] 3342. *The Educational Times (Reprint)*, vol. 16, p. 68.
[1871af]   Solution to [Question No.] 3342. *The Educational Times (Reprint)*, vol. 16, pp. 68-69.
[1871ag]   [Question No.] 3515. *The Educational Times (Reprint)*, vol. 16, p. 93.
[1871ah]   Solution to [Question No.] 3515. *The Educational Times (Reprint)*, vol. 16, p. 93.
[1871ai]   [Question No.] 3433. *The Educational Times (Reprint)*, vol. 16, p. 96.
[1871aj]   Solution to [Question No.] 3433. *The Educational Times (Reprint)*, vol. 16, pp. 96-97.
[1871ak]   Solution to [Question No.] 3408. *The Educational Times (Reprint)*, vol. 16, p. 103.

[1872a]   Solution to [Question No.] 3644. *The Educational Times*, vol. 24, (February 1872), p.268.
[1872b]   Solution to [Question No.] 3561. *The Educational Times*, vol. 24, (February 1872), p. 269.
[1872c]   [Question No.] 3646. *The Educational Times*, vol. 25, p. 244.
[1872d]   Solution to [Question No.] 3644. *The Educational Times (Reprint)*, vol. 17, p.36.
[1872e]   Solution to [Question No.] 3561. *The Educational Times (Reprint)*, vol. 17, p.41.

[1877a]   [Question No.] 5259. *The Educational Times*, vol. 30, p. 21.
[1877b]   [Question No.] 5291. *The Educational Times*, vol. 30, p. 44.
[1877c]   [Question No.] 5318. *The Educational Times*, vol. 30, p. 69.
[1877d]   Symbolical or abbreviated language, with an application to mathematical probability. *The Educational Times*, vol. 29, (July 1877), p. 91-92.
[1877e]   [Question No.] 5339. *The Educational Times*, vol. 30, p. 92.
[1877f]   [Question No.] 5373. *The Educational Times*, vol. 30, pp. 123-124.
[1877g]   [Question No.] 5402. *The Educational Times*, vol. 30, p. 147.
[1877h]   Symbolical language: - Nr. 2. *The Educational Times*, vol. 29, (November 1877), p. 195.
[1877i]   [Question No.] 5450. *The Educational Times*, vol. 30, p. 173.
[1877j]   Solution to [Question No.] 5281. *The Educational Times*, vol. 29, (April 1877), p. 171.
[1877k]   [Question No.] 5469. *The Educational Times*, vol. 30, p.196.
[1877l]   [Question No.] 5511. *The Educational Times*, vol. 30, p. 221.
[1877m]   Solution to [Question No.] *5155. The Educational Times*, vol. 30, p. 300.
[1877n]   [Question No.] 5231. *The Educational Times*, vol. 30, p. 302.
[1877o]   The Calculus of Equivalent Statements and Integration Limits. ***Proceedings of the London Mathematical Society***, (1877-1878), vol. 9, pp. 9-20.
[1877p]   The Calculus of Equivalent Statements (II). ***Proceedings of the London***

	Mathematical Society, vol. 9, pp. 177-186.
[1877q]	Solution to [Question No.] 5155. *The Educational Times (Reprint)*, vol. 27, p.66.
[1877r]	Solution to [Question No.] 5213. *The Educational Times (Reprint)*, vol. 27, p.84.
[1877s]	Solution to [Question No.] 5231. *The Educational Times (Reprint)*, vol. 27, p.85.
[1878a]	[Question No.] 5584. *The Educational Times*, vol. 31, p. 89.
[1878b]	[Question No.] 5645. *The Educational Times*, vol. 31, p. 138.
[1878c]	[Question No.] 5808. *The Educational Times*, vol. 31, p. 294.
[1878d]	The Calculus of Equivalent Statements (III). **Proceedings of the London Mathematical Society**, (1878-1879), vol. 10, pp. 16-28.
[1878e]	Symbolical or Abbreviated Language, with an Application to Mathematical Probability. *The Educational Times (Reprint)*, vol. 28, pp. 20-23.
[1878f]	Symbolical Language:--No. 2. *The Educational Times (Reprint)*, vol. 28, p.100.
[1878g]	[Question No.] 5339. *The Educational Times (Reprint)*, vol. 29, p.66.
[1878h]	Solution to [Question No.] 5339. *The Educational Times (Reprint)*, vol. 29, pp. 66-68.
[1879a]	[Question No.] 5861. *The Educational Times*, vol. 32, p. 22.
[1879b]	[Question No.] 6107. *The Educational Times*, vol. 32, p. 292.
[1879c]	[Question No.] 6133. *The Educational Times*, vol. 32, p. 317.
[1879d]	[Question No.] 6173. *The Educational Times*, vol. 32, p. 341.
[1879e]	The Calculus of Equivalent Statements (IV). **Proceedings of the London Mathematical Society**, vol. 11, pp. 113-21.
[1879f]	A note on Prof. C. S. Peirce's probability notation of 1867. **Proceedings of the London Mathematical Society**, vol. 12, p.102.
[1879g]	[Question No.] 5450. *The Educational Times (Reprint)*, vol. 30, p.50.
[1879h]	[Question No.] 5511. *The Educational Times (Reprint)*, vol. 30, p.78.
[1879i]	Solution to [Question No.] 5511. *The Educational Times (Reprint)*, vol. 30, pp. 79-80.
[1879j]	[Question No.] 5861. *The Educational Times (Reprint)*, vol. 31, p.43.
[1879k]	Solution to [Question No.] 5861. *The Educational Times (Reprint)*, vol. 31, pp. 43-44.
[1880a]	*Letter to W. S. Jevons,* **18. VIII. 1880**.
[1880b]	[Question No.] 6133. *The Educational Times*, vol. 33, pp. 21-22.
[1880c]	Solution to [Question No.] 6133. *The Educational Times*, vol. 33, pp. 21-22.
[1880d]	[Question No.] 6206. *The Educational Times*, vol. 33, p. 23.
[1880e]	[Question No.] 6235. *The Educational Times*, vol. 33, p. 68.
[1880f]	[Question No.] 6258. *The Educational Times and Journal of the College of Preceptors*, vol. 33, p. 97.
[1880g]	Arithmetical Notation. To the Editor of the Educational Times. *The Educational Times and Journal of the College of Preceptors*, vol. 33, p. 111.
[1880h]	[Question No.] 6330. *The Educational Times*, vol. 33, p. 144.
[1880i]	[Question No.] 6173. *The Educational Times*, vol. 33, p. 214.
[1880j]	Solution to [Question No.] 6353. *The Educational Times*, vol. 33, p. 237.

[1880k]   [Question No.] 6461. ***The Educational Times***, vol. 33, p. 288.
[1880l]   Solution to [Question No.] 6235. ***The Educational Times***, vol. 33, p. 262.
[1880m] [Question No.] 6525. ***The Educational Times***, vol. 33, p. 289.
[1880n] On the Diagrammatic and Mechanical Representation of Propositions and Reasoning. ***The London, Edinburgh and Dublin philosophical Magazine and Journal of Science***, vol. 10, pp. 168-171.
[1880o] Implication and Equational Logic. ***The London, Edinburgh and Dublin philosophical Magazine and Journal of Science***, vol. 11, pp. 40-43.
[1880p] Symbolical Reasoning (I). ***Mind***, vol. 5, pp. 45-60.

[1881a] *Letter to W. S. Jevons* , **27. V. 1881**.
[1881b] Solution to [Question No.] 6485. ***The Educational Times***, vol. 34, p.22.
[1881c] Solution to [Question No.] 6560. ***The Educational Times***, vol. 34, p.22.
[1881d] [Question No.] 6575. ***The Educational Times***, vol. 34, p.24.
[1881e] [Question No.] 6615. ***The Educational Times***, vol. 34, p.75.
[1881f]   Solution to [Question No.] 6574. ***The Educational Times***, vol. 34, pp. 146-147.
[1881g] On the growth and use of a symbolical language. ***Proceedings of the Manchester Literary and Philosophical Society***, vol. 20, p. 103.
[1881h] Solution to [Question No.] 6383. ***The Educational Times (Reprint)***, vol. 34, p. 36.
[1881i]   [Question No.] 6173. ***The Educational Times (Reprint)***, vol. 34, p. 40.
[1881j]   Solution to [Question No.] 6353. ***The Educational Times (Reprint)***, vol. 34, p. 51.
[1881k] [Question No.] 6235. ***The Educational Times (Reprint)***, vol. 34, p. 69.
[1881l]   Solution to [Question No.] 6235. ***The Educational Times (Reprint)***, vol. 34, p. 69.
[1881m] [Question No.] 6461. ***The Educational Times (Reprint)***, vol. 34, p. 85.

[1882a] On Probability and Listerism. ***The Educational Times***, vol. 35, p. 103.
[1882b] On Probability and Listerism. No.2. ***The Educational Times***, vol. 35, p. 127.
[1882c] Solution to [Question No.] 5397. ***The Educational Times***, vol. 35, p. 128.
[1882d] On the Growth and Use of a Symbolical Language. ***Memoirs of the Manchester Literary and Philosophical Society***, vol. 7, pp. 225-248.

[1883]   *Letter to C. S. Peirce,* **16. V. 1883**.

[1884a] To the Editor of the Educational Times. ***Memoirs of the Manchester Literary and Philosophical Society***, vol. 37, p. 201.
[1884b] To the Editor of the Educational Times. ***Memoirs of the Manchester Literary and Philosophical Society***, vol. 37, pp. 231-32.
[1884c] [Question No.] 7871. ***The Educational Times***, vol. 37, p. 330.
[1884d] [Question No.] 7897. ***The Educational Times***, vol. 37, p. 356.
[1884e] Solution to [Question No.] 7849. ***The Educational Times***, vol. 37, p. 385.
[1884f]   On the Limits of Multiple Integrals. ***Proceedings of the London Mathematical Society***, vol. 16, pp. 142-48.

[1885a] [Question No.] 7983. ***The Educational Times***, vol. 38, p. 30.
[1885b] [Question No.] 8026. ***The Educational Times***, vol. 38, p. 69.

[1885c]   [Question No.] 8093. *The Educational Times*, vol. 38, p. 160.
[1885d]   [Question No.] 8286. *The Educational Times*, vol. 38, p. 350.
[1885e]   [Question No.] 8387. *The Educational Times*, vol. 38, p. 412.

[1886]   Solution to [Question No.] 8387. *The Educational Times*, vol. 39, p. 134.

[1887]   *Letter to C. S. Peirce* , **17. VIII. 1887.**

[1888a]   [Question No.] 1898 & 4043. *The Educational Times*, vol. 41, p. 134.
[1888b]   A New Angular Trigonometrical Notation, with Applications. ***Proceedings of the London Mathematical Society***, vol. 20, read 14. 03. 1889, not printed, p. 180.

[1889a]   ***Mr. Stranger's Sealed Packet***, Chatto & Windus, London.
[1889b]   [Question No.] 10119. *The Educational Times*, vol. 42, p. 223.
[1889c]   [Question No.] 3646. *The Educational Times (Reprint)*, vol. 51, p.130.
[1889d]   [Question No.] 4043. *The Educational Times (Reprint)*, vol. 51, pp. 135-136.

[1890a]   [Question No.] 10619. *The Educational Times*, vol. 44, p. 268.
[1890b]   [Question No.] 10708. *The Educational Times*, vol. 44, p. 404.
[1890c]   [Question No.] 10750. *The Educational Times*, vol. 44, p. 446.
[1890d]   [Question No.] 10780. *The Educational Times*, vol. 44, p. 485.
[1890e]   [Question No.] 10814. *The Educational Times*, vol. 44, p. 529.
[1890f]   [Question No.] 9996. *The Educational Times (Reprint)*, vol. 52, pp. 33-34.
[1890g]   [Question No.] 10119. *The Educational Times (Reprint)*, vol. 52, pp. 37-38.
[1890h]   Solution to [Question No.] 10274. *The Educational Times (Reprint)*, vol. 52, pp. 103-104.
[1890i]   [Question No.] 4155. *The Educational Times (Reprint)*, vol. 52, pp. 133-34.

[1891a]   ***Ednor Whitlock***, Chatto & Windus, London.
[1891b]   [Question No.] 10937. *The Educational Times*, vol. 44, p. 157.
[1891c]   [Question No.] 11017. *The Educational Times*, vol. 44, p. 239.
[1891d]   [Question No.] 11061. *The Educational Times*, vol. 44, p. 275.
[1891e]   [Question No.] 11104. *The Educational Times*, vol. 44, p. 307.
[1891f]   Solution to [Question No.] 11O61. *The Educational Times*, vol. 44, p. 343.
[1891g]   [Question No.] 11186. *The Educational Times*, vol. 44, p. 345.
[1891h]   [Question No.] 11235. *The Educational Times*, vol. 44, p. 411.
[1891i]   [Question No.] 11104. *The Educational Times*, vol. 44, p. 527.
[1891j]   Solution to [Question No.] 11104. *The Educational Times*, vol. 44, p. 527.
[1891k]   Solution to [Question No.] 5469. *The Educational Times (Reprint)*, vol. 55, p. 136.
[1891l]   Solution to [Question No.] 10536. *The Educational Times (Reprint)*, vol. 55, p. 180.

[1892a]   [Question No.] 11061. *The Educational Times (Reprint)*, vol. 56, pp. 51-54.
[1892b]   [Question No.] 10750. *The Educational Times (Reprint)*, vol. 56, pp. 89-90.
[1892c]   [Question No.] 11104. *The Educational Times (Reprint)*, vol. 56, pp. 115-116.
[1892d]   [Question No.] 6107. *The Educational Times (Reprint)*, vol. 56, p. 134.

[1892e] [Question No.] 11186. *The Educational Times (Reprint)*, vol. 57, p. 38.

[1893a] [Question No.] 7937a. *The Educational Times*, vol. 46, p. 303.
[1893b] [Question No.] 7937b. *The Educational Times*, vol. 46, pp. 303-304.

[1894] [Question No.] 3646. *The Educational Times (Reprint)*, vol. 61, pp. 68-69.

[1895a] [Question No.] 3464. *The Educational Times (Reprint)*, vol. 62, p. 122.
[1895b] [Question No.] 3552. *The Educational Times (Reprint)*, vol. 62, p. 125.
[1895c] [Question No.] 3771. *The Educational Times (Reprint)*, vol. 63, p. 124.

[1896a] [Question No.] 13234. *The Educational Times*, vol. 49, p. 396.
[1896b] The Calculus of Equivalent Statements (V). *Proceedings of the London Mathematical Society*, vol. 28, pp. 156-183.
[1896c] The Calculus of Equivalent Statements (VI). *Proceedings of the London Mathematical Society*, vol. 28, pp. 555-579.

[1897a] Symbolic[al] Reasoning (II). *Mind*, Vol. 6, pp. 493-510.
[1897b] The Calculus of Equivalent Statements (VII). *Proceedings of the London Mathematical Society*, vol. 29, pp. 98-109.
[1897c] The Calculus of Equivalent Statements (VIII). *Proceedings of the London Mathematical Society*, vol. 29, (communicated in abstract 09.06.1898, not printed), p. 545.
[1897d] [Question No.] 13234. *The Educational Times (Reprint)*, vol. 66, pp. 95-97.
[1897e] [Question No.] 13288. *The Educational Times (Reprint)*, vol. 66, pp. 119-120.
[1897f] [Question No.] 5318. *The Educational Times (Reprint)*, vol. 67, p. 125.

[1898a] The Calculus of Equivalent Statements. Explanatory Note and Correction. *Proceedings of the London Mathematical Society*, vol. 30, pp. 330-332.
[1898b] [Question No.] 13449. *The Educational Times (Reprint)*, vol. 68, pp. 30-31.
[1898c] Solution to [Question No.] 13517. *The Educational Times (Reprint)*, vol. 68, p. 68.
[1898d] [Question No.] 13517. *The Educational Times (Reprint)*, vol. 68, p. 68.
[1898e] [Question No.] 5584. *The Educational Times (Reprint)*, vol. 68, p. 123.
[1898f] [Question No.] 5645. *The Educational Times (Reprint)*, vol. 68, p. 124.

[1899a] Review of « A Treatise on Universal Algebra with Applications », Vol. 1. By Alfred North Whitehead, M. A., Fellow and Lecturer of Trinity College, Cambridge, Cambridge University Press, 1898. *Mind*,. Vol. 8, pp. 108-113.
[1899b] [Question No.] 14030. *The Educational Times (Reprint)*, vol. 71, p. 31.
[1899c] [Question No.] 5645. *The Educational Times (Reprint)*, vol. 71, pp. 118-119.

[1900a] [Question No.] 14439. *The Educational Times*, vol. 53, p. 37.
[1900b] Solution to [Question No.] 6330. *The Educational Times*, vol. 53, p. 82.
[1900c] [Question No.] 14469. *The Educational Times*, vol. 53, p. 83.
[1900d] [Question No.] 14498. *The Educational Times*, vol. 53, p. 149.
[1900e] [Question No.] 14557. *The Educational Times*, vol. 53, p. 225.

[1900f]   Solution to [Question No.] 14122. *The Educational Times*, vol. 53, p. 258.
[1900g]   Solution to [Question No.] 14394. *The Educational Times*, vol. 53, pp. 377-378.
[1900h]   [Question No.] 14665. *The Educational Times*, vol. 53, p. 380.
[1900i]   [Question No.] 14721. *The Educational Times*, vol. 53, p. 463.
[1900j]   [Question No.] 14751. *The Educational Times*, vol. 53, p. 503.
[1900k]   Symbolic[al] Reasoning (III). *Mind*, Vol. 9, pp. 75-84.
[1900l]   [Note:] 'I should be glad to have the opinions of logicians ...'. *Mind*, Vol. 9, p. 144.

[1901a]   *Letter to Bertrand Russell*, **28. VI. 1901**.
[1901b]   *Letter to Bertrand Russell*, **19. VII. 1901**.
[1901c]   *Letter to Bertrand Russell*, **10. IX. 1901**.
[1901d]   *Letter to Bertrand Russell*, **11. IX. 1901**.
[1901e]   *Letter to Bertrand Russell*, **6. X. 1901**.
[1901f]   La Logique Symbolique et ses Applications. ***Bibliothèque du 1° Congrès International de Philosophie. Logique et Histoire des Sciences***, pp. 135-183.
[1901g]   [Question No.] 14781. *The Educational Times*, vol. 54, p. 36.
[1901h]   [Question No.] 14807. *The Educational Times*, vol. 54, p. 84.
[1901i]   [Question No.] 14828. *The Educational Times*, vol. 54, p. 154.
[1901j]   [Question No.] 14853. *The Educational Times*, vol. 54, p. 191.
[1901k]   Solution to [Question No.] 14375. *The Educational Times*, vol. 54, p. 500.
[1901l]   [Question No.] 15024. *The Educational Times*, vol. 54, p. 502.
[1901m]   Limits of Logical Statements. ***Proceedings of the London Mathematical Society***, vol. 34, p. 132. (communicated 14. 11. 1901, not printed).

[1902a]   ***Mathematical Questions and Solutions from the "Educational Times" with Many Papers and Solutions in Addition to Those Published in the "Educational Times". New Series.*** Francis Hodgson, London.
[1902b]   Solution to [Question No.] 14751. *The Educational Times*, vol. 55, pp. 37-38.
[1902c]   [Question No.] 8093. *The Educational Times*, vol. 55, p. 40.
[1902d]   Solution to [Question No.] 14665. *The Educational Times*, vol. 55, p.155.
[1902e]   [Question No.] 15064. *The Educational Times*, vol. 55, p. 157.
[1902f]   [Question No.] 15085. *The Educational Times*, vol. 55, p. 197.
[1902g]   Erratum. *The Educational Times*, vol. 55, p. 198.
[1902h]   [Question No.] 15206. *The Educational Times*, vol. 55, p. 437.
[1902i]   Symbolic[al] Reasoning (IV). *Mind*, Vol. 11, pp. 352-368.
[1902j]   On the Validity of Certain Formulae. ***Proceedings of the London Mathematical Society***, vol. 35, p. 459. (Communicated from the Chair 16. 04. 1903, not printed).
[1902k]   Logique tabulaire. ***Revue de Métaphysique et de Morale***, vol. 10, pp. 213-217.
[1902l]   [Question No.] 14807. *The Educational Times (Reprint)*, vol. 1, p. 76.
[1902m]   [Question No.] 14781. *The Educational Times (Reprint)*, vol. 1, p. 81.
[1902n]   Solution to [Question No.] 14781. *The Educational Times (Reprint)*, vol. 1, p. 81-82.
[1902o]   Solution to [Question No.] 14786. *The Educational Times (Reprint)*, vol. 1, p. 88.
[1902p]   [Question No.] 14853. *The Educational Times (Reprint)*, vol. 1, p. 97.

[1902q]   Solution to [Question No.] 14853. *The Educational Times (Reprint)*, vol. 1, pp. 97-98.
[1902r]   [Question No.] 14375. *The Educational Times (Reprint)*, vol. 1, p. 111.
[1902s]   Solution to [Question No.] 14375. *The Educational Times (Reprint)*, vol. 1, pp. 111-113.
[1902t]   [Question No.] 14751. *The Educational Times (Reprint)*, vol. 2, p. 29.
[1902u]   Solution to [Question No.] 14751. *The Educational Times (Reprint)*, vol. 1, p. 29-30.

[1903a]   Symbolic Logic. *The Athenaeum*, p. 137.
[1903b]   Symbolic Logic II. *The Athenaeum*, pp. 321-322.
[1903c]   Symbolic Logic III. *The Athenaeum*, pp. 354-355.
[1903d]   Symbolic Logic III. *The Athenaeum*, p. 385.
[1903e]   La logique symbolique. *L'Enseignement mathématique*, vol. 5, pp. 415-430.
[1903f]   Solution to [Question No.] 15206. *The Educational Times*, vol. 56, pp. 90-91.
[1903g]   [Question No.] 15299. *The Educational Times*, vol. 56, p. 157.
[1903h]   [Question No.] 15429. *The Educational Times*, vol. 56, p. 441.
[1903i]   Symbolic[al] Reasoning (V). *Mind*, Vol. 12, pp. 355-364.
[1903j]   Paradoxes in Symbolic Logic. *The Educational Times (Reprint)*, vol. 4, pp. 34-36.
[1903k]   [Question No.] 15206. *The Educational Times (Reprint)*, vol. 4, pp. 47-48.
[1903l]   Solution to [Question No.] 15206. *The Educational Times (Reprint)*, vol. 4, p. 48.

[1904a]   *Letter to Bertrand Russell*, **26. VIII. 1904**.
[1904b]   Symbolic Logic. A correction *The Athenaeum*.
[1904c]   Symbolic Logic. *The Athenaeum*, pp. 13-16.
[1904d]   Symbolic Logic VI. *The Athenaeum*, pp. 149-151.
[1904e]   Symbolic Logic VII. *The Athenaeum*, pp. 213-214.
[1904f]   Symbolic Logic IV. *The Athenaeum*, pp. 244-245.
[1904g]   Symbolic Logic V. *The Athenaeum*, pp. 468-469.
[1904h]   Symbolic Logic VIII. *The Athenaeum*, pp. 879-880.
[1904i]   Symbolic Logic IX. *The Athenaeum*, p. 911.
[1904j]   La logique symbolique (II). *L'Enseignement mathématique*, vol. 6, pp. 372-375.
[1904k]   Solution to [Question No.] 15429. *The Educational Times*, vol. 57, pp. 159-160.
[1904l]   [Question No.] 15588. *The Educational Times*, vol. 57, p. 313.
[1904m]   [Question No.] 10050. *The Educational Times*, vol. 57, p. 450. [Old Questions as Yet Unsolved].
[1904n]   The Calculus of Limits. *The Educational Times (Reprint)*, vol. 5, pp. 66-70.
[1904o]   [Question No.] 15429. *The Educational Times (Reprint)*, vol. 6, p. 58.
[1904p]   Solution to [Question No.] 15429. *The Educational Times (Reprint)*, vol. 6, pp. 58-59.

[1905a]   *Letter to Bertrand Russell*, **26. I. 1905**.
[1905b]   *Letter to Bertrand Russell*, **03. II. 1905**.
[1905c]   *Letter to Bertrand Russell*, **03. III. 1905**.
[1905d]   *Letter to Bertrand Russell*, **05. IV. 1905**.

[1905e]	*Letter to Bertrand Russell*, **03. V. 1905**.
[1905f]	*Letter to Bertrand Russell*, **10. V. 1905**.
[1905g]	*Letter to Bertrand Russell*, **15. V. 1905**.
[1905h]	*Letter to Bertrand Russell*, **16. V. 1905**.
[1905i]	*Letter to Bertrand Russell*, **17. V. 1905**.
[1905j]	*Letter to Bertrand Russell*, **28. V. 1905**.
[1905k]	*Letter to Bertrand Russell*, **29 VI. 1905**.
[1905l]	*Letter to Bertrand Russell*, **22. VII. 1905**.
[1905m]	[Question No.] 7983. *The Educational Times*, vol. 58, p. 494. [Old Questions as Yet Unsolved].
[1905n]	[Question No.] 10780. *The Educational Times*, vol. 58, p. 534. [Old Questions as Yet Unsolved].
[1905o]	Symbolic[al] Reasoning (VI). *Mind,* Vol. 14, pp. 74-81.
[1905p]	Existential Import. *Mind,* Vol. 14, pp. 295-296.
[1905q]	Symbolic[al] Reasoning (VII). *Mind,* Vol. 14, pp. 390-397.
[1905r]	The Existential Import of Propositions [A Reply to Bertrand Russell]. *Mind,* Vol. 14, pp. 401-402.
[1905s]	The Existential Import of Propositions. *Mind,* Vol. 14, pp. 578-579.
[1906a]	***Symbolic Logic and its Applications***, Longmans, Green and Co., London.
[1906b]	*Letter to Bertrand Russell*, **24. II. 1906**.
[1906c]	*Letter to Bertrand Russell*, **05. III. 1906**.
[1906d]	*Letter to Bertrand Russell*, **12. III. 1906**.
[1906e]	*Letter to Bertrand Russell*, **02. IV. 1906**.
[1906f]	Solution to [Question No.] 15957. *The Educational Times*, vol. 59, pp. 267-268.
[1906g]	[Question No.] 16059. *The Educational Times*, vol. 59, p. 413.
[1906h]	Solution to [Question No.] *16059*. *The Educational Times*, vol. 59, p .493.
[1906i]	[Question No.] 11285. *The Educational Times*, vol. 59, [Old Questions as Yet Unsolved], p. 540.
[1906j]	Chance or Purpose. *The Hibbert Journal*, vol. 5, pp. 384-396.
[1906k]	Symbolic Reasoning (VIII). *Mind,* Vol. 15, pp. 504-518.
[1907a]	*Letter to Bertrand Russell*, **18. I. 1907**.
[1907b]	[Question No.] 10708. *The Educational Times*, vol. 59, p. 82.
[1907c]	[Question No.] 16292. *The Educational Times*, vol. 59, p.458.
[1907d]	Solution to [Question No.] *16292*. *The Educational Times*, vol. 59, p.544.
[1907e]	What and Where is the Soul? *The Hibbert Journal*, vol. 6, pp. 158-170.
[1907f]	Symbolic Logic. A reply. *Mind,* Vol. 16, pp. 470-473.
[1908a]	What and Where is the Soul? *The Hibbert Journal*, vol. 6, pp. 673-674.
[1908b]	'If' and 'Imply'. *Mind,* Vol. 17, pp. 151-152.
[1908c]	'If' and 'Imply'. *Mind,* Vol. 17, pp. 453-455.
[1909a]	***Man's Origin, Destiny and Duty***, Williams and Norgate, London.
[1909b]	*Letter to Bertrand Russell*, **04. X. 1909**.
[1909c]	*Letter to Bertrand Russell*, **18. XII. 1909**.
[1909d]	[Question No.] 16734. *The Educational Times*, vol. 62, p.429.

[1909e] Mathematics and Theology *The Hibbert Journal*, vol. 7, pp. 916-918.
[1909f] Ptolemaic and Copernian Views of the Place of Mind in the Universe (II) *The Hibbert Journal*, vol. 8, pp. 429-430.

[1910a] Solution to [Question No.] 16751. *The Educational Times*, vol. 63, p. 35.
[1910b] [Question No.] 16788. *The Educational Times*, vol. 63, p. 37.
[1910c] Linguistic Misunderstandings (I). *Mind,* Vol. 19, pp. 186-199.
[1910d] Linguistic Misunderstandings (II). *Mind,* Vol. 19, pp. 337-355.

[1911] [Question No.] 12359 [Old Questions as Yet Unsolved]. *The Educational Times*, vol. 64, p. 138.

[1973a] Symbolic Reasoning (VI). **Lackey, D. (ed.): Essays in Analysis, George Allen and Unwin Ltd., London**, pp. 308-316.
[1973b] Existential Import. **Lackey, D. (ed.): Essays in Analysis, George Allen and Unwin Ltd., London**, p.317.
[1973c] The Existential Import of Propositions (A Reply to Chapter 4 above) **Lackey, D. (ed.): Essays in Analysis, George Allen and Unwin Ltd., London**, pp. 317-319.
[1973d] The Existential Import of Propositions. **Lackey, D. (ed.): Essays in Analysis, George Allen and Unwin Ltd., London**, pp. 319-322.

[2007a] Marriage Certificate for Hortense Lina Marchal and Hugh MacColl: 17.08.1887.
[2007b] Death Certificate for Hugh MacColl: 27.12.1909.

Références

Angell, R. B.: 1962, 'A Propositional Logic with Subjunctive Conditionals', *Journal of Symbolic Logic* 27, pp. 327-343.

Angell, R. B.: 2002, *A-Logic*, Lanham: University Press of America.

Aristotle: 1928, *The Works of Aristotle Translated into English*, vol. I, Oxford University Press, Oxford.

Astroh, M.: 1993, 'Der Begriff der Implikation in einigen frühen Schriften von Hugh McColl', in W. Stelzner, S. W. Stelzner *Philosophie und Logik, Frege-Kolloquien Jena 1989/1991*, Walter de Gruyter, Berlin, New York, 1993, pp. 128-144.

Astroh, M.: 1995, 'Subjekt, Prädikat und Modalität. Ein Beitrag zur logischen Rekonstruktion einer kategorischen Syllogistik', Saarbrücken: Typoskript.

Astroh, M.: 1996, 'Präsupposition und Implikatur', in M. Dascal/D. Gerhardus/K. Lorenz/G. Meggle, *Sprachphilosophie, Philosophy of Language, La philosophie du langage*, vol. II, 1996, pp. 1391-1407.

Astroh, M.: 1999a, 'Connexive Logic', *Nordic Journal of Philosophical Logic*, vol. 4, pp. 31-71.

Astroh, M.: 1999b, 'MacColl's evolutionary design of language', in: Astroh, M., Read, S. (eds.): [1999], pp. 141-173.

Astroh, M., Read, S. (eds.): 1999, *Proceedings of the Conference "Hugh MacColl and the Tradition of Logic." at Greifswald* (1998), *Nordic Journal of Philosophical Logic*, vol 3/1/dec.1998.

Astroh, M. Grattan-Guiness, I. Read, S.: 2001, 'A survey of the life of Hugh MacColl (1837-1909)', *History and Philosophy of Logic*. vol. 22, Num 2 (June).

Boethius, A. M. T. S.: 1969, *De hypotheticis syllogismis*, Paideia, Brescia.

Christie, A.: 'Hugh MacColl in der wissenschaftlichen Literatur seit 1970: Zur Methodik'. Typoscript, 1-5.

Christie, A.: 'Nature as a source in the history of logic1870-1910', *History and Philosophy of Logic*, vol. 11, pp. 1-3.

Couturat L.: 1899, 'La Logique mathématique de M. Peano', *Revue de Métaphysique et de Morale*, vol. 7, pp. 616-646.

Cuypers, S. E.: 1999, 'The Metaphysical foundations of Hugh MacColl's religious ethics', in: Astroh, M., Read, S. (eds.): [1999], pp. 175-196.

Frege G.: 1882, *Über den Zweck der Begriffsschrift*. In: I. Angelelli (ed.)., *Gottlob Frege, Begriffssschrift und andere Aufsätze*. Darmstadt: Wissenschaftliche Buchgesellschaft, 1964, pp. 97-105.

Frege G.: 1884, *Grundlagen der Arithmetik*. Breslau: Keubner, 1884.

Gabbay, D. M.: 1987, *Modal Provability Foundations for Negation by Failure*, ESPRIT, Technical Report TI 8, Project 393, ACORD.

Gardner, M.: 1996, *The Universe in a Handkerchief. Lewis Carroll's Mathematical Recreations, Games, Puzzles and Word Plays*, Copernicus (Springer-Verlag), New York.

Grattan-Guiness, I.: 1999, 'Are other logics possible? MacColl's logic and some English reactions, 1905-1912', in: Astroh, M., Read, S. (eds.): [1999], pp. 1-16.

Grice, H. P.: 1967, *Conditionals. Privately Circulated Notes*, University of California, Berkeley.

Grice, H. P.: 1989, *Studies in the Way of Words*, MIT-Press, Cambridge, MA.

Hailperin, T.: 1996, *Sentential Probability Logic*, Lehigh UP/Associated UP, Bethlehem/London.

Hoepelman, J. P. and A. J. M. van Hoof: 1988, 'The Success of Failure', *Proceedings of COLING*, Budapest, pp. 250-254.

Jevons W.S.: 1881, 'Recent mathematico-logical memoirs', *Nature 23*, pp. 485-7.

Ladd-Franklin, C.: 1889, 'On Some Characteristics of Symbolic Logic', *American Journal of Psychology*, (Neudruck 1966), pp. 543-567.

McCall, S.: 1963, *Aristotle's Modal Syllogisms*, North-Holland, Amsterdam.

McCall, S.: 1964, 'A New Variety of Implication', *Journal of Symbolic Logic*, vol. 29, pp. 151-152.

McCall, S.: 1966, 'Connexive conditional', *Journal of Symbolic Logic*, vol. 31, pp. 415-432.

McCall, S.: 1967a, 'Connexive conditional and the Syllogism', *Mind*, vol. 76, pp. 346-356.

McCall, S.: 1967b, 'MacColl', in P. Edwards (ed.): 1975, *Encyclopedia of Philosophy*, Macmillan, London. vol. IV, pp. 545-546.

McCall, S.: 1990, 'Connexive conditional', in A. R. Anderson, and N. D. Belnap, *Entailment* I, Princeton University Press, Princeton, NJ, pp. 432-441.

McCall, S.: 2007, 'Sequent calculus for Connexive Logic', Manuscript.

Merridew, H. M.: 1866, *Merridews illustrated Guide to Boulogne and Environs with plan and map*. Boulogne-sur-Mer: Merridew - English Booksellers.

Olsen, S. H.: 1999, 'Hugh MacColl-Victorian', in: Astroh, M., Read, S. (eds.) [1999], pp. pp. 197-229.

Peckhaus, V.: 1996, 'Case studies towards the establishment of a social history of logic'. *History and Philosophy of Logic*, vol. 7, pp. 185-186.

Peckhaus, V.: 1999, 'Hugh MacColl and the German Algebra of Logic', in: Astroh, M., Read, S. (eds.): [1999], pp. 141-173.

Peirce, C. S. :1867, 'On an improvement in Boole's Calculus of Logic', in *Collected Papers of Charles Sanders Peirce*, vol. III, 1974, Cambridge-Massachusetts, Harvard University Press, 1974, p. 9.

Pizzi, C.: 1977, 'Boethius' Thesis and Conditional Logic', *Journal of Philosophical Logic*, 6: pp. 283-302.

Pizzi, C.: 1991, 'Decision Procedures for Logics of Consequential Implication', *Notre Dame Journal of Formal Logic*, 32: pp. 618-636.

Pizzi, C.: 1993, 'Consequential Implication: A Correction', *Notre Dame Journal of Formal Logic*, 34: pp. 621-624.
Pizzi, C.: 1996, 'Weak vs. Strong Boethius' Thesis: A Problem in the Analysis of Consequential Implication', in: A. Ursini and P. Aglinanó (eds.), *Logic and Algebra*, New York: Marcel Dekker, pp. 647-654.
Pizzi, C. and Williamson, T.: 1997, 'Strong Boethius' Thesis and Consequential Implication', *Journal of Philosophical Logic*, 26: pp. 569-588.
Poggendorff, J.C., Feddersen, B.W and Von Oettingen, A.J. (eds). 1863-1938. *Biographisch-literarisches Handwörterbuch zur Geschichte der exacten Wissenschaften enthaltend Nachweisungen über Lebensverhältnisse und Leistungen von Mathematikern, Astronomen, Physikern, Chemikern, Mineralogen, Geologen, Geographen u.s.w. aller Voelker und Zeiten*, vol. 1-6. Leipzig: Verlag von Johann Ambrosius Barth.
Priest, G.: 1999, 'Negation as Cancellation and Connexive Logic', *Topoi*, vol. 18, pp. 141-148.
Priest, G.: 2005, *Towards Non-Being. The logic and Metaphysics of Intentionality*. Oxford, Clarendon Press, Oxford.
Rahman, S.: 1997a, *Die Logik der zusammenhängenden Behauptungen im frühen Werk von Hugh MacColl*, 'Habilitationsschrift', Universität des Saarlandes, Saarbrücken.
Rahman, S.: 1997b 'Hugh MacColl: eine bibliographische Erschließung seiner Hauptwerke und Notizen zu ihrer Rezeptionsgeschichte'. *History and Philosophy of Logic*, vol. 18, 1997, pp. 165-83)
Rahman, S.: 1998, *Redundanz und Wahrheitswertbestimmung bei Hugh MacColl*, FR 5.1 Philosophie, Universität des Saarlandes, Memo Nr. 23.
Rahman, S.: 1999, 'Ways of understanding Hugh MacColl's concept of Symbolic Existence', in: Astroh, M., Read, S. (eds.): [1999], pp. 35-58.
Rahman, S.: 2000, 'Hugh MacColl's criticism of Boole's formalization of traditional hypotheticals. In J. Gasser (ed.): *A Boole Anthology. Recent and Classical Studies in the Logic of George Boole*. Dordrecht / Boston / London: Kluwer, Synthese Library, pp. 287-310.
Rahman, S.: 2001, 'On Frege's Nightmare', H. Wansing (ed.): *Essays on Non-Classical Logic*. London: World Scientific, pp. 61-85.
Rahman S., Redmond, J., 2005, Talk: *Hugh MacColl: Reasoning about No-Thing*, in Workshop: 'Que prouve la science fiction? Raison, machines, corps et mondes', organised by Université Lille III, Maison de la recherche, Vendredi 1 et samedi 2 avril 2005.
Rahman, S., Rückert, H.: 2001, 'Dialogical Connexive Logic', *Synthese,* vol 127, Nos 1-2, 2001, pp. 105-139.
Read, S.: 1994, *Thinking About Logic*, Oxford University Press, Oxford, New York.
Read, S.: 1999, *Hugh MacColl and the Algebra of Strict Implication*, in: Astroh, M., Read, S. (eds.): [1999], pp. 59-83.

Routley, R. and H. Montgomery.: 1968, 'On Systems Containing Aristotle's Thesis', *The Journal of Symbolic Logic* 3, pp. 82-96.
Russell, G.W.E. (ed.): 1914, 'M. MacColl, Memoirs and Correspondence', London: Smith Elder & Co.
Schröder, E.: 1890-95, *Vorlesungen über die Algebra der Logik*. Leipzig: B. G. Teubner, 3 vol: vol. I (1890), vol. II (1891-1905), vol. III. (1895).
Simons, P.: 1999, 'MacColl and Many-Valued Logic: An Exclusive Conjunction', in: Astroh, M., Read, S. (eds.): [1999], pp. 85-90.
Stelzner, W.: 1999, 'Context-Sensitivity and the Truth-Operator in Hugh MacColl's Modal Distinctions', in: Astroh, M., Read, S. (eds.): [1999], pp. 91-118.
Sundholm, G: 1999, 'MacColl on Judgement and Inference', in: Astroh, M., Read, S. (eds.): [1999], pp. 141-173.
Thiel, C.: 1996, 'Reasearch of the history of logic at Erlangen', in I. Angelelli and M. Cerzo (eds.); Studies on the History of Logic, De Gruter, Berlin, N. York, 1996, pp. 397-401.
Vailati, G.: 1899, 'La logique mathématique et sa nouvelle phase de développement dans les écrits de M. J. Peano', *Revue de Métaphysique et de Morale*, vol. 7, pp. 86-102.
Wansing, H.: 2005, 'Connexive Modal Logic', in: R. Schmidt et al. (eds.), *Advances in Modal Logic. Volume 5*, London: King's College Publications, pp. 367-383.
Wansing, H.: 2006, 'Connexive Logic', in: Online entry in the *Stanford Encyclopaedia of Philosophy* : http://plato.stanford.edu/entries/logic-connexive
Woleński, J.: 1999, 'MacColl on Modalities', in: Astroh, M., Read, S. (eds.): [1999], pp. 133-141.

Index des noms

A

Angell · 37
Aristote · xv, 32, 34
Astroh · xiv, xv, 3, 37

B

Boèce · 31, 32, 33

C

Carroll · 30, 32
Christie · xiv
Couturat · xii, xix, 21
Cuypers · 60, 62

F

Frege · xii, xiii, xviii, 1, 7, 8, 10, 12, 13, 14, 21

G

Gabbay · 40
Gardner · 78
Grattan-Guinness · xv
Grice · 30

H

Hailperin · 63
Hoepelmann · 40

J

Jevons · xii, xvii, xviii, 24, 69, 70

L

Lewis · 20, 37

M

MacCall · 26, 32, 37
McCall · xiii

O

Olsen · 60, 62

P

Peckhaus · xiv
Peirce · xii, xiii, xvii, xviii, 63, 69, 70, 71
Pizzi · 37
Poggendorff · 79
Priest · 51

R

Rahman · xiv, xviii, 4, 10, 20, 21, 35, 36, 37, 38, 40, 51, 57, 66
Read · xiv, xv, 20, 21, 24, 25, 26, 30, 37
Routley · 33, 51
Russell · xii, xiii, xvi, xvii, 6, 8, 10, 11, 12, 13, 14, 21, 22, 61, 73, 74, 75, 417

S

Schröder · xii, xviii, 24
Simons · xiv, 24
Stelzner · 24, 25
Sundholm · xiv, 4

T

Thiel · xiv

V

Vailati · xix
Venn · xii, xviii, 24, 32

W

Wansing · 37, 57
Woleński · xiv, 24

EXTRAITS MAJEURS DE L'OEUVRE DE HUGH MACCOLL

[1906a]: Symbolic Logic and its Applications, Longmans, Green and Co., London.

SYMBOLIC LOGIC
AND ITS APPLICATIONS

BY

HUGH MacCOLL
B.A. (London)

LONGMANS, GREEN, AND CO.
39 PATERNOSTER ROW, LONDON
NEW YORK AND BOMBAY
1906

All rights reserved

PREFACE

This little volume may be regarded as the final concentrated outcome of a series of researches begun in 1872 and continued (though with some long breaks) until to-day. My article entitled "Probability Notation No. 2," which appeared in 1872 in the *Educational Times*, and was republished in the mathematical "Reprint," contains the germs of the more developed method which I afterwards explained in the *Proceedings of the London Mathematical Society* and in *Mind*. But the most important developments from the logical point of view will be found in the articles which I contributed within the last eight or nine years to various magazines, English and French. Among these I may especially mention those in *Mind* and in the *Athenæum*, portions of which I have (with the kind permission of these magazines) copied into this brief epitome.

Readers who only want to obtain a clear general view of symbolic logic and its applications need only attend to the following portions: §§ 1 to 18, §§ 22 to 24, §§ 46 to 53, §§ 76 to 80, §§ 112 to 120, §§ 144 to 150.

Students who have to pass elementary examinations in ordinary logic may restrict their reading to §§ 1 to 18, §§ 46 to 59, §§ 62 to 66, §§ 76 to 109, § 112.

Mathematicians will be principally interested in the last five chapters, from § 114 to § 156; but readers

PREFACE

who wish to obtain a complete mastery of my symbolic system and its applications should read the whole. They will find that, in the elastic adaptability of its notation, it bears very much the same relation to other systems (including the ordinary formal logic of our text-books) as algebra bears to arithmetic. It is mainly this notational adaptability that enables it to solve with ease and simplicity many important problems, both in pure logic and in mathematics (see § 75 and § 157), which lie wholly beyond the reach of any other symbolic system within my knowledge.

HUGH MacCOLL.

August 17th, 1905.

CONTENTS

INTRODUCTION

SECS.   PAGE

1–3. General principles—Origin of language . . .   1

CHAPTER I

4–12. Definitions of symbols—Classification of propositions—Examples and formulæ   4

CHAPTER II

13–17. Logic of Functions—Application to grammar . . .   9

CHAPTER III

18–24. Paradoxes—Propositions of the second, third, and higher degrees   12

CHAPTER IV

25–32. Formulæ of operations with examples worked—Venn's problem   20

CHAPTER V

33–38. Elimination—Solutions of implications and equations—Limits of statements   27

CHAPTER VI

39–43. Jevons's "Inverse Problem"; its complete solution on the principle of limits, with examples . . .   33

viii CONTENTS

CHAPTER VII
SECS. PAGE
44–53. Tests of Validity—Symbolic Universe, or Universe of Discourse—No syllogism valid as usually stated . 39

CHAPTER VIII
54–63. The nineteen traditional syllogisms all deducible from one simple formula—Criticism of the technical words 'distributed' and 'undistributed'—The usual syllogistic 'Canons' unreliable; other and simpler tests proposed 49

CHAPTER IX
64–66 (a). Enthymemes—Given one premise of a syllogism and the conclusion, to find the missing premise—Strongest conclusion from given premises . . 66

CHAPTER X
67–75. To find the weakest data from which we can prove a given complex statement, and also the strongest conclusion deducible from the statement—Some contested problems—'Existential Import of Propositions'—Comparison of symbolic methods . 70

CHAPTER XI
76–80. The nature of *inference*—The words *if*, *therefore*, and *because*—Causation and discovery of causes . . 80

CHAPTER XII
81–89. Solutions of some questions set at recent examinations 86

CHAPTER XIII
90–113. Definitions and explanations of technical terms often used in logic—Meaningless symbols and their uses; mathematical examples — *Induction:* inductive reasoning not absolutely reliable; a curious case in mathematics—'*Infinite*' and '*infinitesimal*' . . 91

CALCULUS OF LIMITS

CHAPTER XIV

SECS.   PAGE
114–131   Application to elementary algebra, with examples   .   106

CHAPTER XV

132–140. Nearest limits—Table of Reference   117

CHAPTER XVI

141–143. Limits of two variables—Geometrical illustrations   .   123

CHAPTER XVII

144–150. Elementary probability—Meaning of 'dependent' and 'independent' in probability, with geometrical illustrations   128

CHAPTER XVIII

151–157. Notation for Multiple Integrals—Problems that require the integral calculus   132

ALPHABETICAL INDEX

(The numbers indicate the **sections**, *not the pages.)*

Alternative, 7, 41
Ampliative, 108
Antecedent, 28
Cause, 79
Complement, 46
Connotation, 93
Consequent, 28
Contraposition, 97
Contrary, 94
Conversion, 98
Couturat's notation, 132 (footnote)
Dichotomy, 100
Dilemma, 101–103
Elimination, 33-38
Enthymeme, 64
Equivalence, 11, 19
Essential, 108
Excluded Middle, 92
Existential import of propositions, 72, 73
Factor, 7, 28
Formal, 109
Functions, 13-17
Grammar, 17
Illicit process, 63 (footnote)
Immediate inference, 91
Implication, 10, 18

Induction, 112
Inference, nature of, 76–80
Infinite and infinitesimal, 113
Jevons's 'inverse problem,' 39–43
Limits of statements, 33
Limits of variable ratios, 114–143
Major, middle, minor, 54
Material, distinguished from Formal, 109
Meaningless symbols, 110
Mediate inference, 91
Modality, 99
Multiple, 28
Particulars, 49
Ponendo ponens, &c., 104–107
Product, 7
Sorites, 90
Strong statements, 33, 34
Subalterns and subcontraries, 95, 96
Syllogisms, 54
Transposition, 56
Universals, 49
Universe of discourse, 46–50
Venn's problem, 32
Weak statements, 33, 34

SYMBOLIC LOGIC

INTRODUCTION

1. IN the following pages I have done my best to explain in clear and simple language the principles of a useful and widely applicable method of research. Symbolic logic is usually thought to be a hard and abstruse subject, and unquestionably the Boolian system and the more modern methods founded on it *are* hard and abstruse. They are, moreover, difficult of application and of no great utility. The symbolic system explained in this little volume is, on the contrary, so simple that an ordinary schoolboy of ten or twelve can in a very short time master its fundamental conceptions and learn to apply its rules and formulæ to practical problems, especially in elementary mathematics (see §§ 114, 118). Nor is it less useful in the higher branches of mathematics, as my series of papers published in the *Proceedings of the London Mathematical Society* abundantly prove. There are two leading principles which separate my symbolic system from all others. The first is the principle that there is nothing sacred or eternal about symbols; that all symbolic conventions may be altered when convenience requires it, in order to adapt them to new conditions, or to new classes of problems. The symbolist has a right, in such circumstances, to give a new meaning to any old symbol, or arrangement of symbols, provided the change of sense be accompanied by a fresh definition, and provided the nature of the

A

problem or investigation be such that we run no risk of confounding the new meaning with the old. The second principle which separates my symbolic system from others is the principle that the complete statement or proposition is the real *unit* of all reasoning. Provided the complete statement (alone or in connexion with the context) convey the meaning intended, the words chosen and their arrangement matter little. Every intelligible argument, however complex, is built up of individual statements; and whenever a simple elementary symbol, such as a letter of the alphabet, is sufficient to indicate or represent any statement, it will be a great saving of time, space, and brain labour thus to represent it.

2. The words *statement* and *proposition* are usually regarded as synonymous. In my symbolic system, however, I find it convenient to make a distinction, albeit the distinction may be regarded as somewhat arbitrary. I define a *statement* as any sound, sign, or symbol (or any arrangement of sounds, signs, or symbols) employed to give information; and I define a *proposition* as a statement which, *in regard to form,* may be divided into two parts respectively called *subject* and *predicate*. Thus every proposition is a statement; but we cannot affirm that every statement is a proposition. A nod, a shake of the head, the sound of a signal gun, the national flag of a passing ship, and the warning "Caw" of a sentinel rook, are, by this definition, statements but not propositions. The nod may mean "I see him"; the shake of the head, "I do not see him"; the warning "Caw" of the rook, "A man is coming with a gun," or "Danger approaches"; and so on. These propositions express more specially and precisely what the simpler statements express more vaguely and generally. In thus taking statements as the ultimate constituents of symbolic reasoning I believe I am following closely the gradual evolution of human language from its primitive

INTRODUCTION

prehistoric forms to its complex developments in the languages, dead or living, of which we have knowledge now. There can be little doubt that the language or languages of primeval man, like those of the brutes around him, consisted of simple elementary statements, indivisible into subject and predicate, but differing from those of even the highest order of brutes in being uninherited—in being more or less conventional and therefore capable of indefinite development. From their grammatical structure, even more than from their community of roots, some languages had evidently a common origin; others appear to have started independently; but all have sooner or later entered the propositional stage and thus crossed the boundary which separates all brute languages, like brute intelligence, from the human.

3. Let us suppose that amongst a certain prehistoric tribe, the sound, gesture, or symbol S was the understood representation of the general idea *stag*. This sound or symbol might also have been used, as single words are often used even now, to represent a complete statement or proposition, of which *stag* was the central and leading idea. The symbol S, or the word *stag*, might have vaguely and varyingly done duty for " It is a stag," or " I see a stag," or " A stag is coming," &c. Similarly, in the customary language of the tribe, the sound or symbol B might have conveyed the general notion of *bigness*, and have varyingly stood for the statement " It is *big*," or " I see a *big* thing coming," &c. By degrees primitive men would learn to combine two such sounds or signs into a compound statement, but of varying form or arrangement, according to the impulse of the moment, as SB, or BS, or S_B, or S^B, &c., any of which might mean " I see a *big stag*," or " The *stag* is *big*," or " A *big stag* is coming," &c. In like manner some varying arrangement, such as SK, or S^K, &c., might mean " The *stag* has been *killed*," or " I have *killed* the *stag*," &c.

SYMBOLIC LOGIC [§§ 3, 4]

Finally, and after many tentative or haphazard changes, would come the grand chemical combination of these linguistic atoms into the compound linguistic molecules which we call *propositions*. The arrangement S^B (or some other) would eventually crystallize and *permanently* signify "The *stag* is *big*," and a similar form S^K would *permanently* mean "The *stag* is *killed*." These are two complete propositions, each with distinct subject and predicate. On the other hand, S_B and S_K (or some other forms) would *permanently* represent "The *big stag*" and "The *killed stag*." These are *not* complete propositions; they are merely qualified subjects waiting for their predicates. On these general ideas of linguistic development I have founded my symbolic system.

CHAPTER I

4. THE symbol A^B denotes a proposition of which the individual A is the subject and B the predicate. Thus, if A represents *my aunt*, and B represents *brown-haired*, then A^B represents the proposition "My *aunt* is *brown-haired*." Now the word *aunt* is a class term; a person may have several *aunts*, and any one of them may be represented by the symbol A. To distinguish between them we may employ numerical suffixes, thus A_1, A_2, A_3, &c., Aunt No. 1, Aunt No. 2, &c.; or we may distinguish between them by attaching to them different attributes, so that A_B would mean *my brown-haired aunt*, A_R *my red-haired aunt*, and so on. Thus, when A is a class term, A_B denotes the individual (or an individual) of whom or of which the proposition A^B is true. For example, let H mean "*the horse*"; let w mean "*it won the race*"; and let s mean "*I sold it*," or "*it has been sold by me*." Then H_w^s, which is short for $(H_w)^s$, represents the complex proposition "The *horse* which *won* the race has been *sold* by me," or "I have sold the horse which

§§ 4–6]   EXPLANATIONS OF SYMBOLS   5

won the race." Here we are supposed to have a series of horses, H_1, H_2, H_3, &c., of which H_w is one; and we are also supposed to have a series, S_1, S_2, S_3, &c., of things which, at some time or other, I sold; and the proposition H_w^s asserts that the individual H_w, of the first series H, belongs also to the second series S. Thus the suffix w is *adjectival;* the exponent s *predicative.* If we interchange suffix and exponent, we get the proposition H_s^w, which asserts that "the horse which I have sold won the race." The symbol H^w, without an adjectival suffix, merely asserts that a horse, or the horse, won the race without specifying which horse of the series $H_1, H_2,$ &c.

5. A small *minus* before the predicate or exponent, or an acute accent affecting the whole statement, indicates denial. Thus if H^c means "The *horse* has been *caught*"; then H^{-c} or $(H^c)'$ means "The *horse* has *not* been *caught.*" In accordance with the principles of notation laid down, the symbol H_{-c} will, on this understanding, mean "The *horse* which has *not* been *caught,*" or the "*uncaught horse*"; so that a minus suffix, like a suffix without a minus, is *adjectival.* The symbol H_c ("The *caught horse*") assumes the statement H^c, which *asserts* that "The *horse* has been *caught.*" Similarly H_{-c} assumes the statement H^{-c}.

6. The symbol 0 denotes *non-existence,* so that $0_1, 0_2, 0_3,$ &c., denote a series of names or symbols which correspond to nothing in our universe of admitted realities. Hence, if we give H and C the same meanings as before, the symbol H_c^0 will assert that "The *horse caught does not exist,*" which is equivalent to the statement that "No horse has been caught." The symbol H_c^{-0}, which denies the statement H_c^0, may therefore be read as "The *horse caught does* exist," or "*Some* horse has been caught." Following the same principle of notation, the symbol H_{-c}^0 may be read "An *uncaught horse does not exist,*" or "Every horse has been caught." The context would, of course, indicate the particular totality of horses

referred to. For example, H^0_{-c} may mean "Every horse *that escaped* has been caught," the words in italics being understood. On the same principle H^{-0}_{c} denies H^0_{-c}, and may therefore be read "*Some uncaught horse does exist*," or "*Some horse has not been caught*."

7. The symbol $A^B \times C^D$, or its usually more convenient synonym $A^B \cdot C^D$, or (without a point) $A^B C^D$, asserts two things—namely, that A *belongs to the class* B, and that C *belongs to the class* D; or, as logicians more briefly express it, that "A is B" and that "C is D." The symbol $A^B + C^D$ asserts an alternative—namely, that "*Either* A *belongs to the class* B, *or else* C *to the class* D"; or, as it is more usually and briefly expressed, that "Either A is B, or C is D." The alternative $A^B + C^D$ does not necessarily imply that the propositions A^B and C^D are mutually exclusive; neither does it imply that they are not. For example, if A^B means "*Alfred is a barrister*," and C^D means "*Charles is a doctor*"; then $A^B C^D$ asserts that "*Alfred* is a *barrister*, and *Charles* a *doctor*," while $A^B + C^D$ asserts that "Either *Alfred* is a *barrister*, or *Charles* a *doctor*," a statement which (apart from context) does not necessarily exclude the possibility of $A^B C^D$, that *both* A^B and C^D are true.* Similar conventions hold good for $A^B C^D E^F$ and $A^B + C^D + E^F$, &c. From these conventions we get various self-evident formulæ, such as (1) $(A^B C^D)' = A^{-B} + C^{-D}$; (2) $(A^B + C^D)' = A^{-B} C^{-D}$; (3) $(A^B C^{-D})' = A^{-B} + C^D$; (4) $(A^B + C^{-D})' = A^{-B} C^D$.

8. In *pure* or *abstract* logic statements are represented by single letters, and we classify them according to attributes as *true, false, certain, impossible, variable*, respectively denoted by the five Greek letters τ, ι, ϵ, η, θ. Thus the symbol $A^\tau B^\iota C^\epsilon D^\eta E^\theta$ asserts that A is *true*, that B is *false*, that C is *certain*, that D is *impossible*, that

---

* To preserve mathematical analogy, A^B and C^D may be called *factors* of the *product* $A^B C^D$, and *terms* of the sum $A^B + C^D$; though, of course, these words have quite different meanings in logic from those they bear in mathematics.

§§ 8–10]   EXPLANATIONS OF SYMBOLS   7

E is *variable* (possible but uncertain). The symbol A^v only asserts that A is true in a particular case or instance. The symbol A^c asserts more than this: it asserts that A is *certain*, that A is *always* true (or true in *every case*) within the limits of our data and definitions, that its probability is 1. The symbol A^i only asserts that A is false in a particular case or instance; it says nothing as to the truth or falsehood of A in other instances. The symbol A^η asserts more than this; it asserts that A contradicts some datum or definition, that its probability is 0. Thus A^v and A^i are simply *assertive;* each refers only to one case, and raises no general question as to data or probability. The symbol A^θ (A is a *variable*) is equivalent to $A^{-\eta}A^{-c}$; it asserts that A is neither *impossible* nor *certain*, that is, that A is *possible* but *uncertain*. In other words, A^θ asserts that the probability of A is neither 0 nor 1, but some proper fraction between the two.

9. The symbol A^{bc} means $(A^b)^c$; it asserts * that the statement A^b belongs to the class C, in which C may denote *true*, or *false*, or *possible*, &c. Similarly A^{bcd} means $(A^{bc})^d$, and so on. From this definition it is evident that A^{rs} is not necessarily or generally equivalent to A^{sr}, nor A^{ss} equivalent to A^s.

10. The symbol $A^b : C^d$ is called an *implication*, and means $(A^b C^{-d})^\eta$, or its synonym $(A^{-b} + C^d)^c$. It may be read in various ways, as (1) A^b implies C^d; (2) If A belongs to the class B, then C belongs to the class D; (3) It is impossible that A can belong to the class B without C belonging to the class D; (4) It is certain that either A does not belong to the class B or else C belongs to the class D. Some logicians consider these four propositions equivalent, while others do not; but all ambiguity may be avoided by the convention, adopted

---

* The symbol A^{bc} must not be confounded with the symbol $A^{b,c}$, which I sometimes use as a convenient abbreviation for $A^b A^c$; nor with the symbol $A^{b.c}$, which I use as short for $A^b + A^c$.

here, that they are synonyms, and that each, like the symbol $A^B:C^D$, means $(A^B C^{-D})^\eta$, or its synonym $(A^{-B}+C^D)^\epsilon$. Each therefore usually asserts more than $(A^B C^{-D})^\iota$ and than $(A^{-B}+C^D)^\tau$, because A^η and A^ϵ (for any statement A) asserts more than A^ι and A^τ respectively (see § 8).

11. Let the proposition A^B be denoted by a single letter α; then α' will denote its denial A^{-B} or $(A^B)'$. When each letter denotes a statement, the symbol $A:B:C$ is short for $(A:B)(B:C)$. It asserts that A implies B and that B implies C. The symbol $(A=B)$ means $(A:B)(B:A)$. The symbol $A!B$ (which may be called an *inverse implication*) asserts that A is *implied in* B; it is therefore equivalent to $B:A$. The symbol $A!B!C$ is short for $(A!B)(B!C)$; it is therefore equivalent to $C:B:A$. When we thus use single letters to denote statements, we get numberless self-evident or easily proved formulæ, of which I subjoin a few. To avoid an inconvenient multiplicity of brackets in these and in other formulæ I lay down the convention that the sign of equivalence $(=)$ is of longer reach than the sign of implication $(:)$, and that the sign of implication $(:)$ is of longer reach than the sign of disjunction or alternation $(+)$. Thus the equivalence $\alpha=\beta:\gamma$ means $\alpha=(\beta:\gamma)$, *not* $(\alpha=\beta):\gamma$, and $A+B:x$ means $(A+B):x$, *not* $A+(B:x)$.

(1) $x(\alpha+\beta) = x\alpha + x\beta$; (2) $(\alpha\beta)' = \alpha' + \beta'$;
(3) $(\alpha+\beta)' = \alpha'\beta'$; (4) $\alpha:\beta = \beta':\alpha'$;
(5) $(x:\alpha)(x:\beta) = x:\alpha\beta$; (6) $\alpha+\beta:x = (\alpha:x)(\beta:x)$;
(7) $(A:B:C):(A:C)$; (8) $(A!B!C):(A!C)$;
(9) $(A!C)!(A!B!C)$; (10) $(A:C)!(A:B:C)$;
(11) $(A+A')^\iota$; (12) $(A^\tau+A^\iota)^\tau$; (13) $(AA')^\eta$,
(14) $(A^\epsilon+A^\eta+A^\theta)^\epsilon$; (15) $A^\iota:A^\tau$;
(16) $A^\eta:A^\iota$; (17) $A^\epsilon = (A')^\eta$; (18) $A^\eta = (A')^\epsilon$;
(19) $A^\theta = (A')^\theta$; (20) $\epsilon:A = A^\epsilon$; (21) $A:\eta = A^\eta$;
(22) $A\epsilon = A$; (23) $A\eta = \eta$.

§§ 11–14]   LOGIC OF FUNCTIONS   9

These formulæ, like all valid formulæ in symbolic logic, hold good whether the individual letters represent certainties, impossibilities, or variables.

12. The following examples will illustrate the working of this symbolic calculus in simple cases.

(1) $(A + B'C')' = A'(B'C')' = A'(B + C') = A'B + A'C'$.
(2) $(A^\epsilon + B^{-\epsilon}C^\epsilon)' = A^{-\epsilon}(B^{-\epsilon}C^\epsilon)' = A^{-\epsilon}(B^\epsilon + C^{-\epsilon})$
      $= (A^\eta + A^\theta)(B^\epsilon + C^\eta + C^\theta)$.
(3) $(A^{-\theta} + A^\theta B^\theta)' = A^\theta(A^\theta B^\theta)' = A^\theta(A^{-\theta} + B^{-\theta})$
      $= A^\theta A^{-\theta} + A^\theta B^{-\theta} = A^\theta(B^\epsilon + B^\eta)$;

for $A^\theta A^{-\theta} = \eta$ (an impossibility), and $B^{-\theta} = B^\epsilon + B^\eta$.

CHAPTER II

13. Symbols of the forms $F(x)$, $f(x)$, $\phi(x)$, &c., are called *Functions of x*. A *function of x* means *an expression containing the symbol x*. When a symbol $\phi(x)$ denotes a function of x, the symbols $\phi(a)$, $\phi(\beta)$, &c., respectively denote what $\phi(x)$ becomes when a is put for x, when β is put for x, and so on. As a simple mathematical example, let $\phi(x)$ denote $5x^2 - 3x + 1$. Then, by definition, $\phi(a)$ denotes $5a^2 - 3a + 1$; and any tyro in mathematics can see that $\phi(4) = 69$, that $\phi(1) = 3$, that $\phi(0) = 1$, that $\phi(-1) = 9$, and so on. As an example in symbolic logic, let $\phi(x)$ denote the complex implication $(A : B) : (A^x : B^x)$. Then $\phi(\epsilon)$ will denote $(A : B) : (A^\epsilon : B^\epsilon)$, which is easily seen to be a valid* formula; while $\phi(\theta)$ will denote $(A : B) : (A^\theta : B^\theta)$, which is *not* valid.

14. Symbols of the forms $F(x, y)$, $\phi(x, y)$, &c., are called *functions of x and y*. Any of the forms may be employed to represent *any expression that contains both the symbols x and y*. Let $\phi(x, y)$ denote any function of x and y; then the symbol $\phi(a, \beta)$ will denote what $\phi(x, y)$

* Any formula $\phi(x)$ is called *valid* when it is true for all admissible values (or meanings), x_1, x_2, x_3, &c., of x.

10   SYMBOLIC LOGIC   [§§ 14–16

becomes when α is put for x and β for y. Hence, $\phi(y, x)$ will denote what $\phi(x, y)$ becomes when x and y interchange places. For example, let B = *boa-constrictor*, let R = *rabbit*, and let $\phi(B, R)$ denote the statement that "The *boa-constrictor* swallowed the *rabbit*." It follows that the symbol $\phi(R, B)$ will assert that "The *rabbit* swallowed the *boa-constrictor*."

15. As another example, let τ (as usual) denote *true*, and let p denote *probable*. Also let $\phi(\tau, p)$ denote the implication $(A^\tau B^\tau)^p : (A^p B^p)^\tau$, which asserts that "If it is *probable* that A and B are both *true*, it is *true* that A and B are both *probable*." Then $\phi(p, \tau)$ will denote the converse (or inverse) implication, namely, "If it is *true* that A and B are both *probable*, it is *probable* that A and B are both *true*." A little consideration will show that $\phi(\tau, p)$ is always true, but not always $\phi(p, \tau)$.

16. Let ϕ denote any function of one or more constituents; that is to say, let ϕ be short for $\phi(x)$, or for $\phi(x, y)$, &c. The symbol ϕ^ϵ asserts that ϕ is *certain*, that is, true for all admissible values (or meanings) of its constituents; the symbol ϕ^η asserts that ϕ is *impossible*, that is, false for all admissible values (or meanings) of its constituents; the symbol ϕ^θ means $\phi^{-\epsilon}\phi^{-\eta}$, which asserts that ϕ is *neither certain nor impossible*; while ϕ^0 asserts that ϕ is a *meaningless* statement which is neither true nor false. For example, let w = *whale*, h = *herring*, c = *conclusion*. Also let $\phi(w, h)$ denote the statement that "A small *whale* can swallow a large *herring*." We get

$$\phi^\epsilon(w, h) \cdot \phi^\eta(h, w) \cdot \phi^0(w, c),$$

a three-factor statement which asserts (1) that it is *certain* that a small *whale* can swallow a large *herring*, (2) that it is *impossible* that a small *herring* can swallow a large whale, and (3) that it is *unmeaning* to say that a small *whale* can swallow a large *conclusion*. Thus we see that $\phi(x, y)$, $F(x, y)$, &c., are really *blank forms* of more or less

§§ 16, 17]   APPLICATION TO GRAMMAR   11

complicated expressions or statements, the blanks being represented by the symbols x, y, &c., and the symbols or words to be substituted for or in the blanks being a, β, $a\beta$, $a+\beta$, &c., as the case may be.

17. Let $\phi(x, y)$ be any proposition containing the *words* x and y; and let $\phi(x, z)$, in which z is substituted for y, have the same meaning as the proposition $\phi(x, y)$. Should we in this case consider y and z as necessarily of the same part of speech? In languages which, like English, are but little inflected, the rule generally holds good and may be found useful in teaching grammar to beginners; but from the narrow conventional view of grammarians the rule would not be accepted as absolute. Take, for example, the two propositions "He talks nonsense" and "He talks foolishly." They both mean the same thing; yet grammarians would call *non-sense* a noun, while they would call *foolishly* an adverb. Here conventional grammar and strict logic would appear to part company. The truth is that so-called "general grammar," or a collection of rules of construction and classification applicable to all languages alike, is hardly possible. The *complete proposition* is the unit of all reasoning; the manner in which the separate words are combined to construct a proposition varies according to the particular bent of the language employed. In no two languages is it exactly the same. Consider the following example. Let S = His *son*, let A = in *Africa*, let K = has been *killed*, and let $\phi(S, K, A)$ denote the proposition "His *son* has been *killed* in *Africa*." By our symbolic conventions, the symbol $\phi(S, A, K)$, in which the symbols K and A have interchanged places, denotes the proposition "His *son* in *Africa* has been *killed*." Do these two propositions differ in meaning? Clearly they do. Let S_A denote his *son in Africa* (to distinguish him, say, from S_C, his *son in China*), and let K_A denote *has been killed in Africa* (as distinguished from K_C, *has been killed in China*). It follows that

$\phi(S, A, K)$ must mean S_A^K, whereas $\phi(S, K, A)$ must mean S^{KA}. In the former, A has the force of an *adjective* referring to the noun S, whereas, in the latter, A has the force of an *adverb* referring to the verb K. And in general, as regards symbols of the form A_x, A_y, A_z, &c., the letter A denotes the leading or *class* idea, the point of *resemblance*, while the subscripta x, y, z, &c., denote the points of *difference* which distinguish the separate members of the general or class idea. Hence it is that when A denotes a noun, the subscripta denote adjectives, or adjective-equivalents; whereas when A denotes a verb, the subscripta denote adverbs, or adverb-equivalents. When we look into the matter closely, the inflections of verbs, to indicate moods or tenses, have really the force of adverbs, and, from the logical point of view, may be regarded as adverb-equivalents. For example, if S denote the word *speak*, S_x may denote *spoke*, S_y may denote *will speak*, and so on; just as when S denotes *He spoke*, S_x may denote *He spoke well*, or *He spoke French*, and S_y may denote *He spoke slowly*, or *He spoke Dutch*, and so on. So in the Greek expression οἱ τότε ἄνθρωποι (the *then* men, or the men *of that time*), the adverb τοτε has really the force of an adjective, and may be considered an *adjective-equivalent*.

CHAPTER III

18. THE main cause of symbolic paradoxes is the ambiguity of words in current language. Take, for example, the words *if* and *implies*. When we say, "*If* in the centigrade thermometer the mercury falls below zero, water will freeze," we evidently assert a general law which is true in all cases within the limits of tacitly understood conditions. This is the sense in which the word *if* is used throughout this book (see § 10). It is understood to refer to a *general law* rather than to

§§ 18, 19] PARADOXES AND AMBIGUITIES

a particular case. So with the word *implies*. Let M^z denote "The *mercury* will fall below *zero*," and let W^F denote "*Water* will *freeze*." The preceding conditional statement will then be expressed by $M^z : W^F$, which asserts that the proposition M^z *implies* the proposition W^F. But this convention forces us to accept some paradoxical-looking formulæ, such as $\eta : x$ and $x : \epsilon$, which hold good whether the statement x be true or false. The former asserts that *if* an *impossibility* be true *any* statement x is true, or that an *impossibility implies any* statement. The latter asserts that the statement x (whether true or false) *implies* any certainty ϵ, or (in other words) that if x is true ϵ is true. The paradox will appear still more curious when we change x into ϵ in the first formula, or x into η in the second. We then get the formula $\eta : \epsilon$, which asserts that any *impossibility* implies any *certainty*. The reason why the last formula appears paradoxical to some persons is probably this, that they erroneously understand $\eta : \epsilon$ to mean $Q^\eta : Q^\epsilon$, and to assert that if any statement Q is *impossible* it is also *certain*, which would be absurd. But $\eta : \epsilon$ does *not* mean this (see § 74); by definition it simply means $(\eta \epsilon')^\eta$, which asserts that the statement $\eta \epsilon'$ is an impossibility, as it evidently is. Similarly, $\eta : x$ means $(\eta x')^\eta$, and asserts that $\eta x'$ is an impossibility, which is true, since the statement $\eta x'$ contains the impossible factor η. We prove $x : \epsilon$ as follows:

$$x : \epsilon = (x\epsilon')^\eta = (x\eta)^\eta = \eta^\eta = \epsilon.$$

For $\epsilon' = \eta$, since the denial of any certainty is some impossibility (see § 20). That, on the other hand, the implication $Q^\eta : Q^\epsilon$ is not a valid formula is evident; for it clearly fails in the case Q^η. Taking $Q = \eta$, we get

$$Q^\eta : Q^\epsilon = \eta^\eta : \eta^\epsilon = \epsilon : \eta = (\epsilon \eta')^\eta = (\epsilon \epsilon)^\eta = \eta.$$

19. Other paradoxes arise from the ambiguity of the sign of equivalence ($=$). In this book the statement

($\alpha = \beta$) does not necessarily assert that α and β are *synonymous*, that they have the *same meaning*, but only that they are *equivalent* in the sense that each *implies* the other, using the word 'implies' as defined in § 10. In this sense any two *certainties*, ϵ_1 and ϵ_2, are *equivalent*, however different in meaning; and so are any two *impossibilities*, η_1 and η_2; but not necessarily two different *variables*, θ_1 and θ_2. We prove this as follows. By definition, we have

$$(\epsilon_1 = \epsilon_2) = (\epsilon_1 : \epsilon_2)(\epsilon_2 : \epsilon_1) = (\epsilon_1 \epsilon'_2)^\eta (\epsilon_2 \epsilon'_1)^\eta$$
$$= (\epsilon_1 \eta_1)^\eta (\epsilon_2 \eta_2)^\eta = \eta_3^\eta \eta_4^\eta = \epsilon_3 \epsilon_4 = \epsilon_5 ;$$

for the denial of any *certainty* ϵ_x is some *impossibility* η_y. Again we have, by definition,

$$(\eta_1 = \eta_2) = (\eta_1 : \eta_2)(\eta_2 : \eta_1) = (\eta_1 \eta'_2)^\eta (\eta_2 \eta'_1)^\eta$$
$$= \eta_3^\eta \eta_4^\eta = \epsilon_1 \epsilon_2 = \epsilon_3.$$

But we cannot assert that any two *variables*, θ_1 and θ_2, are necessarily equivalent. For example, θ_2 might be the denial of θ_1, in which case we should get

$$(\theta_1 = \theta_2) = (\theta_1 = \theta'_1) = (\theta_1 : \theta'_1)(\theta'_1 : \theta_1) = (\theta_1 \theta_1)^\eta (\theta'_1 \theta'_1)^\eta$$
$$= \theta_1^\eta (\theta'_1)^\eta = \eta_1 \eta_2 = \eta_3.$$

The symbol used to assert that any two statements, α and β, are not only *equivalent* (in the sense of each implying the other) but also *synonymous*, is $(\alpha \equiv \beta)$; but this being an awkward symbol to employ, the symbol $(\alpha = \beta)$, though it asserts less, is generally used instead.

20. Let the symbol π temporarily denote the word *possible*, let p denote *probable*, let q denote *improbable*, and let u denote *uncertain*, while the symbols $\epsilon, \eta, \theta, \tau, \iota$ have their usual significations. We shall then, by definition, have $(A^\pi = A^{-\eta})$ and $(A^u = A^{-\epsilon})$, while A^p and A^q will respectively assert that the chance of A is greater than $\frac{1}{2}$, that it is less than $\frac{1}{2}$. These conventions give us the nine-factor formula

$$(\tau')^\iota (\iota')^\tau (p')^q (q')^p \ (\epsilon')^\eta (\eta')^\epsilon (\theta')^\theta (\pi')^u (u')^\pi,$$

§ 20]   PARADOXES AND AMBIGUITIES   15

which asserts (1, 2) that the denial of a *truth* is an *untruth*, and conversely; (3, 4) that the denial of a *probability* is an *improbability*, and conversely; (5, 6) that the denial of a *certainty* * is an *impossibility*, and conversely; (7) that the denial of a *variable* is a *variable*; (8, 9) that the denial of a *possibility* is an *uncertainty*, and conversely. The first four factors are pretty evident; the other five are less so. Some persons might reason, for example, that instead of $(\pi')^u$ we should have $(\pi')^\eta$; that the denial of a *possibility* * is not merely an *uncertainty* but an *impossibility*. A single concrete example will show that the reasoning is not correct. The statement "It will rain to-morrow" may be considered a *possibility*; but its denial "It will *not* rain to-morrow," though an *uncertainty* is not an *impossibility*. The formula $(\pi')^u$ may be proved as follows: Let Q denote any statement taken at random out of a collection of statements containing *certainties*, *impossibilities*, and *variables*. To prove $(\pi')^u$ is equivalent to proving $Q^\pi : (Q')^u$. Thus we get

$$(\pi')^u = Q^\pi : (Q')^u = Q^\epsilon + Q^\theta : (Q')^\eta + (Q')^\theta$$
$$= Q^\epsilon + Q^\theta : Q^\epsilon + Q^\theta = \epsilon ;$$

for $(Q')^\eta = Q^\epsilon$, and $(Q')^\theta = Q^\theta$, whatever be the statement Q. To prove that $(\pi')^\eta$, on the other hand, is *not* valid, we have only to instance a single case of failure. Giving Q the same meaning as before, a case of failure is Q^θ; for we then get, putting $Q = \theta_1$,

$$(\pi')^\eta = Q^\pi : (Q')^\eta = \theta_1^\pi : (\theta'_1)^\eta = \theta_1^\pi : \theta_1^\epsilon$$
$$= \epsilon_1 : \eta_1 = (\epsilon_1 \eta'_1)^\eta = (\epsilon_1 \epsilon_2)^\eta = \eta_2$$

---

* By the " denial of a certainty " is not meant $(A^\epsilon)'$, or its synonym $A^{-\epsilon}$, which denies that a particular statement A is certain, but $(A_\epsilon)'$ or its synonym A'_ϵ, the denial of the *admittedly certain* statement A_ϵ. This statement A_ϵ (since a suffix or subscriptum is adjectival and not predicative) *assumes* A to be certain; for both A_x and its denial A'_x *assume* the truth of A^x (see §§ 4, 5). Similarly, "the denial of a possibility" does not mean $A^{-\pi}$ but A'_π, or its synonym $(A_\pi)'$, the denial of the *admittedly possible* statement A_π.

21. It may seem paradoxical to say that the proposition A is not quite synonymous with A^τ, nor A' with A^ι; yet such is the fact. Let $A = \textit{It rains}$. Then $A' = \textit{It does not rain}$; $A^\tau = \textit{it is true that it rains}$; and $A^\iota = \textit{it is false that it rains}$. The two propositions A and A^τ are *equivalent* in the sense that each *implies* the other; but they are not *synonymous*, for we cannot always substitute the one for the other. In other words, the equivalence $(A = A^\tau)$ does not necessarily imply the equivalence $\phi(A) = \phi(A^\tau)$. For example, let $\phi(A)$ denote A^ϵ; then $\phi(A^\tau)$ denotes $(A^\tau)^\epsilon$, or its synonym $A^{\tau\epsilon}$ (see § 13). Suppose now that A denotes θ_τ, a variable that turns out true, or happens to be true in the case considered, though it is not true in all cases. We get

$$\phi(A) = A^\epsilon = \theta_\tau^\epsilon = (\theta_\tau)^\epsilon = \eta;$$

for a *variable* is never a *certainty*, though it may turn out true in a particular case.

Again, we get

$$\phi(A^\tau) = (A^\tau)^\epsilon = (\theta_\tau^\tau)^\epsilon = \epsilon^\epsilon = \epsilon;$$

for θ_τ^τ means $(\theta_\tau)^\tau$, which is a formal certainty. In this case, therefore, though we have $A = A^\tau$, yet $\phi(A)$ is *not* equivalent to $\phi(A^\tau)$. Next, suppose A denotes θ_ι, a variable that happens to be false in the case considered, though it is not false always. We get

$$\phi(A') = (A')^\epsilon = A'^\eta = \theta_\iota^\eta = \eta;$$

for no *variable* (though it may turn out false in a particular case) can be an *impossibility*. On the other hand, we get

$$\phi(A^\iota) = (A^\iota)^\epsilon = A^{\iota\epsilon} = \theta_\iota^{\iota\epsilon} = (\theta_\iota^\iota)^\epsilon = \epsilon^\epsilon = \epsilon;$$

for θ_ι^ι means $(\theta_\iota)^\iota$, which is a formal certainty. In this case, therefore, though we have $A' = A^\iota$, yet $\phi(A')$ is *not* equivalent to $\phi(A^\iota)$. It is a remarkable fact that nearly all civilised languages, in the course of their evolution, as if impelled by some unconscious instinct, have drawn

§§ 21, 22]   DEGREES OF STATEMENTS   17

this distinction between a simple affirmation A and the statement A^τ, that A is *true;* and also between a simple denial A' and the statement A^ι, that A is *false.* It is the first step in the *classification of statements*, and marks a faculty which man alone of all terrestrial animals appears to possess (see §§ 22, 99).

22. As already remarked, my system of logic takes account not only of statements of the second degree, such as $A^{\alpha\beta}$, but of statements of higher degrees, such as $A^{\alpha\beta\gamma}$, $A^{\alpha\beta\gamma\delta}$, &c. But, it may be asked, what is *meant* by statements of the second, third, &c., degrees, when the primary subject is itself a statement? The statement $A^{\alpha\beta\gamma}$, or its synonym $(A^{\alpha\beta})^\gamma$, is a statement of the *first* degree as regards its immediate subject $A^{\alpha\beta}$; but as it is synonymous with $(A^\alpha)^{\beta\gamma}$, it is a statement of the *second* degree as regards A^α, and a statement of the *third* degree as regards A, the *root* statement of the series. Viewed from another standpoint, A^α may be called a *revision* of the judgment A, which (though here it is the *root* statement, or root judgment, of the series) may itself have been a revision of some previous judgment here unexpressed. Similarly, $(A^\alpha)^\beta$ may be called a revision of the judgment A^α, and so on. To take the most general case, let A denote any complex statement (or judgment) of the nth degree. If it be neither a *formal certainty* (see § 109), like $(\alpha\beta : \alpha)^\epsilon$, nor a *formal impossibility*, like $(\alpha\beta : \alpha)^\theta$, it may be a *material certainty, impossibility,* or *variable*, according to the special data on which it is founded. If it follows necessarily from these data, it is a *certainty*, and we write A^ϵ; if it is incompatible with these data, it is an *impossibility*, and we write A^η; if it neither follows from nor is incompatible with our data, it is a *variable*, and we write A^θ. But whether this new or revised judgment be A^ϵ or A^η or A^θ, it must necessarily be a judgment (or statement) of the $(n+1)^{\text{th}}$ degree, since, by hypothesis, the statement A is of the nth degree. Suppose, for example, A denotes a functional statement $\phi(x, y, z)$ of

B

the nth degree, which may have m different meanings (or values) ϕ_1, ϕ_2, ϕ_3, &c., depending upon the different meanings x_1, x_2, x_3, &c., y_1, y_2, y_3, &c., z_1, z_2, z_3, &c., of x, y, z. Of these m different meanings of A, or its synonym ϕ, let one be taken at random. If A, or its synonym $\phi(x, y, z)$, be true for r meanings out of its m possible meanings, then the chance of A is r/m, and the chance of its denial A′ is $(m-r)/m$. When $r = m$, the chance of A is *one*, and the chance of A′ is *zero*, so that we write $A^\epsilon (A')^\eta$. When $r = o$, the chance of A is *zero*, and the chance of A′ is *one*, so that we write $A^\eta (A')^\epsilon$. When r is some number less than m and greater than o, then r/m and $(m-r)/m$ are two proper fractions, so we write $A^\theta (A')^\theta$. But, as before, whether we get A^ϵ or A^η or A^θ, this revised judgment, though it is a judgment of the *first* degree as regards its *expressed* root A, is a judgment of the $(n+1)^{\text{th}}$ degree as regards some *unexpressed* root $\psi(x, y, z)$. For instance, if A denote $\psi^{\epsilon\eta\theta}$, then A^θ will denote $\psi^{\epsilon\eta\theta\theta}$, so that it will be a judgment (or statement) of the *fourth* degree as regards ψ.

23. It may be remarked that any statement A and its denial A′ are always of the same degree, whereas A^τ and A^ι, their respective *equivalents* but not *synonyms* (see §§ 19, 21), are of one degree higher. The statement A^τ is a revision and *confirmation* of the judgment A; while A^ι is a revision and *reversal* of the judgment A. We suppose two incompatible alternatives, A and A′, to be placed before us with fresh data, and we are to decide which is true. If we pronounce in favour of A, we *confirm* the previous judgment A and write A^τ; if we pronounce in favour of A′, we *reverse* the previous judgment A and write A^ι.

24. Some logicians say that it is not correct to speak of any statement as "sometimes true and sometimes false"; that if true, it must be true always; and if false, it must be false always. To this I reply, as I did in my seventh paper published in the *Proceedings of the London*

§ 24]   VARIABLE STATEMENTS   19

Mathematical Society, that when I say "A is sometimes true and sometimes false," or "A is a *variable,*" I merely mean that the symbol, word, or collection of words, denoted by A sometimes represents a truth and sometimes an untruth. For example, suppose the symbol A denotes the statement "Mrs. Brown is not at home." This is not a formal certainty, like $3 > 2$, nor a formal impossibility, like $3 < 2$, so that *when we have no data but the mere arrangement of words,* "Mrs. Brown is not at home," we are justified in calling this *proposition,* that is to say, *this intelligible arrangement of words,* a *variable,* and in asserting A^θ. If at the moment the servant tells me that "Mrs. Brown is not at home" I happen to see Mrs. Brown walking away in the distance, then *I have fresh data* and form the judgment A^e, which, of course, implies A^τ. In this case I say that "A is *certain,*" because its denial A' ("Mrs. Brown is at home") would contradict my data, the evidence of my eyes. But if, instead of seeing Mrs. Brown walking away in the distance, I see her face peeping cautiously behind a curtain through a corner of a window, I obtain fresh data of an opposite kind, and form the judgment A^η, which implies A^ι. In this case I say that "A is impossible," because the statement represented by A, namely, "Mrs. Brown is not at home," this time *contradicts* my data, which, as before, I obtain through the medium of my two eyes. To say that the proposition A is a *different proposition* when it is *false* from what it is when it is *true,* is like saying that Mrs. Brown is a *different person* when she is *in* from what she is when she is *out.*

CHAPTER IV

25. THE following three rules are often useful:—

(1) $A^\epsilon \phi(A) = A^\epsilon \phi(\epsilon)$.
(2) $A^\eta \phi(A) = A^\eta \phi(\eta)$.
(3) $A^\theta \phi(A) = A^\theta \phi(\theta_x)$.

In the last of these formulæ, θ_x denotes the first variable of the series $\theta_1, \theta_2, \theta_3$, &c., that comes *after the last-named in our argument*. For example, if the last variable that has entered into our argument be θ_3, then θ_x will denote θ_4. In the first two formulæ it is not necessary to state which of the series $\epsilon_1, \epsilon_2, \epsilon_3$, &c., is represented by the ϵ in $\phi(\epsilon)$, nor which of the series η_1, η_2, η_3, &c., is represented by the η in $\phi(\eta)$; for, as proved in § 19, we have always $(\epsilon_x = \epsilon_y)$, and $(\eta_x = \eta_y)$, whatever be the certainties ϵ_x and ϵ_y, and whatever the impossibilities η_x and η_y. Suppose, for example, that ψ denotes

$$A^\epsilon B^\eta C^\theta (\theta_1 C : AB + CA).$$

We get

$$\psi = A^\epsilon B^\eta C^\theta (\theta_1 \theta_2 : \epsilon\eta + \theta_2 \epsilon) = A^\epsilon B^\eta C^\theta (\theta_1 \theta_2 : \theta_2)$$
$$= A^\epsilon B^\eta C^\theta \epsilon = A^\epsilon B^\eta C^\theta;$$

so that the fourth or bracket factor of ψ may be omitted without altering the value or meaning of ψ. In this operation we assumed the formulæ

(1) $(a\eta = \eta)$; (2) $(a\epsilon = a)$; (3) $(\eta + a = a)$.

Other formulæ frequently required are

(4) $(AB)' = A' + B'$; (5) $(A + B)' = A'B'$;
(6) $\epsilon + A = \epsilon$; (7) $AA' = \eta$; (8) $A + A' = \epsilon$;
(9) $\epsilon' = \eta$; (10) $\eta' = \epsilon$; (11) $A + AB = A$;
(12) $(A + B)(A + C) = A + BC$.

§§ 26, 27]   FORMULÆ OF OPERATION

26. For the rest of this chapter we shall exclude the consideration of variables, so that A, A^r, A^ϵ will be considered mutually equivalent, as will also A', A^ι, A^η. On this understanding we get the formulæ

(1) $A\phi(A) = A\phi(\epsilon)$; (2) $A'\phi(A) = A'\phi(\eta)$;
(3) $A\phi(A') = A\phi(\eta)$; (4) $A'\phi(A') = A'\phi(\epsilon)$.

From these formulæ we derive others, such as

(5) $AB'\phi(A, B) = AB'\phi(\epsilon, \eta)$;
(6) $AB'\phi(A', B) = AB'\phi(\eta, \eta)$;
(7) $AB'\phi(A', B') = AB'\phi(\eta, \epsilon)$,

and so on; like signs, as in $A\phi(A)$ or $A'\phi(A')$, in the same letter, producing $\phi(\epsilon)$; and unlike signs, as in $B'\phi(B)$ or $B\phi(B')$, producing $\phi(\eta)$. The following examples will show the working of these formulæ:—

Let $\phi(A, B) = AB'C + A'BC'$.   Then we get
$AB'\phi(A, B) = AB'(AB'C + A'BC')$
$= AB'(\epsilon\epsilon C + \eta\eta C')$
$= AB'(C + \eta) = AB'C$.

$A'B\phi'(A, B) = A'B(AB'C + A'BC')'$
$= A'B(\eta\eta C + \epsilon\epsilon C')'$
$= A'B(C'\eta)' = A'BC$.

Next, let $\phi(B, D) = (CD' + C'D + B'C')'$.
Then, $B'D'\phi(B, D) = B'D'(CD' + C'D + B'C')'$
$= B'D'(C\epsilon + C'\eta + \epsilon C')'$
$= B'D'(C + C')'$
$= B'D'\epsilon' = B'D'\eta = \eta$.

The application of Formulæ (4), (5), (11) of §·25 would, of course, have obtained the same result, but in a more troublesome manner.

27. If in any product ABC any statement-factor is implied in any other factor, or combination of factors, the *implied factor* may be omitted. If in any sum (*i.e.*,

alternative) A+B+C, any term implies any other, or the sum of any others, the *implying term* may be omitted. These rules are expressed symbolically by the two formulæ—

(1) (A:B):(AB=A); (2) (A:B):(A+B=B).

By virtue of the formula $(x:\alpha)(x:\beta)=x:\alpha\beta$, these two formulæ may be combined into the single formula—

(3) (A:B):(AB=A)(A+B=B).

As the converse of each of these three formulæ also holds good, we get

(4) A:B=(AB=A)=(A+B=B).

Hence, we get A+AB=A, omitting the term AB, because it implies the term A; and we also get A(A+B)=A, omitting the factor A+B, because it is implied by the factor A.

28. Since A:B is equivalent to (AB=A), and B is a factor of AB, it follows that the consequent B may be called a *factor* of the antecedent A, in any implication A:B, and that, for the same reason, the antecedent A may be called a *multiple* of the consequent B. The equivalence of A:B and (A=AB) may be proved as follows:—

$(A=AB)=(A:AB)(AB:A)=(A:AB)\epsilon$
$=A:AB=(A:A)(A:B)=\epsilon(A:B)=A:B.$

The equivalence of A:B and (A+B=B) may be proved as follows:—

$(A+B=B)=(A+B:B)(B:A+B)=(A+B:B)\epsilon$
$=A+B:B=(A:B)(B:B)=A:B.$

The formulæ assumed in these two proofs are

$(x:\alpha\beta)=(x:\alpha)(x:\beta)$, and $\alpha+\beta:x=(\alpha:x)(\beta:x)$,

both of which may be considered axiomatic. For to assert that "If x is true, then α and β are both true" is equivalent to asserting that "If x is true α is true, and

§§ 28, 29]   REDUNDANT TERMS   23

if x is true β is true." Also, to assert that "If either a or β is true x is true" is equivalent to asserting that "If a is true x is true, and if β is true x is true."

29. To discover the redundant terms of any logical sum, or alternative statement.

These redundant terms are easily detected by mere inspection when they evidently imply (or are multiples of) single co-terms, as in the case of the terms underlined in the expression

$$\underline{a'\beta\gamma} + a'\gamma + \underline{a\beta\gamma'} + \beta\gamma',$$

which therefore reduces to $a'\gamma + \beta\gamma'$. But when they do not imply single co-terms, but the sum of two or more co-terms, they cannot generally be thus detected by inspection. They can always, however, be discovered by the following rule, which includes all cases.

Any term of a logical sum or alternative may be omitted as redundant when this term multiplied by the denial of the sum of all its co-terms gives an impossible product η; but if the product is not η, the term must not be omitted. Take, for example, the alternative statement

$$CD' + C'D + B'C' + B'D'.$$

Beginning with the first term we get

$$CD'(C'D + B'C' + B'D')' = CD'(\eta\eta + B'\eta + B'\epsilon)'$$
$$= CD'(B')' = BCD'.$$

Hence, the first term CD' must *not* be omitted. Taking next the second term $C'D$, we get

$$C'D(CD' + B'C' + B'D')' = C'D(\eta\eta + B'\epsilon + B'\eta)'$$
$$= C'D(B')' = BC'D.$$

Hence, the second term $C'D$ must not be omitted. We next take the third term $B'C'$, getting

$$B'C'(CD' + C'D + B'D')' = B'C'(\eta D' + \epsilon D + \epsilon D')'$$
$$= B'C'(D + D')' = B'C'\eta = \eta.$$

This shows that the third term $B'C'$ *can* be omitted as

redundant. Omitting the third term, we try the last term B'D', thus

$$B'D'(CD' + C'D)' = B'D'(C\epsilon + C'\eta)' = B'D'C'.$$

This shows that the fourth term B'D' cannot be omitted as redundant *if we omit the third term*. But if we retain the third term B'C', we may omit the fourth term B'D', for we then get

$$B'D'(CD' + C'D + B'C')' = B'D'(C\epsilon + C'\eta + \epsilon C')'$$
$$= B'D'(C + C')' = B'D'\eta = \eta.$$

Thus, we may omit either the third term B'C', or else the fourth term B'D', as redundant, *but not both*.

30. A complex alternative may be said to be in its simplest form* when it contains no redundant terms, and none of its terms (or of the terms left) contains any redundant factor. For example, $a + ab + m + m'n$ is reduced to its simplest form when we omit the redundant term ab, and out of the last term strike out the unnecessary factor m'. For $a + ab = a$, and $m + m'n = m + n$, so that the simplest form of the expression is $a + m + n$. (See § 31.)

31. To reduce a complex alternative to its simplest form, apply the formula $(a + \beta)' = a'\beta'$ to the *denial* of the alternative. Then apply the formula $(a\beta)' = a' + \beta'$ to the negative compound factors of the result, and omit the redundant terms in this new result. Then develop the denial of this product by the same formulæ, and go through the same process as before. The final result will be the simplest equivalent of the original alternative. Take, for example, the alternative given in § 30, and denote it by ϕ. We get

$$\phi = a + ab + m + m'n = a + m + m'n.$$
$$\phi' = (a + m + m'n)' = a'm'(m'n)' = a'm'(m + n') = a'm'n'.$$
$$\phi = (\phi')' = (a'm'n')' = a + m + n.$$

---

* What I here call its "simplest form" I called its "primitive form" in my third paper in the *Proceedings of the London Mathematical Society;* but the word "primitive" is hardly appropriate.

§§ 31, 32]   METHODS OF SIMPLIFICATION   25

As another example take the alternative

$$AB'C' + ABD + A'B'D' + ABD' + A'B'D,$$

and denote it by ϕ. Then, omitting, as we go along, all terms which mere inspection will show to be redundant, we get

$$\phi = AB'C' + AB(D + D') + A'B'(D' + D)$$
$$= AB'C' + AB\epsilon + A'B'\epsilon = AB'C' + AB + A'B'.$$
$$\phi' = (AB'C')'(AB)'(A'B')'$$
$$= (A' + B + C)(A' + B')(A + B)$$
$$= (A' + B'C)(A + B) = A'B + AB'C.$$
$$\phi = (\phi')' = (A'B + AB'C)' = (A + B')(A' + B + C')$$
$$= AB + AC' + A'B' + B'C'.$$

Applying the test of § 29 to discover redundant terms, we find that the second or fourth term (AC' or B'C') may be omitted as redundant, but not both. We thus get

$$\phi = AB + A'B' + B'C' = AB + AC' + A'B',$$

either of which may be taken as the simplest form of ϕ.

32. We will now apply the preceding principles to an interesting problem given by Dr. Venn in his "Symbolic Logic" (see the edition of 1894, page 331).

Suppose we were asked to discuss the following set of rules, in respect to their mutual consistency and brevity.

α. The Financial Committee shall be chosen from amongst the General Committee.

β. No one shall be a member both of the General and Library Committees unless he be also on the Financial Committee.

γ. No member of the Library Committee shall be on the Financial Committee.

Solution.

Speaking of a member taken at random, let the symbols F, G, L, respectively denote the statements "He will be on the *Financial* Committee," "He will be on the *General* Committee," "He will be on the *Library* Committee." Putting η, as usual, for any statement that contradicts our data, we have

$$\alpha = (F:G); \quad \beta = (GLF':\eta); \quad \gamma = (LF:\eta);$$

so that

$$\begin{aligned}\alpha\beta\gamma &= (F:G)(GLF':\eta)(FL:\eta) \\ &= (FG':\eta)(GLF':\eta)(FL:\eta) \\ &= FG' + GLF' + FL:\eta.\end{aligned}$$

Putting ϕ for the antecedent $FG' + GLF' + FL$, we get

$$\phi' = (F' + G)(G' + L' + F)(F' + L')$$

(See § 25, Formulæ (4) and (5))

$$= (F' + GL')(G' + L' + F) = F'G' + F'L' + GL' + FGL'$$
$$= F'G' + GL';$$

for the term FGL', being a multiple of the term GL', is redundant by inspection, and $F'L'$ is also redundant, because, by § 29,

$$F'L'(F'G' + GL')' = F'L'(\epsilon G' + G\epsilon)' = F'L'(G' + G)' = \eta.$$

Hence, finally, we get (omitting the redundant term FL)

$$\phi = (\phi')' = (F'G' + GL')' = FG' + GL,$$

and therefore

$$\alpha\beta\gamma = \phi:\eta = (FG' + GL:\eta) = (F:G)(G:L').$$

That is to say, the three club rules, α, β, γ, may be replaced by the two simple rules $F:G$ and $G:L'$, which assert, firstly, that "If any member is on the *Financial* Committee, he must be also on the *General* Committee," which is rule α in other words; and, secondly, that "If any member is on the *General* Committee, he is not to be on the *Library* Committee."

CHAPTER V

33. FROM the formula

$$(a:b)(c:d) = ab' + cd' : \eta$$

the product of any number of implications can always be expressed in the form of a single implication,

$$a + \beta + \gamma + \&c. : \eta,$$

of which the antecedent is a logical sum (or alternative), and the consequent an impossibility. Suppose the implications forming the data of any problem that contains the statement x among its constituents to be thus reduced to the form

$$Ax + Bx' + C : \eta,$$

in which A is the coefficient or co-factor of x, B the coefficient of x', and C the term, or sum of the terms, which contain neither x nor x'. It is easy to see that the above data may also be expressed in the form

$$(B:x)(x:A')(C:\eta),$$

which is equivalent to the form

$$(B:x:A')(C:\eta).$$

When the data have been reduced to this form, the given implication, or product of implications, is said to be *solved with respect to* x; and the statements B and A' (which are generally more or less complex) are called the *limits of* x; the antecedent B being the *strong* * or *superior* limit; and the consequent A', the *weak* or *inferior* limit. Since the

---

* When from our data we can infer $a:\beta$, but have no data for inferring $\beta:a$, we say that a is *stronger* than β. For example, since we have $AB:A:A+B$, we say that AB is stronger than A, and A stronger than $A+B$.

factor $(B:x:A')$ implies $(B:A')$, and our data also imply the factor $(C:\eta)$, it follows that our data imply

$$(B:A')(C:\eta),$$

which is equivalent to $AB+C:\eta$. Thus we get the formula of elimination

$$(Ax+Bx'+C:\eta):(AB+C:\eta),$$

which asserts that the strongest conclusion deducible from our data, *and making no mention of* x, is the implication $AB+C:\eta$. As this conclusion is equivalent to the two-factor statement $C^\eta(AB)^\eta$, it asserts that the statement C and the combination of statements AB are both impossible.

34. From this we deduce the solution of the following more general problem. Let the functional symbol $\phi(x,y,z,a,b)$, or simply the symbol ϕ, denote data which refer to any number of constituent statements x, y, z, a, b, and which may be expressed (as in the problem of § 33) in the form of a single implication $\alpha+\beta+\gamma+\&c.:\eta$, the terms α, β, γ, &c., being more or less complex, and involving more or less the statements x, y, z, a, b. It is required, firstly, to find successively in any desired order the limits (*i.e.*, the weakest antecedent and strongest consequent) of x, y, z; secondly, to eliminate x, y, z in the same order; and, thirdly, to find the strongest implicational statement (involving a or b, but neither x nor y nor z) that remains after this elimination.

Let the assigned order of limits and elimination be z, y, x. Let A denote the sum of the terms containing the factor z; let B denote the sum of the terms containing the factor z', and let C denote the sum of the terms containing neither z nor z'. Our data being ϕ, we get

$$\phi = Az+Bz'+C:\eta = (B:z)(z:A')(C:\eta)$$
$$= (B:z:A')(C:\eta) = (B:z:A')(B:A')(C:\eta).$$

§ 34]   SOLUTIONS, ELIMINATIONS, LIMITS   29

The expression represented by $Az + Bz' + C$ is understood to have been reduced to its simplest form (see §§ 30, 31), before we collected the coefficients of z and z'. The limits of z are therefore B and A'; and the result after the elimination of z is

$$(B : A')(C : \eta), \text{ which} = AB + C : \eta.$$

To find the limits of y from the implication $AB + C : \eta$, we reduce $AB + C$ to its simplest form (see §§ 30, 31), which we will suppose to be $Dy + Ey' + F$. We thus get, as in the previous expression in z,

$$AB + C : \eta = Dy + Ey' + F : \eta = (E : y : D')(E : D')(F : \eta).$$

The limits of y are therefore E and D', and the result after the successive elimination of z and y is

$$(E : D')(F : \eta), \text{ which} = ED + F : \eta.$$

To find the limits of x from the implication $ED + F : \eta$, we proceed exactly as before. We reduce $ED + F$ to its simplest form, which we will suppose to be $Gx + Hx' + K$, and get

$$ED + F : \eta = Gx + Hx' + K : \eta = (H : x : G')(H : G')(K : \eta).$$

The limits of x are therefore H and G', and the result after the successive elimination of z, y, x is

$$(H : G')(K : \eta), \text{ which} = HG + K : \eta.$$

The statements z, y, x having thus been successively eliminated, there remains the implication $GH + K : \eta$, which indicates the relation (if any) connecting the remaining constituent statements a and b. Thus, we finally get

$$\phi = (B : z : A')(E : y : D')(H : x : G')(GH + K : \eta).$$

in which A and B do not contain z (that is, they make no mention of z); D and E contain neither z nor y; G and H contain neither z nor y nor x; and the expression K in the last factor will also be destitute of (*i.e.*, will make no mention of) the constitutents x, y, z, though, like G and H, it may contain the constituent statements a and b.

In the course of this process, since $\eta : a$ and $a : \epsilon$ are certainties whatever the statement a may be (see § 18), we can supply η for any missing antecedent, and ϵ for any missing consequent.

35. To give a concrete example of the general problem and solution discussed in § 34, let ϕ denote the data

$$\epsilon : xyza' + xyb' + xy'z' + y'z'a'.$$

We get, putting ϕ for these data,

$$\phi = (xyza' + xyb' + xy'z' + y'z'a')' : \eta$$
$$= x'y + byz' + y'z + abz + ax' : \eta,$$

when the antecedent of this last implication has been reduced to its simplest form by the process explained in § 31. Hence we get

$$\phi = (y' + ab)z + (by)z' + (x'y + ax') : \eta$$
$$= Az + Bz' + C : \eta,$$

putting A for $y' + ab$, B for by, and C for $x'y + ax'$. As in § 34, we get

$$(B : z : A')(AB + C : \eta),$$

so that the limits of z are B and A', and the result after the elimination of z is $AB + C : \eta$. Substituting their values for A, B, C, this last implication becomes

$$(ab + x')y + ax' : \eta,$$

which we will denote by $Dy + Ey' + F : \eta$, putting D for $ab + x'$, E for η, and F for ax'. Thus we get

$$\phi = (B : z : A')(Dy + Ey' + F : \eta)$$
$$= (B : z : A')(E : y : D')(ED + F : \eta).$$

Having thus found the limits (*i.e.*, the weakest ante-

§§ 35, 36]   SOLUTIONS, ELIMINATIONS, LIMITS   31

cedents and strongest consequents) of z and y, we proceed to find the limits of x from the implication $ED + F : \eta$, which is the strongest implication that remains after the elimination of z and y. Substituting for D, E, F the values which they represent, we get

$$DE + F : \eta = (ab + x')\eta + ax' : \eta = Gx + Hx' + K : \eta,$$

in which G, H, K respectively denote η, a, η. We thus get

$$DE + F : \eta = (H : x : G')(HG + K : \eta) ;$$

so that our final result is

$$\begin{aligned}\phi &= (B : z : A')(E : y : D')(H : x : G')(HG + K : \eta) \\ &= (by : z : a'y + b'y)(\eta : y : a'x + b'x)(a : x : \epsilon)(\eta : \eta) \\ &= (by : z : a'y + b'y)(y : a'x + b'x)(a : x).\end{aligned}$$

To obtain this result we first substituted for A, B, D, E, G, H, K the values we had assigned to them; then we omitted the redundant antecedent η in the second factor, the redundant consequent ϵ in the third factor, and the redundant certainty $(\eta : \eta)$, which constituted the fourth factor. The fact that the fourth factor $(HG + K : \eta)$ reduces to the form $(\eta : \eta)$, which is a formal certainty (see § 18), indicates that, in this particular problem, nothing can be implicationally affirmed in terms of a or b (without mentioning either x or y or z) except formal certainties such as $(ab : a)$, $(aa' : \eta)$, $ab(a' + b') : \eta$, &c., which are true always and independently of our data ϕ.

36. If in the preceding problem we had not reduced the alternative represented by $Az + Bz' + C$ to its simplest form (see §§ 30, 31), we should have found for the inferior limit or consequent of z, not $a'y + b'y$, but $x(a'y + b'y)$. From this it might be supposed that the strongest conclusion deducible from z (in conjunction with, or within the limits of, our data) was not A' but xA'. But though xA' is *formally* stronger than A', that

is to say, stronger than A' *when we have no data but our definitions*, here we *have* other data, namely, ϕ; and ϕ implies (as we shall prove) that A' is in this case *equivalent* to xA', so that *materially* (that is to say, within the limits of our particular data ϕ) neither of the two statements can be called stronger or weaker than the other. This we prove as follows:—

$$\phi : (z : A' : y : D' : x) : (A' : x) : (A' = xA');$$

a proof which becomes evident when for A' and D' we substitute their respective values $a'y + b'y$ and $a'x + b'x$; for it is clear that y is a factor of the former, and x a factor of the latter.

37. In the problem solved in § 35, in which ϕ denoted our data, namely, the implication

$$\epsilon : xyza' + xyb' + xy'z' + y'z'a',$$

we took z, y, x as the order of limits and of elimination. Had we taken the order y, x, z, our final result would have been

$$\phi = (z : y : b'x + xz)(z + a : x)(z : a' + b').$$

38. The preceding method of finding what I call the "limits" of logical statements is closely allied to, and was suggested by, my method (published in 1877, in the *Proc. of the Lond. Math. Soc.*) for successively finding the limits of integration for the variables in a multiple integral (see § 138). In the next chapter the method will be applied to the solution (so far as solution is possible) of Professor Jevons's so-called "Inverse Problem," which has given rise to so much discussion, not only among logicians but also among mathematicians.

CHAPTER VI

39. BRIEFLY stated, the so-called "inverse problem" of Professor Jevons is this. Let ϕ denote any alternative, such as $abc + a'bc + ab'c'$. It is required to find an implication, or product of implications,* that implies this alternative.

Now, any implication whatever (or any product of implications) that is equivalent to ϕ^ϵ, or is a multiple of ϕ^ϵ, as, for example, $\epsilon : \phi$, or $\phi' : \eta$, or $(abc : ab)(\epsilon : \phi)$, or $(a : b)(\phi' : \eta)$, &c., must necessarily imply the given alternative ϕ, so that the number of possible solutions is really unlimited. But though the problem as enunciated by Professor Jevons is thus indeterminate, the number of possible solutions may be restricted, and the problem rendered far more interesting, as well as more useful and instructive, by stating it in a more modified form as follows:—

Let ϕ denote any alternative involving any number of constituents, a, b, c, &c. It is required to resolve the implication $\epsilon : \phi$ into factors, so that it will take the form
$$(M : a : N)(P : b : Q)(R : c : S), \&c.,$$
in which the limits M and N (see § 33) may contain b, c, &c.; but not a; the limits P and Q may contain c, d, &c., but neither a nor b; the limits R and S may contain d, e, &c., but neither a nor b nor c; and so on to the last constituent. When no nearer limits of a constituent can be found we give it the limits η and ϵ; the former being its antecedent, and the latter its consequent (see §§ 18, 34).

---

* Professor Jevons calls these implications "laws," because he arrives at them by a long tentative inductive process, like that by which scientific investigators have often discovered the so-called "laws of nature" (see § 112).

As a simple example, suppose we have *

$$\phi = abc + a'bc + ab'c',$$

the terms of which are mutually exclusive. Reducing ϕ to its simplest form (see §§ 30, 31), we get $\phi = bc + ab'c'$, and therefore

$$\epsilon : \phi = \phi' : \eta = (bc)'(ab'c')' : \eta = (b' + c')(a' + b + c) : \eta$$
$$= a'b' + b'c + a'c' + bc' : \eta.$$

This alternative equivalent of ϕ' may be simplified (see § 31) by omitting either the first or the third term, but not both; so that we get

$$\epsilon : \phi = b'c + a'c' + bc' : \eta = a'b' + b'c + bc' : \eta.$$

Taking the first equivalent of $\epsilon : \phi$, and (in order to find the limits of a) arranging it in the form $Aa + Ba' + C : \eta$, we get (see §§ 33, 34)

$$\epsilon : \phi = \eta a + c'a' + (b'c + bc') : \eta$$
$$= (c' : a : \epsilon)(c : b : c)(\eta : c : \epsilon).$$

Thus, we have successively found the limits of a, b, c (see §§ 34, 35). But since $(a : \epsilon)$, $(\eta : c)$, and $(c : \epsilon)$ are all formal certainties, they may be omitted as factors, so that we get

$$\epsilon : \phi = (c' : a)(c : b : c) = (c' : a)(c = b).$$

The first of these two factors asserts that any term of the given alternative ϕ which contains c' must also contain a. The second asserts that any term which contains c must also contain b, and, conversely, that any term which contains b must also contain c. A glance at the given alternative ϕ will verify these assertions.

* Observe that here and in what follows the symbol ϕ denotes an *alternative*. In §§ 34, 35 the symbol ϕ denotes a given *implication*, which may take either such a form as $\epsilon : a + \beta + \gamma + \&c.$, or as $a + \beta + \gamma + \&c. : \eta$.

§§ 39, 40] JEVONS'S "INVERSE PROBLEM"

We will now take the second equivalent of $\epsilon : \phi$, namely,
$$a'b' + b'c + bc' : \eta,$$
and resolve it into three factors by successively finding the limits of a, b, c. Proceeding as before, we get
$$\epsilon : \phi = (b' : a)(c = b).$$
At first sight it might be supposed that the two ways of resolving $\epsilon : \phi$ into factors gave different results, since the factor $(c' : a)$ in the former result is replaced by the factor $(b' : a)$ in the latter. But since the second factor $(c = b)$, common to both results, informs us that b and c are equivalent, it follows that the two implications $c' : a$ and $b' : a$ are equivalent also.

If we had taken the alternative equivalent of ϕ', namely, $a'b' + b'c + a'c' + bc'$, in its unsimplified form, we should have found
$$\epsilon : \phi = \phi' : \eta = (b' + c' : a)(c = b) = (b' : a)(c' : a)(c = b),$$
in which either the factor $(b' : a)$ or the factor $(c' : a)$ may be omitted as redundant, but not both. For though the factor $(c = b)$ *alone* neither implies $(b' : a)$ nor $(c' : a)$, yet $(b' : a)(c = b)$ implies $(c' : a)$, and $(c' : a)(c = b)$ implies $(b' : a)$. This redundancy of factors in the result is a necessary consequence of the redundancy of *terms* in the alternative equivalent of ϕ' at the starting. For the omission of the term $a'b'$ in the alternative leads to the omission of the implicational factor $(a'b' : \eta)$, or its equivalent $(b' : a)$, in the result; and the omission of the term $a'c'$ in the alternative leads, in like manner, to the omission of the factor $(a'c' : \eta)$, or its equivalent $(c' : a)$, in the result.

40. I take the following alternative from Jevons's "Studies in Deductive Logic" (edition of 1880, p. 254, No. XII.), slightly changing the notation,
$$abcd + abc'd + ab'cd' + a'bcd' + a'b'c'd'.$$
Let ϕ denote this alternative, and let it be required to

find successively the limits of a, b, c, d. In other words, we are required to express $\epsilon : \phi$ in the form

$$(M : a : N)(P : b : Q)(R : c : S)(T : d : U),$$

in which M and N are not to contain a; P and Q are neither to contain a nor b; R and S are neither to contain a nor b nor c; and T and U must be respectively η and ϵ. By the process of §§ 34, 35, we get

$$M = d + bc' + b'c, \; N = bd + b'c, \; P = d, \; Q = c + d, \; R = \eta,$$
$$S = \epsilon, \; T = \eta, \; U = \epsilon.$$

Omitting the last two factors $R : c : S$ and $T : d : U$ because they are formal certainties, we get

$$\epsilon : \phi = (d + bc' + b'c : a : bd + b'c)(d : b : c + d).$$

A glance at the given alternative ϕ will verify this result, which asserts (1) that whenever we have either d or bc' or $b'c$, then we have a; (2) that whenever we have a, then we have either bd or $b'c$; (3) that whenever we have d, then we have b; (4) that whenever we have b, then we have either c or d; and (5) that from the implication $\epsilon : \phi$ we can infer no relation connecting c with d without making mention of a or b; or, in other words, that c cannot be expressed in terms of d alone, since the factor $\eta : c : \epsilon$ is a formal certainty and therefore true from our definitions alone apart from any special data. The final factor is only added for form's sake, for it *must* always have η for antecedent and ϵ for consequent. In other words, when we have n constituents, if x be the nth or last in the order taken, the last factor must necessarily be $\eta : x : \epsilon$, and therefore a formal certainty which may be left understood. Others of the factors may (as in the case of $\eta : c : \epsilon$ here) turn out to be formal certainties also, but not necessarily.

We have found the limits of the constituents a, b, c, d, taken successively in alphabetic order. If we take the reverse order d, c, b, a, our result will be

$$\epsilon : \phi = (ab + ac' + bc' : d : ab)(ab' + a'b : c : a + b),$$

§§ 40, 41]   ALTERNATIVES   37

omitting the third and fourth factors $\eta : b : \epsilon$ and $\eta : a : \epsilon$ because they are formal certainties. There is one point in this result which deserves notice. Since every double implication $a : x : \beta$ always implies $a : \beta$, it follows that (in the first bracket) $ab + ac' + bc'$ implies ab. Now, the latter is *formally* stronger than the former, since any statement x is formally stronger than the alternative $x + y$. But the formally stronger statement x, though it can never be weaker, either formally or materially, than $x + y$, may be *materially equivalent* to $x + y$; and it must be so whenever y *materially* (i.e., by the special data of the problem) implies x, but not otherwise. Let us see whether our special data, in the present case, justifies the inferred implication $ab + ac' + bc' : ab$. Call this implication ψ. By virtue of the formula $a + \beta + \gamma : x = (a : x)(\beta : x)(\gamma : x)$, we get (putting ab for a and for x, ac' for β, and bc' for γ)

$$\psi = (ab : ab)(ac' : ab)(bc' : ab) = \epsilon(ac' : ab)(bc' : ab)$$
$$= (ac' : a)(ac' : b)(bc' : a)(bc' : b)$$
$$= \epsilon(ac' : b)(bc' : a)\epsilon = (ac' : b)(bc' : a).$$

This asserts that (within the limits of our data in this problem) whenever we have ac' we have also b, and that whenever we have bc' we have also a. A glance at the given fully developed alternative ϕ will show that this is a fact (see § 41). Hence, the inferred implication $ab + ac' + bc' : ab$ is, *in this problem*, legitimate, in spite of the fact that its antecedent is *formally* weaker than its consequent.

41. An alternative is said to be *fully developed* when, and only when, it satisfies the following conditions: Firstly, every single-letter constituent, or its denial, must be a factor of every term; secondly, no term must be a formal certainty nor a formal impossibility; thirdly, all the terms must be *mutually incompatible*, which means that no two terms can be true at the same time. This last condition implies that no term is redundant or repeated.

For example, the fully developed form of $a+\beta$ is $a\beta+a\beta'+a'\beta$. To obtain this we multiply the two factors $a+a'$ and $\beta+\beta'$, and strike out the term $a'\beta'$, because it is equivalent to $(a+\beta)'$, the denial of the given alternative $a+\beta$. As another example, let it be required to find the fully developed form of $a+\beta'\gamma$. Here we first find the product of the three factors $a+a'$, $\beta+\beta'$, and $\gamma+\gamma'$. We next find that $(a+\beta'\gamma)'$ is equivalent to $a'(\beta'\gamma)'$, which is equivalent to $a'(\beta+\gamma')$, and therefore, finally, to $a'\beta+a'\gamma'$. Then, out of the eight terms forming the product we strike out the three terms $a'\beta\gamma$, $a'\beta\gamma'$, $a'\beta'\gamma'$, because each of these contains either $a'\beta$ or $a'\gamma'$, which are the two terms of $a'\beta+a'\gamma'$, the denial of the given alternative $a+\beta'\gamma$. The result will be

$$a\beta'\gamma+a'\beta'\gamma+a\beta\gamma+a\beta'\gamma'+a\beta\gamma',$$

which is, therefore, the fully developed form of the given alternative $a+\beta'\gamma$.

42. Let ϕ denote $a'cde+b'cd+cd'e'+a'd'e'$. Here we have 5 elementary constituents a, b, c, d, e; so that the product of the five factors $(a+a'), (b+b')$, &c., will contain 2^5 (or 32) terms. Of these 32 terms, 11 terms will constitute the fully developed form of ϕ, and the remaining 21 will constitute the fully developed form of its denial ϕ'. Let ψ denote the fully developed form of ϕ. Then the alternatives ϕ and ψ will, of course, only differ in *form*; they will be logically equivalent. Suppose the alternative ψ to be given us (as in Jevons's "inverse problem"), and we are required to find the limits of the 5 constituents in the alphabetic order a, b, c, d, e, from the data $\epsilon : \psi$. When we have reduced the alternative ψ to its simplest form, we shall find the result to be ϕ. Thus we get

$$\epsilon : \psi = \epsilon : \phi = \phi' : \eta = ac' + bde' + c'd + d'e + abe : \eta$$
$$= (\eta : a : b'c + ce')(\eta : b : d' + e)(d : c : \epsilon)(e : d : \epsilon)(\eta : e : \epsilon).$$

This is the final result *with every limit expressed*. Omit-

§§ 42–44]   UNRESTRICTED FUNCTIONS   89

ting the superior limit η and the inferior limit ϵ wherever they occur, and also the final factor $\eta : e : \epsilon$ because it is a formal certainty (see § 18), we get

$$\epsilon : \psi = (a : b'c + ce')(b : d' + c)(d : c)(e : d).$$

Suppose next we are required to find the limits in the order d, e, c, a, b. Our final result in this case will be

$$\epsilon : \psi = (e : d : b'c + ce)(\eta : e : a'c + b'c)(a : c : \epsilon)(\eta : a : \epsilon)(\eta : b : \epsilon)$$
$$= (e : d : b'c + ce)(e : a'c + b'c)(a : c).$$

43. When an alternative ϕ contains n constituents, the number of possible permutations in the order of constituents when all are taken is $1.2.3.4...n$. In an alternative of 5 constituents, like the one in § 42, the number of possible solutions cannot therefore exceed $1.2.3.4.5$, which $=120$. For instance, in the example of § 42, the solution in the order d, e, c, a, b (the last given), is virtually the same as the solution in the order d, e, c, b, a; the only difference being that the last two factors in the first case are (as given), $\eta : a : \epsilon$ and $\eta : b : \epsilon$; while in the second case they are $\eta : b : \epsilon$ and $\eta : a : \epsilon$; that is to say, the order changes, and both, being certainties, may be omitted. It will be observed that when the order of limits is prescribed, the exact solution is prescribed also: no two persons can (without error) give different solutions, though they may sometimes appear different in *form* (see §§ 39, 40).

CHAPTER VII

44. Let $F_u(x, y, z)$, or its abbreviated synonym F_u, represent the functional proposition $F(x, y, z)$, when the values or meanings of its constituents x, y, z are *unrestricted*; while the symbol $F_r(x, y, z)$, or its abbreviated synonym F_r, represents the functional proposition $F(x, y, z)$ when the values of x, y, z are *restricted*. For example, if x can have only four values x_1, x_2, x_3, x_4; y

40   SYMBOLIC LOGIC   [§§ 44, 45

the four values y_1, y_2, y_3, y_4; and z the three values z_1, z_2, z_3; then we write F_r, and not F_u. But if each of the three symbols x, y, z may have any value (or meaning) whatever out of the infinite series x_1, x_2, x_3, &c., y_1, y_2, y_3, &c., z_1, z_2, z_3, &c.; then we write F_u, and not F_r. The suffix r is intended to suggest the adjective *restricted*, and the suffix u the adjective *unrestricted*. The symbols F^ϵ, F^η, F^θ, as usual, assert respectively that F is *certain*, that F is *impossible*, that F is *variable*; but here the word *certain* is understood to mean *true for all the admissible values of x, y, z* in the functional statement $F(x, y, z)$; *impossible* means *false for every admissible value of x, y, z* in the statement $F(x, y, z)$; and *variable* means *neither certain nor impossible*. Thus F^θ asserts that $F(x, y, z)$ is neither always true nor always false; it is synonymous with $F^{-\epsilon}F^{-\eta}$, which is synonymous with $(F^\epsilon)'(F^\eta)'$.

45. From these symbolic conventions we get the three formulæ :

(1) $(F_u^\epsilon : F_r^\epsilon)$; (2) $(F_u^\eta : F_r^\eta)$; (3) $(F_r^\theta : F_u^\theta)$;

but the converse (or inverse) implications are not necessarily true, so that the three formulæ would lose their validity if we substituted the sign of equivalence $(=)$ for the sign of implication $(:)$. The first two formulæ need no proof; the third is less evident, so we will prove it as follows. Let ϕ_1, ϕ_2, ϕ_3 denote the above three formulæ respectively. The first two being self-evident, we assume $\phi_1 \phi_2$ to be a *certainty*, so that we get the deductive sorites

$\epsilon : \phi_1\phi_2 : (F_u^\epsilon : F_r^\epsilon)(F_u^\eta : F_r^\eta)$
$\quad\quad\quad : (F_r^{-\epsilon} : F_u^{-\epsilon})(F_r^{-\eta} : F_u^{-\eta})$ [for $a : \beta = \beta' : a'$]
$\quad\quad\quad : (F_r^{-\epsilon}F_r^{-\eta} : F_u^{-\epsilon}F_u^{-\eta})$ [for $(A : a)(B : b) : (AB : ab)$]
$\quad\quad\quad : (F_r^\theta : F_u^\theta)$ [for $A^{-\epsilon}A^{-\eta} = A^\theta$, by definition].

This proves the third formula ϕ_3, when we assume the first two ϕ_1 and ϕ_2. To give a concrete illustration of the difference between F_u and F_r, let the symbol H

§§ 45, 46]   SYLLOGISTIC REASONING   41

represent the word *horse*, and let F(H) denote the statement "The *horse* has been caught." Then $F^\epsilon(H)$ asserts that every horse of the series H_1, H_2, &c., has been caught; the symbol $F^\eta(H)$ asserts that not one horse of the series H_1, H_2, &c., has been caught; and the symbol $F^\theta(H)$ denies both the statements $F^\epsilon(H)$ and $F^\eta(H)$, and is therefore equivalent to $F^{-\epsilon}(H) . F^{-\eta}(H)$, which may be more briefly expressed by $F^{-\epsilon}F^{-\eta}$, the symbol (H) being left understood. But what *is* the series H_1, H_2, &c.? This universe of horses may mean, for example, *all the horses owned by the horse-dealer;* or it may mean a *portion* only of these horses, as, for example, *all the horses that had escaped.* If we write F^ϵ_u we assert that *every horse owned by the horse-dealer has been caught;* if we write F^ϵ_r we only assert that *every horse of his that escaped has been caught.* Now, it is clear that the first statement implies the second, but that the second does not necessarily imply the first; so that we have $F^\epsilon_u : F^\epsilon_r$, but not necessarily $F^\epsilon_r : F^\epsilon_u$. The last implication $F^\epsilon_r : F^\epsilon_u$ is not necessarily true; for the fact that all the horses that had escaped were caught would not necessarily imply that all the horses owned by the horse-dealer had been caught, since some of them may not have escaped, and of these it would not be correct to say that they had been caught. The symbol F_u may refer to the series $F_1, F_2, F_3, \ldots, F_{60}$, while F_r may refer only to the series $F_1, F_2, F_3, \ldots, F_{10}$. The same concrete illustration will make evident the truth of the implications $F^\eta_u : F^\eta_r$ and $F^\theta_r : F^\theta_u$, and also that the converse implications $F^\eta_r : F^\eta_u$ and $F^\theta_u : F^\theta_r$ are not necessarily true.

46. Let us now examine the special kind of reasoning called *syllogistic*. Every valid syllogism, as will be shown, is a particular case of my general formula

$$(\alpha : \beta)(\beta : \gamma) : (\alpha : \gamma),$$

or, as it may be more briefly expressed,

$$(\alpha : \beta : \gamma) : (\alpha : \gamma).$$

Let S denote our *Symbolic Universe*, or "Universe of Discourse," consisting of all the things S_1, S_2, &c., *real, unreal, existent*, or *non-existent*, expressly mentioned or tacitly understood, in our argument or discourse. Let X denote any class of individuals X_1, X_2, &c., forming a *portion* of the Symbolic Universe S; then $`X$ (with a grave accent) denotes the class of individuals $`X_1$, $`X_2$, &c., that do not belong to the class X; so that the individuals $X_1$, $X_2$, &c., of the class X, *plus* the individuals $`X_1$, $`X_2$, &c., of the class $`X$, always make up the total Symbolic Universe S_1, S_2, &c. The class $`X$ is called the *complement* of the class X, and *vice versa*. Thus, any class A and its complement $`A$ make up together the whole Symbolic Universe S; each forming a portion only, and both forming the whole.

47. Now, there are two mutually complementary classes which are so often spoken of in logic that it is convenient to designate them by special symbols; these are the class of individuals which, in the given circumstances, have a *real existence*, and the class of individuals which, in the given circumstances, have *not* a real existence. The first class is the class e, made up of the individuals e_1, e_2, &c. To this class belongs every individual of which, in the given circumstances, one can truly say "It *exists*"—that is to say, not merely symbolically but really. To this class therefore may belong *horse, town, triangle, virtue, vice*. We may place *virtue* and *vice* in the class e, because the statement "Virtue exists" or "Vice exists" really asserts that virtuous *persons*, or vicious *persons*, exist; a statement which every one would accept as true.

The second class is the class 0, made up of the individuals 0_1, 0_2, &c. To this class belongs every individual of which, in the given circumstances, we can truly say "It does not exist"—that is to say, "It does not exist *really*, though (like everything else named) it exists *symbolically*." To this class necessarily belong

§§ 47–49] REALITIES AND UNREALITIES   43

centaur, mermaid, round square, flat sphere. The Symbolic Universe (like any class A) may consist wholly of realities e_1, e_2, &c.; or wholly of unrealities 0_1, 0_2, &c., or it may be a mixed universe containing both. When the members A_1, A_2, &c., of any class A consist wholly of realities, or wholly of unrealities, the class A is said to be a *pure* class; when A contains at least one reality and also at least one unreality, it is called a *mixed* class. Since the classes e and 0 are mutually complementary, it is clear that 'e is synonymous with 0, and '0 with e.

48. In no case, however, in fixing the limits of the class e, must the *context* or given circumstances be overlooked. For example, when the symbol H_c^0 is read "*The horse caught does not exist*," or " No horse has been caught" (see §§ 6, 47), the *understood* universe of realities, e_1, e_2, &c., may be a limited number of horses, H_1, H_2, &c., *that had escaped*, and in that case the statement H_c^0 merely asserts that to that limited universe the individual H_c, the *horse caught*, or a *horse caught*, does not belong; it does not deny the possibility of a horse being caught at some other time, or in some other circumstances. Symmetry and convenience require that the admission of any class A into our symbolic universe must be always understood to imply the existence also in the same universe of the complementary class 'A. Let A and B be any two classes that are *not mutually complementary* (see § 46); if A and B are mutually exclusive, their respective complements 'A and 'B, overlap; and, conversely, if 'A and 'B are mutually exclusive, A and B overlap.

49. Every statement that enters into a syllogism of the traditional logic has one or other of the following four forms:

(1) *Every* X *is* Y ; (2) *No* X *is* Y ;
(3) *Some* X *is* Y ; (4) *Some* X *is not* Y.

It is evident that (3) is simply the denial of (2), and (4

the denial of (1). From the conventions of §§ 6, 47, we get

(1) X^0_{-Y} = Every X is Y ; (2) X^0_Y = No X is Y ;
(3) $X^e_Y = X^{-0}_{-Y}$ = Some X is Y ;
(4) $X^e_{-Y} = X^{-0}_Y$ = Some X is not Y.

The first two are, in the traditional logic, called *universals*; the last two are called *particulars*; and the four are respectively denoted by the letters A, E, I, O, for reasons which need not be here explained, as they have now only historical interest. The following is, however, a simpler and more symmetrical way of expressing the above four standard propositions of the traditional logic; and it has the further advantage, as will appear later, of showing how all the syllogisms of the traditional logic are only particular cases of more general formulæ in the logic of pure statements.

50. Let S be any individual taken at random out of our Symbolic Universe, or Universe of Discourse, and let x, y, z respectively denote the three propositions S^x, S^y, S^z. Then x', y', z' must respectively denote S^{-x}, S^{-y}, S^{-z}. By the conventions of § 46, the three propositions x, y, z, like their denials x', y', z', are all possible but uncertain; that is to say, all six are *variables*. Hence, we must always have x^θ, y^θ, z^θ, $(x')_\theta$, $(y')_\theta$, $(z')_\theta$; and never x^η nor y^η nor z^η nor x^ϵ nor y^ϵ nor z^ϵ. Hence, when x, y, z respectively denote the propositions S^x, S^y, S^z, the propositions $(x:\eta)'$, $(y:\eta)'$, $(z:\eta)'$ (which are respectively synonymous with $x^{-\eta}$, $y^{-\eta}$, $z^{-\eta}$) must always be considered to form part of our data, whether expressed or not; and their denials, $(x:\eta)$, $(y:\eta)$, $(z:\eta)$, must be considered impossible. With these conventions we get—

(A) Every (or all) X is Y = $S^x : S^y = (x:y) = (xy')^\eta$
(O) Some X is not Y = $(S^x : S^y)' = (x:y)' = (xy')^{-\eta}$
(E) No X is Y = $S^x : S^{-y} = x : y' = (xy)^\eta$
(I) Some X is Y = $(S^x : S^{-y})' = (x:y')' = (xy)^{-\eta}$.

§ 50]   GENERAL AND TRADITIONAL LOGIC   45

In this way we can express every syllogism of the traditional logic in terms of x, y, z, which represent three propositions having the same subject S, but different predicates X, Y, Z. Since none of the propositions x, y, z (as already shown) can in this case belong to the class η or ϵ, the values (or meanings) of x, y, z are *restricted*. Hence, every traditional syllogism expressed in terms of x, y, z must belong to the class of *restricted* functional statements $F_r(x, y, z)$, or its abbreviated synonym F_r, and *not* to the class of *unrestricted* functional statements $F_u(x, y, z)$, or its abbreviated synonym F_u, as this last statement assumes that the values (or meanings) of the propositions x, y, z are wholly unrestricted (see § 44). The proposition $F_u(x, y, z)$ assumes not only that each constituent statement x, y, z may belong to the class η or ϵ, as well as to the class θ, but also that the three statements x, y, z need not even have the same subject. For example, let $F(x, y, z)$, or its abbreviation F, denote the formula

$$(x : y)(y : z) : (x : z).$$

This formula asserts that "If x implies y, and y implies z, then x implies z." The formula holds good whatever be the statements x, y, z; whether or not they have (as in the traditional logic) the same subject S; and whether or not they are certainties, impossibilities, or variables. Hence, with reference to the above formula, it is always correct to assert F^ϵ whether F denotes F_u or F_r. When x, y, z have a common subject S, then F^ϵ will mean F^ϵ_r and will denote the syllogism of the traditional logic called *Barbara* ;* whereas when x, y, z are wholly unrestricted, F^ϵ will mean F^ϵ_u and will therefore be a more general formula, of which the traditional *Barbara* will be a particular case.

* *Barbara* asserts that "If every X is Y, and every Y is Z, then every X is Z," which is equivalent to $(S^x : S^y)(S^y : S^z) : (S^x : S^z)$.

But now let F, or F(x, y, z), denote the implication

$$(y:z)(y:x):(x:z')'.$$

If we suppose the propositions x, y, z to be limited by the conventions of §§ 46, 50, the traditional syllogism called *Darapti* will be represented by F_r and not by F_u. Now, by the first formula of § 45, we have $F_u^{\epsilon}:F_r^{\epsilon}$, and, consequently, $F_r^{-\epsilon}:F_u^{-\epsilon}$, but *not* necessarily $F_u^{-\epsilon}:F_r^{-\epsilon}$. Thus, if F_u be valid, the traditional Darapti must be valid also. We find that F_u is not valid, for the above implication represented by F fails in the case $y^\eta(xz)^\eta$, as it then becomes

$$(\eta:z)(\eta:x):(xz)^{-\eta},$$

which is equivalent to $\epsilon\epsilon:\eta^{-\eta}$, and consequently to $\epsilon:\eta$, which $= (\epsilon\eta')^\eta = (\epsilon\epsilon)^\eta = \eta$. But since (as just shown) $F_u^{-\epsilon}$ does not necessarily imply $F_r^{-\epsilon}$, *this discovery does not justify us in concluding that the traditional Darapti is not valid.* The only case in which F fails is $y^\eta(xz)^\eta$, and *this case cannot occur* in the limited formula F_r (which here represents the traditional *Darapti*), because in F_r the propositions x, y, z are always variable and therefore possible. In the *general* and *non-traditional* implication F_u, the case $x^\eta y^\eta z^\eta$, since it implies $y^\eta(xz)^\eta$, is also a case of failure; but it is not a case of failure in the traditional logic.

51. The traditional *Darapti*, namely, "*If every* Y *is* Z, *and every* Y *is also* X, *then some* X *is* Z," is thought by some logicians (I formerly thought so myself) to fail when the class Y is non-existent, while the classes X and Z are real but mutually exclusive. But this is a mistake, as the following concrete example will show. Suppose we have

$$Y = (0_1, 0_2, 0_3),\ Z = (e_1, e_2, e_3),\ X = (e_4, e_5, e_6).$$

Let P denote the first premise of the given syllogism, Q the second, and R the conclusion. We get

$$P = \text{Every Y is Z} = \eta_1;\ Q = \text{Every Y is X} = \eta_2;$$

and $R = $ Some X is $Z = \eta_3$; three statements, η_1, η_2, η_3,

§§ 51, 52] TRADITIONAL SYLLOGISMS 47

each of which contradicts our data, since, by our data in this case, the three classes X, Y, Z are mutually exclusive. Hence in this case we have

$$PQ : R = (\eta_1 \eta_2 : \eta_3) = (\eta_4 : \eta_3) = (\eta_4 \eta'_3)^\eta = \epsilon_1 ;$$

so that, *when presented in the form of an implication*, Darapti does *not* fail in the case supposed. (But see § 52.)

52. Startling as it may sound, however, it is a demonstrable fact that *not one syllogism* of the traditional logic—neither Darapti, nor Barbara, nor any other—is valid in the form in which it is usually presented in our text-books, and in which, I believe, it has been always presented ever since the time of Aristotle. In this form, every syllogism makes four positive assertions: it *asserts* the first premise; it *asserts* the second; it *asserts* the conclusion; and, by the word 'therefore,' it *asserts* that the conclusion follows necessarily from the premises, *i.e.* that *if* the premises be true, the conclusion must be true also. Of these four assertions the first three may be, and often are, false; the fourth, and the fourth alone, is a formal certainty. Take the standard syllogism *Barbara*. Barbara (in the usual text-book form) says this:

"Every A is B; every B is C; *therefore* every A is C."

Let $\psi(A, B, C)$ denote this syllogism. If valid it must be true whatever values (or meanings) we give to A, B, C. Let A=*ass*, let B=*bear*, and let C=*camel*. If $\psi(A, B, C)$ be valid, the following syllogism must therefore be true: "Every *ass* is a *bear*; every *bear* is a *camel*; therefore, every *ass* is a *camel*." Is this concrete syllogism really true? Clearly not; it contains three false statements. Hence, in the above form, Barbara (here denoted by ψ) is *not* valid; for have we not just adduced a case of failure? And if we give random values to A, B, C out of a large number of classes taken haphazard (*kings, queens, sailors, doctors, stones, cities, horses, French, Europeans, white things, black things,* &c., &c.), we shall find that the cases in which this syllogism will turn out false enormously outnumber the cases in which it will turn out true. *But it is always true in the following form, whatever values we give to* A, B, C:—

"*If* every A is B, and every B is C, *then* every A is C."

Suppose as before that A = *ass*, that B = *bear*, and that C = *camel*. Let P denote the combined premises, "Every *ass* is a *bear*, and every *bear* is a *camel*," and let Q denote the conclusion, "Every *ass* is a *camel*." Also, let the symbol ∴, as is customary, denote the word *therefore*. The first or *therefore*-form asserts P ∴ Q, which is equivalent* to the two-factor statement P(P : Q); the second or *if*-form asserts only the second factor P : Q. The *therefore*-form vouches for the truth of P and Q, which are both false; the *if*-form vouches only for the truth of the implication P : Q, which, by definition, means $(PQ')^\eta$, and is a formal certainty. (See § 10.)

53. Logicians may say (as some *have* said), in answer to the preceding criticism, that my objection to the usual form of presenting a syllogism is purely verbal; that the premises are always understood to be merely hypothetical, and that therefore the syllogism, in its *general* form, is not supposed to guarantee either the truth of the premises or the truth of the conclusion. This is virtually an admission that though (P ∴ Q) is *asserted*, the weaker statement (P : Q) is the one really *meant*—that though logicians assert "P *therefore* Q," they only mean "*If* P then Q." But why depart from the ordinary common-sense linguistic convention? In ordinary speech, when we say "P is true, *therefore* Q is true," we vouch for the truth of P; but when we say "*If* P is true, then Q is true," we do not. As I said in the *Athenæum*, No. 3989:—

"Why should the linguistic convention be different in logic? . . . Where is the necessity? Where is the advantage? Suppose a general, whose mind, during his past university days, had been over-imbued with the traditional logic, were in war time to say, in speaking of an untried and possibly innocent prisoner, 'He is a spy; *therefore* he

---

* I pointed out this equivalence in *Mind*, January 1880.

§§ 53, 54]   TRADITIONAL SYLLOGISMS   49

must be shot,' and that this order were carried out to the letter. Could he afterwards exculpate himself by saying that it was all an unfortunate mistake, due to the deplorable ignorance of his subordinates; that if these had, like him, received the inestimable advantages of a logical education, they would have known at once that what he really meant was '*If he is a spy, he must be shot*'? The argument in defence of the traditional wording of the syllogism is exactly parallel."

It is no exaggeration to say that nearly all fallacies are due to neglect of the little conjunction, *If*. Mere hypotheses are accepted as if they were certainties.

CHAPTER VIII

54. In the notation of § 50, the following are the nineteen syllogisms of the traditional logic, in their usual order. As is customary, they are arranged into four divisions, called *Figures*, according to the position of the "middle term" (or middle constituent), here denoted by y. This constituent y always appears in both premises, but not in the conclusion. The constituent z, in the traditional phraseology, is called the "major term," and the constituent x the "minor term." Similarly, the premise containing z is called the "major premise," and the premise containing x the "minor premise." Also, since the conclusion is always of the form "All X is Z," or "Some X is Z," or "No X is Z," or "Some X is not Z," it is usual to speak of X as the 'subject' and of Z as the 'predicate.' As usual in text-books, the major premise precedes the minor.

Figure 1

Barbara $= (y:z)(x:y) : (x:z)$
Celarent $= (y:z')(x:y) : (x:z')$
Darii $= (y:z)(x:y')' : (x:z')'$
Ferio $= (y:z')(x:y')' : (x:z)'$

Figure 2

Cesare $= (z : y')(x : y) : (x : z')$
Camestres $= (z : y)(x : y') : (x : z')$
Festino $= (z : y')(x : y')' : (x : z)'$
Baroko $= (z : y)(x : y)' : (x : z)'$

Figure 3

Darapti $= (y : z)(y : x) : (x : z')'$
Disamis $= (y : z')'(y : x) : (x : z')'$
Datisi $= (y : z)(y : x')' : (x : z')'$
Felapton $= (y : z')(y : x) : (x : z)'$
Bokardo $= (y : z)'(y : x) : (x : z)'$
Ferison $= (y : z')(y : x')' : (x : z)'$

Figure 4

Bramantip $= (z : y)(y : x) : (x : z')'$
Camenes $= (z : y)(y : x') : (x : z')$
Dismaris $= (z : y')'(y : x) : (x : z')'$
Fesapo $= (z : y')(y : x) : (x : z)'$
Fresison $= (z : y')(y : x')' : (x : z)'$

Now, let the symbols (Barbara)$_u$, (Celarent)$_u$, &c., denote, in conformity with the convention of § 44, these nineteen functional statements respectively, when the values of their constituent statements x, y, z are *unrestricted*; while the symbols (Barbara)$_r$, (Celarent)$_r$, &c., denote the same functional statements when the values of x, y, z are *restricted* as in § 50. The syllogisms (Barbara)$_r$, (Celarent)$_r$, &c., with the suffix r, indicating restriction of values, are the real syllogisms of the traditional logic; and all these, without exception, are valid—*within the limits of the understood restrictions*. The nineteen syllogisms of *general* logic, that is to say, of the pure logic of *statements*,

§§ 54-56]   GENERAL LOGIC   51

namely, (Barbara)$_u$, (Celarent)$_u$, &c., in which x, y, z are *unrestricted* in values, are more general than and imply the traditional nineteen in which x, y, z are *restricted* as in § 50; and four of these unrestricted syllogisms, namely, (Darapti)$_u$, (Felapton)$_u$, (Bramantip)$_u$, and (Fesapo)$_u$, fail in certain cases. (Darapti)$_u$ fails in the case $y^\eta(xz)^\eta$, (Felapton)$_u$ and (Fesapo)$_u$ fail in the case $y^\eta(xz')^\eta$, and (Bramantip)$_u$ fails in the case $z^\eta(x'y)^\eta$.

55. It thus appears that there are *two* Barbaras, *two* Celarents, *two* Darii, &c., of which, in each case, the one belongs to the traditional logic, with restricted values of its constituents x, y, z; while the other is a more general syllogism, of which the traditional syllogism is a particular case. Now, as shown in § 45, when a general law F_u, with *unrestricted* values of its constituents, implies a general law F_r, with *restricted* values of its constituents, if the former is true absolutely and never fails, the same may be said of the latter. This is expressed by the formula $F_u^\epsilon : F_r^\epsilon$. But an exceptional case of failure in F_u does not necessarily imply a corresponding case of failure in F_r; for though $F_u^\epsilon : F_r^\epsilon$ is a valid formula, the implication $F_u^{-\epsilon} : F_r^{-\epsilon}$ (which is equivalent to the converse implication $F_r^\epsilon : F_u^\epsilon$) is *not* necessarily valid. For example, the general and non-traditional syllogism (Darapti)$_u$ implies the less general and traditional syllogism (Darapti)$_r$. The former fails in the exceptional case $y^\eta(xz)^\eta$; but in the traditional syllogism this case cannot occur because of the restrictions which limit the statement y to the class θ (see § 50). Hence, though this case of failure necessitates the conclusion (Darapti)$_u^{-\epsilon}$, we cannot, from this conclusion, infer the further, but incorrect, conclusion (Darapti)$_r^{-\epsilon}$. Similar reasoning applies to the unrestricted non-traditional and restricted traditional forms of Felapton, Bramantip, and Fesapo.

56. All the preceding syllogisms, with many others not recognised in the traditional logic, may, by means of the formulæ of transposition $\alpha : \beta = \beta' : \alpha'$ and $\alpha\beta' : \gamma' = \alpha\gamma : \beta$,

be shown to be only particular cases of the formula $(x:y)(y:z):(x:z)$, which expresses Barbara. Two or three examples will make this clear. Let $\phi(x, y, z)$ denote this standard formula. Referring to the list in § 54, we get

$$\text{Baroko} = (z:y)(x:y)':(x:z)';\text{ which, by transposition,}$$
$$= (x:z)(z:y):(x:y) = \phi(x, z, y).$$

Thus Baroko is obtained from the general standard formula $\phi(x, y, z)$ by interchanging y and z.

Next, take the syllogism Darii. Transposing as before, we get

$$\text{Darii} = (y:z)(x:y)':(x:z)' = (y:z)(x:z'):(x:y')$$
$$= (y:z)(z:x'):(y:x') = \phi(y, z, x').$$

Next, take (Darapti)$_r$. We get (see § 54)

$$(\text{Darapti})_r = (y:z)(y:x):(x:z')' = (y:zx):(xz:\eta)'$$
$$= (y:xz)(xz:\eta):\eta = (y:xz)(xz:\eta):(y:\eta);$$

for, in the *traditional* logic, $(y:\eta) = \eta$, since, by the convention of § 50, y must always be a *variable*, and, therefore, always *possible*. Thus, finally (Darapti)$_r = \phi(y, xz, \eta)$.

Lastly, take (Bramantip)$_r$. We get

$$(\text{Bramantip})_r = (z:y)(y:x):(x:z')' = (z:y)(y:x)(x:z'):\eta$$
$$= (z:y)(z:x')(y:x):\eta = (z:yx')(y:x):\eta$$
$$= (z:yx')(yx':\eta):\eta = (z:yx')(yx':\eta):(z:\eta);$$

for, in the *traditional* logic, $(z:\eta) = \eta$, since z must be a *variable* and therefore *possible*. Hence, finally, we get

$$(\text{Bramantip})_r = \phi(z, yx', \eta).$$

57. By similar reasoning the student can verify the following list (see §§ 54–56):

$\phi(x, y, z) = $ Barbara ; $\phi(x, y, z') = $ Celarent = Cesare ;
$\phi(y, z, x') = $ Darii = Datisi ; $\phi(x, z, y') = $ Ferio = Festino
= Ferison = Fresison ; $\phi(z, y, x') = $ Camestres
= Camenes ;

§§ 57–59] TESTS OF SYLLOGISTIC VALIDITY 53

$\phi(y, x, z')$ = Disamis = Dismaris; $\phi(x, z, y)$ = Baroko;
$\phi(y, x, z)$ = Bokardo; $\phi(y, xz, \eta)$ = (Darapti)$_r$;
$\phi(y, xz', \eta)$ = (Felapton)$_r$ = (Fesapo)$_r$;
$\phi(z, yx', \eta)$ = (Bramantip)$_r$.

58. It is evident (since $x:y = y':x'$) that $\phi(x, y, z) = \phi(z,' y', x')$ in the preceding list; so that all these syllogisms remain valid when we change the order of their constituents, provided we, at the same time, change their signs. For example, Camestres and Camenes may each be expressed, not only in the form $\phi(z, y, x')$, as in the list, but also in the form $\phi(x, y', z')$.

59. Text-books on logic usually give rather complicated rules, or "canons," by which to test the validity of a supposed syllogism. These we shall discuss further on (see §§ 62, 63); meanwhile we will give the following rules, which are simpler, more general, more reliable, and more easily applicable.

Let an accented capital letter denote a *non-implication* (or "particular"), that is to say, the denial of an implication; while a capital without an accent denotes a simple implication (or "universal"). Thus, if A denote $x:y$, then A' will denote $(x:y)'$. Now, let A, B, C denote any syllogistic implications, while A', B', C' denote their respective denials. Every valid syllogism must have one or other of these three forms:

(1) AB : C; (2) AB' : C'; (3) AB : C';

that is to say, either the two premises and the conclusion are all three implications (or "universals") as in (1); or one premise only and the conclusion are both non-implications (or "particulars") as in (2); or, as in (3), both premises are implications (or "universals"), while the conclusion is a non-implication (or "particular"). If any supposed syllogism does not come under form (1) nor under form (2) nor under form (3), it is not *valid*; that is to say, there will be cases in which it will fail.

The second form may be reduced to the first form by transposing the premise B′ and the conclusion C′, and changing their signs; for AB′ : C′ is equivalent to AC : B, each being equivalent to AB′C′ : η. When thus transformed the validity of AB′ : C′, that is, of AC : B, may be tested in the same way as the validity of AB : C. The test is easy. Suppose the conclusion C to be $x:z$, in which z may be affirmative or negative. If, for example, $z = He\ is\ a\ soldier$; then $z' = He\ is\ not\ a\ soldier$. But if $z = He\ is\ not\ a\ soldier$; then $z' = He\ is\ a\ soldier$. The conclusion C being, by hypothesis, $x:z$, the syllogism AB : C, if valid, becomes (see § 11) either

$$(x:y:z):(x:z), \text{ or else } (x:y':z):(x:z),$$

in which the statement y refers to the middle class (or "term") Y, not mentioned in the conclusion $x:z$. If any supposed syllogism AB : C cannot be reduced to either of these two forms, it is not valid; if it *can* be reduced to either form, it is valid. To take a concrete example, let it be required to test the validity of the following implicational syllogism:

If no *Liberal* approves of *Protection*, though some Liberals approve of fiscal *Retaliation*, it follows that some person or persons who approve of fiscal *Retaliation* do not approve of *Protection*.

Speaking of a person taken at random, let L = He is a *Liberal*; let P = He approves of *Protection*; and let R = He approves of fiscal *Retaliation*. Also, let Q denote the syllogism. We get

$$Q = (L:P')(L:R')':(R:P)'.$$

To get rid of the non-implications, we transpose them (see § 56) and change their signs from negative to affirmative, thus transforming them into implications. This transposition gives us

$$Q = (L:P')(R:P):(L:R').$$

§§ 59, 60] TESTS OF SYLLOGISTIC VALIDITY   55

Since in this form of Q, the syllogistic propositions are all three implications (or "universals"), the combination of premises, $(L:P')(R:P)$, must (if Q be valid) be equivalent

either to $L:P:R'$ or else to $L:P':R'$;

in which P is the letter left out in the new consequent or conclusion $L:R'$. Now, the factors $L:P$ and $P:R'$ of $L:P:R'$ are not equivalent to the premises $L:P'$ and $R:P$ in the second or transposed form of the syllogism Q; but the factors $L:P'$ and $P':R'$ (which is equivalent to $R:P$) of $L:P':R'$ are equivalent to the premises in the second or transformed form of the syllogism Q. Hence Q is valid.

As an instance of a *non-valid* syllogism of the form $AB:C$, we may give

$$(x:y')(y:z'):(x:z');$$

for since the y's in the two premises have different signs, the one being negative and the other affirmative, the combined premises can neither take the form $x:y:z'$ nor the form $x:y':z'$, which are respective abbreviations for $(x:y)(y:z')$ and $(x:y')(y':z')$. The syllogism is therefore not valid.

60. The preceding process for testing the validity of syllogisms of the forms $AB:C$ and $AB':C'$ apply to all syllogisms without exception, whether the values of their constituents x, y, z be restricted, as in the traditional logic, or unrestricted, as in my general logic of statements. But as regards syllogisms in *general* logic of the form $AB:C'$ (a form which includes Darapti, Felapton, Fesapo, and Bramantip in the traditional logic), with two implicational premises and a non-implicational conclusion, they can only be true *conditionally;* for in *general* logic (as distinguished from the *traditional* logic) no syllogism of this type is a formal certainty. It therefore becomes an interesting and important problem to deter-

mine the *conditions* on which syllogisms of this type can be held valid. We have to determine two things, firstly, the *weakest premise* (see § 33, footnote) which, when joined to the two premises given, would render the syllogism a *formal certainty*; and, secondly, the weakest condition which, when assumed throughout, would render the syllogism a *formal impossibility*. As will be seen, the method we are going to explain is a general one, which may be applied to other formulæ besides those of the syllogism.

The given implication $AB:C'$ is equivalent to the implication $ABC:\eta$, in which A, B, C are three implications (see § 59) involving three constituents x, y, z. Eliminate successively x, y, z as in § 34, not as in finding the successive limits of x, y, z, but taking each variable independently. Let α denote the strongest conclusion deducible from ABC and containing no reference to the eliminated x. Similarly, let β and γ respectively denote the strongest conclusions after the elimination of y alone (x being left), and after the elimination of z alone (x and y being left). Then, if we join the factor α' or β' or γ' to the premises (*i.e.* the antecedent) of the given implicational syllogism $AB:C'$, the syllogism will become a formal certainty, and therefore valid. That is to say, $AB\alpha':C'$ will be a formal certainty; and so will $AB\beta':C'$ and $AB\gamma':C'$. Consequently, $AB(\alpha'+\beta'+\gamma'):C'$ is a formal certainty; so that, on the one hand, the weakest premise needed to be joined to AB to render the given syllogism $AB:C'$ *valid* (*i.e.* a *formal certainty*) is the alternative $\alpha'+\beta'+\gamma'$, and, on the other, the weakest datum needed to make the syllogism $AB:C'$ a *formal impossibility* is the *denial* of $\alpha'+\beta'+\gamma'$, that is, $\alpha\beta\gamma$.

61. Take as an example the syllogism Darapti. Here we have an implication $AB:C'$ in which A, B, C respectively denote the implications $(y:x)$, $(y:z)$, $(x:z')$. By the method of § 34 we get

$$ABC = yx' + yz' + xz : \eta = Mx + Nx' + P : \eta, \text{ say,}$$

§ 61]   CONDITIONS OF VALIDITY   57

in which M, N, P respectively denote the co-factor of x, the co-factor of x', and the term not containing x. The strongest consequent not involving x is $MN + P : \eta$, in which here $M = z$, $N = y$, and $P = yz'$; so that we have

$$MN + P : \eta = zy + yz' : \eta = y(z + z') : \eta$$
$$= y\epsilon : \eta = y : \eta.$$

Thus we get $a = y : \eta$, so that the premise required when we eliminate x is $(y : \eta)'$; and therefore

$$(y : x)(y : z)(y : \eta)' : (x : z')'$$

should be a formal certainty, which is a fact; for, getting rid of the non-implications by transposition, this complex implication becomes

$$(y : x)(y : z)(x : z') : (y : \eta),$$
$$\text{which} = (y : xz)(xz : \eta) : (y : \eta);$$

and this is a formal certainty, being a particular case of the standard formula $\phi(x, y, z)$, which represents Barbara both in general and in the traditional logic (see § 55). Eliminating y alone in the same manner from $AB : C'$, we find that $\beta = xz : \eta = x : z'$; so that the complex implication

$$(y : x)(y : z)(x : z')' : (x : z')'$$

should be a formal certainty. That it is so is evident by inspection, on the principle that the implication $PQ : Q$, for all values of P and Q, is a formal certainty. Finally, we eliminate z, and find that $\gamma = y : \eta$. This is the same result as we obtained by the elimination of x, as might have been foreseen, since x and z are evidently interchangeable.

Thus we obtain the information sought, namely, that $a' + \beta' + \gamma'$, the weakest premise to be joined to the premises of Darapti to make this syllogism a formal certainty in *general* logic is

$$(y : \eta)' + (xz : \eta)' + (y : \eta)', \text{ which } = y^{-\eta} + (xz)^{-\eta};$$

SYMBOLIC LOGIC [§§ 61, 62

and that $\alpha\beta\gamma$, the weakest presupposed condition that would render the syllogism Darapti a logical impossibility, is therefore

$$\{y^{-\eta} + (xz)^{-\eta}\}', \text{ which} = y^\eta(xz)^\eta.$$

Hence, the Darapti of *general* logic, with unrestricted values of its constituents x, y, z, fails in the case $y^\eta(xz)^\eta$: but in the traditional logic, as shown in § 50, this case cannot arise. The preceding reasoning may be applied to the syllogisms Felapton and Fesapo by simply changing z into z'.

Next, take the syllogism Bramantip. Here we get

$$ABC = yx' + zy' + xz : \eta,$$

and giving α, β, γ the same meanings as before, we get $\alpha = z^\eta$, $\beta = z^\eta$, $\gamma = (x'y)^\eta$. Hence, $\alpha\beta\gamma = z^\eta(x'y)^\eta$, and $\alpha' + \beta' + \gamma' = z^{-\eta} + (x'y)^{-\eta}$. Thus, in general logic, Bramantip is a formal certainty when we assume $z^{-\eta} + (x'y)^{-\eta}$, and a formal impossibility when we assume $z^\eta(x'y)^\eta$; but in the traditional logic the latter assumption is inadmissible, since z^η is inadmissible by § 50, while the former is obligatory, since it is implied in the necessary assumption $z^{-\eta}$.

62. The validity tests of the traditional logic turn mainly upon the question whether or not a syllogistic 'term' or class is '*distributed*' or '*undistributed*.' In ordinary language these words rarely, if ever, lead to any ambiguity or confusion of thought; but logicians have somehow managed to work them into a perplexing tangle. In the proposition "All X is Y," the class X is said to be 'distributed,' and the class Y 'undistributed.' In the proposition "No X is Y," the class X and the class Y are said to be both 'distributed.' In the proposition "Some X is Y," the class X and the class Y are said to be both 'undistributed.' Finally, in the proposition "Some X is not Y," the class X is said to be 'undistributed,' and the class Y 'distributed.'

§ 62]   'DISTRIBUTED'—'UNDISTRIBUTED'   59

Let us examine the consequences of this tangle of technicalities. Take the leading syllogism Barbara, the validity of which no one will question, provided it be expressed in its *conditional* form, namely, "*If* all Y is Z, and all X is Y, then all X is Z." Being, in this form (see § 52), admittedly valid, this syllogism must hold good whatever values (or meanings) we give to its constituents X, Y, Z. It must therefore hold good when X, Y, and Z are synonyms, and, therefore, all denote *the same class*. In this case also the two premises and the conclusion will be three truisms which no one would dream of denying. Consider now one of these truisms, say "All X is Y." Here, by the usual logical convention, the class X is said to be 'distributed,' and the class Y 'undistributed.' But when X and Y are synonyms they denote the *same class*, so that the same class may, at the same time and in the same proposition, be both '*distributed*' and '*undistributed*.' Does not this sound like a contradiction? Speaking of a certain concrete collection of apples in a certain concrete basket, can we consistently and in the same breath assert that "All the apples are already *distributed*" and that "All the apples are still *undistributed*"? Do we get out of the dilemma and secure consistency if on every apple in the basket we stick a ticket X and also a ticket Y? Can we then consistently assert that all the X apples are *distributed*, but that all the Y apples are *undistributed*? Clearly not; for every X apple is also a Y apple, and every Y apple an X apple. In ordinary language the classes which we can respectively qualify as *distributed* and *undistributed* are mutually exclusive; in the logic of our text-books this is evidently not the case. Students of the traditional logic should therefore disabuse their minds of the idea that the words 'distributed' and 'undistributed' necessarily refer to classes mutually exclusive, as they do in everyday speech; or that there is anything but a forced and fanciful connexion between the 'distributed' and 'undistributed' of current English and the technical 'distributed' and 'undisturbed' of logicians.

Now, how came the words 'distributed' and 'undistributed' to be employed by logicians in a sense which plainly does not coincide with that usually given them? Since the statement "No X is Y" is equivalent to the statement "All X is 'Y," in which (see §§ 46–50) the class 'Y (or non-Y) contains all the individuals of the Symbolic Universe excluded from the class Y, and since "Some X is not Y" is equivalent to "Some X is 'Y," the definitions of 'distributed' and 'undistributed' in textbooks virtually amount to this: that a class X is *distributed* with regard to a class Y (or 'Y) when every individual of the former is synonymous or identical with some individual or other of the latter; and that when this is not the case, then the class X is *undistributed* with regard to the class Y (or 'Y). Hence, when in the statement "All X is Y" we are told that X is *distributed* with regard to Y, but that Y is *undistributed* with regard to X, this ought to imply that X and Y cannot denote exactly the same class. In other words, the proposition that "All X is Y" ought to imply that "Some Y is not X." But as no logician would accept this implication, it is clear that the technical use of the words 'distributed' and 'undistributed' to be found in logical treatises is lacking in linguistic consistency. In answer to this criticism, logicians introduce psychological considerations and say that the proposition "All X is Y" gives us information about *every* individual, X_1, X_2, &c., of the class X, but *not* about *every* individual, Y_1, Y_2, &c., of the class Y; and that this is the reason why the term X is said to be 'distributed' and the term Y 'undistributed.' To this explanation it may be objected, firstly, that formal logic should not be mixed up with psychology—that its formulæ are independent of the varying mental attitude of individuals; and, secondly, that if we accept this 'information-giving' or 'non-giving' definition, then we should

§ 62]   'DISTRIBUTED'—'UNDISTRIBUTED'   61

say, not that X *is* distributed, and Y undistributed, but that X is *known* or *inferred* to be distributed, while Y is *not* known to be distributed—that the inference requires further data.

To throw symbolic light upon the question we may proceed as follows. With the conventions of § 50 we have

(1) All X is Y $= x:y$;   (2) No X is Y $= x:y'$
(3) Some X is Y $= (x:y')'$;   (4) Some X is not Y $= (x:y)'$.

The positive class (or 'term') X is usually spoken of by logicians as the 'subject'; and the positive class Y as the 'predicate.' It will be noticed that, in the above examples, the *non-implications* in (3) and (4) are the respective denials of the *implications* in (2) and (1). The definitions of 'distributed' and 'undistributed' are as follows.

(a) The class (or 'term') referred to by the *antecedent* of an *implication* is, in text-book language, said to be '*distributed*'; and the class referred to by the *consequent* is said to be 'undistributed.'

(β) The class referred to by the *antecedent* of a *non-implication* is said to be '*undistributed*'; and the class referred to by the *consequent* is said to be '*distributed*.'

Definition (a) applies to (1) and (2); definition (β) applies to (3) and (4). Let the symbol X^d assert that X is '*distributed*,' and let X^u assert that X is 'undistributed.' The class 'X being the complement of the class X, and *vice versa* (see § 46), we get $('X)^d = X^u$, and $('X)^u = X^d$. From the definitions (a) and (β), since $('Y)^d = Y^u$, and $('Y)^u = Y^d$, we therefore draw the following four conclusions:—

In (1) $X^d Y^u$; in (2) $X^d Y^d$; in (3) $X^u Y^u$; in (4) $X^u Y^d$. For in (2) the definition (a) gives us $X^d ('Y)^u$, and $('Y)^u = Y^d$. Similarly, in (3) the definition (β) gives us $X^u ('Y)^d$, and $('Y)^d = Y^u$.

If we change y into x in proposition (1) above, we

62 SYMBOLIC LOGIC [§§ 62, 63

get "All X is X" $=x:x$. Here, by definition (a), we have $X^d X^u$; which shows that there is no necessary antagonism between X^d and X^u; that, in the text-book sense, the same class may be both 'distributed' and 'undistributed' at the same time.

63. The six canons of syllogistic validity, as usually given in text-books, are:—

(1) Every syllogism has three and only three terms, namely, the 'major term,' the 'minor term,' and the 'middle term' (see § 54).

(2) Every syllogism consists of three and only three propositions, namely, the 'major premise,' the 'minor premise,' and the 'conclusion' (see § 54).

(3) The middle term must be distributed at least once in the premises; and it must not be ambiguous.

(4) No term must be distributed in the conclusion, unless it is also distributed in one of the premises.*

(5) We can infer nothing from two negative premises.

(6) If one premise be negative, the conclusion must be so also; and, *vice versa*, a negative conclusion requires one negative premise.

Let us examine these traditional canons. Suppose $\psi(x, y, z)$ to denote any valid syllogism. The syllogism being valid, it must hold good whatever be the classes to which the statements x, y, z refer. It is therefore valid when we change y into x, and also z into x; that is to say, $\psi(x, x, x,)$ is valid (§ 13, footnote). Yet this is a case which Canon (1) appears arbitrarily and needlessly to exclude. Canon (2) is simply a definition, and requires no comment. The second part of Canon (3) applies to all arguments alike, whether syllogistic or not.

---

* Violation of Canon (4) is called "Illicit Process." When the term illegitimately distributed in the conclusion is the major term, the fallacy is called "Illicit Process of the Major"; when the term illegitimately distributed in the conclusion is the minor term, the fallacy is called "Illicit Process of the Minor" (see § 54).

§ 63]   'CANONS' OF TRADITIONAL LOGIC   63

It is evident that if we want to avoid fallacies, we must also avoid ambiguities. The first part of Canon (3) cannot be accepted without reservation. The rule about the necessity of middle-term distribution does not apply to the following perfectly valid syllogism, " If every X is Y, and every Z is also Y, then something that is not X is not Z." Symbolically, this syllogism may be expressed in either of the two forms

$$(x:y)(z:y):(x':z)' \qquad \ldots \qquad (1)$$
$$(xy')^\eta(zy')^\eta:(x'z')^{-\eta} \qquad \ldots \qquad (2)$$

Conservative logicians who still cling to the old logic, finding it impossible to contest the validity of this syllogism, refuse to recognise it as a syllogism at all, on the ground that it has *four* (instead of the regulation *three*) terms, namely, X, Y, Z, 'Z, the last being the class containing all the individuals excluded from the class X. Yet a mere change of the three constituents, x, y, z, of the syllogism Darapti (which they count as valid) into their denials x', y', z' makes Darapti equivalent to the above syllogism. For Darapti is

$$(y:x)(y:z):(x:z')' \qquad \ldots \qquad (3);$$

and by virtue of the formula $a:\beta = \beta':a'$, the syllogism (1) in question becomes

$$(y':x')(y':z'):(x':z)' \qquad \ldots \qquad (4).$$

Thus, if $\psi(x, y, z)$ denote Darapti, then $\psi(x', y', z')$ will denote the contested syllogism (1) in its form (4); and, *vice versa*, if $\psi(x, y, z)$ denote the contested syllogism, namely, (1) or (4), then $\psi(x', y', z')$ will denote Darapti. To assert that any individual is *not* in the class X is equivalent to asserting that it *is* in the complementary class 'X. Hence, if we call the class 'X the non-X class, the syllogism in question, namely,

$$(y':x')(y':z'):(x':z)' \qquad \ldots \qquad (4),$$

may be read, " If every non-Y is a non-X, and every non-

Y is also a non-Z, then some non-X is a non-Z." For $(x':z)'$ is equivalent to $(x'\,z')^{-\eta}$, which asserts that it is possible for an individual to belong at the same time both to the class non-X and to the class non-Z. In other words, it asserts that some non-X is non-Z. Thus read, the contested syllogism becomes a case of Darapti, the classes X, Y, Z being replaced by their respective complementary classes 'X, 'Y, 'Z. It is evident that when we change any constituent x into x' in any syllogism, the words 'distributed' and 'undistributed' interchange places.

Canon (4) of the traditional logic asserts that "No term must be distributed in the conclusion, unless it is also distributed in one of the premises." This is another canon that cannot be accepted unreservedly. Take the syllogism Bramantip, namely,

$$(z:y)(y:x):(x:z')',$$

and denote it by $\psi(z)$. Since the syllogism is valid within the restrictions of the traditional logic (see § 50), it should be valid when we change z into z', and consequently z' into z. We should then get

$$\psi(z')=(z':y)(y:x):(x:z)'.$$

Here (see § 62) we get Z^u in the first premise, and Z^d in the conclusion, which is a flat contradiction to the canon. Upholders of the traditional logic, unable to deny the validity of this syllogism, seek to bring it within the application of Bramantip by having recourse to distortion of language, thus:—

"If every non-Z is Y, and every Y is X, then some X is non-Z."

Thus treated, the syllogism, instead of having Z^u in the first premise and Z^d in the conclusion, *which would contradict the canon*, would have $('Z)^d$ in the first premise and $('Z)^u$ in the conclusion, which, *though it means exactly the same thing*, serves to "save the face" of the canon and to hide its real failure and inutility.

§ 63] TESTS OF SYLLOGISTIC VALIDITY 65

Canon (5) asserts that "We can infer nothing from two negative premises." A single instance will show the unreliability of the canon. The example is

$$(y:x')(y:z'):(x':z)',$$

which is obtained from Darapti by simply changing z into z', and x into x'. It may be read, "If no Y is X, and no Y is Z, then something that is not X is not Z." Of course, logicians may "save the face" of this canon also by throwing it into the Daraptic form, thus: "If all Y is non-X, and all Y is also non-Z, then some non-X is non-Z." But in this way we might rid logic of *all* negatives, and the canon about negative premises would then have no *raison d'être*.

Lastly, comes Canon (6), which asserts, firstly, that "if one premise be negative, the conclusion must be negative; and, secondly, that a negative conclusion requires one negative premise." The objections to the preceding canons apply to this canon also. In order to give an appearance of validity to these venerable syllogistic tests, logicians are obliged to have recourse to distortion of language, and by this device they manage to make their negatives look like affirmatives. But when logic has thus converted all real negatives into seeming affirmatives the canons about negatives must disappear through want of negative matter to which they can refer. The following three simple formulæ are more easily applicable and will supersede all the traditional canons:—

(1) $(x:y:z):(x:z)$ . . . . . Barbara.
(2) $(z:y:x):(x:z')'$ . . . . . Bramantip.
(3) $(y:x)(y:z):(x:z')'$ . . . . Darapti.

The first of these is valid both in general logic and in the traditional logic; the second and third are only valid in the traditional logic. Apart from this limitation, they all three hold good whether any constituent be affirma-

E

tive or negative, and in whatever order we take the letters. Any syllogism that cannot, directly or by the formulæ of transposition, $\alpha:\beta=\beta':\alpha'$ and $\alpha\beta':\gamma'=\alpha\gamma:\beta$, be brought to one or other of these forms is invalid.

CHAPTER IX

Given one Premise and the Conclusion, to find the missing Complementary Premise.*

64. When in a valid syllogism we are given one premise and the conclusion, we can always find the *weakest* complementary premise which, with the one given, will imply the conclusion. When the given conclusion is an implication (or "universal") such as $x:z$ or $x:z'$, the complementary premise required is found readily by mere inspection. For example, suppose we have the conclusion $x:z'$ and the given major premise $z:y$ (see § 54). The syllogism required must be

either $(x:y:z'):(x:z')$ or $(x:y':z'):(x:z')$,

the middle term being either y or y'. The major premise of the first syllogism is $y:z'$, which is *not* equivalent to the given major premise $z:y$. Hence, the first syllogism is not the one wanted. The major premise of the second syllogism is $y':z'$, and this, by transposition and change of signs, is equivalent to $z:y$, which is the given major premise. Hence, the second syllogism is the one wanted, and the required minor premise is $x:y'$.

When the conclusion, but not the given premise, is a non-implication (or "particular"), we proceed as follows. Let P be the given implicational (or "universal") premise, and C' the given non-implicational (or "particular") conclusion. Let W be the required weakest premise which,

* A syllogism with one premise thus left understood is called an *enthymeme*.

§§ 64, 65]   TO FIND A MISSING PREMISE   67

joined to P, will imply C'. We shall then have PW : C', which, by transposition, becomes PC : W'. Let S be the strongest conclusion deducible from PC. We shall then have both PC : S and PC : W'. These two implications having the same antecedent PC, we suppose their consequents S and W' to be equivalent. We thus get S = W', and therefore W = S'. *The weakest premise required is therefore the denial of the strongest conclusion deducible from* PC (*the given premise and the denial of the given conclusion*).

For example, let the given premise be $y : x$, and the given conclusion $(x : z')'$. We are to have

$$(y : x)\mathrm{W} : (x : z')'.$$

Transposing and changing signs, this becomes

$$(y : x)(x : z') : \mathrm{W}'.$$

But, by our fundamental syllogistic formula, we have also (see § 56)

$$(y : x)(x : z') : (y : z').$$

We therefore assume $\mathrm{W}' = y : z'$, and, consequently, $\mathrm{W} = (y : z')'$. The weakest premise required* is therefore $(y : z')'$, and the required syllogism is

$$(y : x)(y : z')' : (x : z')'.$$

65. The only formulæ needed in finding the weakest complementary premise are

(1) $a : \beta = \beta' : a'.$
(2) $(a : \beta)(\beta : \gamma) : (a : \gamma).$
(3) $(a : \beta)(a : \gamma) : (\beta\gamma)^{-\eta}.$

The first two are true universally, whatever be the statements a, β, γ; the third is true on the condition $a^{-\eta}$, that a is possible—a condition which exists in the

---

* The implication $y : z$, since in the traditional logic it implies $(y : z')'$, would also answer as a premise ; but it would not be the *weakest* (see § 33, footnote, and § 73).

traditional logic, as here any of the statements a, β, γ may represent any of the three statements x, y, z, or any of their denials x', y', z', every one of which six statements is possible, since they respectively refer to the six classes X, Y, Z, 'X, 'Y 'Z, every one of which is understood to exist in our Universe of Discourse.

Suppose we have the major premise $z:y$ with the conclusion $(x:z')'$, and that we want to find the weakest complementary minor premise W. We are to have

$$(z:y)W:(x:z')',$$

which, by transposition and change of signs, becomes

$$(z:y)(x:z'):W'.$$

This, by the formula $a:\beta = \beta':a'$, becomes

$$(z:y)(z:x'):W'.$$

But by Formula (3) we have also

$$(z:y)(z:x'):(yx')^{-\eta}.$$

We therefore assume $W' = (yz')^{-\eta}$, and consequently $W = (yx')^\eta = y:x$. The weakest minor premise required is therefore $y:x$; and the required syllogism is

$$(z:y)(y:x):(x:z')',$$

which is the syllogism Bramantip. As the weakest premise required turns out in this case to be an *implication*, and not a *non-implication*, it is not only the weakest complementary premise required, but no other complementary premise is possible. (See § 64, second footnote.)

66. When the conclusion and given premise are both non-implications (or " particulars "), we proceed as follows. Let P' be the given non-implicational premise, and C' the non-implicational conclusion, while W denotes the required weakest complementary premise. We shall then have P'W:C' or its equivalent WC:P, which we obtain by transposition. The consequent P of the second

§§ 66, 66 (a)] THE STRONGEST CONCLUSION   69

implication being an implication (or "universal") we have only to proceed as in § 64 to find W. For example, let the given non-implicational premise be $(y:z)'$, and the given non-implicational conclusion $(x:z)'$. We are to have

$$(y:z)'\,\mathrm{W}:(x:z)'.$$

By transposition this becomes

$$\mathrm{W}(x:z):(y:z).$$

The letter missing in the consequent $y:z$ is x. The syllogism WC : P must therefore be

either $(y:x:z):(y:z)$ or else $(y:x':z):(y:z)$;

one or other of which must contain the implication C, of which the given non-implicational conclusion C′, representing $(x:z)'$, is the denial. The syllogism WC : P must therefore denote the first of these two syllogisms, and not the second; for it is the first and not the second that contains the implication C, or its synonym $x:z$. Hence $\mathrm{W}=y:x$. Now, WC : P is equivalent, by transposition, to WP′ : C′, which is the syllogism required. Substituting for W, P′, C′, we find the syllogism sought to be

$$(y:x)(y:z)' : (x:z)',$$

and the required missing minor premise to be $y:x$.

66 (a). By a similar process we find the strongest conclusion derivable from two given premises. One example will suffice. Suppose we have the combination of premises $(z:y)(x:y)'$. Let S denote the strongest conclusion required. We get

$(z:y)(x:y)'$: S, which, by transposition, is $(z:y)\mathrm{S}' : (x:y)$.

The letter missing in the implicational consequent of the second syllogism is z, so that its antecedent $(z:y)\mathrm{S}'$ must be

either $x:z:y$ or else $x:z':y$.

The first antecedent is the one that contains the factor $z:y$, so that its other factor $x:z$ must be the one denoted by S'. Hence, we get $S' = x:z$, and $S = (x:z)'$. The strongest * conclusion required is therefore $(x:z)'$.

CHAPTER X

67. WE will now introduce three new symbols, $W\phi$, $V\phi$, $S\phi$, which we define as follows. Let $A_1, A_2, A_3, \ldots A_m$ be m statements which are all possible, but of which one only is true. Out of these m statements let it be understood that $A_1, A_2, A_3, \ldots A_r$ imply (each separately) a conclusion ϕ; that $A_{r+1}, A_{r+2}, A_{r+3}, \ldots A_s$ imply ϕ'; and that the remaining statements, $A_{s+1}, A_{s+2}, \ldots A_m$ neither imply ϕ nor ϕ'. On this understanding we lay down the following definitions:—

(1) $W\phi = A_1 + A_2 + A_3 + \ldots + A_r$.
(2) $W\phi' = A_{r+1} + A_{r+2} + \ldots + A_s$.
(3) $V\phi = V\phi' = A_{s+1} + A_{s+2} + \ldots + A_m$.
(4) $S\phi = W\phi + V\phi = W\phi + V\phi'$.
(5) $S\phi' = W\phi' + V\phi' = W\phi' + V\phi$.
(6) $W'\phi$ means $(W\phi)'$, the denial of $W\phi$.
(7) $S'\phi$ means $(S\phi)'$, the denial of $S\phi$.

The symbol $W\phi$ denotes the *weakest statement* that implies ϕ; while $S\phi$ denotes the *strongest statement* that ϕ implies (see § 33, footnote). As A is stronger formally than $A + B$, while $A + B$ is formally stronger than $A + B + C$, and so on, we are justified in calling $W\phi$ the *weakest statement that implies* ϕ, and in calling $S\phi$ *the strongest statement that ϕ implies*. Generally $W\phi$ and $S\phi$

* Since here the strongest conclusion is a *non-implication*, there is no other and weaker conclusion. An *implicational* conclusion $x:z$ would also admit of the weaker conclusion $(x:z')'$.

§§ 67, 68]   EXPLANATIONS OF SYMBOLS   71

present themselves as *logical sums* or *alternatives*; but, in exceptional cases, they may present themselves as single terms. From the preceding definitions we get the formulæ, (1) $W\phi' = S'\phi$; (2) $S\phi' = W'\phi$; (3) $V^0\phi = (W\phi = S\phi = \phi)$. The last of these three formulæ asserts that to deny the existence of $V\phi$ in our arbitrary universe of admissible statements, A_1, A_2, &c., is equivalent to affirming that $W\phi$, $S\phi$, and ϕ are all three equivalent, each implying the others. The statement $V^0\phi$, which means $(V\phi)^0$, is not synonymous with $V^\eta\phi$; the former asserts that $V\phi$ is *absent from a certain list* A_1, A_2, ... A_m, which constitutes our universe of intelligible statements; whereas $V^\eta\phi$, which means $(V\phi)^\eta$, assumes the existence of the statement $V\phi$ in this list, and asserts that *it is an impossibility*, or, in other words, that it contradicts our data or definitions. The statement $V^0\phi$ *may* be true; the statement $V^\eta\phi$ *cannot* be true. The statement $V^0\phi$ is true when, as sometimes happens, every term of the series A_1, A_2, ... A_m either implies ϕ or implies ϕ'. The statement $V^\eta\phi$ is necessarily false, because it asserts that $V\phi$, *which by definition neither implies ϕ nor ϕ'*, is a statement of the class η; whereas *every statement of the class η implies both ϕ and ϕ'*, since (as proved in § 18) the implication $\eta : a$ is always true, whatever be the statement represented by a. The statement $V^\eta\phi$ also contradicts the convention laid down that all the statements A_1, A_2, ... A_m are possible. Similarly, we may have $W^0\phi$ or $W^0\phi'$.

68. The following examples will illustrate the meanings of the three symbols $W\phi$, $V\phi$, $S\phi$. Suppose our total (or "universe") of possible hypotheses to consist of the nine terms resulting from the multiplication of the two certainties $A^\epsilon + A^\eta + A^\theta$ and $B^\epsilon + B^\eta + B^\theta$. The product is

$$A^\epsilon B^\epsilon + A^\epsilon B^\eta + A^\epsilon B^\theta + A^\eta B^\epsilon + A^\eta B^\eta + A^\eta B^\theta \\ + A^\theta B^\epsilon + A^\theta B^\eta + A^\theta B^\theta.$$

Let ϕ denote $(AB)^\theta$. We get

(1) $W(AB)^\theta = A^\epsilon B^\theta + A^\theta B^\epsilon$.
(2) $S(AB)^\theta = A^\epsilon B^\theta + A^\theta B^\epsilon + A^\theta B^\theta = A^{-\eta}B^\theta + A^\theta B^{-\eta}$.
(3) $W(AB)^{-\theta} = S'(AB)^\theta = A^\eta + B^\eta + A^\epsilon B^\epsilon$. (See § 69.)
(4) $S(AB)^{-\theta} = W'(AB)^\theta = A^\eta + B^\eta + A^\epsilon B^\epsilon + A^\theta B^\theta$. (See § 69.)

The first of the above formulæ asserts that the weakest data from which we can conclude that AB is a *variable* is the alternative $A^\epsilon B^\theta + A^\theta B^\epsilon$, which affirms that *either* A *is certain and* B *variable, or else* A *variable and* B *certain*. The second formula asserts that the strongest conclusion we can draw from the statement that AB is a variable is the alternative $A^{-\eta}B^\theta + A^\theta B^{-\eta}$, which asserts that *either* A *is possible and* B *variable, or else* A *variable and* B *possible*. Other formulæ which can easily be proved, when not evident by inspection, are the following:—

(5) $W\phi : \phi : S\phi$.
(6) $(W\phi = S\phi) = (W\phi = \phi)(S\phi = \phi)$.
(7) $W(AB)^\epsilon = A^\epsilon B^\epsilon = S(AB)^\epsilon$.
(8) $W(A+B)^\epsilon = A^\epsilon + B^\epsilon$.
(9) $S(A+B)^\epsilon = A^\epsilon + B^\epsilon + A^\theta B^\theta$.
(10) $W(A+B)^\eta = A^\eta B^\eta = S(A+B)^\eta = (A+B)^\eta$.
(11) $W(A+B)^\theta = A^\eta B^\theta + A^\theta B^\eta$.
(12) $S(A+B)^\theta = A^{-\epsilon}B^\theta + A^\theta B^{-\epsilon}$.
(13) $W(AB)^\eta = A^\eta + B^\eta$.
(14) $S(AB)^\eta = A^\eta + B^\eta + A^\theta B^\theta$.
(15) $W(A:B) = W(AB')^\eta = A^\eta + B^\epsilon$.
(16) $S(A:B) = S(AB')^\eta = A^\eta + B^\epsilon + A^\theta B^\theta$.
(17) $W(A:B)' = S'(A:B) = A^\epsilon B^{-\epsilon} + A^{-\eta}B^\eta$.
(18) $S(A:B)' = W'(A:B) = A^{-\eta}B^{-\epsilon}$.

The formulæ (15) and (16) may evidently be deduced from (13) and (14) by changing B into B′. Formula (17) asserts that the weakest data from which we can

§§ 68, 69]   APPLICATIONS OF SYMBOLS   73

conclude that A does not imply B is the alternative that *either* A *is certain and* B *uncertain, or else* A *possible and* B *impossible*. The formula may be proved as follows:

$$W(A:B)' = S'(A:B) = (A^\eta + B^\epsilon + A^\theta B^\theta)' = (A^\eta)'(B^\epsilon)'(A^\theta B^\theta)'$$
$$= A^{-\eta} B^{-\epsilon}(A^{-\theta} + B^{-\theta}) = A^\epsilon B^{-\epsilon} + A^{-\eta} B^\eta;$$

for, evidently, $A^{-\eta} A^{-\theta} = A^\epsilon$, and $B^{-\epsilon} B^{-\theta} = B^\eta$.

69. All the formulæ of § 68 may be proved from first principles, though some may be deduced more readily from others. Take, for example, (1), (2), (3). We are required to find $W(AB)^\theta$, $S(AB)^\theta$, $W(AB)^{-\theta}$. We first write down the nine terms which constitute the product of the two certainties $A^\epsilon + A^\eta + A^\theta$ and $B^\epsilon + B^\eta + B^\theta$, as in § 68. This done, we *underdot* every term that implies $(AB)^\theta$, which asserts that AB is a variable; we *underline* every term that implies $(AB)^{-\theta}$, which asserts that AB is *not* a variable; and we enclose in brackets every term that neither implies $(AB)^\theta$ nor $(AB)^{-\theta}$. We thus get

$$\underline{A^\epsilon B^\epsilon} + \underline{A^\epsilon B^\eta} + \dot{A^\epsilon B^\theta} + \underline{A^\eta B^\epsilon} + \underline{A^\eta B^\eta} + A^\eta B^\theta$$
$$+ \dot{A^\theta B^\epsilon} + A^\theta B^\eta + (A^\theta B^\theta).$$

By our definitions in § 67 we thus have

$$W(AB)^\theta = A^\epsilon B^\theta + A^\theta B^\epsilon \quad \ldots \ldots \quad (1)$$

By definition also we have $V(AB)^\theta = A^\theta B^\theta$, and therefore

$$S(AB)^\theta = W(AB)^\theta + V(AB)^\theta = A^\epsilon B^\theta + A^\theta B^\epsilon + A^\theta B^\theta$$
$$= A^\epsilon B^\theta + A^\theta B^\epsilon + A^\theta B^\theta, \text{ for } a = a + a$$
$$= (A^\epsilon + A^\theta) B^\theta + A^\theta (B^\epsilon + B^\theta) = A^{-\eta} B^\theta$$
$$+ A^\theta B^{-\eta} \quad \ldots \ldots \quad (2).$$

We may similarly deduce (3) and (4) from first principles, but they may be deduced more easily from the two formulæ

$$W(\phi + \psi) = W\phi + W\psi \quad \ldots \ldots \quad (\alpha)$$
$$S(\phi + \psi) = S\phi + S\psi \quad \ldots \ldots \quad (\beta),$$

as follows:

$$W(AB)^{-\theta} = W\{(AB)^\epsilon + (AB)^\eta\} = W(AB)^\epsilon + W(AB)^\eta$$
$$= A^\epsilon B^\epsilon + A^\eta + B^\eta, \text{ from § 68, Formulæ 7, 13.}$$
$$S(AB)^{-\theta} = S\{(AB)^\epsilon + (AB)^\eta\} = S(AB)^\epsilon + S(AB)^\eta$$
$$= A^\epsilon B^\epsilon + A^\eta + B^\eta + A^\theta B^\theta, \text{ from § 68, Formulæ 7, 14.}$$

70. The following is an example of inductive, or rather inverse, implicational reasoning (see §§ 11, 112).

The formula $(A:x)+(B:x):(AB:x)$ is always true; when (if ever) is the converse, implication $(AB:x):(A:x)+(B:x)$, false? Let ϕ denote the first and valid formula, while ϕ_c denotes its converse formula to be examined. We get

$$\phi_c = (ABx')^\eta : (Ax')^\eta + (Bx')^\eta$$
$$= (Ax' . Bx')^\eta : (Ax')^\eta + (Bx')^\eta$$
$$= (\alpha\beta)^\eta : \alpha^\eta + \beta^\eta, \text{ putting } \alpha \text{ for } Ax', \text{ and } \beta \text{ for } Bx'.$$

Hence (see § 11), we get

$$\phi_c' \,!\, (\alpha\beta)^\eta(\alpha^\eta + \beta^\eta)' \,!\, (\alpha\beta)^\eta \alpha^{-\eta}\beta^{-\eta} \,!\, (\alpha\beta)^\eta \alpha^\theta \beta^\theta$$
$$!\, (Ax' . Bx')^\eta (Ax')^\theta (Bx')^\theta \,!\, (ABx')^\eta (Ax')^\theta (Bx')^\theta$$

Thus, the converse implication ϕ_c fails in the case $(\alpha\beta)^\eta \alpha^{-\eta}\beta^{-\eta}$, which represents the statement

$$(ABx')^\eta (Ax')^{-\eta}(Bx')^{-\eta} \quad \ldots \quad (1);$$

and it therefore also fails in the case $(\alpha\beta)^\eta \alpha^\theta \beta^\theta$, which represents the statement

$$(ABx')^\eta (Ax')^\theta (Bx')^\theta \quad \ldots \quad (2);$$

for the second statement implies the first. The failure of ϕ_c in the second may be illustrated by a diagram as on opposite page.

Out of the total ten points marked in this diagram, take a point P at random, and let the three symbols A, B, x assert respectively (as propositions) that the

§§ 70, 71]   CERTAIN DISPUTED PROBLEMS   75

point P will be in the circle A, that P will be in the circle B, that P will be in the ellipse x. It is evident that the respective chances of the four propositions A, B, x, AB are $\frac{5}{10}$, $\frac{5}{10}$, $\frac{6}{10}$, $\frac{2}{10}$; so that they are all variables. It is also clear that the respective chances of the three statements ABx', Ax', Bx', are 0, $\frac{2}{10}$, $\frac{2}{10}$; so that we also have $(ABx')^\eta(Ax')^\theta(Bx')^\theta$, which, by pure symbolic reasoning, we found to be a case of failure. We may also show this by direct appeal to the diagram, as follows. The implication AB: x asserts that the point P cannot be in both the circles A and B without being also in the ellipse x, a statement which is a *material certainty*, as it follows necessarily from the 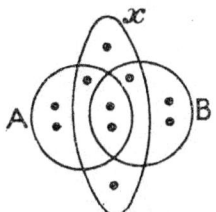 special data of our diagram (see § 109). The implication A: x asserts that P cannot be in A without being in x, a statement which is a *material impossibility*, as it is inconsistent with the data of our diagram; and B: x is *impossible* for the same reason. Thus we have AB: $x=\epsilon$, A: $x=\eta$, B: $x=\eta$, so that we get

$$\phi = (A:x) + (B:x) : (AB:x) = \eta + \eta : \epsilon = \epsilon$$
$$\phi_c = (AB:x) : (A:x) + (B:x) = \epsilon : \eta + \eta = \eta.$$

The Boolian logicians consider ϕ and ϕ_c equivalent, because they draw no distinction between the *true* (τ) and the *certain* (ϵ), nor between the *false* (ι) and the *impossible* (η). Every proposition is with them either *certain* or *impossible*, the propositions which I call *variables* (θ) being treated as non-existent. The preceding illustration makes it clear that this is a serious and fundamental error.

71. The diagram above will also illustrate two other propositions which by most logicians are considered equivalent, but which, according to my interpretation of the word *if*, are *not* equivalent. They are the complex

conditional, "*If* A *is true, then if* B *is true x is true*," and the simple conditional, "*If* A *and* B *are both true, then* x *is true.*" Expressed in my notation, and with my interpretation of the conjunction *if* (see § 10), these conditionals are respectively A : (B : x) and AB : x. Giving to the propositions A, B, x, AB the same meanings as in § 70 (all having reference to the same subject, the random point P), it is evident that B : x, which asserts that the random point P cannot be in the circle B without being also in the ellipse x, *contradicts our data*, and is therefore *impossible*. The statement A, on the other hand, does *not* contradict our data; neither does its denial A′, for both, in the given conditions, are possible though uncertain. Hence, A is a *variable*, and B : x being *impossible*, the complex conditional A : (B : x) becomes $\theta : \eta$, which is equivalent to θ^η, and therefore an impossibility. But the simple conditional AB : x, instead of being impossible, is, in the given conditions, a *certainty*, for it is clear from the figure that P cannot be in both A and B without being also in x. Hence, though A : (B : x) always implies AB : x, the latter does not always imply the former, so that the two are not, in all cases, equivalent.

72. A question much discussed amongst logicians is the "Existential Import of Propositions." When we make an affirmation A^B, or a denial A^{-B}, do we, at the same time, implicitly affirm the existence of A? Do we affirm the existence of B? Do the four technical propositions of the traditional logic, namely, "All A is B," "No A is B," "Some A is B," "Some A is not B," taking each separately, necessarily imply the existence of the class A? Do they necessarily imply the existence of the class B? My own views upon this question are fully explained in *Mind* (see vol. xiv., N.S., Nos. 53–55); here a brief exposition of them will suffice. The convention of a "Symbolic Universe" (see §§ 46–50) necessarily leads to the following conclusions:—

§§ 72, 73] EXISTENTIAL IMPORT 77

Firstly, when any symbol A denotes an *individual;* then, any intelligible statement $\phi(A)$, containing the symbol A, implies that the individual represented by A has a *symbolic* existence; but whether the statement $\phi(A)$ implies that the individual represented by A has a *real* existence depends upon the context.

Secondly, when any symbol A denotes a *class*, then, any intelligible statement $\phi(A)$ containing the symbol A implies that the whole class A has a *symbolic* existence; but whether the statement $\phi(A)$ implies that the class A is *wholly real*, or *wholly unreal*, or *partly real and partly unreal*, depends upon the context.

As regards this question of "Existential Import," the one important point in which I appear to differ from other symbolists is the following. The null class 0, which they define as *containing no members*, and which I, for convenience of symbolic operations, define as consisting of the null or unreal members $0_1, 0_2, 0_3$, &c., is understood by them to be *contained in every class*, real or unreal; whereas I consider it to be *excluded from every real class*. Their convention of universal inclusion leads to awkward paradoxes, as, for example, that "Every round square is a triangle," because round squares form a *null* class, which (by them) is understood to be *contained in every class*. My convention leads, in this case, to the directly opposite conclusion, namely, that "No round square is a triangle," because I hold that every purely *unreal* class, such as the class of round squares, is necessarily excluded from every purely *real* class, such as the class of figures called triangles.

73. Another paradox which results from this convention of universal inclusion as regards the null class 0, is their paradox that the two universals "All X is Y" and "No X is Y" are mutually compatible; that it is possible for both to be true at the same time, and that this is necessarily the case when the class X is null or non-existent. My convention of a "Symbolic Universe"

leads, on the contrary, to the common-sense conclusion of the traditional logic that the two propositions "All X is Y" and "No X is Y" are *incompatible*. This may be proved formally as follows. Let ϕ denote the proposition to be proved. We have

$$\phi = (x:y)(x:y') : \eta = (xy')^{\eta}(xy)^{\eta} : \eta$$
$$= (xy' + xy : \eta) : \eta = \{x(y' + y) : \eta\} : \eta$$
$$= (x\epsilon : \eta) : \eta = (x : \eta) : \eta = (\theta : \eta) : \eta$$
$$= \eta : \eta = (\eta\eta')^{\eta} = \epsilon.$$

In this proof the statement x is assumed to be a variable by the convention of § 46. See also § 50. It will be noticed that ϕ, the proposition just proved, is equivalent to $(x:y):(x:y')'$, which asserts that "All X is Y" implies "Some X is Y."

74. Most symbolic logicians use the symbol A⤙B, or some other equivalent (such as Schroeder's A⊈B), to assert that *the class* A *is wholly included in the class* B; and they imagine that this is virtually equivalent to my symbol A:B, which asserts that *the statement* A *implies the statement* B. That this is an error may be proved easily as follows. If the statement A:B be always equivalent to the statement A⤙B, the equivalence must hold good when A denotes η, and B denotes ϵ. Now, the statement $\eta:\epsilon$, by definition, is synonymous with $(\eta\epsilon')^{\eta}$, which only asserts the truism that the impossibility $\eta\epsilon'$ is an impossibility. (For the compound statement ηa, whatever a may be, is clearly an impossibility because it has an impossible factor η.) But by their definition the statement $\eta \prec \epsilon$ asserts that the class η is wholly included in the class ϵ; that is to say, it asserts that every individual *impossibility*, η_1, η_2, η_3, &c., of the class η is also an individual (either ϵ_1, or ϵ_2, or ϵ_3, &c.) of the class of *certainties* ϵ; which is absurd. Thus, $\eta:\epsilon$ is a formal certainty, whereas $\eta \prec \epsilon$ is a formal impossibility. (See § 18.)

§ 75]   CLASS INCLUSION AND IMPLICATION   79

75. Some logicians (see § 74) have also endeavoured to drag my formula

$$(A:B)(B:C):(A:C) \quad . \quad . \quad . \quad . \quad . \quad (1)$$

into their systems under some disguise, such as

$$(A \prec B)(B \prec C) \prec (A \prec C) \quad . \quad . \quad . \quad . \quad (2).$$

The meaning of (1) is clear and unambiguous; but how can we, without having recourse to some distortion of language, extract any sense out of (2)? The symbol $A \prec B$ (by virtue of their definition) asserts that every individual of the class A is also an individual of the class B. Consistency, therefore, requires that the complex statement (2) shall assert that every individual of the class $(A \prec B)(B \prec C)$ is also an individual of the class $(A \prec C)$. But how can the double-factor compound *statement* $(A \prec B)(B \prec C)$ be intelligibly spoken of as a *class* contained in the single-factor *statement* $(A \prec C)$? It is true that the compound statement $(A \prec B)(B \prec C)$ *implies* the single statement $(A \prec C)$, an implication expressed, *not* by their formula (2) but by

$$(A \prec B)(B \prec C):(A \prec C) \quad . \quad . \quad . \quad . \quad (3);$$

but that is quite another matter. The two formulæ (1) and (3) are both valid, though not synonymous; whereas their formula (2) cannot, without some arbitrary departure from the accepted conventions of language, be made to convey any meaning whatever.

The inability of other systems to express the new ideas represented by my symbols A^{xy}, A^{xyz}, &c., may be shown by a single example. Take the statement $A^{\theta\theta}$. This (unlike *formal certainties*, such as ϵ^τ and $AB:A$, and unlike *formal impossibilities*, such as θ^ϵ and $\theta:\eta$) may, in my system, be a *certainty*, an *impossibility*, or a *variable*, according to the special data of our problem or investigation (see §§ 22, 109). But how could the proposition $A^{\theta\theta}$ be expressed in other systems? In these it could not be expressed at all, for its recognition would involve the abandonment of their erroneous and unworkable hypothesis (assumed always) that *true* is synonymous with *certain*, and *false* with *impossible*. If they ceased to consider their A (when it denotes a proposition) as equivalent to their (A=1), and their A' (or their corresponding symbol for a denial) as equivalent to their (A=0), and if they employed their symbol (A=1) in the sense of my symbol A^ε, and their symbol (A=0) in the sense of my symbol A^η, they *might* then express my statement $A^{\theta\theta}$ in their notation; but the expression would be extremely long and intricate. Using A≠B (in accordance with usage) as the denial of (A=B), my statement A^θ would then be expressed by (A≠0)(A≠1), and my $A^{\theta\theta}$ by

$$\{(A\neq 0)(A\neq 1)\neq 0\}\{(A\neq 0)(A\neq 1)\neq 1\}.$$

This example of the difference of notations speaks for itself.

CHAPTER XI

76. LET A denote the premises, and B the conclusion, of any argument. Then A ∴ B ("A is true, *therefore* B is true"), or its synonym B ∵ A ("B is true *because* A is true"), each of which synonyms is equivalent to $A(AB')^\eta$, denotes the argument. That is to say, the argument asserts, firstly, that the statement (or collection of statements) A is true, and, secondly, that the affirmation of A coupled with the denial of B constitutes an *impossibility*, that is to say, a statement that is incompatible with our data or definitions. When the person to whom the argument is addressed believes in the truth of the statements A and $(AB')^\eta$, he considers the argument valid; if he disbelieves either, he considers the argument invalid. This does not necessarily imply that he dis-

§§ 76, 77]   'BECAUSE' AND 'THEREFORE'   81

believes either the premises A or the conclusion B; he may be firmly convinced of the truth of both without accepting the validity of the argument. For the truth of A coupled with the truth of B does not necessarily imply the truth of the proposition $(AB')^\eta$, though it does that of $(AB')^\iota$. The statement $(AB')^\iota$ is *equivalent* to $(AB')'$ (see § 23) and therefore to $A' + B$. Hence we have

$$A(AB')^\iota = A(A' + B) = AB = A^\tau B^\tau.$$

But A∴B, like its synonym $A(AB')^\eta$, asserts more than $A^\tau B^\tau$. Like $A(AB')^\iota$, it asserts that A is true, but, unlike $A(AB')^\iota$, it asserts not only that AB' is *false*, but that it is *impossible*—that it is incompatible with our data or definitions. For example, let $A = He\ turned\ pale$, and let $B = He\ is\ guilty$. Both statements may happen to be true, and then we have $A^\tau B^\tau$, which, as just shown, is equivalent to $A(AB')^\iota$; yet the argument A∴B ("He turned pale; *therefore* he is guilty") is not valid, for though the weaker statement $A(AB')^\iota$ happens on this occasion to be true, the stronger statement $A(AB')^\eta$ is not true, because of its false second factor $(AB')^\eta$. I call this factor false, because it asserts not merely $(AB')^\iota$, that it is *false* that he turned pale without being guilty, an assertion which may be true, but also $(AB')^\eta$, that it is impossible he should turn pale without being guilty, an assertion which is *not* true.

77. The convention that A∴B shall be considered equivalent to $A(A:B)$, and to its synonym $A(AB')^\eta$, obliges us however to accept the argument A∴B as valid, even when the only bond connecting A and B is the fact that they are both certainties. For example, let A denote the statement $13 + 5 = 18$, and let B denote the statement $4 + 6 = 10$. It follows from our symbolic conventions that in this case A∴B and B∴A are both valid. Yet here it is not easy to discover any bond of connexion between the two statements A and B; we know the truth of each statement independently of

F

all consideration of the other. We might, it is true, give the *appearance* of logical deduction somewhat as follows :—

By our data, $13+5=18$. From each of these equals take away 9. This gives us (subtracting the 9 from the 13) $4+5=9$. To each of these equals add 1 (adding the 1 to the 5). We then, finally, get $4+6=10$; *quod erat demonstrandum*.

Every one must feel the unreality (from a psychological point of view) of the above argument; yet much of our so-called 'rigorous' mathematical demonstrations are on lines not very dissimilar. A striking instance is Euclid's demonstration of the proposition that any two sides of a triangle are together greater than the third—a proposition which the Epicureans derided as patent even to asses, who always took the shortest cut to any place they wished to reach. As marking the difference between $A \therefore B$ and its implied factor $A:B$, it is to be noticed that though $A:\epsilon$ and $\eta:A$ are formal certainties (see § 18), neither of the two other and stronger statements, $A \therefore \epsilon$ and $\eta \therefore A$, can be accepted as valid. The first evidently fails when $A = \eta$, and the second is always false; for $\eta \therefore x$, like its synonym $\eta(\eta:x)$, is false, because, though its second factor $\eta:x$ is necessarily true, its first factor η is necessarily false by definition.

78. Though in purely formal or symbolic logic it is generally best to avoid, when possible, all psychological considerations, yet these cannot be wholly thrust aside when we come to the close discussion of first principles, and of the exact meanings of the terms we use. The words *if* and *therefore* are examples. In ordinary speech, when we say, "*If* A is true, then B is true," or "A is true, *therefore* B is true," we suggest, if we do not positively affirm, that the knowledge of B depends in some way or other upon previous knowledge of A. But in formal logic, as in mathematics, it is convenient, if not absolutely necessary, to work with symbolic statements

CAUSE AND EFFECT

whose truth or falsehood in no way depends upon the mental condition of the person supposed to make them. Let us take the extreme case of crediting him with absolute omniscience. On this hypothesis, the word *therefore*, or its symbolic equivalent $\therefore$, would, from the *subjective* or *psychological* standpoint, be as meaningless, in no matter what argument, as we feel it to be in the argument $(7 \times 9 = 63)$ *therefore* $(2 + 1 = 3)$; for, to an omniscient mind all true theorems would be equally self-evident or axiomatic, and proofs, arguments, and logic generally would have no *raison d'être*. But when we lay aside psychological considerations, and define the word 'therefore,' or its synonym $\therefore$, as in § 76, it ceases to be meaningless, and the seemingly meaningless argument, $(7 \times 9 = 63) \therefore (2 + 1 = 3)$, becomes at once clear, definite, and a formal certainty.

79. In order to make our symbolic formulæ and operations as far as possible independent of our changing individual opinions, we will arbitrarily lay down the following definitions of the word 'cause' and 'explanation.' Let A, as a statement, be understood to assert the existence of the circumstance A, or the occurrence of the event A, while V asserts the posterior or simultaneous occurrence of the event V; and let both the statement A and the implication A:V be true. In these circumstances A is called a *cause* of V; V is called the *effect* of A; and the symbol A(A:V), or its synonym A$\therefore$V, is called an *explanation* of the event or circumstance V. To possess an explanation of any event or phenomenon V, we must therefore be in possession of two pieces of knowledge: we must know the existence or occurrence of some cause A, and we must know the law or implication A:V. The product or combination of these two factors constitute the argument A$\therefore$V, which is an explanation of the event V. We do not call A *the* cause of V, nor do we call the argument A$\therefore$V *the* explanation of V, because we may have also

B ∴ V, in which case B would be another sufficient cause of V, and the argument B ∴ V another sufficient explanation of V.

80. Suppose we want to discover the cause of an event or phenomenon x. We first notice (by experiment or otherwise) that x is invariably found in each of a certain number of circumstances, say A, B, C. We therefore *provisionally* (till an exception turns up) regard each of the circumstances A, B, C as a sufficient cause of x, so that we write $(A:x)(B:x)(C:x)$, or its equivalent $A+B+C:x$. We must examine the different circumstances A, B, C to see whether they possess some circumstance or factor in common which might *alone* account for the phenomena. Let us suppose that they do have a common factor f. We thus get (see § 28)

$$(A:f)(B:f)(C:f), \text{ which} = A+B+C:f.$$

We before possessed the knowledge $A+B+C:x$, so that we have now

$$A+B+C:fx.$$

If f be not posterior to x, we may suspect it to be *alone* the real cause of x. Our next step should be to seek out some circumstance a which is consistent with f, but not with A or B or C; that is to say, some circumstance a which is sometimes found associated with f, but not with the co-factors of f in A or B or C. If we find that fa is invariably followed by x—that is to say, if we discover the implication $fa:x$—then our suspicion is confirmed that the reason why A, B, C are each a sufficient cause of x is to be found in the fact that each contains the factor f, which may therefore be provisionally considered as alone, and independently of its co-factors, a sufficient cause of x. If, moreover, we discover that while on the one hand fa implies x, on the other $f'a$ implies x'; that is to say, if we discover $(fa:x)(f'a:x')$, our suspicion that f alone is the cause of x is confirmed

§ 80] CAUSE AND EFFECT

still more strongly. To obtain still stronger confirmation we vary the circumstances, and try other factors, β, γ, δ, consistent with f, but inconsistent with A, B, C and with each other. If we similarly find the same result for these as for α; so that

$$(f\alpha : x)(f'\alpha : x'), \text{ which} = f\alpha : x : f + \alpha'$$
$$(f\beta : x)(f'\beta : x'), \text{ which} = f\beta : x : f + \beta'$$
$$(f\gamma : x)(f'\gamma : x'), \text{ which} = f\gamma : x : f + \gamma'$$
$$(f\delta : x)(f'\delta : x'), \text{ which} = f\delta : x : f + \delta'$$

our conviction that f alone is a sufficient cause of x receives stronger and stronger confirmation. But by no inductive process can we reach absolute *certainty* that f is a sufficient cause of x, when (as in the investigation of natural laws and causes) the number of hypotheses or possibilities logically consistent with f are unlimited; for, eventually, some circumstance q may turn up such that fq does *not* imply x, as would be proved by the actual occurrence of the combination fqx'. Should this combination ever occur—and in natural phenomena it is always *formally possible*, however antecedently improbable—the supposed law $f : x$ would be at once disproved. For, since, by hypothesis, the unexpected combination fqx' has actually occurred, we may add this fact to our data ϵ_1, ϵ_2, ϵ_3, &c.; so that we get

$$\epsilon : fqx' : (fqx')^{-\eta} : (fx')^{-\eta} : (f : x)'.$$

This may be read, "It is certain that fqx' has occurred. The occurrence fqx' implies that fqx' is possible. The possibility of fqx' implies the possibility of fx'; and the *possibility* of fx' implies the denial of the implication $f : x$."

The inductive method here described will be found, upon examination, to include all the essential principles of the methods to which Mill and other logicians have given the names of 'Method of Agreement' and 'Method of Difference' (see § 112).

CHAPTER XII

We will now give symbolic solutions of a few miscellaneous questions mostly taken from recent examination papers.

81. Test the validity of the reasoning, "All fairies are mermaids, for neither fairies nor mermaids exist."

Speaking of anything S taken at random out of our symbolic universe, let $f =$ "It is a *fairy*," let $m =$ "it is a mermaid," and let $e =$ "it exists." The *implication* of the argument, in symbolic form, is

$$(f:e')(m:e'):(f:m)$$
$$\text{which} = (f:e')(e:m'):(f:m).$$

Since the conclusion $f:m$ is a "universal" (or implication), the premises of the syllogism, if valid, must (see § 59) be either $f:e:m$ or $f:e':m$. This is not the case, so that the syllogism is not valid. Of course, 0 may replace e'.

Most symbolic logicians, however, would consider this syllogism valid, as they would reason thus: "By our data, $f = 0$ and $m = 0$; therefore $f = m$. Hence, all fairies are mermaids, and all mermaids are fairies" (see § 72).

82. Examine the validity of the argument: "It is not the case that any metals are compounds, and it is incorrect to say that every metal is heavy; it may therefore be inferred that some elements are not heavy, and also that some heavy substances are not metals."

Let $e =$ "it is an *element*" = "it is *not* a compound"; let $m =$ "it is a *metal*"; and let $h =$ "it is *heavy*."

The above argument, or rather implication (always supposing the word "*If*" understood before the premises) is

$$(m:e)(m:h)':(e:h)'(h:m)'.$$

Let $A = m:e$, let $B = m:h$, let $C = e:h$, let $D = h:m$, and

§§ 82, 83]   MISCELLANEOUS EXAMPLES   87

let ϕ denote the implication of the given argument. We then get

$$\phi = AB' : C'D' = (AB' : C')(AB' : D'),$$
since $x : yz = (x : y)(x : z)$.

In order that ϕ may be valid, the two implications $AB' : C'$ and $AB' : D'$ must both be valid. Now, we have (see § 59)

$$AB' : C' = AC : B = (m : e)(e : h) . (m : h),$$

which is valid by § 56. Hence, C′, which asserts $(e : h)'$, that "some elements are not heavy" is a legitimate conclusion from the premises A and B′. We next examine the validity of the implication $AB' : D'$. We have

$$AB' : D' = (m : e)(m : h)' : (h : m)'.$$

Now, this is not a syllogism at all, for the middle term m, which appears in the two premises, appears also in the conclusion. Nor is it a valid implication, as the subjoined figure will show. Let the eight points in the circle m constitute the class m; let the twelve points in the circle e constitute the class e; and let the five points in the circle h constitute the class h. Here, the premises "Every m is e, and some m is not h" are both true; yet the conclusion, "Some h is not m," is false.

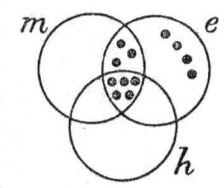

Hence, though the conclusion C′ is legitimate, the conclusion D′ is not.

83. Examine the argument, "No young man is wise; for only experience can give wisdom, and experience comes only with age."

Let $y =$ "he is *young*"; let $w =$ "he is *wise*"; and let $e =$ "he has had *experience*." Also, let ϕ denote the implication factor of the given argument. We have

$$\phi = (e' : w')(y : e') : (y : w') = (y : e' : w') : (y : w').$$

88   SYMBOLIC LOGIC   [§§ 83–85

The given implication is therefore valid (see §§ 11, 56, 59).

84. Examine the argument, " His reasoning was correct, but as I knew his conclusion to be false, I was at once led to see that his premises must be false also."

Let P = " his *premises* were true," and let C = " his *conclusion* was true." Then P : C = " his reasoning (or rather implication) was valid." Let ϕ denote the *implication* of the argument to be examined. We get (see § 105)

$$\phi = (P:C)C' : P'$$
= the valid form of the *Modus tollendo tollens*.

Thus interpreted ϕ is valid. But suppose the word " premises " means P *and* Q, and not a *single compound statement* P. We then get

$$\phi = (PQ:C)C' : P'Q' \, ;$$

an interpretation which fails in the case $C^\eta P^\epsilon Q^\eta$, and also in the case $C^\eta P^\eta Q^\epsilon$. To prove its failure in the latter case, we substitute for C, P, Q their respective exponential values η, η, ϵ, and thus get

$$\phi = (\eta\epsilon : \eta)\eta' : \eta'\epsilon' = (\eta : \eta)\epsilon : \epsilon\eta = \epsilon\epsilon : \eta = \eta.$$

85. Supply the missing premise in the argument: " Not all mistakes are culpable; for mistakes are sometimes quite unavoidable."

Let m = " it is a *mistake*," let c = " it is *culpable*," let u = " it is *unavoidable*," and let ϕ denote the implication of the argument. Putting Q for the missing premise, we get (see §§ 59, 64)

$$\phi = (m:u')'Q : (m:c)' = (m:c)Q : (m:u').$$

For this last implication to be valid (see § 64), we must have its premises (or antecedent) either in the form

$m : c : u'$, or else in the form $m : c' : u'$.

The first form contains the antecedent premise $m : c$; the

§§ 85-87]   MISCELLANEOUS EXAMPLES   89

second form does not. The first form is therefore the one to be taken, and the complete syllogism is

$$(m : c : u') : (m : u'),$$

the missing premise Q being $c : u'$, which asserts that "nothing *culpable* is *unavoidable.*" The original reasoning in its complete form should therefore be, "Since mistakes are sometimes unavoidable, and nothing culpable is unavoidable, some mistakes are not culpable."

86. Supply the missing premise in the argument, "Comets must consist of heavy matter; for otherwise they would not obey the law of gravitation."

Let $c =$ "it is a *comet*," let $h =$ "it consists of *heavy matter*," and let $g =$ "it obeys the law of *gravitation.*" Putting ϕ for the implication of the argument, and Q for the missing premise understood, we get

$$\phi = (h' : g')Q : (c : h) = (c : g : h) : (c : h),$$

by application of § 64; for $g : h = h' : g'$, so that the missing minor premise Q understood is $c : g$, which asserts that "all comets obey the law of gravitation." The full reasoning is therefore (see § 11)

$$(c : h) \,!\, (c : g)(g : h),$$

or its equivalent (see § 11)

$$(c : g)(g : h) : (c : h).$$

In the first form it may be read, "*Comets* consist of *heavy* matter; for all *comets* obey the law of *gravitation*, and everything that obeys the law of *gravitation* consists of *heavy* matter."

87. Supply the missing proposition which will make the following enthymeme* into a valid syllogism: "Some professional men are not voters, for every voter is a householder."

Let P = "he is a *professional* man," let V = "he is a

---
* An *enthymeme* is a syllogism incompletely stated.

voter," and let H = "he is a *householder*." Let ϕ denote the implication of the argument, and W the weakest additional premise required to justify the conclusion. We have (see § 11)

$$\phi = (P:V)'\,!\,(V:H)W = (V:H)W : (P:V)'$$
$$= (P:V)(V:H) : W' = (P:V:H) : W'.$$

The strongest conclusion deducible from P : V : H is P : H. We therefore assume P : H = W', and consequently W = (P : H)', which is therefore the weakest premise required. The complete argument is therefore this: "Some professional men are not voters, for every voter is a householder, and some professional men are not householders."

88. Put the following argument into syllogistic form, and examine its validity: "The absence of all trace of paraffin and matches, the constant accompaniments of arson, proves that the fire under consideration was not due to that crime."

Let F = "it was the *fire* under consideration"; let A = "it was due to *arson*"; let T = "it left a *trace* of paraffin and matches"; and let ϕ denote the implication of the given argument. We get

$$\phi = (F:T')(A:T) : (F:A') = (F:T')(T':A') : (F:A')$$
$$= (F:T':A') : (F:A').$$

The implication of the given argument is therefore valid.

The argument might also be expressed unsyllogistically (in the *modus tollendo tollens*) as follows (see § 105). Let T = "the fire left a *trace* of paraffin and matches"; let A = "the fire was due to the crime of *arson*"; and let ϕ denote the implication of the argument. We get (see § 105)

$$\phi = T'(A:T) : A'$$

which is the valid form of the *Modus tollendo tollens*.

89. Put the following argument into syllogistic form: "How can any one maintain that pain is always an evil,

§§ 89, 90]   TECHNICAL WORDS EXPLAINED   91

who admits that remorse involves pain, and yet may sometimes be a real good?"

Let R = "It is *remorse*"; let P = "it causes *pain*"; let E = "it is an *evil*"; and let ϕ denote the implication of the argument. We get (as in Figure 3, Bokardo)

$$\phi = (R : P)(R : E)' : (P : E)'$$
$$= (R : P)(P : E) : (R : E),$$

which is a syllogism of the Barbara type. But to reduce the reasoning to syllogistic form we have been obliged to consider the premise, "Remorse may sometimes be a real *good*," as equivalent to the weaker premise $(R : E)'$, which only asserts that "Remorse is not necessarily an *evil*." As, however, the reasoning is valid when we take the weaker premise, it must remain valid when we substitute the stronger premise; only in that case it will not be strictly syllogistic.

CHAPTER XIII

IN this chapter will be given definitions and explanations of some technical terms often used in treatises on logic.

90. *Sorites.*—This is an extension of the syllogism Barbara. Thus, we have

$$\text{Barbara} = (A : B : C) : (A : C)$$
$$(\text{Sorites})_1 = (A : B : C : D) : (A : D)$$
$$(\text{Sorites})_2 = (A : B : C : D : E) : (A : E)$$
$$\&c., \&c.$$

Taken in the reverse order (see § 11) we get what may be called *Inverse Sorites*, thus:—

$$\text{Barbara} = (A \:!\: C) \:!\: (A \:!\: B \:!\: C)$$
$$(\text{Sorites})_1 = (A \:!\: D) \:!\: (A \:!\: B \:!\: C \:!\: D).$$
$$\&c.$$

91. *Mediate* and *Immediate Inferences*. When from a proposition $\phi(x, y, z)$ we infer another proposition $\psi(x, z)$ in which one or more constituents of the first proposition are left out (or "eliminated"), we call it *Mediate Inference*. If all the constituents of the first proposition are also found in the second, none being eliminated, we have what is called *Immediate Inference*. For example, in Barbara we have *mediate* inference, since from $x:y:z$ we infer $x:z$; the middle term y being eliminated. On the other hand, when from $x:y$ we infer $y':x'$, or $\alpha x:y$, we have *immediate inference*, since there is no elimination of any constituent.

92. *Law of Excluded Middle*. This is the name given to the certainty $A^B + A^{-B}$, or its equivalent $a + a'$. The individual A either belongs to the class B or it does not belong to the class B—an alternative which is evidently a formal certainty.

93. *Intension* and *Extension*, or *Connotation* and *Denotation*. Let the symbols (AB), (ABC), &c., with brackets, as in § 100, denote the collection of individuals, $(AB)_1$, $(AB)_2$, &c., or $(ABC)_1$, $(ABC)_2$, &c., common to the classes inside the brackets; so that $S^{(AB)}$ will not be synonymous with S^{AB}, nor $S^{(ABC)}$ with S^{ABC} (see § 9). With this interpretation of the symbols employed, let S be any individual taken at random out of our universe of discourse, and let $S^x = S^{(AB)}$ be our definition of the term or class X. The term X is said to *connote* the properties A and B, and to *denote* the individuals X_1, X_2, &c., or $(AB)_1$, $(AB)_2$, &c., possessing the properties A and B. As a rule the greater the number of properties, A, B, C, &c., ascribed to X, the fewer the individuals possessing them; or, in other words, the greater the *connotation* (or *intension*), the smaller the *denotation* (or *extension*). In A^a the symbol a *connotes* as *predicate*, and in A_a it *denotes* as *adjective*.

94. *Contrary* and *Contradictory*. The two propositions "All X is Y" (or $x:y$) and "No X is Y" (or $x:y'$) are

§§ 94–98]   TECHNICAL WORDS EXPLAINED   93

called *contraries*, each being the *contrary* of the other. The propositions "All X is Y" and "Some X is not Y," respectively represented by the implication $x:y$ and its denial $(x:y)'$ are called *Contradictories*, each being the contradictory or denial of the other (see § 50). Similarly "No X is Y" and "Some X is Y," respectively represented by the implication $x:y'$ and its denial $(x:y')'$, are called *Contradictories*.

95. *Subcontraries*. The propositions "Some X is Y" and "Some X is not Y," respectively represented by the non-implications $(x:y')'$ and $(x:y)'$, are called *Subcontraries*. It is easily seen that both may be true, but that both cannot be false (see § 73).

96. *Subalterns*. The universal proposition "All X is Y," or $x:y$, implies the particular "Some X is Y," or $(x:y')'$; and the universal "No X is Y," or $x:y'$, implies the particular "Some X is not Y," or $(x:y)'$. In each of these cases the implication, or universal, is called the *Subalternant*, and the non-implication, or particular, is called the *Subalternate* or *Subaltern*. That $x:y$ implies $(x:y')'$ is proved in § 73; and by changing y into y' and *vice versa*, this also proves that $x:y'$ implies $(x:y)'$.

97. *Contraposition*. This is the name given by some logicians to the formula $x:y=y':x'$, which, with the conventions of §§ 46, 50, asserts that the proposition "All X is Y" is equivalent to the proposition "All non-Y is non-X." But other logicians define the word differently.

98. *Conversion*. Let $\phi(x, y)$ denote any proposition, A, E, I, or O, of the traditional logic (see § 50); and let $\psi(y, x)$ denote any other proposition which the first implies, *the letters x and y being interchanged*. The implication $\phi(x, y) : \psi(y, x)$ is called *Conversion*. When the two implications $\phi(x, y)$ and $\psi(y, x)$ are *equivalent*, each implying the other, as in $x:y'=y:x'$, and in $(x:y')'=(y:x')'$, the conversion is called *Simple Conversion*. When the proposition $\phi(x, y)$ implies but is not

implied by $\psi(y, x)$, as in the case of $(x:y):(y:x')'$, the conversion is called *Conversion by Limitation* or *Per accidens*. In all these cases, the antecedent $\phi(x, y)$ is called the *Convertend;* and the consequent $\psi(y, x)$ is called the *Converse*.

99. *Modality.* In the traditional logic any proposition A^B of the *first degree* is called a *pure* proposition, while any of my propositions A^{BC} or A^{BCD}, &c., of a *higher degree* would generally be considered a *modal proposition;* but upon this point we cannot speak with certainty, as logicians are not agreed as to the meaning of the word 'modal.' For example, let the pure proposition A^B assert that "*Alfred* will go to *Belgium*"; then A^{Bc} might be read "*Alfred* will *certainly* go to *Belgium*," which would be called a *modal* proposition. Again, the proposition A^{-B}, which asserts that "*Alfred* will *not* go to *Belgium*," would be called a *pure* proposition; whereas $A^{B\iota}$, or its synonym $(A^B)^\iota$, which asserts that A^B is false, would, by most logicians, be considered a *modal* proposition (see §§ 21, 22, and note 2, p. 105).

100. *Dichotomy.* Let the symbols (AB), (AB'), (ABC), &c., with *brackets*, be understood to denote *classes* (as in Boolean systems) and not the *statements* AB, AB', ABC, &c. We get*

$$A = A(B + B') = A(B + B')(C + C') = \&c.$$
$$= (AB) + (AB') = (ABC) + (ABC') + (AB'C) + (AB'C')$$
$$= \&c.$$

Thus any class A in our universe of discourse may be divided, first, into *two* mutually exclusive divisions; then, by similar subdivision of each of these, into *four* mutually exclusive divisions; and so on. This process of division into *two, four, eight,* &c., mutually exclusive

---

* The symbol (AB) denotes the total of individuals common to A and B; the symbol (AB') denotes the total number in A but not in B; and so on.

§§ 100–105] TECHNICAL WORDS EXPLAINED 95

divisions is called *Dichotomy*. The celebrated *Tree of Porphyry*, or *Ramean Tree*, affords a picture illustration of this division by Dichotomy. Jeremy Bentham wrote enthusiastically of " the matchless beauty of the Ramean Tree."

101. *Simple Constructive Dilemma.* This, expressed symbolically, is the implication

$$(A:x)(B:x)(A+B):x.$$

It may be read, " If A implies x, and B implies x, and either A or B is true, then x is true."

102. *Complex Constructive Dilemma.* This is the implication

$$(A:x)(B:y)(A+B):x+y.$$

103. *Destructive Dilemma.* This is

$$(A:x)(B:y)(x'+y'):A'+B'.$$

It may be read, " If A implies x, and B implies y, and either x or y is false, then either A or B is false."

104. *Modus ponendo ponens* (see Dr. Keynes's " Formal Logic "). There are two forms of this, the one valid, the other not, namely,

$$(A:B)A:B \text{ and } (A:B)B:A.$$

The first form is self-evident; the second form fails in the case $A^\eta B^{-\eta}$ and in the case $A^{-\epsilon}B^\epsilon$; for, denoting the second form by ϕ, we get (see §§ 67–69)

$$W\phi' = A^\eta B^{-\eta} + A^{-\epsilon}B^\epsilon.$$

105. *Modus tollendo tollens.* Of this also there are two forms; the first valid, the second not, namely,

$$(A:B)B':A' \text{ and } (A:B)A':B'.$$

The first is evident; the second fails, as before, in the case $A^\eta B^{-\eta}$, and in the case $A^{-\epsilon}B^\epsilon$. For, denoting the

96   SYMBOLIC LOGIC   [§§ 105-108

second form by ϕ, we get $W\phi' = A^\eta B^{-\eta} + A^{-\epsilon}B^\epsilon$. (See §§ 67-69.)

106. *Modus tollendo ponens.* This also has two forms; the first valid, the other not. They are

$$(A+B)A' : B \text{ and } (AB)'B' : A.$$

The first may be proved formally as follows:—

$$(A+B)A' : B = A'B'(A+B) : \eta = (\eta+\eta) : \eta$$
$$= \eta : \eta = \epsilon.$$

The second is not valid, for

$$(AB)'B' : A = A'B'(AB)' : \eta = A'B' : \eta$$
$$= (A+B)^\epsilon;$$

which fails both in the case $(A+B)^\eta$ and in the case $(A+B)^\theta$. To prove its failure in the last case, let ϕ denote the given implication. We get

$$\phi = (AB)'B' : A = (A+B)^\epsilon,$$

as already proved. Therefore, putting $A+B=\theta$, we get $\phi = \theta^\epsilon = \eta$.

107. *Modus ponendo tollens.* This also has a valid and an invalid form, namely,

$$(AB)'A : B' \text{ and } (A+B)B : A'.$$

The first is valid, for

$$(AB)'A : B' = AB(AB)' : \eta = \eta : \eta = \epsilon.$$

The second is not valid, for

$$(A+B)B : A' = AB(A+B) : \eta = AB : \eta,$$

which fails both in the case $(AB)^\epsilon$ and in the case $(AB)^\theta$. In the first case the given implication becomes $\epsilon : \eta$, which $= \eta$; and in the second case it becomes $\theta : \eta$, which also $= \eta$.

108. *Essential* (or *Explicative*) and *Ampliative.* Let x be any word or symbol, and let $\phi(x)$ be any proposition

§§ 108–110] TECHNICAL WORDS EXPLAINED 97

containing x (see § 13). When $\phi(x)$ is, or follows necessarily from, a *definition* which explains the *meaning* of the word (or collection of words) x; then the proposition $\phi(x)$ is called an *essential*, or an *explicative*, proposition. *Formal certainties* are *essential* propositions (see § 109). When we have a proposition, such as x^a, or x^{-a}, or $x^a + x^\beta$, which gives information about x not contained in any definition of x; such a proposition is called *ampliative*.

109. *Formal* and *Material*. A proposition is called a *formal certainty* when it follows necessarily from our definitions, or our understood linguistic conventions, without further data; and it is called a *formal impossibility*, when it is inconsistent with our definitions or linguistic conventions. It is called a *material certainty* when it follows necessarily from some special data not necessarily contained in our definitions. Similarly, it is called a *material impossibility* when it contradicts some special datum or data not contained in our definitions. In this book the symbols ϵ and η respectively denote certainties and impossibilities without any necessary implication as to whether formal or material. When no special data are given beyond our definitions, the certainties and impossibilities spoken of are understood to be formal; when special data are given then ϵ and η respectively denote material certainties and impossibilities.

110. *Meaningless Symbols*. In logical as in mathematical researches, expressions sometimes turn up to which we cannot, for a time, or in the circumstances considered, attach any meaning. Such expressions are not on that account to be thrown aside as useless. The meaning and the utility may come later; the symbol $\sqrt{-1}$ in mathematics is a well-known instance. From the fact that a certain simple or complex symbol x happens to be meaningless, it does not follow that every statement or expression containing it is also meaningless. For example, the logical statement $A^x + A^{-x}$, which

G

asserts that A either belongs to the class x or does *not* belong to it, is a formal certainty whether A be meaningless or not, and also whether x be meaningless or not. Suppose A meaningless and x a certainty. We get

$$A^x + A^{-x} = 0^\epsilon + 0^{-\epsilon} = \eta + \epsilon = \epsilon.$$

Next, suppose A a certainty and x meaningless. We get

$$A^x + A^{-x} = \epsilon^0 + \epsilon^{-0} = \eta + \epsilon = \epsilon.$$

Lastly, suppose A and x both meaningless. We get

$$A^x + A^{-x} = 0^0 + 0^{-0} = \epsilon + \eta = \epsilon.$$

Let A_x denote any function of x, that is, any expression containing the symbol x; and let $\phi(A_x)$ be any statement containing the symbol A_x; so that the statement $\phi(A_x)$ is a *function of a function* of x (see § 13). Suppose now that the symbol A_x, though intelligible for most values (or meanings) of x, happens to be meaningless when x has a particular value a, and also when x has a particular value β. Suppose also that the statement $\phi(A_x)$ is true (and therefore intelligible) for all values of x except the values a and β, but that for these two values of x the statement $\phi(A_x)$ becomes meaningless, and therefore neither true nor false. Suppose, thirdly, that $\phi(A_x)$ becomes true (and therefore intelligible) also for the exceptional cases $x = a$ and $x = \beta$ provided we lay down the convention or definition that the hitherto meaningless symbol A_a shall have a certain intelligible meaning m_1, and that, similarly, the hitherto meaningless symbol A_β shall have a certain intelligible meaning m_2. Then, the hitherto meaningless symbols A_a and A_β will henceforth be synonyms of the intelligible symbols m_1 and m_2, and the general statement or formula $\phi(A_x)$, which was before meaningless in the cases $x = a$ and $x = \beta$, will now be true and intelligible for all values of x without exception. It is on this principle that

§§ 110, 111]   MEANINGLESS SYMBOLS   99

mathematicians have been led to give meanings to the originally meaningless symbols a^0 and a^{-n}, the first of which is now synonymous with 1, and the second with $\dfrac{1}{a^n}$.

Suppose we have a formula, $\phi(x)=\psi(x)$, which holds good for all values of x with the exception of a certain meaningless value ς. For this value of x we further suppose $\phi(x)$ to become meaningless, while $\psi(x)$ remains still intelligible. In this case, since $\phi(\varsigma)$ is, by hypothesis, meaningless, we are at liberty to give it any convenient meaning that does not conflict with any previous definition or established formula. In order, therefore, that the formula $\phi(x)=\psi(x)$ may hold good for all values of x without exception (not excluding even the meaningless value ς), we may legitimately lay down the convention or definition that the hitherto meaningless expression $\phi(\varsigma)$ shall henceforth be synonymous with the always intelligible expression $\psi(\varsigma)$. With this convention, the formula, $\phi(x)=\psi(x)$, which before had only a restricted validity, will now become true in all cases.

111. Take, for example, the formula, $\sqrt{x}\sqrt{x}=x$ in mathematics. This is understood to be true for all *positive* values of x; but the symbol $\sqrt{x}$, and consequently also the symbol $\sqrt{x}\sqrt{x}$, become meaningless when x is negative, for (unless we lay down further conventions) the square roots of negative numbers or fractions are non-existent. Mathematicians, therefore, have arrived tacitly, and, as it were, unconsciously, at the understanding that when x is negative, then, whatever meaning may be given to the symbol $\sqrt{x}$ itself, the *combination* $\sqrt{x}\sqrt{x}$, like its synonym $(\sqrt{x})^2$, shall be synonymous with x; and, further, that whatever meaning it may in future be found convenient to give to $\sqrt{-1}$, that meaning must not conflict with any previous formula

or definition. These remarks bear solely on the *algebraic* symbol $\sqrt{-1}$, which we have given merely as a concrete illustration of the wider general principles discussed previously. In *geometry* the symbol $\sqrt{-1}$ now conveys by itself a clear and intelligible meaning, and one which in no way conflicts with any algebraic formula of which it is a constituent.

112. *Induction.*—The reasoning by which we infer, or rather suspect, the existence of a general law by observation of particular cases or instances is called *Induction*. Let us imagine a little boy, who has but little experience of ordinary natural phenomena, to be sitting close to a clear lake, picking up pebbles one after another, throwing them into the lake, and watching them sink. He might reason *inductively* as follows: "This is a stone" (a); "I throw it into the water" (β); "It sinks" (γ). These three propositions he repeats, or rather tacitly and as it were mechanically *thinks*, over and over again, until finally he discovers (as he imagines) the universal law $a\beta : \gamma$, that $a\beta$ implies γ, that *all stones thrown into water sink*. He continues the process, and presently, to his astonishment, discovers that the inductive law $a\beta : \gamma$ is *not* universally true. An exception has occurred. One of the pebbles which he throws in happens to be a pumice-stone and does *not* sink. Should the lake happen to be in the crater of an extinct volcano, the pebbles might be all pumice-stones, and the little boy might then have arrived inductively at the general law, *not* that *all stones sink*, but that *all stones float*. So it is with every so-called "law of nature." The whole collective experience of mankind, even if it embraced millions of ages and extended all round in space beyond the farthest stars that can ever be discovered by the most powerful telescope, must necessarily occupy but an infinitesimal portion of infinite time, and must ever be restricted to a mere infinitesimal portion of infinite space. Laws founded upon data thus confined, as it were, within the limits of an

§ 112]   "LAWS OF NATURE"   101

infinitesimal can never be regarded (like most formulæ in logic and in mathematics) as absolutely certain; they should not therefore be extended to the infinite universe of time and space beyond—a universe which must necessarily remain for ever beyond our ken. This is a truth which philosophers too often forget (see § 80).

Many theorems in mathematics, like most of the laws of nature, were discovered inductively before their validity could be rigorously deduced from unquestionable premises. In some theorems thus discovered further researches have shown that their validity is restricted within narrower limits than was at first supposed. Taylor's Theorem in the Differential Calculus is a well-known example. Mathematicians used to speak of the "failure cases" of Taylor's Theorem, until Mr. Homersham Cox at last investigated and accurately determined the exact conditions of its validity. The following example of a theorem discovered inductively by successive experiments may not be very important; but as it occurred in the course of my own researches rather more than thirty years ago, I venture to give it by way of illustration.

Let C be the centre of a square. From C draw in a random direction a straight line CP, meeting a side of the square at P. What is the average area of the circle whose variable radius is CP?

The question is very easy for any one with an elementary knowledge of the integral calculus and its applications, and I found at once that the average area required is equal to that of the given square. I next took a rectangle instead of a square, and found that the average area required (*i.e.* that of the random circle) was equal to that of the rectangle. This led me to suspect that the same law would be found to hold good in regard to all symmetrical areas, and I tried the ellipse. The result was what I had expected: taking C as the centre of the ellipse, and CP in a random direction meeting the curve at P, I found that the average area of the variable

102   SYMBOLIC LOGIC   [§ 112

circle whose radius is CP must be equal to that of the ellipse. Further trials with other symmetrical figures confirmed my opinion as to the universality of the law. Next came the questions: Need the given figure be symmetrical? and might not the law hold good for *any* point C in *any* area, regular or irregular? Further trials again confirmed my suspicions, and led me to the discovery of the general theorem, that if there be any given areas in the same plane, and we take any point C anywhere in the plane (whether in one of the given areas or not), and draw any random radius CP meeting the boundary of any given area at a variable point P, the average area of the circle whose radius is CP is always equal to the sum of the given areas, provided we consider the variable circle as *positive* when P is a point of *exit* from any area, *negative* when P is a point of entrance, and *zero* when P is *non-existent*, because the random radius meets none of the given boundaries.

Next came the question: Might not the same general theorem be extended to any number of given *volumes* instead of *areas*, with an average *sphere* instead of *circle*? Experiment again led to an affirmative answer—that is to say, to the discovery of the following theorem which (as No. 3486) I proposed in the *Educational Times* as follows:

> Some shapeless solids lie about—
>   No matter where they be;
> Within such solid, or without,
>   Let's take a centre C.
> From centre C, in countless hosts,
>   Let random radii run,
> And meet a surface each at P,
>   Or, may be, meet with none.
> Those shapeless solids, far or near,
>   Their total prove to be
> The average volume of the sphere
>   Whose radius is CP.

§§ 112, 113]   FINITE, INFINITE, ETC.   103

>   The sphere, beware, is *positive*
>   When *out* at P they fly;
>   But, changing sign, 'tis *negative*
>   When *entrance* there you spy.
>   One caution more, and I have done:
>   The sphere is *naught* when P there's *none*.

In proposing the question in verse instead of in plain prose, I merely imitated the example of more distinguished contributors. Mathematicians, like other folk, have their moments of exuberance, when they burst forth into song just to relieve their feelings. The theorem thus discovered inductively was proved deductively by Mr. G. S. Carr. A fuller and therefore clearer proof was afterwards given by Mr. D. Biddle, who succeeded Mr. Miller as mathematical editor of the *Educational Times*.

113. *Infinite* and *Infinitesimal*. Much confusion of ideas is caused by the fact that each of those words is used in different senses, especially by mathematicians. Hence arise most of the strange and inadmissible paradoxes of the various non-Euclidean geometries. To avoid all ambiguities, I will define the words as follows. The symbol a denotes any *positive* quantity or ratio *too large to be expressible in any recognised notation*, and any such ratio is called a *positive infinity*. As we may, in the course of an investigation, have to speak of several such ratios, the symbol a denotes a *class* of ratios called *infinities*, the respective individuals of which may be designated by a_1, a_2, a_3, &c. An immensely large number is not necessarily *infinite*. For example, let M denote a million. The symbol M^M, which denotes the millionth power of a million, is a number so inconceivably large that the ratio which a million miles has to the millionth part of an inch would be negligible in comparison; yet this ratio M^M is too small to be reckoned among the infinities a_1, a_2, a_3, &c., of the class a, because, *though inconceivably*

large, its exact value is still expressible in our decimal notation; for we have only to substitute 10^6 or 1,000,000 for M, and we get the exact expression at once. The symbol β, or its synonym $-a$, denotes any *negative infinity;* so that β_1, β_2, β_3, &c., denote different negative ratios, each of which is *numerically* too large to be expressible in any recognised notation. Mathematicians often use the symbols ∞ and $-\infty$ pretty much in the sense here given to a and β; but unfortunately they also employ ∞ and $-\infty$ indifferently to denote expressions such as $\frac{1}{0}$, $\frac{3}{0}$, &c., *which are not ratios at all, but pure non-existences of the class* 0 (see § 6). Mathematicians consider ∞ and $-\infty$ equivalent when they are employed in this sense; but it is clear that a and $-a$ are *not* equivalent. They speak of all parallel straight lines meeting at a *point at infinity;* but this is only an abbreviated way of saying that all straight lines which meet at any infinite distance a_1, or a_2, or a_3, &c., or β_1, or β_2, or β_3, &c., can never be distinguished by any possible instrument from parallel straight lines; and may, therefore, for all practical purposes, be considered parallel.

The symbol h, called a *positive infinitesimal*, denotes any *positive* quantity or ratio *too small numerically to be expressible in any recognised notation;* and the symbol k, called a *negative infinitesimal*, denotes any *negative* quantity or ratio *too small numerically to be expressible in any recognised notation.* Let c temporarily denote any positive *finite* number or ratio—that is to say, a ratio neither too large nor too small to be expressible in our ordinary notation; and let symbols of the forms xy, $x+y$, $x-y$, &c., have their customary mathematical meanings. From these conventions we get various self-evident formulæ, such as

§ 113]   FINITE, INFINITE, ETC.   105

(1) $(ca)^\alpha$, $(c\beta)^\beta$;   (2) $(ch)^h$, $(ck)^k$;   (3) $(a-c)^\alpha_\iota$;

(4) $(c\pm h)^c$;   (5) $(\beta+c)^\beta$;   (6) $\left(\dfrac{c}{h}\right)^\alpha$, $\left(\dfrac{c}{k}\right)^\beta$;

(7) $\left(\dfrac{h}{c}\right)^h$, $\left(\dfrac{k}{c}\right)^k$;   (8) $(a^2)^\alpha$, $(\beta^2)^\alpha$;   (9) $(a\beta)^\beta$;

(10) $x^c : x^{-\alpha} x^{-\beta}$;   (11) $x^\alpha + x^\beta : x^{-c}$;   (12) $(hk)^k$.

The first formula asserts that the product of a *positive finite* and a *positive infinite* is a *positive infinite;* the tenth formula asserts that if any ratio x is a *positive finite*, it is neither a *positive* nor a *negative* infinite. The third formula asserts that the difference between a *positive infinite* and a *positive finite* is a *positive infinite*.

NOTE 1.—A fuller discussion of the finite, the infinite, and the infinitesimal will be found in my eighth article on " Symbolic Reasoning" in *Mind.* The article will probably appear next April.

NOTE 2.—The four " Modals" of the traditional logic are the four terms in the product of the two certainties $A^r + A^\iota$ and $A^\epsilon + A^\eta + A^\theta$. This product is $A^\epsilon + A^\eta + A^r A^\theta + A^\iota A^\theta$; it asserts that every statement A is either *necessarily true* (A^ϵ), or *necessarily false* (A^η), or *true in the case considered but not always* ($A^r A^\theta$), or *false in the case considered but not always* ($A^\iota A^\theta$). See § 99.

CALCULUS OF LIMITS

CHAPTER XIV

114. We will begin by applying this calculus to simple problems in elementary algebra. Let A denote any number, ratio, or fraction. The symbol A^x asserts that A belongs to the class x, the symbol x denoting some such word as *positive*, or *negative*, or *zero*,* or *imaginary*, &c. The symbols $A^x B^y$, $A^x + B^y$, $A^x : B^y$, A^{-x}, &c., are to be understood in the same sense as in §§ 4–10. For example, let $P = positive$, let $N = negative$, and let $0 = zero^*$; while all numbers or ratios not included in one or other of these three classes are excluded from our Universe of Discourse—that is to say, left entirely out of consideration. Thus we get $(6-4)^P$, $(4-6)^N$, $(3-3)^0$, $\left(\dfrac{0}{3}\right)^0$, $(3 \times 0)^0$, $(3P_1)^P$, $(P_1 P_2)^P$, $(P_1 N_1)^N$, $(N_1 N_2)^P$, $(P_1 + P_2)^P$, $(N_1 + N_2)^N$, and many other self-evident formulæ, such as

(1) $(AB)^P = A^P B^P + A^N B^N$.
(2) $(AB)^N = A^N B^P + A^P B^N$.
(3) $(AB)^0 = A^0 + B^0$.
(4) $(Ax - B)^P = \left\{ A\left(x - \dfrac{B}{A}\right) \right\}^P = A^P \left(x - \dfrac{B}{A}\right)^P + A^N \left(x - \dfrac{B}{A}\right)^N$.

---

* In this chapter and after, the symbol 0, representing *zero*, denotes not simple *general* non-existence, as in § 6, but that particular non-existence through which a variable passes when it changes from a positive infinitesimal to a negative infinitesimal, or *vice versa*. (See § 113.)

§§ 114, 115]   CALCULUS OF LIMITS

(5) $(Ax-B)^N = \left\{A\left(x-\dfrac{B}{A}\right)\right\}^N = A^N\left(x-\dfrac{B}{A}\right)^P + A^P\left(x-\dfrac{B}{A}\right)^N.$

(6) $(Ax-B)^0 = \left\{A\left(x-\dfrac{B}{A}\right)\right\}^0 = A^0 + \left(x-\dfrac{B}{A}\right)^0.$

(7) $(ax=ab)=(ax-ab)^0 = \{a(x-b)\}^0 = a^0 + (x-b)^0.$

115. The words *greater* and *less* have a wider meaning in algebra than in ordinary speech. In algebra, when we have $(x-a)^P$, we say that "x is *greater* than a," whether a is positive or negative, and whether x is positive or negative. Also, without any regard to the sign of x or a, when we have $(x-a)^N$, we say that "x is *less* than a." Thus, in algebra, whether x be positive or negative, and whether a be positive or negative, we have

$$(x-a)^P = (x>a), \text{ and } (x-a)^N = (x<a).$$

From this it follows, by changing the sign of a, that

$$(x+a)^P = (x>-a), \text{ and } (x+a)^N = (x<-a);$$

the symbols $>$ and $<$ being used in their customary algebraic sense.

For example, let $a=3$. We get

$$(x-3)^P = (x>3), \text{ and } (x-3)^N = (x<3).$$

In other words, to assert that $x-3$ is *positive* is equivalent to asserting that x is *greater* than 3; while to assert that $x-3$ is *negative* is equivalent to asserting that x is less than 3.

Next, let $a=-3$. We get

$$(x-a)^P = (x+3)^P = (x>-3)$$
$$(x-a)^N = (x+3)^N = (x<-3).$$

Let $x=6$, we get

$$(x>-3) = (x+3)^P = (6+3)^P = \epsilon \text{ (a certainty)}.$$

Let $x=0$, we get

$$(x>-3) = (x+3)^P = (0+3)^P = \epsilon \text{ (a certainty)}.$$

Let $x = -1$, we get

$(x > -3) = (x+3)^P = (-1+3)^P = \epsilon$ (a certainty).

Let $x = -4$, we get

$(x > -3) = (x+3)^P = (-4+3)^P = \eta$ (an impossibility).

It is evident that $(x > -3)$ is a certainty (ϵ) for all positive values of x, and for all negative values of x between 0 and -3; but that $x > -3$ is an impossibility (η) for all negative values of x not comprised between 0 and -3. With $(x < -3)$ the case is reversed. The statement $(x < -3)$ is an impossibility (η) for all positive values of x and for all negative values between 0 and -3; while $(x < -3)$ is a certainty (ϵ) for all negative values of x not comprised between 0 and -3. Suppose, for example, that $x = -8$; we get

$(x < -3) = (x+3)^N = (-8+3)^N = \epsilon$ (a certainty).

Next, suppose $x = -1$; we get

$(x < -3) = (-1+3)^N = \eta$ (an impossibility).

116. From the conventions explained in § 115, we get the formulæ

$(A > B) = (-A) < (-B)$, and $(A < B) = (-A) > (-B)$;

for $\{(-A) < (-B)\} = \{(-A) - (-B)\}^N = (-A+B)^N$
$= (A-B)^P = (A > B)$,

and $\{(-A) > (-B)\} = \{(-A) - (-B)\}^P = (-A+B)^P$
$= (A-B)^N = (A < B)$.

117. Let x be a variable number or fraction, while a is a constant of fixed value. When we have $(x-a)^P$, or its synonym $(x > a)$, we say that a is an *inferior limit* of x; and when we have $(x-a)^N$, or its synonym $(x < a)$, we say that a is a *superior limit* of x. And this definition holds good when we change the sign of a. Thus $(x+a)^P$ asserts that $-a$ is an *inferior limit* of x, and $(x+a)^N$ asserts that $-a$ is a *superior limit* of x.

§§ 118, 119] CALCULUS OF LIMITS

118. For example, let it be required to find the superior or inferior limit of x from the given inequality

$$3x - \frac{x-3}{2} > x + \frac{x+6}{3}$$

Let A denote this given statement of inequality. We get

$$A = \left(3x - \frac{x-3}{2} - x - \frac{x+6}{3}\right)^P = \left(2x - \frac{x-3}{2} - \frac{x+6}{3}\right)^P$$

$$= \left\{6\left(2x - \frac{x-3}{2} - \frac{x+6}{3}\right)\right\}^P = (7x-3)^P = \left(x - \frac{3}{7}\right)^P$$

Hence, $\frac{3}{7}$ is an *inferior* limit of x. In other words, the given statement A is impossible for any positive value of x lower than $\frac{3}{7}$, and also impossible for all negative values of x.

119. Given the statements A and B, in which

A denotes $3x - \frac{5-x}{2} < \frac{1}{4}$, and B denotes $\frac{6-x}{3} - 3x < \frac{1}{4}$.

Find the limits of x. We have

$$A = \left(3x - \frac{5-x}{2} - \frac{1}{4}\right)^N = (12x - 10 + 2x - 1)^N$$

$$= (14x - 11)^N = \left(x - \frac{11}{14}\right)^N = \left(x < \frac{11}{14}\right).$$

$$B = \left(\frac{6-x}{3} - 3x - \frac{1}{4}\right)^N = (24 - 4x - 36x - 3)^N$$

$$= (21 - 40x)^N = (40x - 21)^P = \left(x - \frac{21}{40}\right)^P = \left(x > \frac{21}{40}\right).$$

Hence we get $AB = \left(\frac{11}{14} > x > \frac{21}{40}\right).$

Thus x may have any value between the superior limit $\frac{11}{14}$ and the inferior limit $\frac{21}{40}$; but any value of x not comprised within these limits would be incompatible with our data. For example, suppose $x=1$. We get

$$A : \left(3 - \frac{5-1}{2} - \frac{1}{4}\right)^N : \left(3 - 2 - \frac{1}{4}\right)^N : \left(\frac{3}{4}\right)^N : \eta \text{ (an impossibility).}$$

$$B : \left(\frac{6-1}{3} - 3 - \frac{1}{4}\right)^N : \left(\frac{5}{3} - 3\frac{1}{4}\right)^N ; \epsilon \text{ (a certainty).}$$

Thus the supposition $(x=1)$ is incompatible with A though not with B.

Next, suppose $x=0$. We get

$$A : \left(0 - \frac{5}{2} - \frac{1}{4}\right)^N : \epsilon \text{ (a certainty).}$$

$$B : \left(\frac{6}{3} - 0 - \frac{1}{4}\right)^N : \eta \text{ (an impossibility).}$$

Thus, the supposition $(x=0)$ is incompatible with B though not with A.

120. Next, suppose our data to be AB, in which

$$A \text{ denotes } 5x - \frac{3}{4} > 4x + \frac{1}{3}.$$

$$B \text{ denotes } 6x - \frac{1}{2} < 4x + \frac{3}{4}.$$

We get

$$A = \left(5x - \frac{3}{4} - 4x - \frac{1}{3}\right)^P = \left(x - \frac{13}{12}\right)^P = \left(x > \frac{13}{12}\right)$$

$$B = \left(6x - \frac{1}{2} < 4x + \frac{3}{4}\right) = \left(6x - \frac{1}{2} - 4x - \frac{3}{4}\right)^N$$

$$= \left(2x - \frac{5}{4}\right)^N = \left(x - \frac{5}{8}\right)^N = x < \frac{5}{8}.$$

§§ 120, 121]   CALCULUS OF LIMITS   111

Hence we get

$$AB = \frac{5}{8} > x > \frac{13}{12} = \eta \text{ (an impossibility)}$$

$$\text{for } \left(\frac{5}{8} > x > \frac{13}{12}\right) : \left(\frac{5}{8} > \frac{13}{12}\right) : \eta.$$

In this case therefore our data AB are mutually incompatible. Each datum, A or B, is possible taken by itself; but the combination AB is impossible.

121. Find for what positions of x the ratio F is positive, and for what positions negative, when F denotes $\dfrac{2x-1}{x-3} - \dfrac{28}{x}$.

$$F = \frac{2x^2 - 29x + 84}{x(x-3)} = \frac{2(x-4)(x-10\frac{1}{2})}{x(x-3)}.$$

As in § 113, let a denote *positive infinity*, and let β denote *negative infinity*. Also let the symbol (m, n) assert as a statement that x lies between the superior limit m and the inferior limit n, so that the three symbols (m, n), $(m > x > n)$, and $(m-x)^{\text{p}}(x-n)^{\text{p}}$ are synonyms. We have to consider six limits, namely, $a, 10\frac{1}{2}, 4, 3, 0, \beta$, in descending order, and the five intervening spaces corresponding to the five statements $(a, 10\frac{1}{2})$, $(10\frac{1}{2}, 4)$, $(4, 3)$, $(3, 0)$, $(0, \beta)$. Since x must lie in one or other of these five spaces, we have

$$\epsilon = (a, 10\frac{1}{2}) + (10\frac{1}{2}, 4) + (4, 3) + (3, 0) + (0, \beta).$$

Taking these statements separately, we get

$(a, 10\frac{1}{2}) : (x - 10\frac{1}{2})^{\text{p}} : (x - 10\frac{1}{2})^{\text{p}}(x-4)^{\text{p}}(x-3)^{\text{p}}x^{\text{p}} : F^{\text{p}}$

$(10\frac{1}{2}, 4) : (x - 10\frac{1}{2})^{\text{n}}(x-4)^{\text{p}} : (x - 10\frac{1}{2})^{\text{n}}(x-4)^{\text{p}}(x-3)^{\text{p}}x^{\text{p}} : F^{\text{n}}$

$(4, 3) : (x-4)^{\text{n}}(x-3)^{\text{p}} : (x - 10\frac{1}{2})^{\text{n}}(x-4)^{\text{n}}(x-3)^{\text{p}}x^{\text{p}} : F^{\text{p}}$

$(3, 0) : (x-3)^{\text{n}}x^{\text{p}} : (x - 10\frac{1}{2})^{\text{n}}(x-4)^{\text{n}}(x-3)^{\text{n}}x^{\text{p}} : F^{\text{n}}$

$(0, \beta) : x^{\text{n}} : x^{\text{n}}(x-3)^{\text{n}}(x-4)^{\text{n}}(x - 10\frac{1}{2})^{\text{n}} : F^{\text{p}}.$

Thus, these five statements respectively imply F^{p}, F^{n}, F^{p},

F^N, F^P, the ratio or fraction F changing its sign four times as x passes downwards through the limits $10\frac{1}{2}$, 4, 3, 0. Hence we get

$$F^P = (a, 10\frac{1}{2}) + (4, 3) + (0, \beta);$$
$$F^N = (10\frac{1}{2}, 4) + (3, 0).$$

That is to say, the statement that F is *positive* is equivalent to the statement that x is either between a and $10\frac{1}{2}$, or between 4 and 3, or between 0 and β; and the statement that F is *negative* is equivalent to the statement that x is either between $10\frac{1}{2}$ and 4 or else between 3 and 0.

122. Given that $\dfrac{2x-1}{x-3} = \dfrac{28}{x}$, to find the value or values of x.

It is evident by inspection that there are two values of x which do *not* satisfy this equation; they are 0 and 3. When $x = 0$, we get $\dfrac{2x-1}{x-3} = \dfrac{1}{3}$, while $\dfrac{28}{x} = \dfrac{28}{0}$; and evidently a real ratio $\dfrac{1}{3}$ cannot be equal to a meaningless ratio or unreality $\dfrac{28}{0}$ (see § 113). Again when $x = 3$, we get $\dfrac{2x-1}{x-3} = \dfrac{5}{0}$, while $\dfrac{28}{x} = \dfrac{28}{3}$; and evidently $\dfrac{5}{0}$ cannot be equal to $\dfrac{28}{3}$. Excluding therefore the suppositions $(x=0)$ and $(x=3)$ from our universe of possibilities, let A denote our data, and let $F = \dfrac{2x-1}{x-3} - \dfrac{28}{x}$. We get

$$A : F^0 : \left(\dfrac{2x-1}{x-3} - \dfrac{28}{x}\right)^0 : \left\{\dfrac{2(x-4)(x-10\frac{1}{2})}{x(x-3)}\right\}^0$$
$$: \{(x-4)(x-10\frac{1}{2})\}^0 : (x-4)^0 + (x-10\frac{1}{2})^0$$
$$: (x=4) + (x=10\frac{1}{2}).$$

§§ 122–124]   CALCULUS OF LIMITS

From our data, therefore, we conclude that x must be either 4 or $10\tfrac{1}{2}$.

123. Suppose we have given $\dfrac{13x}{8} - \dfrac{3}{4} > \dfrac{3x}{4} - \dfrac{6-7x}{8}$ to find the limits of x.

Let A denote the given statement. We have
$$A = \left(\dfrac{13x}{8} - \dfrac{3}{4} - \dfrac{3x}{4} + \dfrac{6-7x}{8}\right)^{\mathrm{P}} = (13x - 6 - 6x + 6 - 7x)^{\mathrm{P}}$$
$$= 0^{\mathrm{P}} = \eta.$$

If in the given statement we substitute the sign $<$ for the sign $>$, we shall get $A = 0^{\mathrm{N}} = \eta$. Thus, the statement that $\dfrac{13x}{8} - \dfrac{3}{4}$ is *greater* than $\dfrac{3x}{4} - \dfrac{6-7x}{8}$ is impossible, and so is the statement that $\dfrac{13x}{8} - \dfrac{3}{4}$ is *less* than $\dfrac{3x}{4} - \dfrac{6-7x}{8}$. Hence $\dfrac{13x}{8} - \dfrac{3}{4}$ must be *equal* to $\dfrac{3x}{4} - \dfrac{6-7x}{8}$, *whatever value we give to* x. This is evident from the fact that $\dfrac{3x}{4} - \dfrac{6-7x}{8}$, when reduced to its simplest form, is $\dfrac{13x-6}{8}$, which, for all values of x, is equivalent to $\dfrac{13x}{8} - \dfrac{3}{4}$. If in the given statement we substitute the sign $=$ for the sign $>$, we shall get
$$A = \left(\dfrac{13x}{8} - \dfrac{3}{4} - \dfrac{3x}{4} + \dfrac{6-7x}{8}\right)^{0} = 0^{0} = \epsilon;$$
so that, in this case, A is a formal certainty, whatever be the value of x.

124. Let A denote the statement $x^2 + 3 > 2x$; to find the limits of x. We have
$$A = (x^2 - 2x + 3)^{\mathrm{P}} = \{(x^2 - 2x + 1) + 2\}^{\mathrm{P}}$$
$$= \{(x-1)^2 + 2\}^{\mathrm{P}} = \epsilon.$$

H

Here A is a formal certainty whatever be the value of x, so that there are no *real* finite limits of x (see § 113). If we put the sign $=$ for the sign $>$ we shall get

$$A = \{(x-1)^2 + 2\}^0 = \eta.$$

Here A is a formal impossibility, so that no *real* value of x satisfies the equation $x^2 + 3 = 2x$. It will be remembered that, by § 114, *imaginary* ratios are excluded from our universe of discourse.

125. Let it be required to find the value or values of x from the datum $x - \sqrt{x} = 2$. We get

$$(x - \sqrt{x} = 2) = (x - \sqrt{x} - 2)^0 = (x^P + x^N + x^0)$$
$$(x - \sqrt{x} - 2)^0 = x^P(x - \sqrt{x} - 2)^0$$
$$= x^P\{(x^{\frac{1}{2}} - 2)(x^{\frac{1}{2}} + 1)\}^0 = x^P(x^{\frac{1}{2}} - 2)^0 = (x = 4);$$

for $(x=4)$ implies x^P, and x^0 and x^N are incompatible with the datum $(x - \sqrt{x} - 2)^0$.

126. Let it be required to find the limits of x from the datum $(x - \sqrt{x} > 2)$.

$$(x - \sqrt{x} > 2) = (x - \sqrt{x} - 2)^P = (x^P + x^N + x^0)(x - \sqrt{x} - 2)^P$$
$$= x^P(x - \sqrt{x} - 2)^P$$
$$= x^P\{(x^{\frac{1}{2}} - 2)(x^{\frac{1}{2}} + 1)\}^P = x^P(x^{\frac{1}{2}} - 2)^P = (x > 4);$$

for $(x > 4)$ implies x^P, and x^0 and x^N are incompatible with the datum $(x - \sqrt{x} - 2)^P$.

127. Let it be required to find the limits of x from the datum $(x - \sqrt{x} < 2)$.

$$(x - \sqrt{x} < 2) = (x - \sqrt{x} - 2)^N = (x^P + x^N + x^0)(x - \sqrt{x} - 2)^N$$
$$= (x^P + x^0)(x - \sqrt{x} - 2)^N = x^P(x - \sqrt{x} - 2)^N + x^0$$
$$= x^P\{(x^{\frac{1}{2}} - 2)(x^{\frac{1}{2}} + 1)\}^N + x^0 = x^P(x^{\frac{1}{2}} - 2)^N + x^0$$
$$= x^P(x^{\frac{1}{2}} < 2) + x^0 = x^P(x < 4) + x^0$$
$$= (4 > x > 0) + (x = 0).$$

Here, therefore, x may have any value between 4 and zero, *including zero, but not including* 4.

128. The symbol gm denotes any number or ratio

§§ 128, 129]   CALCULUS OF LIMITS   115

greater than m, while lm denotes any number or ratio *less* than m (see § 115). The symbols g_1m, g_2m, g_3m, &c., denote a series of different numbers or ratios, *each greater than m*, and collectively forming the class gm. Similarly, the symbols l_1m, l_2m, l_3m, &c., denote a series of different numbers or ratios, *each less than m*, and collectively forming the class lm. The symbol x^{gm} asserts that the number or ratio x belongs to the class gm, while x^{lm} asserts that x belongs to the class lm (see § 4). The symbol $x^{gm \cdot gn}$ is short for $x^{gm} x^{gn}$; the symbol $x^{gm \cdot ln}$ is short for $x^{gm} x^{ln}$; and so on (see § 9, footnote).

These symbolic conventions give us the formulæ

(1) $x^{gm} = (x > m) = (x - m)^P$.
(2) $x^{lm} = (x < m) = (x - m)^N$.
(3) $x^{gm \cdot ln} = x^{gm} x^{ln} = (x > m)(x < n)$
$= (x - m)^P (x - n)^N = (n > x > m)$.

129. Let m and n be two different numbers or ratios. We get the formula

(1) $x^{gm \cdot gn} = x^{gm} m^{gn} + x^{gn} n^{gm}$
$= (x > m > n) + (x > n > m)$.

To prove this we have (since m and n are different numbers)

$x^{gm \cdot gn} = x^{gm \cdot gn}(m^{gn} + n^{gm})$, for $m^{gn} + n^{gm} = e$
$= x^{gm} x^{gn} m^{gn} + x^{gm} x^{gn} n^{gm}$
$= (x^{gm} m^{gn}) x^{gn} + (x^{gn} n^{gm}) x^{gm}$
$= x^{gm} m^{gn} + x^{gn} n^{gm} = (x > m > n) + (x > n > m)$,

for in each term the outside factor may be omitted, because it is implied in the compound statement in the bracket, since $x > m > n$ implies $x > n$, and $x > n > m$ implies $x > m$. Similarly, we get and prove the formula

(2) $x^{lm \cdot ln} = x^{lm} m^{ln} + x^{ln} n^{lm} = (x < m < n) + (x < n < m)$.

This formula may be obtained from (1) by simply substituting l for g; and the proof is obtained by the same substitution.

130. Let m, n, r be the three different numbers or ratios. We get the formulæ

(1) $x^{gm \cdot gn \cdot gr} = x^{gm} m^{gn} n^{gr} + x^{gn} n^{gm} n^{gr} + x^{gr} r^{gm} r^{gn}$.

(2) $x^{lm \cdot ln \cdot lr} = x^{lm} m^{ln} m^{lr} + x^{ln} n^{lm} n^{lr} + x^{lr} r^{lm} r^{ln}$.

These two formulæ are almost self-evident; but they may be formally proved in the same way as the two formulæ of § 129; for since m, n, r are, by hypothesis, different numbers or ratios, we have

$$m^{gn \cdot gr} + n^{gm \cdot gr} + r^{gm \cdot gn} = \epsilon_1$$
$$m^{ln \cdot lr} + n^{lm \cdot lr} + r^{lm \cdot ln} = \epsilon_2;$$

while $x^{gm \cdot gn \cdot gr} = x^{gm \cdot gn \cdot gr} \epsilon_1$, by the formula $a = a\epsilon$, and $x^{lm \cdot ln \cdot lr} = x^{lm \cdot ln \cdot lr} \epsilon_2$, by the same formula. When we have multiplied $x^{gm \cdot gn \cdot gr}$ by the alternative ϵ_1, and omitted implied factors, as in § 129, we get Formula (1). When we have multiplied $x^{lm \cdot ln \cdot lr}$ by the alternative ϵ_2, and omitted implied factors, as in § 129, we get Formula (2). The same principle evidently applies to four ratios, m, n, r, s, and so on to any number.

131. If, in § 130, we suppose m, n, r to be *inferior* limits of x, the three terms of the alternative ϵ_1, namely, $m^{gn \cdot gr}$, $n^{gm \cdot gr}$, $r^{gm \cdot gn}$, respectively assert that m is the nearest inferior limit, that n is the nearest inferior limit, that r is the nearest inferior limit. And if we suppose m, n, r to be *superior* limits of x, the three terms of the alternative ϵ_2, namely, $m^{ln \cdot lr}$, $n^{lm \cdot lr}$, $r^{lm \cdot ln}$, respectively assert that m is the nearest superior limit, that n is the nearest superior limit, that r is the nearest superior limit. For of any number of *inferior* limits of a variable x, the *nearest* to x is the *greatest;* whereas, of any number of *superior* limits, the *nearest* to x is the *least*. And since in each case one or other of the limits m, n, r must be the nearest, we have the certain alternative ϵ_1 in the former case, and the certain alternative ϵ_2 in the latter.

§§ 131–133] CALCULUS OF LIMITS 117

It is evident that m^{gn} may be replaced by $(m-n)^{\text{P}}$, that m^{ln} may be replaced by $(m-n)^{\text{N}}$, that $m^{ln \cdot lr}$ may be replaced by $(m-n)^{\text{N}}(m-r)^{\text{N}}$; and so on.

CHAPTER XV

132. When we have to speak often of several limits, x_1, x_2, x_3, &c., of a variable x, it greatly simplifies and shortens our reasoning to register them, one after another, as they present themselves, in a *table of reference*. The * symbol $x_{m' \cdot n}$ asserts that x_m is a *superior* limit, and x_n an *inferior* limit, of x. The * symbol $x_{m' \cdot n' \cdot r \cdot s}$ asserts that x_m and x_n are superior limits of x, while x_r and x_s are inferior limits of x. Thus

$x_{m' \cdot n}$ means $(x-x_m)^{\text{N}}(x-x_n)^{\text{P}}$ or $(x_m > x > x_n)$,

$x_{m' \cdot n' \cdot r \cdot s}$ means $(x-x_m)^{\text{N}}(x-x_n)^{\text{N}}(x-x_r)^{\text{P}}(x-x_s)^{\text{P}}$,

and so on.

133. The symbol $x_{m'}$ (with an acute accent on the numerical suffix m) always denotes a proposition, and is synonymous with $(x-x_m)^{\text{N}}$, which is synonymous with $(x < x_m)$. It affirms that the mth limit of x registered in our table of reference is a *superior* limit. The symbol x_m (with no accent on the numerical suffix), *when used as a proposition*, asserts that the mth limit of x registered in our table of reference is an *inferior* limit of x. Thus x_m means $(x-x_m)^{\text{P}}$.

* In my memoir on *La Logique Symbolique et ses applications* in the *Bibliothèque du Congrès International de Philosophie*, I adopted the symbol x_n^m (suggested by Monsieur L. Couturat) instead of $x_{m' \cdot n}$, and $x_{r \cdot s}^{m \cdot n}$ instead of $x_{m' \cdot n' \cdot r \cdot s}$. The student may employ whichever he finds the more convenient. From long habit I find the notation of the text easier; but the other occupies rather less space, and has certain other advantages in the process of finding the limits. When, however, the limits have been found and the multiple integrals have to be evaluated, the notation of the text is preferable, as the other might occasionally lead to ambiguity (see §§ 151, 156).

134. The employment of the symbol x_n sometimes to denote the *proposition* $(x - x_m)^{\text{P}}$, and sometimes to denote the simple number or ratio x_m, never leads to any ambiguity; for the context always makes the meaning perfectly evident. For example, when we write

$$\left(x - \frac{5}{8}\right)^{\text{P}} = (x - x_3)^{\text{P}} = x_3,$$

it is clear that the x_3 *inside* the bracket denotes the fraction $\frac{5}{8}$, which is supposed to be marked in the table of reference as the third limit of x; whereas the x_3, *outside* the bracket, is affirmed to be equivalent to the *statement* $(x - x_3)^{\text{P}}$, and is therefore a statement also. Similarly, when we write

$$A = (2x^2 + 84 > 29x) = (x - 10\tfrac{1}{2})^{\text{P}} + (x - 4)^{\text{N}}$$
$$= (x - x_1)^{\text{P}} + (x - x_2)^{\text{N}} = x_1 + x_2,$$

we assert that the *statement* A is equivalent to the alternative statement $x_1 + x_2$, of which the first term x_1 asserts (as a *statement*) that the limit x_1 (denoting $10\tfrac{1}{2}$) is an *inferior* limit of x, and the second term x_2 asserts that the limit x_2 (denoting 4) is a *superior* limit of x. Thus, the alternative statement $x_1 + x_2$ asserts that " either x_1 is an inferior limit of x, or else x_2 is a superior limit of x."

135. The operations of this calculus of limits are mainly founded on the following three formulæ (see §§ 129-131):

(1) $x_{m,n} = x_m(x_m - x_n)^{\text{P}} + x_n(x_n - x_m)^{\text{P}}.$
(2) $x_{m',n'} = x_{m'}(x_m - x_n)^{\text{N}} + x_{n'}(x_n - x_m)^{\text{N}}.$
(3) $x_{m',n} = x_{m'.n}(x_m - x_n)^{\text{P}}.$

In the first of the above formulæ, the symbol $x_{m,n}$ means $x_m x_n$, and asserts that x_m and x_n are both *inferior* limits

§§ 135, 136]   CALCULUS OF LIMITS   119

of x. The statement $(x_m - x_n)^\text{P}$ asserts that x_m is *greater* than x_n and therefore a *nearer* inferior limit of x; while the statement $(x_n - x_m)^\text{P}$ asserts, on the contrary, that x_n and not x_m is the nearer inferior limit (see §§ 129, 131). In the second formula, the symbol $x_{m'.n'}$ asserts that x_m and x_n are both *superior* limits of x. The statement $(x_m - x_n)^\text{N}$ asserts that x_m is *less* than x_n and therefore a *nearer* superior limit of x; while the statement $(x_n - x_m)^\text{N}$ asserts, on the contrary, that x_n and not x_m is the nearer superior limit. The third formula is equivalent to

$$x_{m'.n} : (x_m - x_n)^\text{P},$$

and asserts that if x_m is a superior limit, and x_n an inferior limit, of x, then x_m must be greater than x_n.

136. When we have *three inferior* limits, Formula (1) of § 135 becomes

$$x_{m.n.r} = x_m a + x_n \beta + x_r \gamma,$$

in which a asserts that x_m is the nearest of the three inferior limits, β asserts that x_n is the nearest, and γ asserts that x_r is the nearest. In other words,

$$a = (x_m - x_n)^\text{P}(x_m - x_r)^\text{P}$$
$$\beta = (x_n - x_m)^\text{P}(x_n - x_r)^\text{P}$$
$$\gamma = (x_r - x_m)^\text{P}(x_r - x_n)^\text{P}.$$

When we have *three superior* limits, Formula (2) of § 135 becomes

$$x_{m'.n'.r'} = x_{m'} a + x_{n'} \beta + x_{r'} \gamma,$$

in which, this time, a asserts that x_m is the nearest of the three superior limits, β asserts that x_n is the nearest, and γ asserts that x_r is the nearest. In other words,

$$a = (x_m - x_n)^\text{N}(x_m - x_r)^\text{N}$$
$$\beta = (x_n - x_m)^\text{N}(x_n - x_r)^\text{N}$$
$$\gamma = (x_r - x_m)^\text{N}(x_r - x_n)^\text{N}.$$

Evidently the same principle may be extended to any number of inferior or superior limits.

137. There are certain limits which present themselves so often that (to save the trouble of consulting the Table of Limits) it is convenient to represent them by special symbols. These are *positive infinity*, *negative infinity*, and *zero* (or rather an *infinitesimal*). Thus, when we have any variable x, in addition to the limits x_1, x_2, x_3, &c., registered in the table, we may have always understood the superior limit x_a, which will denote *positive infinity*, the limit x_0, which will denote *zero* (or rather, in strict logic, a positive or negative *infinitesimal*), and the always understood inferior limit x_β, which will denote *negative infinity* (see § 113). Similarly with regard to any other variable y, we may have the three *understood* limits y_a, y_0, y_β, in addition to the *registered* limits y_1, y_2, y_3, &c. Thus, when we are speaking of the *limits* of x and y, we have $x_a = y_a = a$; $x_0 = y_0 = 0$ (or dx or dy); $x_\beta = y_\beta = -a$. On the other hand, the *statement* $x_{a'.m}$ asserts that x lies between *positive infinity* x_a, and the limit x_m registered in the table of reference; whereas $x_{m'.\beta}$ asserts that x lies between the limit x_m and the *negative infinity* x_β. Similarly, $x_{m'.0}$ asserts that x lies between the *superior limit* x_m and the *inferior limit* 0; while $x_{0'.n}$ asserts that x lies between the *superior limit* 0 and the *inferior limit* x_n. Thus, the statement $x_{m'.0}$ implies that x is positive, and $x_{0'.n}$ implies that x is negative. Also, the *statement* $x_{0'}$ is synonymous with the *statement* x^N; and the *statement* x_0 is synonymous with the statement x^P. As shown in § 134, the employment of the symbol x_0 sometimes to denote a *limit*, and sometimes to denote a *statement*, need not lead to any ambiguity.

138. Just as in finding the limits of *statements* in pure logic (see §§ 33–40) we may supply the superior limit η when no other superior limit is given, and the inferior limit ϵ when no other inferior limit is given, so in finding the limits of *variable ratios* in mathematics, we may supply the positive infinity a (represented by x_a or y_a or z_a, &c., according to the variable in question) when no

§§ 138, 139]   CALCULUS OF LIMITS   121

other superior limit is given, and the negative infinity β (represented by x_β or y_β or z_β, &c.) when no other inferior limit is given. Thus, when x_m denotes a statement, namely, the statement $(x - x_m)^P$, it may be written $x_{a'.m}$; and, in like manner, for the statement $x_{n'}$, which denotes $(x - x_n)^N$, we may write $x_{n'.\beta}$ (see § 137).

139. Though the formulæ of § 135 may generally be dispensed with in easy problems with only one or two variables, we will nevertheless apply them first to such problems, in order to make their meaning and object clearer when we come to apply them afterwards to more complicated problems which cannot dispense with their aid.

Given that $7x - 53$ is positive, and $67 - 9x$ negative; required the limits of x.

Let A denote the first datum, and B the second. We get

TABLE

$$A = (7x - 53)^P = \left(x - \frac{53}{7}\right)^P = x_1 = x_{a'.1}$$

$$B = (67 - 9x)^N = (9x - 67)^P = \left(x - \frac{67}{9}\right)^P = x_{a'.2}$$

$$x_1 = \frac{53}{7} \qquad x_2 = \frac{67}{9}$$

Hence, we get
$$AB = x_{a'.1} x_{a'.2} = x_{a'.1.2}$$

By Formula (1) of § 135, we get

$$x_{1.2} = x_1(x_1 - x_2)^P + x_2(x_2 - x_1)^P$$
$$= x_1\left(\frac{53}{7} - \frac{67}{9}\right)^P + x_2\left(\frac{67}{9} - \frac{53}{7}\right)^P$$
$$= x_1(477 - 469)^P + x_2(469 - 477)^P, \text{ for } Q^P = (63Q)^P$$
$$= x_1 \epsilon + x_2 \eta = x_1 \text{ (see § 11, Formulæ 22, 23).}$$

Thus we get $AB = x_{a'.1.2} = x_{a'.1}$. From the data AB therefore we infer that x lies between x_a and x_1; that is, between positive infinity and $\frac{53}{7}$. In other words, x is greater than $\frac{53}{7}$ or $7\frac{4}{7}$.

Now, here evidently the formula of § 135 was not wanted; for it is evident by mere inspection that x_1 is greater than x_2, so that x_1 being therefore the *nearest* inferior limit, the limit x_2 is superseded and may be left out of account. In fact A implies B, so that we get $AB = A = x_{a'.1}$.

140. Given that $7x - 53$ is negative and $67 - 9x$ positive; required the limits of x.

Let A denote the first datum, and B the second. We get—

$$A = (7x - 53)^{\scriptscriptstyle N} = \left(x - \frac{53}{7}\right)^{\scriptscriptstyle N} = x_{1'} = x_{1'.\beta}$$

$$B = (67 - 9x)^{p} = (9x - 67)^{\scriptscriptstyle N} = \left(x - \frac{67}{9}\right)^{\scriptscriptstyle N}$$

$$= x_{2'.\beta}.$$

$$x_1 = \frac{53}{7}$$

$$x_2 = \frac{67}{9}$$

Hence, we get

$$AB = x_{1'.\beta} x_{2'.\beta} = x_{1'.2'.\beta}.$$

By Formula 2 of § 135 we get—

$$x_{1'.2'} = x_1(x_1 - x_2)^{\scriptscriptstyle N} + x_2(x_2 - x_1)^{\scriptscriptstyle N}$$
$$= x_1\left(\frac{53}{7} - \frac{67}{9}\right)^{\scriptscriptstyle N} + x_2\left(\frac{7}{9} - \frac{53}{7}\right)^{\scriptscriptstyle N}$$
$$= x_1(477 - 469)^{\scriptscriptstyle N} + x_2(469 - 477)^{\scriptscriptstyle N}$$
$$= x_1 \eta + x_2 \epsilon = x_{2'} \text{ (see § 11, Formulæ 22, 23).}$$

This shows that the nearer superior limit x_2 supersedes the more distant superior limit x_1; so that we get $AB = x_{1'.2'.\beta} = x_{2'.\beta}$. Thus x lies between the superior limit x_2 $\left(\text{or } \dfrac{67}{9}\right)$ and negative infinity.

CHAPTER XVI

141. WE will now consider the limits of two variables, and first with only *numerical* constants (see § 156).

Suppose we have given that the variables x and y are both positive, while the expressions $2y - 3x - 2$ and $3y + 2x - 6$ are both negative; and that from these data we are required to find the limits of y and x in the order y, x.

Let A denote our whole data. We have

$$A = y^P x^P (2y - 3x - 2)^N (3y + 2x - 6)^N.$$

Beginning with the first bracket factor, we get*

TABLE OF LIMITS.

$y_1 = \dfrac{3}{2}x + 1$	$x_1 = \dfrac{6}{13}$
$y_2 = 2 - \dfrac{2}{3}x$	$x_2 = -\dfrac{2}{3}$
	$x_3 = 3$

$$(2y - 3x - 2)^N = \left(y - \frac{3}{2}x - 1\right)^N = (y - y_1)^N = y_{1'}.$$

Then, taking the second bracket factor, we get

$$(3y + 2x - 6)^N = \left(y + \frac{2}{3}x - 2\right)^N = (y - y_2)^N = y_{2'}.$$

Also $y^P x^P = y_{a'.0} x_{a'.0}$ (see §§ 137, 139), so that

$$A = y_{a'.0} x_{a'.0} y_{1'} y_{2'} = y_{a'.1'.2'.0} x_{a'.0} = y_{1'.2'.0} x_{a'.0};$$

for the nearer superior limits y_1 and y_2 supersede the more distant limit y_a. Applying Formula (2) of § 135 to the statement $y_{1'.2'}$, we get

$$y_{1'.2'} = y_{1'}(y_1 - y_2)^N + y_{2'}(y_2 - y_1)^N = y_{1'}\left(\frac{13}{6}x - 1\right)^N$$
$$+ y_{2'}\left(\frac{13}{6}x - 1\right)^P = y_{1'}\left(x - \frac{6}{13}\right)^N + y_{2'}\left(x - \frac{6}{13}\right)^P$$
$$= y_1 x_{1'} + y_2 x_1.$$

---

* The limits are registered in the table, one after another, as they are found, so that the table *grows* as the process proceeds.

Substituting this alternative for $y_{1'.2'}$ in the expression for A, we get

$$A = (y_1 x_{1'} + y_{2'} x_1) y_0 x_{a'.0} = (y_{1'.0} x_{1'} + y_{2'.0} x_1) x_{a'.0}$$
$$= y_{1'.0} x_{a'.1'.0} + y_{2'.0} x_{a'.1.0} = y_{1'.0} x_{1'.0} + y_{2'.0} x_{a'.1};$$

omitting in the first term the superior limit x_a because it is superseded by the nearer superior limit x_1; and omitting in the second term the limit x_0, because it is superseded by the nearer limit x_1. The next step is to apply Formula (3) of § 135 to the y-factors $y_{1'.0}$ and $y_{2'.0}$. We get

$$y_{1'.0} = y_{1'.0}(y_1 - y_0)^{\text{P}} = y_{1'.0}(y_1)^{\text{P}} = y_{1'.0}\left(\frac{3}{2}x + 1\right)^{\text{P}}$$
$$= y_{1'.0}(3x+2)^{\text{P}} = y_{1'.0}\left(x + \frac{2}{3}\right)^{\text{P}} = y_{1'.0}(x - x_2)^{\text{P}}$$
$$= y_{1'.0} x_2;$$
$$y_{2'.0} = y_{2'.0}(y_2 - y_0)^{\text{P}} = y_{2'.0}(y_2)^{\text{P}} = y_{2'.0}\left(2 - \frac{2}{3}x\right)^{\text{P}}$$
$$= y_{2'.0}(6 - 2x)^{\text{P}} = y_{2'.0}(3 - x)^{\text{P}} = y_{2'.0}(x - 3)^{\text{N}}$$
$$= y_{2'.0} x_{3'}.$$

Substituting these equivalents of $y_{1'.0}$ and $y_{2'.0}$ in A, we get

$$A = y_{1'.0} x_{1'.2.0} + y_{2'.0} x_{a'.3'.1} = y_{1'.0} x_{1'.0} + y_{2'.0} x_{3'.1};$$

for evidently x_0 is a nearer inferior limit than x_2, and therefore supersedes x_2; while x_3 is a nearer superior limit than x_a (which denotes positive infinity), and therefore supersedes x_a. We have now done with the y-statements, and it only remains to apply Formula (3) of § 135 to the x-statements $x_{1'.0}$ and $x_{3'.1}$. It is evident, however, by mere inspection of the table, that this is needless, as it would introduce no new factor, nor discover any inconsistency, since x_1 is evidently greater than x_0, that is, than zero, and x_3 is evidently greater than x_1. The process therefore here terminates, and the limits are fully deter-

§ 141] CALCULUS OF LIMITS 125

mined. We have found that either x varies between x_1 and zero, and y between y_1 and zero; or else x varies between x_3 and x_1, and y between y_2 and zero.

The figure below will illustrate the preceding process and table of reference. The symbol x denotes the distance of any point P (taken at random out of those in the shaded figure) from the line x_0, and the symbol y denotes the distance of the point P from the line y_0. The first equivalent of the data A is the statement

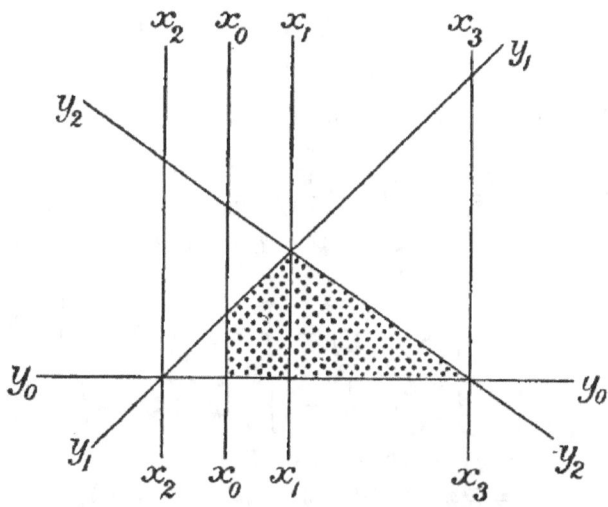

$y_{1'.2'.0}x_0$, which asserts that y_1 and y_2 are superior limits of y, that y_0 (or zero) is an inferior limit of y, and that x_0 (or zero) is an inferior limit of x. It is evident that this compound statement A is true for every point P in the shaded portion of the figure, and that it is *not* true for any point outside the shaded portion. The final equivalent of the data A is the alternative $y_{1'.0}x_{1'.0}$ $+ y_{2'.0}x_{3'.1}$, the first term of which is true for every point P in the quadrilateral contained by the lines y_1, y_0, x_1, x_0; and the second term of which is true for the triangle contained by the lines y_2, y_0, x_1.

142. Given that $y^2 - 4x$ is negative and $y + 2x - 4$ positive; required the limits of y and x.

Let A denote our data. We get

$$A = (y^2 - 4x)^N (y + 2x - 4)^P$$
$$= (y^2 - 4x)^N (y - y_1)^P ;$$
$$(y^2 - 4x)^N = \{(y - 2\sqrt{x})(y + 2\sqrt{x})\}^N$$
$$= (y - 2\sqrt{x})^N (y + 2\sqrt{x})^P ;$$

TABLE OF LIMITS.

$y_1 = 4 - 2x$	$x_1 = 4$
$y_2 = 2\sqrt{x}$	$x_2 = 1$
$y_3 = -2\sqrt{x}$	

for $(y - 2\sqrt{x})^P (y + 2\sqrt{x})^N$ is impossible. We therefore get

$$(y^2 - 4x)^N = (y - y_2)^N (y - y_3)^P = y_{2'.3}$$
$$A = y_{2'.3}(y - y_1)^P = y_{2'.3} y_1 = y_{2'.3.1}$$

By Formula (1) of § 135 we get

$$y_{3.1} = y_3 (y_3 - y_1)^P + y_1 (y_1 - y_3)^P$$
$$= y_3 (2x - 2\sqrt{x} - 4)^P + y_1 (2x - 2\sqrt{x} - 4)^N$$
$$= y_3 (x - \sqrt{x} - 2)^P + y_1 (x - \sqrt{x} - 2)^N \text{ (see §§ 126, 127)}$$
$$= y_3 \left\{ \left(\sqrt{x} - \frac{1}{2}\right)^2 - \frac{9}{4} \right\}^P + y_1 \left\{ \left(\sqrt{x} - \frac{1}{2}\right)^2 - \frac{9}{4} \right\}^N$$
$$= y_3 \left\{ \left(\sqrt{x} - \frac{1}{2}\right) - \frac{3}{2} \right\}^P + y_1 \left\{ \left(\sqrt{x} - \frac{1}{2}\right) - \frac{3}{2} \right\}^N$$
$$= y_3 (x - 4)^P + y_1 (x - 4)^N$$
$$= y_3 (x - x_1)^P + y_1 (x - x_1)^N = y_3 x_1 + y_1 x_{1'}.$$

Therefore
$$A = y_{2'.3} x_1 + y_{2'.1} x_{1'}.$$

We now apply Formula (3) of § 135, thus

$$y_{2'.3} = y_{2'.3} (y_2 - y_3)^P = y_{2'.3} (2\sqrt{x} + 2\sqrt{x})^P = y_{2'.3} \epsilon$$
$$y_{2'.1} = y_{2'.1} (y_2 - y_1)^P = y_{2'.1} (2x + 2\sqrt{x} - 4)^P$$
$$= y_{2'.1} (x + \sqrt{x} - 2)^P = y_{2'.1} \left\{ \left(\sqrt{x} + \frac{1}{2}\right)^2 - \left(\frac{3}{2}\right)^2 \right\}^P$$
$$= y_{2'.1} (x - 1)^P = y_{2'.1} (x - x_2)^P = y_{2'.1} x_2.$$

§§ 142, 143]　　CALCULUS OF LIMITS　　127

Thus the application of Formula (3) of § 135 to $y_{2'.3}$ introduces no new factor, but its application to the other compound statement $y_{2'.1}$ introduces the new statement x_2, and at the same time the new limit x_2. Hence we finally get (since Form 3 of § 135 applied to $x_{a'.1}$ and $x_{1'.2}$ makes no change)·

$$A = y_{2'.3} x_{a'.1} + y_{2'.1} x_{1'.2} \quad \text{(see §§ 137, 138)}.$$

This result informs us that "either x lies between x_a (positive infinity) and x_1, and y between the superior

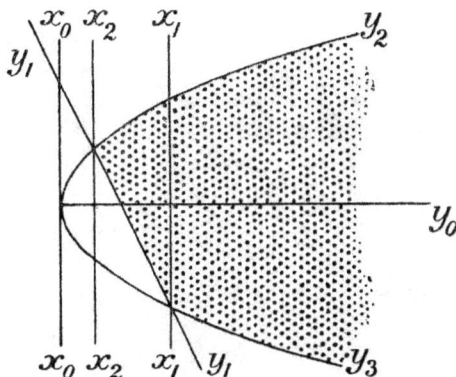

limit y_2 and the inferior limit y_3; or else x lies between x_1 and x_2, and y between y_2 and y_1. The above figure will show the position of the limits. With this geometrical interpretation of the symbols x, y, &c., all the points marked will satisfy the conditions expressed by the statement A, and so will all other points bounded by the upper and lower branches of the parabole, with the exception of the blank area cut off by the line y_1.

143. Given that $y^2 - 4x$ is negative, and $y + 2x - 4$ also negative; required the limits of y and x.

Here the required limits (though they may be found

independently as before) may be obtained at once from the diagram in § 142. The only difference between this problem and that of § 142 is that in the present case $y+2x-4$ is negative, instead of being, as before, positive. Since y^2-4x is, as before, negative, y_2 will be, as before, a superior limit, and y_3 an inferior limit of y; so that, as before, all the points will be restricted within the two branches of the parabola. But since $y+2x-4$ has now changed sign, all the admissible points, while still keeping between the two branches of the parabola, *will cross the line* y_1. The result will be that the only admissible points will now be restricted to the *blank* portion of the parabola cut off by the line y_1, instead of being, as before, restricted to the shaded portion within the two branches and extending indefinitely in the positive direction towards positive infinity. A glance at the diagram of § 142 will show that the required result now is

$$y_{2'.3}x_{2'.0} + y_{1'.3}x_{1'.2'}$$

with, of course, the same table of limits.

CHAPTER XVII

144. The symbol $\dfrac{A}{B}$, *when the numerator and denominator denote statements*, expresses *the chance that* A *is true on the assumption that* B *is true;* B being some statement compatible with the data of our problem, but not necessarily implied by the data.

145. The symbol $\dfrac{A}{\epsilon}$ denotes the chance that A is true *when nothing is assumed but the data of our problem*. This is what is usually meant when we simply speak of the "*chance of* A."

§§ 146, 147]   CALCULUS OF LIMITS   129

146. The symbol $\delta\frac{A}{B}$, or its synonym $\delta(A, B)$, denotes $\frac{A}{B} - \frac{A}{\epsilon}$; and this is called *the dependence** of the statement A *upon the statement* B. It indicates the increase, or (when negative) the decrease, undergone by the absolute chance $\frac{A}{\epsilon}$ when the supposition B is added to our data.

The symbol $\delta^0\frac{A}{B}$, or its synonym $\delta^0(A, B)$, asserts that the dependence of A upon B is *zero*. In this case the state-

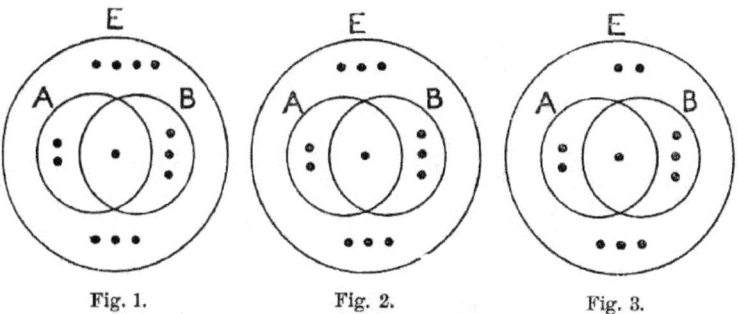

Fig. 1.   Fig. 2.   Fig. 3.

ment A is said to be *independent of the statement* B; which implies, as will be seen further on (see § 149), that B is independent of A.

147. The symbols a, b, c, &c. (small italics) respectively represent the chances $\frac{A}{\epsilon}$, $\frac{B}{\epsilon}$, $\frac{C}{\epsilon}$, &c. (see § 145); and the symbols a', b', c', &c., respectively denote the chances $\frac{A'}{\epsilon}$, $\frac{B'}{\epsilon}$, $\frac{C'}{\epsilon}$, &c., so that we get

$$1 = a + a' = b + b' = c + c' = \&c.$$

---

* Obscure ideas about 'dependence' and 'independence' in probability have led some writers (including Boole) into serious errors. The definitions here proposed are, I believe, original.

I

148. The diagrams on p. 129 will illustrate the preceding conventions and definitions.

Let the symbols A, B assert respectively as *propositions* that a point P, taken at random out of the total number of points in the circle E, will be in the circle A, that it will be in the circle B. Then AB will assert that P will be in both circles A and B; AB' will assert that P will be in the circle A, but not in the circle B; and similarly for the statements A'B and A'B'.

In Fig. 1 we have

$$\frac{A}{\epsilon}=a=\frac{3}{13};\ \frac{A'}{\epsilon}=a'=\frac{10}{13};\ \frac{AB}{\epsilon}=\frac{1}{13};\ \frac{AB'}{\epsilon}=\frac{2}{13}.$$

In Fig. 2 we have

$$\frac{A}{\epsilon}=a=\frac{3}{12};\ \frac{A'}{\epsilon}=a'=\frac{9}{12};\ \frac{AB}{\epsilon}=\frac{1}{12};\ \frac{AB'}{\epsilon}=\frac{2}{12}.$$

In Fig. 3 we have

$$\frac{A}{\epsilon}=a=\frac{3}{11};\ \frac{A'}{\epsilon}=a'=\frac{8}{11};\ \frac{AB}{\epsilon}=\frac{1}{11};\ \frac{AB'}{\epsilon}=\frac{2}{11}.$$

It is evident also that

in Fig. 1, $\delta(A, B)=\frac{A}{B}-\frac{A}{\epsilon}=\frac{AB}{B}-\frac{A}{\epsilon}=\frac{1}{4}-\frac{3}{13}=+\frac{1}{52}$;

in Fig. 2, $\delta(A, B)=\frac{A}{B}-\frac{A}{\epsilon}=\frac{AB}{B}-\frac{A}{\epsilon}=\frac{1}{4}-\frac{3}{12}=0$;

in Fig. 3, $\delta(A, B)=\frac{A}{B}-\frac{A}{\epsilon}=\frac{AB}{B}-\frac{A}{\epsilon}=\frac{1}{4}-\frac{3}{11}=-\frac{1}{44}$.

Similarly, we get

in Fig. 1, $\qquad \delta(B, A)=+\frac{1}{39}$;

in Fig. 2, $\qquad \delta(B, A)=0$;

in Fig. 3, $\qquad \delta(B, A)=-\frac{1}{33}$

§§ 149, 150] CALCULUS OF LIMITS

149. The following formulæ are easily verified:—

(1) $\dfrac{A}{B} = \dfrac{a}{b} \cdot \dfrac{B}{A}$; (2) $\delta\dfrac{A}{B} = \dfrac{a}{b} \cdot \delta\dfrac{B}{A}$;

(3) $\dfrac{A'}{B} = 1 - \dfrac{A}{B}$; (4) $\delta\dfrac{A'}{B} = -\delta\dfrac{A}{B}$;

(5) $\dfrac{A}{B'} = \dfrac{a}{b'} - \dfrac{b}{b'} \cdot \dfrac{A}{B}$; (6) $\delta\dfrac{A}{B'} = -\dfrac{b}{b'}\delta\dfrac{A}{B}$;

(7) $\dfrac{A'}{B'} = 1 - \dfrac{A}{B'}$; (8) $\delta\dfrac{A'}{B'} = -\delta\dfrac{A}{B'} = \dfrac{b}{b'}\delta\dfrac{A}{B}$.

The second of the above eight formulæ shows that if any statement A is independent of another statement B, then B is independent of A; for, by Formula (2), it is clear that $\delta^0(A, B)$ implies $\delta^0(B, A)$. To the preceding eight formulæ may be added the following:—

(9) $\dfrac{AB}{\epsilon} = \dfrac{A}{\epsilon} \cdot \dfrac{B}{A} = \dfrac{B}{\epsilon} \cdot \dfrac{A}{B}$; (10) $\dfrac{AB}{Q} = \dfrac{A}{Q} \cdot \dfrac{B}{AQ} = \dfrac{B}{Q} \cdot \dfrac{A}{BQ}$;

(11) $\dfrac{A+B}{\epsilon} = \dfrac{A}{\epsilon} + \dfrac{B}{\epsilon} - \dfrac{AB}{\epsilon}$; (12) $\dfrac{A+B}{Q} = \dfrac{A}{Q} + \dfrac{B}{Q} - \dfrac{AB}{Q}$.

150. Let A be any statement, and let x be any positive proper fraction; then A^x is short for the statement $\left(\dfrac{A}{\epsilon} = x\right)$, which asserts that the chance of A is x. Similarly, $(AB)^x$ means $\left(\dfrac{AB}{\epsilon} = x\right)$; and so on. This convention gives us the following formulæ, in which a and b (as before) are short for $\dfrac{A}{\epsilon}$ and $\dfrac{B}{\epsilon}$.

(1) $A^x B^y : \left(\dfrac{A}{B} = \dfrac{x}{y} \cdot \dfrac{B}{A}\right)$; (2) $A^x B^y (AB)^z : (A+B)^{x+y-z}$;

(3) $(AB)^x (A+B)^y : (x+y = a+b)$; (4) $\delta^0(A, B) = (AB)^{ab}$;

(5) $(AB)^n = (A+B)^{a+b}$;

$$(6)\ \left(\frac{A}{B}=\frac{A}{B'}\right)=\left(\frac{A}{B}=\frac{A}{\epsilon}\right)=\delta^0(A, B);$$

$$(7)\ \left(\frac{A}{B}=\frac{B}{A}\right):\left(\frac{A}{B}=0\right)+(a=b):(AB)^\eta+(a=b).$$

It is easy to prove all these formulæ, of which the last may be proved as follows:

$$\left(\frac{A}{B}=\frac{B}{A}\right):\left(\frac{A}{B}=\frac{b}{a}\cdot\frac{A}{B}\right):\left(\frac{A}{B}-\frac{b}{a}\frac{A}{B}\right)^0:\left\{\frac{A}{B}\left(1-\frac{b}{a}\right)\right\}^0$$

$$:\left\{\frac{A}{B}(a-b)\right\}^0:\left(\frac{A}{B}=0\right)+(a-b)^0:(AB)^\eta+(a=b).$$

The following chapter requires some knowledge of the integral calculus.

CHAPTER XVIII

151. In applying the Calculus of Limits to multiple integrals, it will be convenient to use the following notation, which I employed for the first time rather more than twenty years ago in a paper on the "Limits of Multiple Integrals" in the *Proc. of the Math. Society*.

The symbols $\phi(x)x_{m'.n}$ and $x_{m'.n}\phi(x)$, which differ in the relative positions of $\phi(x)$ and $x_{m'.n}$, differ also in meaning. The symbol $\phi(x)x_{m'.n}$ is short for the integration $\int\phi(x)dx$, taken between the superior limit x_m and the inferior limit x_n; an integration which would be commonly expressed either in the form $\int_{x_n}^{x_m}dx\phi(x)$ or $\int_{x_n}^{x_m}\phi(x)dx$. The symbol $x_{m'.n}\phi(x)$, with the symbol $x_{m'.n}$ to the *left*, is short for $\phi(x_m)-\phi(x_n)$.

For example, suppose we have $\int\phi(x)dx=\psi(x)$. Then, by substitution of notation, we get $\int_{x_n}^{x_m}\phi(x)dx=\phi(x)x_{m'.n}$ $=x_{m'.n}\psi(x)=\psi(x_m)-\psi(x_n)$; so that we can thus entirely

§§ 151–153]   CALCULUS OF LIMITS   133

dispense with the symbol of integration, $\int$, as in the following concrete example.

Let it be required to evaluate the integral

$$\int_{z_2}^{z_1} dz \int_{y_2}^{y_1} dy \int_{x_0}^{x_1} dx,$$

TABLE OF LIMITS.

| $z_1 = y$ | $y_1 = x$ | $x_1 = a$ |
| $z_2 = c$ | $y_2 = b$ | $x_0 = 0$ |

the limits being as in the given table. The full process is as follows, the order of variation being z, y, x.

Integral $z_{1'.2}y_{1'.2}x_{1'.0} = (z_1 - z_2)y_{1'.2}x_{1'.0} = (y - c)y_{1'.2}x_{1'.0}$
$= y_{1'.2}(\tfrac{1}{2}y^2 - cy)x_{1'.0} = \{(\tfrac{1}{2}y_1^2 - cy_1) - (\tfrac{1}{2}y_2^2 - cy_2)\}x_{1'.0}$
$= \{(\tfrac{1}{2}x^2 - cx) - (\tfrac{1}{2}b^2 - cb)\}x_{1'.0} = (\tfrac{1}{2}x^2 - cx - \tfrac{1}{2}b^2 + bc)x_{1'.0}$
$= x_{1'.0}(\tfrac{1}{6}x^3 - \tfrac{1}{2}cx^2 - \tfrac{1}{2}b^2x + bcx) = \tfrac{1}{6}a^3 - \tfrac{1}{2}ca^2 - \tfrac{1}{2}b^2a + bca.$

152. The following formulæ of integration are self-evident:—

(1) $x_{m'.n} = -x_{n'.m}$;   (2) $\phi(x)x_{m'.n} = -\phi(x)x_{n'.m}$;
(3) $x_{m'.n}\phi(x) = -x_{n'.m}\phi(x)$;   (4) $x_{m'.n} + x_{n'.r} = x_{m'.r}$;
(5) $\phi(x)(x_{m'.n} + x_{n'.r}) = \phi(x)x_{m'.r}$;
(6) $(x_{m'.n} + x_{n'.r})\phi(x) = x_{m'.r}\phi(x)$;
(7) $y_{m'.n}x_{r'.s} = -y_{n'.m}x_{r'.s} = -y_{m'.n}x_{s'.r} = y_{n'.m}x_{s'.r}$;
(8) $x_{m'.n} + x_{r'.s} = x_{m'.s} + x_{r'.n}$;
(9) $(x_{m'.n} + x_{r'.s})\phi(x) = (x_{m'.s} + x_{r'.n})\phi(x)$;
(10) $\phi(x)(x_{m'.n} + x_{r'.s}) = \phi(x)(x_{m'.s} + x_{r'.n}).$

153. As already stated, the symbol $\dfrac{A}{B}$, when A and B are propositions, denotes the chance that A is true on the assumption that B is true. Now, let x and y be any numbers or ratios. The symbol $\dfrac{xA}{yB}$ means $\dfrac{x}{y} \times \dfrac{A}{B}$; and when either of these two numbers is missing, we may suppose the number 1 understood.

Thus, $\dfrac{xA}{B}$ means $\dfrac{x}{1} \times \dfrac{A}{B}$; and $\dfrac{A}{xB}$ means $\dfrac{1}{x} \times \dfrac{A}{B}$.

154. The symbol *Int* A(x, y, z) denotes the integral $\int dx \int dy \int dz$, *subject to the restrictions of the statement* A, the order of variation being x, y, z. The symbol *Int* A, or sometimes simply A, may be used as an abbreviation for *Int* A(x, y, z) when the context leaves no doubt as to the meaning of the abbreviation.

155. Each of the variables x, y, z is taken at random between 1 and 0; what is the chance that the fraction $\dfrac{z(1-x-y)}{1-y-yz}$ will also be between 1 and 0?

TABLE OF LIMITS.

$x_1 = 1$	$y_1 = 1$	$z_1 = 1$
$x_2 = 1-y$	$y_2 = \dfrac{1}{1+z}$	
$x_3 = \dfrac{y+z-1}{z}$	$y_3 = 1-z$	

$$\epsilon = A = x_{1'\cdot 0} y_{1'\cdot 0} z_{1'\cdot 0}$$

Let the symbol Q, as a proposition, assert that the value of the fraction in question will lie between 1 and 0; and let A denote our data $x_{1'\cdot 0} y_{1'\cdot 0} z_{1'\cdot 0}$. We have to find $\dfrac{Q}{A}$, which here $= \dfrac{Q}{\epsilon}$ (see § 145). Also, let N denote the *numerator* $z(1-x-y)$, and D the *denominator* $1-y-yz$ of the fraction in question; while, this time, to avoid ambiguity, the letter n will denote *negative*, and p *positive* (small italics instead of, as before, capitals). We get

$$Q = N^p D^p (N-D)^n + N^n D^n (N-D)^p.$$

Taking the order of variation x, y, z, as in the table, we get, since z is given positive,

$$N^p = (1-x-y)^p = \{x-(1-y)\}^n = x_{2'}$$
$$N^n = (1-x-y)^n = \{x-(1-y)\}^p = x_2$$
$$D^p = (1-y-yz)^p = \{y(1+z)-1\}^n = y_{2'}$$
$$D^n = (1-y-yz)^n = \{y(1+z)-1\}^p = y_2$$
$$(N-D)^n = (z-zx+y-1)^n = (zx-y-z+1)^p$$
$$= \left(x - \dfrac{y+z-1}{z}\right)^p = x_3$$
$$(N-D)^p = (z-zx+y-1)^p = (zx-y-z+1)^n = x_{3'}.$$

§ 155]   CALCULUS OF LIMITS   135

Substituting these results in our expression for Q, we shall have

$$Q = x_2 y_2 x_3 + x_2 y_2 x_{3'} = x_{2'.3} y_{2'} + x_{3'.2} y_2.$$

Multiplying by the given certainty $x_{1'.0}$ (see table), we get

$$x_{1'.0} Q = x_{2'.1'.3.0} y_{2'} + x_{3'.1'.2.0} y_2.$$

Applying Formulæ (1) and (2) of § 135, we get (see § 137)

$$x_{2'.1'} = x_{2'}(x_2 - x_1)^n + x_1(x_1 - x_2)^n = x_{2'}\epsilon + x_1\eta = x_{2'}$$
$$x_{3.0} = x_3(x_3 - x_0)^p + x_0(x_0 - x_3)^p = x_3 y_3 + x_0 y_{3'}$$
$$x_{3'.1'} = x_{3'}(x_3 - x_1)^n + x_1(x_1 - x_3)^n = x_{3'}\epsilon + x_1\eta = x_{3'}$$
$$x_{2.0} = x_2(x_2 - x_0)^p + x_0(x_0 - x_2)^p = x_2\epsilon + x_0\eta = x_2.$$

Substituting these results in our expression for $x_{1'.0} Q$, we get

$$x_{1'.0} Q = x_{2'}(x_3 y_3 + x_0 y_{3'}) y_{2'} + x_{3'.2} y_2$$
$$= x_{2'.3} y_{2'.3} + x_{2'.0} y_{3'.2'} + x_{3'.2} y_2.$$

We now apply Formula (3) of § 135 to the statements $x_{2'.3}$, $x_{2'.0}$, $x_{3'.2}$, thus

$$x_{2'.3} = x_{2'.3}(x_2 - x_3)^p = x_{2'.3} y_{2'}$$
$$x_{2'.0} = x_{2'.0}(x_2 - x_0)^p = x_{2'.0}\epsilon$$
$$x_{3'.2} = x_{3'.2}(x_3 - x_2)^p = x_{3'.2} y_2.$$

This shows that the application of § 135, Form 3, introduces no new statement in y; so that we have finished with the limits of x, and must now apply the formulæ of § 135 to find the limits of y. Multiplying the expression found for $x_{1'.0} Q$ by the datum $y_{1'.0}$, we get

$$x_{1'.0} y_{1'.0} Q = x_{2'.3} y_{2'.1'.3.0} + x_{2'.0} y_{3'.2'.1'.0} + x_{3'.2} y_{1'.2.0}.$$

By applying the formulæ of § 135, or by simple inspection of the table, we get $y_{2'.1'} = y_{2'}$; $y_{3.0} = y_3$; $y_{3'.2'.1'} = y_{3'}$; $y_2 = y_{2.0}$ and substituting these results in the right-hand side of the last equivalence, we get

$$x_{1'.2} y_{1'.0} Q = x_{2'.3} y_{2'.3} + x_{2'.0} y_{3'.0} + x_{3'.2} y_{1'.2}.$$

The application of § 135, Form 3, to the y-statements will introduce no fresh statements in z, nor destroy any term by showing that it contains an impossible factor η. We have therefore found the nearest limits of y; and it only remains to find the limits of z. Multiplying the last expression by the datum $z_{1'.0}$ we get

$$QA = Q x_{1'.0} y_{1'.0} z_{1'.0} = (x_{2'.3} y_{2'.3} + x_{2'.0} y_{3'.0} + x_{3'.2} y_{1'.2}) z_{1'.0}$$

The application of § 135, Form 3, to the factor $z_{1'.0}$ will effect no change, since $(z_1 - z_0)^p$ is a certainty. The process of finding the limits is therefore over; and it only remains to evaluate the integrals. We get

$$\frac{Q}{A} = \frac{Int\ QA}{Int\ A} = Int\ QA$$

$$= Int(x_{2'.3} y_{2'.3} + x_{2'.0} y_{3'.0} + x_{3'.2} y_{1'.2}) z_{1'.0}$$

for $Int\ A = Int\ x_{1'.0} y_{1'.0} z_{1'.0} = 1$. The integrations are easy, and the result is $\frac{5}{4} - \log 2$ (Naperian base), which is a little above $\frac{5}{9}$.

156. Given that a is positive, that n is a positive whole number, and that the variables x and y are each taken at random between a and $-a$, what is the chance that $\{(x+y)^n - a\}$ is negative and $\{(x+y)^{n+1} - a\}$ positive?

Let A denote our data $y_{1'.2} x_{1'.2}$ (see Table); let Q denote the proposition $\{(x+y)^n - a\}^N$, and let R denote the proposition $\{(x+y)^{n+1} - a\}^P$, in which the exponent N denotes *negative*, and the exponent P *positive*.

We have to find the chance $\dfrac{QR}{A}$, which $= \dfrac{Int\ QRA}{Int\ A}$.

In this problem we have only to find the limits of integration (or variation) for the numerator from the compound statement QRA, the limits of integration for the denominator being already known, since $A = y_{1.2} x_{1.2}$.

§ 156]   CALCULUS OF LIMITS

Table of Limits

$y_1 = a$	$x_1 = a$	$a_1 = 1$
$y_2 = -a$	$x_2 = -a$	$a_2 = (\tfrac{1}{2})^{\tfrac{n}{n-1}}$
$y_3 = a^{\tfrac{1}{n}} - x$	$x_3 = a^{\tfrac{1}{n}} - a$	
$y_4 = -a^{\tfrac{1}{n}} - x$	$x_4 = a - a^{\tfrac{1}{n}}$	$a_3 = (\tfrac{1}{2})^{\tfrac{n+1}{n}}$
$y_5 = a^{\tfrac{1}{n+1}} - x$	$x_5 = a + a^{\tfrac{1}{n+1}}$	
$y_6 = -a^{\tfrac{1}{n+1}} - x$	$x_6 = a + a^{\tfrac{1}{n}}$	$A = \epsilon = y_{1'.2} x_{1'.2}$
	$x_7 = a^{\tfrac{1}{n+1}} - a$	
	$x_8 = -a - a^{\tfrac{1}{n+1}}$	
	$x_9 = a - a^{\tfrac{1}{n+1}}$	

We take the order of integration y, x. The limits being registered in the table, one after another, as they are found, the table grows as the process goes on. For convenience of reference the table should be on a separate slip of paper.

We will first suppose n to be even. Then

$$Q = \left\{(x+y) - a^{\tfrac{1}{n}}\right\}^N \left\{(x+y) + a^{\tfrac{1}{n}}\right\}^P$$
$$= \left\{y - (a^{\tfrac{1}{n}} - x)\right\}^N \left\{y + (a^{\tfrac{1}{n}} + x)\right\}^P = y_{3'.4}$$
$$R = \left\{(x+y) - a^{\tfrac{1}{n+1}}\right\}^P = \left\{y - (a^{\tfrac{1}{n+1}} - x)\right\}^P = y_5.$$

Hence $QR = y_{3'.4} y_5 = y_{3'.4.5}$; and multiplying by the datum $y_{1'.2}$, we get

$$QR y_{1'.2} = y_{3'.1'.2.4.5} = (y_3 x_3 + y_1 x_{3'})(y_2 x_{4.5} + y_5 x_{5'})$$
$$= (y_3 x_3 + y_1 x_{3'})(y_2 x_5 + y_5 x_{5'}) = y_{3'.2} x_5 + y_{3'.5} x_{5'.3} + y_{1'.5} x_{3'};$$

for by application of the formulæ of § 135, $x_{4.5} = x_5$

$x_{5.3} = x_5$; $x_{3'.5} = \eta$ (an impossibility); and $x_{3'.5'} = x_{3'}$. For when $a > 1$ we have $\left(2a - a^{\frac{1}{n}}\right)^P$, and when $a < 1$ we have $\left(a^{\frac{1}{n+1}} - a^{\frac{1}{n}}\right)^P$; so that $x_5 - x_3$ is always positive.

We must now apply § 135, Form 3, to the statements in y. We get $y_{3'.2} = y_{3'.2}x_{6'}$; $y_{3'.5} = y_{3'.5}a_1$; $y_{1'.5} = y_{1'.5}x_7$. Substituting these results, we get

$$QRy_{1'.2} = y_{3'.2}x_{6'.5} + y_{3'.5}x_{5'.3}a_1 + y_{1'.5}x_{3'.7}.$$

Having found the limits of the variable y, we must apply the three formulæ of § 135 to the statements in x. Multiplying by the datum $x_{1'.2}$, we get

$$QRy_{1'.2}x_{1'.2} = y_{3'.2}x_{1'.6'.5.2} + y_{3'.5}x_{1'.5'.3.2}a_1 + y_{1'.5}x_{3'.1'.7.2}$$
$$= y_{3'.5}x_{1'.3}a_1 + y_{1'.5}x_{3'.1'.7};$$
for $x_{1'.5} = \eta$; $x_{1'.5'} = x_{1'}$; $x_{3.2} = x_3$; $x_{7.2} = x_7$.

We obtain these results immediately by simple inspection of the Table of Limits, without having recourse to the formulæ of § 135. Applying the formulæ of § 135 to the statements in x which remain, we get

$$x_{3'.1'} = x_{3'}a_2 + x_1 a_{2'}; \quad x_{1'.3} = x_{1'.3}a_2;$$
$$x_{3'.7} = x_{3'.7}a_1; \quad x_{1'.7} = x_{1'.7}a_3.$$

Substituting these values, we get

$$QRy_{1'.2}x_{1'.2} = QRA = y_{3'.5}x_{1'.3}a_{1.2} + y_{1'.5}(x_{3'.7}a_{2.1} + x_{1'.7}\eta_{2'.3})$$
$$= y_{3'.5}x_{1'.3}a_1 + y_{1'.5}x_{3'.7}a_1$$
$$= (y_{3'.5}x_{1'.3} + y_{1'.5}x_{3'.7})a_1;$$

for $a_{1.2} = a_1 = a_{2.1}$; and $a_{2'.3} = \eta$ (an impossibility).

This is the final step in the process of finding the limits, and the result informs us that, *when n is even*, QRA is only possible when a_1 (which $= 1$) is an inferior limit of a. In other words, when n is even and a is not greater than 1, the chance of QR is *zero*. To find the chance when n is even and a is greater than 1, we have

§ 156] CALCULUS OF LIMITS 139

only to evaluate the integrals, employing the abbreviated notation of § 151. Thus

Integral $A = Int\ y_{1'.2}x_{1'.2} = (y_1 - y_2)x_{1'.2} = (2a)x_{1'.2}$
$= x_{1'.2}(2ax) = 2ax_1 - 2ax_2 = 4a^2$

Integral $QRA = y_{3'.5}x_{1'.3} + y_{1'.5}x_{3'.7}$
$= (y_3 - y_5)x_{1'.3} + (y_1 - y_5)x_{3'.7}$
$= \left(a^{\frac{1}{n}} - a^{\frac{1}{n+1}}\right)x_{1'.3} + \left(a - a^{\frac{1}{n+1}} + x\right)x_{3'.7}$
$= x_{1'.3}\left(a^{\frac{1}{n}} - a^{\frac{1}{n+1}}\right)x + x_{3'.7}\left(ax - a^{\frac{1}{n+1}}x + \tfrac{1}{2}x^2\right)$
$= \left(a^{\frac{1}{n}} - a^{\frac{1}{n+1}}\right)(x_1 - x_3) + \left(a - a^{\frac{1}{n+1}}\right)(x_3 - x_7) + \tfrac{1}{2}(x_3^2 - x_7^2)$
$= \left(a^{\frac{1}{n}} - a^{\frac{1}{n+1}}\right)\left(2a - \tfrac{1}{2}a^{\frac{1}{n}} - \tfrac{1}{2}a^{\frac{1}{n+1}}\right)$

$$\frac{QR}{A} = \frac{Int\ QRA}{Int\ A} = \frac{Int\ QRA}{4a^2}$$
$$= \frac{1}{8a^2}\left(a^{\frac{1}{n}} - a^{\frac{1}{n+1}}\right)\left(4a - a^{\frac{1}{n}} - a^{\frac{1}{n+1}}\right).$$

We have now to find the chance when n is odd. By the same process as before we get

$$QRA = (y_{3'.5}x_{1'.3} + y_{1'.5}x_{3'.7})a_1 + y_{6'.2}x_{9'.2}a_3.$$

Here we have *two* inferior limits of a, namely, a_1 and a_3, so that the process is not yet over. To separate the different possible cases, we must multiply the result obtained by the certainty $(a_1 + a_{1'})(a_3 + a_{3'})$, which here reduces to $a_1 + a_{1'.3} + a_{3'}$, since a_1 is greater than a_3.

For shortness sake let M_1 denote the bracket co-efficient (or co-factor) of a_1 in the result already obtained for QRA; and let M_3 denote $y_{6'.2}x_{9'.2}$, the coefficient of a_3. We get

$$QRA = (M_1 a_1 + M_3 a_3)(a_1 + a_{1'.3} + a_{3'})$$
$$= (M_1 + M_3)a_1 + M_3 a_{1'.3};$$

for $a_{1.3} = a_1$, and $a_{3'.1} = \eta$ (an impossibility). Hence, there are only two possible cases when n is an odd number, the case a_1 (that is to say, $a > a_1$, which here means $a > 1$) and the case $a_{1'.3}$. For the latter, $a_{1'.3}$, we get

$$\frac{QR}{A} = \frac{Int\ M_3}{Int\ A} = \frac{1}{8a^2}\left(2a - a^{\frac{1}{n+1}}\right)^2.$$

For the first case, namely, the case $a > 1$, we get

$$\frac{QR}{A} = \frac{Int(M_1 + M_3)}{Int\ A}.$$

When the integrals in this case are worked out, the result will be found to be

$$\frac{QR}{A} = \frac{1}{4a^2}\left(a^{\frac{1}{n}} - a^{\frac{1}{n+1}}\right)\left(2a - a^{\frac{1}{n}}\right) + \frac{1}{8a^2}\left(a^{\frac{1}{n}} - a^{\frac{1}{n+1}}\right)^2$$
$$+ \frac{1}{8a^2}\left(2a - a^{\frac{1}{n+1}}\right)^2.$$

The expression for the chance $\frac{QR}{A}$ in the case $a > 1$ and the expression for it in the case $a < 1$ evidently ought to give the same result when we suppose $a = 1$. This is easily seen to be the fact; for when we put $a = 1$, each expression gives $\frac{1}{8}$ as the value of the chance $\frac{QR}{A}$.

157. The great advantage of this "Calculus of Limits" is that it is *independent of all diagrams*, and can therefore be applied not only to expressions of two or three variables, but also to expressions of four or several variables. Graphic methods are often more expeditious when they only require straight lines or easily traced and well-known curves; but graphic methods of finding the limits of integration are, in general, difficult when there are three variables, because this involves the perspective representation of the intersections of curved surfaces.

§ 157]   CALCULUS OF LIMITS   141

When there are four or more variables, graphic methods cannot be employed at all. For other examples in probability I may refer the student to my sixth paper in the *Proceedings of the London Mathematical Society* (June 10th, 1897), and to recent volumes of *Mathematical Questions and Solutions from the Educational Times*. It may interest some readers to learn that as regards the problems worked in §§ 155, 156, I submitted my results to the test of actual experiment, making 100 trials in each case, and in the latter case taking $a=10$ and $n=3$. The *theoretical* chances (to two figures) are respectively ·56 and ·43, while the experiments gave the close approximations of ·53 and ·41 respectively.

THE END

Printed by BALLANTYNE, HANSON & Co.
Edinburgh & London

[1877o]: The Calculus of Equivalent Statements and Integration Limits. Proceedings of the London Mathematical Society, (1877-1878), vol. 9, pp. 9-20.

The Calculus of Equivalent Statements and Integration Limits.
By HUGH MCCOLL, B.A.

[*Read Nov. 8, 1877.*]

The above title seems to be the most suitable for an analytical method which I discovered a few months ago, and to which a short introduction was published in the "Educational Times" for last July, under the name of "Symbolical Language." The chief use of the method, as far as I have yet carried it, is to determine the new limits of integration when we change the order of integration or the variables in a multiple integral, and also to determine the limits of integration in questions relating to probability. This object it will accomplish with perfect certainty, and by a process almost as simple and mechanical as the ordinary operations of elementary algebra. The fundamental principles of the method are as follows:

DEFINITION 1.—Let any symbols, say A, B, C, &c., denote *statements* (or propositions) registered for convenience of reference in a table. Then the equation $A = 1$ asserts that the statement A is *true*; the equation $A = 0$ asserts that the statement A is *false*; and the equation $A = B$ asserts that A and B are equivalent statements.

DEF. 2.—The symbol $A \times B \times C$ or ABC denotes a *compound statement*, of which the statements A, B, C may be called the *factors*. The equation $ABC = 1$ asserts that *all the three statements are true;* the equation $ABC = 0$ asserts that all the three statements are *not* true, *i.e.*, that at least *one* of the three is false. Similarly a compound statement of any number of factors may be defined.

DEF. 3.—The symbol $A + B + C$ denotes an *indeterminate* statement, of which the statements A, B, C may be called the *terms*. The equation $A + B + C = 0$ asserts that all the three statements are *false;* the equation $A + B + C = 1$ asserts that all the three are *not* false, *i.e.*, that at least *one* of the three is true. Similarly an indeterminate statement of any number of terms may be defined.

DEF. 4.—The symbol A' is the *denial* of the statement A. The two statements A and A' are so related that they satisfy the two equations $A + A' = 1$ and $AA' = 0$; that is to say, one of the two statements (either A or A') must be true and the other false. The same symbol (*i.e.*, a dash) will convert any complex statement into its denial. For example, $(AB)'$ is the denial of the compound statement AB.

Note.—The statements A and A' are what logicians call "contradictories"; and the two equations $A + A' = 1$ and $AA' = 0$ combined express the principle known in logic as the "Law of Excluded Middle."

DEF. 5.—When only *one* of the terms of an indeterminate statement $A+B+C+\ldots$ can be true, or when no two terms can be true at the same time, the terms are said to be *mutually inconsistent* or *mutually exclusive*.

RULE 1.—The rule of ordinary algebraical multiplication applies to the multiplication of indeterminate statements, thus:

$$A(B+C) = AB+AC; \quad (A+B)(C+D) = AC+AD+BC+BD;$$

and so on for any number of factors, and whatever be the number of terms in the respective factors.

Note.—It is evident that if the terms of every indeterminate factor be mutually inconsistent, the terms of the product will also be mutually inconsistent.

RULE 2.—Let A be any statement whatever, and let B be any statement which is implied in A (and which must therefore be true when A is true, and false when A is false); or else let B be any statement which is admitted to be true independently of A; then (in either case) we have the equation $A = AB$. As particular cases of this we have $A = AA = AAA = $ &c., as repetition neither strengthens nor weakens the logical value of a statement. Also,

$$A = A(B+B') = A(B+B')(C+C') = \&c.,$$

for $B+B' = 1 = C+C' = $ &c. (See Def. 4.)

RULE 3.
$$\begin{aligned}(AB)' &= AB'+A'B+A'B' \\ &= AB'+A'(B+B') = AB'+A' \\ &= A'B+B'(A+A') = A'B+B',\end{aligned}$$

for $A+A' = 1$ and $B+B' = 1$. Similarly we may obtain various equivalents (with mutually inconsistent terms) for $(ABC)'$, $(ABCD)'$, &c.

RULE 4. $\quad (A+B)' = A'B'; \quad (A+B+C)' = A'B'C';$

and so on.

RULE 5. $\quad \begin{aligned}A+B &= \{(A+B)'\}' = (A'B')' \\ &= AB'+A'B+AB \\ &= AB'+(A'+A)B = AB'+B \\ &= A'B+A(B'+B) = A'B+A.\end{aligned}$

Similarly we get equivalents (with mutually inconsistent terms) for $A+B+C$, $A+B+C+D$, &c.

The foregoing principles constitute the elementary *basis* of the method. We now come to the more important part of the subject, namely, the application of the method to multiple integrals and probability.

1877.]   *Equivalent Statements and Integration Limits.*   11

Def. 6.—The symbol p prefixed to any algebraical (or arithmetical) expression converts the expression into a statement, namely, that the expression is *real and positive*; the symbol p' in like manner asserts that the expression affected by it is *real and negative*. For example, if we know that x and y are both real, we have the equations:

$$p(xy) = px\,py + p'x\,p'y,$$
$$p'(xy) = p'x\,py + px\,p'y.$$

Again, suppose we know that x, y, a are all three real and positive, it is easy to see the identities,

$$p(x^2+y^2-a^2) = p\{y-\sqrt{(a^2-x^2)}\}\,p'(x-a) + p(x-a),$$
$$p'(x^2+y^2-a^2) = p'\{y-\sqrt{(a^2-x^2)}\}\,p'(x-a).$$

Note.—We might also use a symbol q to denote the statement that the expression affected by it was imaginary, but I do not think that the need for the symbol would often arise.

Def. 7.—The symbols px, $p'x$, py, $p'y$, &c. occur so frequently that it is convenient to replace them respectively by x_0, $x_{0'}$, y_0, $y_{0'}$, &c. Another reason for the employment of these last symbols will appear later.

Def. 8.—The symbols x_1, x_2, x_3, &c. denote the 1st, 2nd, 3rd, &c. limits of x registered in any convenient order in a table of reference. The limits of the other variables of the expression (or expressions) under consideration are denoted similarly.

Note.—Among the limits of the variables thus registered, the limit *zero* is not included (see Defs. 7 and 9); but we may denote it either by the usual symbol 0, or (for the sake of uniformity in the notation) by any of the symbols x_0, y_0, z_0, &c.

Def. 9.—The symbols x_1, x_2, x_3, &c. also denote *statements*, namely, the statements that the *limits* x_1, x_2, x_3, &c. are *inferior* limits of x. Similarly y_1, y_2, &c., z_1, z_2, &c. are to be interpreted.

Note.—The symbol x_m has thus two meanings: it denotes the mth *limit* of x, and it also denotes the *statement* that this limit is an *inferior limit*; in other words, x_m is an abbreviation for $p(x-x_m)$. The context will always prevent any confusion of ideas resulting from this double signification of the same symbol.

Def. 10.—The symbols $x_{1'}$, $x_{2'}$, $x_{3'}$, &c. denote the *statements* that the limits x_1, x_2, x_3, &c. are *superior* limits of x. Similarly $y_{1'}$, $y_{2'}$, &c., $z_{1'}$, $z_{2'}$, &c. are to be interpreted.

Note.—The symbol $x_{m'}$ is thus an abbreviation for the statement $p'(x-x_m)$.

Def. 11.—The symbol $x_{m'n'r's}$ is an abbreviation for the symbol $x_{m'}\,x_{n'}\,x_{r'}\,x_{s}$, and denotes the compound statement

$$p'(x-x_m)\ p'(x-x_n)\ p(x-x_r)\ p(x-x_s).$$

Similarly a compound statement of any number of factors, and having reference to the limits of x or of any other variable, may be abbreviated.

Rule 6.—The compound statement

$$x_{m'n'r's'\ldots} = x_{m'}\,\alpha + x_{n'}\,\beta + x_{r'}\,\gamma + x_{s'}\,\delta + \ldots,$$

in which $x_{m'}$, $x_{n'}$, &c. are abbreviations for the statements that the mth, n^{th}, &c. limits of x are all *superior* limits; while α, β, γ, &c. respectively denote the statements that amongst these x_m is the nearest superior limit of x, that x_n is the nearest superior limit, that x_r is the nearest superior limit, and so on. In other words, α is an abbreviation for the compound statement

$$p'(x_m-x_n)\ p'(x_m-x_r)\ p'(x_m-x_s)\ldots.$$

The value of β is obtained from this expression by simply interchanging m and n; the value of γ is obtained from the expression for β by interchanging n and r; and so on.

Rule 7.—This is obtained from the preceding Rule by simply copying all the words in it (except *superior*, for which we must write *inferior*), and *omitting all the accents*, both on the numbers m, n, r, s, ... and on the symbol p.

Note.—Rule 6 may be illustrated by a plane figure as follows:—

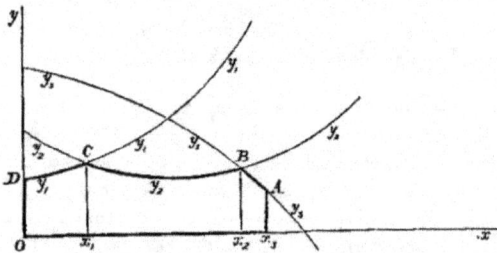

Let $y = \phi(x)$, $y = \psi(x)$, $y = \chi(x)$ be the three equations for the curves marked respectively $y_1\,y_1\,y_1$, $y_2\,y_2\,y_2$, $y_3\,y_3\,y_3$. Then all the points contained within the thick boundary DOx_3ABCD will be expressed by the statement $y_{1'.1'.3'.0}\,x_{3'.0}$;

and this statement is evidently equivalent to the statement

$$y_{1'.0}\,x_{1'.0} + y_{2'.0}\,x_{2'.1} + y_{3'.0}\,x_{3'.2},$$

1877.]   *Equivalent Statements and Integration Limits.*   13

x and y being the Cartesian coordinates of any point whatever within the boundary.

A similar geometrical illustration may be given of Rule 7.

RULE 8.—The statement $x_{m'n} = x_{m'n} a$, in which $a = p(x_m - x_n)$.

Note.—This is a particular case of Rule 2.

Rules 6 and 7 may also be brought under Rule 2 as follows:—Attaching the same meaning to a, β, γ, &c. as in Rule 6, it is evident that $a + \beta + \gamma + \delta + \ldots = 1$. Hence, by Rule 2, we get

$$x_{m'n'r'r'\ldots} = x_{m'n'r'r'\ldots}(a + \beta + \gamma + \delta + \ldots);$$

and since $x_{m'} a = x_{m'n'r'r'\ldots} a$, and so on for the β, γ, δ, &c. terms, we get Rule 6 by the suppression of implied factors. We may similarly show Rule 7 to be a particular case of Rule 2.

The last three Rules, 6, 7, 8 (combined with Rule 1), constitute the pivot on which the whole process turns, whether in its application to the transformation of multiple integrals or to probability. By repeated application of these three Rules to the several variables in succession, any compound statement of the form

$$x_{1'.2'.0'\ldots 5.4.6\ldots} \, y_{5'.7'.3'\ldots 0.3.4\ldots} \, z_{3'.4'.0'\ldots 1.2.6\ldots} \ldots,$$

with any number of variables, and any number of factors for each variable, will finally be reduced either to a single *elementary*[*] term of the form $x_{m'n} y_{r's} z_{t'u} \ldots$, in which x_m and x_n are the *nearest* limits (superior and inferior) of x, and so on for the variables y, z, &c.; or (as will generally happen in complicated cases) it will be reduced to an indeterminate statement consisting of several such terms. The work may generally be much abbreviated by dropping *zero terms* (*i.e.*, terms with inconsistent factors) as we go along, when mere inspection of the table of limits (without having recourse to Rule 8) will suffice to detect them. But if we overlook these zero terms, they will eventually disappear of themselves in the subsequent evolutions of the process. We may also shorten the process by cancelling factors which mere inspection of the table (instead of having recourse to Rule 6 or 7) will show to be implied in their co-factors *of the same variable.*

To give a practical illustration of the method, we will take a problem of some complexity. Suppose we are required to *reverse the order of integration* in the multiple integral

$$\int_{-a}^{2a} du \int_{-u}^{2u} dx \int_{-x}^{2x} dy \int_{-2z}^{\frac{y^2}{2x}} dz \, \phi(u, x, y, z).$$

---

[*] A term of this form may be called *elementary* when the application of Rule 8 will introduce no fresh factor except unity, or such factors as Rules 6 or 7 would afterwards reject as unnecessary.

For greater facility of reference throughout the process, the annexed table of limits may be conveniently made out on a card or moveable slip of paper. The values severally entered in the table during the course of the operations are found to arise spontaneously in reducing the various factors to con-

Table of Limits.			
$u_1 = -a$	$x_1 = -y$	$y_1 = \frac{1}{2}z$	$z_1 = -8a$
$u_2 = 2a$	$x_2 = \frac{1}{2}y$	$y_2 = -z$	$z_2 = 2a$
$u_3 = -x$	$x_3 = -\frac{1}{2}z$	$y_3 = \sqrt{8az}$	$z_3 = 8a$
$u_4 = \frac{1}{2}x$	$x_4 = \dfrac{y^2}{2z}$	$y_4 = -\sqrt{8az}$	
	$x_5 = a$	$y_5 = -2z$	
	$x_6 = -2a$	$y_6 = -4a$	
	$x_7 = 4a$	$y_7 = z$	
		$y_8 = 8a$	

venient symbols. Each limit is registered in the table as soon as it is ascertained, so that the table *grows* as the process proceeds. Sometimes a limit which has already been registered as x_m may again inadvertently be registered as x_n; when this happens the oversight will be detected later by the appearance of an anomalous statement, such as

$$p(x_m - x_n) = p(0).$$

From the integral we get a compound statement of 8 factors (2 for each variable); so that, if we denote the compound statement by A, and the 8 factors by A_1, A_2, A_3, &c., we have

$$A_1 = p(u+a) = u_1,$$
$$A_2 = p'(u-2a) = u_2,$$
$$A_3 = p(x+u) = u_3,$$
$$A_4 = p'(x-2u) = p(u-\tfrac{1}{2}x) = u_4,$$
$$A_5 = p(y+x) = x_1,$$
$$A_6 = p'(y-2x) = p(x-\tfrac{1}{2}y) = x_2,$$
$$A_7 = p(z+2x) = p(x+\tfrac{1}{2}z) = x_3,$$
$$A_8 = p'\left(z-\frac{y^2}{2x}\right) = p'\left\{\frac{(x-x_4)z}{x}\right\}.$$

But evidently $\quad p'\left(\dfrac{x-x_4}{x}\right) = x_{4'.0} + x_{0'.4}; \quad p'z = z_0;$

$p\left(\dfrac{x-x_4}{x}\right) = x_{4.0} + x_{4'.0'}; \quad p'z = z_{0'};$

and therefore $p'\left\{\dfrac{x-x_4}{x}z\right\} = (x_{4'.0} + x_{0'.4})z_0 + (x_{4.0} + x_{4'.0'})z_{0'}$

$\qquad\qquad\qquad = x_{4'.0}z_0 + x_0 z_{0'} + x_{4'} z_{0'},$

for by inspection of the table of limits we see that $x_{0'.4} z_0 = 0$, $x_{4.0} z_{0'} = x_0 z_{0'}, \; x_{4'.0'} z_{0'} = x_{4'} z_{0'}.$ Hence we have

$$A = u_{1'.1.3.4} x_{1.2.3} (x_{4'.0} z_0 + x_0 z_{0'} + x_{4'} z_{0'}).$$

1877.]   *Equivalent Statements and Integration Limits.*   15

Now the simplest and generally the most advantageous order in the application of Rules 6, 7, 8 would be to apply these rules first to the u-factors; then (after multiplying) to the x-factors; then to the y-factors which would arise; and lastly to the z-factors. But a little saving of labour will be effected in this instance by slightly departing from this order, which is never an absolutely necessary order of application. We will first apply Rule 7 to the compound statements $u_{1.3.4}$ and $x_{1.2.3}$, and then Rule 8 to the u-factors, thus:

$$u_{1.3.4} = u_1 a + u_3 \beta + u_4 \gamma,$$

in which
$$\begin{aligned}
a &= p(u_1 - u_3)\ p(u_1 - u_4) \\
&= p(-a+x)\ p(-a-\tfrac{1}{2}x) = x_{5.6'} = 0, \\
\beta &= p(u_3 - u_1)\ p(u_3 - u_4) = x_{5'.0'} = x_{0'}, \\
\gamma &= p(u_4 - u_1)\ p(u_4 - u_3) = x_{6.0} = x_0.
\end{aligned}$$

Expanding the compound statement $x_{1.2.3}$ in the same way into an indeterminate statement, and substituting, we get

$$A = u_{2'}(u_3 x_{0'} + u_4 x_0)(x_1 y_{0'.1'} + x_2 y_{0.2} + x_3 y_{2'.1})(x_{4'.0} z_0 + x_0 z_{0'} + x_{4'} z_{0'}).$$

Multiplying the three indeterminate factors, omitting the zero terms in the result, and cancelling those factors in each term which mere inspection of the table will show to be implied in their co-factors of the same variable, we get (see Appendix, Note a)

$$A = u_{2'.4}(x_{4'.1} y_{0'} + x_{4'.2} y_0) z_0 + u_{2'.4}(x_1 y_{1'} + x_2 y_2 + x_3 y_{2'.1}) z_{0'}.$$

Applying now Rule 8 to the compound factor $u_{2'.4}$, we get

$$u_{2'.4} = u_{2'.4} x_{7'},$$

so that a fresh factor $x_{7'}$ is introduced among the factors in x (see Appendix, Note β).

We have now done with the limits of the variable u, so we apply our rules (when mere inspection of the table is not sufficient) to the x-statements, and we get (after cancelling implied factors in y, see Appendix, Note β)

$$\begin{aligned}
A = &\ u_{2'.4}(x_{4'.1} y_{5'.4} + x_{7'.1} y_{4'.6} + x_{4'.2} y_{5'.7} + x_{7'.2} y_{5'.3}) z_0 \\
&+ u_{2'.4}(x_{7'.1} y_{1'.6} + x_{7'.2} y_{8'.2} + x_{7'.3} y_{2'.1} z_1) z_{0'}.
\end{aligned}$$

Having now done with the limits of u and x, we apply Rule 8 (Rules 6 and 7, not being required as implied factors in y, have already been cancelled) to the y statements, when we shall get finally

$$\begin{aligned}
A = &\ z_{2'.0}(y_{5'.4} x_{4'.1} + y_{4'.6} x_{7'.1}) u_{2'.4} \\
&+ z_{3'.0}(y_{5'.7} x_{4'.2} + y_{8'.3} x_{7'.2}) u_{2'.4} \\
&+ z_{0'.1}(y_{1'.6} x_{7'.1} + y_{8'.2} x_{7'.2} + y_{2'.1} x_{7'.3}) u_{2'.4},
\end{aligned}$$

altogether 7 elementary terms. The first is $x_{1'.0}y_{9'.4}x_{4'.1}u_{5'.4}$, the corresponding term of the transformed integral being

$$\int_0^{z_2} dz \int_{y_4}^{y_5} dy \int_{x_1}^{x_4} dx \int_{u_4}^{u_5} du \, \phi(u, x, y, z);$$

and so on for the remaining 6 terms and the corresponding 6 terms of the transformed integral.

The mode of applying the process to find the limits of integration when we change the *variables* in a multiple integral is so obvious from the mode of applying it in finding the limits when we change the *order of integration*, that it is unnecessary to illustrate it by a separate example. We shall therefore end this article by applying the method to an easy question in probability.

In the quadratic equation $x\theta^2 - y\theta + z = 0$, if the coefficients x, y, z be each taken at random between a and 0, what is the chance that the roots of the equation will be real, all values of x, y, z between the given limits being equally probable?

Let A denote the statement whose truth is taken for granted, namely, the statement $x_{1'.0}y_{1'.0}z_{1'.0}$ (see the table); and let Q denote the statement which may be true or false, namely, $p(y^2 - 4xz)$. Then the required chance is

$$\frac{AQ \iiint dx\,dy\,dz}{A \iiint dx\,dy\,dz},$$

the statements A and AQ fixing the limits of integration for the denominator and numerator respectively.

The denominator of the above fraction is evidently

$$\int_0^{x_1} dx \int_0^{y_1} dy \int_0^{z_1} dz = a^3,$$

TABLE OF LIMITS.

$x_1 = a$	$y_1 = a$	$x_1 = a$
$x_2 = \dfrac{y^2}{4x}$	$y_2 = \sqrt{4ax}$	$x_2 = \dfrac{a}{4}$

the statement A being elementary, since the application of rule 8 will introduce no fresh factors. It remains to find the limits of integration for the numerator from the statement AQ.

Now $Q = p'(4xz - y^2)$, and since x is positive, this

$$= p'\left(z - \frac{y^2}{4x}\right) = z_{2'}.$$

Hence $\qquad AQ = x_{1'.0}y_{1'.0}z_{2'.1'.0}.$

But $\qquad z_{2'.1'} = z_{2'}y_{2'} + z_{1'}y_2,\;$ by Rule 6.

Hence $\qquad AQ = x_{1'.0}y_{1'.0}(z_{2'}y_{2'} + z_{1'}y_2)z_0$

$\qquad\qquad\quad = x_{1'.0}(y_{2'.1'.0}z_{2'.0} + y_{1'.0.2}z_{1'.0}).$

1877.]   *Equivalent Statements and Integration Limits.*   17

But $y_{2'.1'} = y_{2'} x_{2'} + y_{1'} x_2$ by Rule 6, and $y_{0.2} = y_2$ by Rule 7 or mere inspection of the table.

Hence $AQ = x_{2'.1'.0}\, y_{2'.0}\, z_{2'.0} + x_{1'.0.2}\, y_{1'.0}\, z_{2'.0} + x_{1'.0}\, y_{1'.2}\, z_{1'.0}.$

But by mere inspection of the table we get $x_{2'.1'} = x_{2'}$ and $x_{0.2} = x_2$; and in the third term we get (by Rule 8) $y_{1'.2} = y_{1'.2}\, x_{2'}.$

Hence, finally,

$$AQ = x_{2'.0}\, y_{2'.0}\, z_{2'.0} + x_{1'.2}\, y_{1'.0}\, z_{2'.0} + x_{2'.0}\, y_{1'.2}\, z_{1'.0}.$$

These three terms are elementary, since rule 8 will introduce no fresh factors. Hence

$$AQ \iiint dx\, dy\, dz = \int_0^{x_2} dx \int_0^{y_2} dy \int_0^{z_2} dz + \int_{x_2}^{x_1} dx \int_0^{y_1} dy \int_0^{z_2} dz$$
$$+ \int_0^{x_2} dx \int_{y_2}^{y_1} dy \int_0^{z_1} dz.$$

The integrations are easy, and the result is $(\tfrac{5}{36} + \tfrac{1}{6} \log_e 2)\, a^3$.

The required chance is therefore $\tfrac{5}{36} + \tfrac{1}{6} \log_e 2$.

With reference to the preceding solution, the referees of the Mathematical Society have kindly made the following valuable suggestion, which may also be extended to other problems:

"The process will be considerably abbreviated and simplified if from the outset the statements x_0, y_0, z_0, $x_{1'}$, $y_{1'}$, $z_{1'}$ are severally regarded as *unit*-factors, and therefore omitted when not wanted. Thus the whole working would be as follows:

$$AQ = z_{2'} = z_{2'.1'} = z_{2'}\, y_{2'} + z_{1'}\, y_2$$
$$= z_{2'}\, y_{2'.1'} + y_2 = z_{2'}\, (y_{2'}\, x_{2'} + y_{1'}\, x_2) + y_2.$$

And, by restoring or supplying the proper unit-factors, the final result is at once obtained."

In accordance with this suggestion I would propose the following convention:—

When we are analyzing any factor A of any compound statement $ABC...$, the truth of its co-factors B, C, ... may for the time be taken for granted, so that as *unit-factors* they may be introduced or suppressed at pleasure. The equation $A = a$, according to this convention, will assert, *not* that a is *always* an equivalent for A, but *that a may replace A in the particular compound statement of which A is a factor*. We may then have such equations as $A = AB = aB = a = aC = $ &c. According to this convention, it is evident that $A = 1$ asserts either that A is *implied* in some co-factor (which co-factor may be true or false), or else that A is *true absolutely* and independently of any co-factor. Also $A = 0$ asserts either that A is *inconsistent* with some co-factor (which co-factor may be true or false), or else that A is *false absolutely* and independently of any co-factor.

These abbreviations, however, necessitate more caution in the working of the process: for, on the one hand, care is needed lest any factor, left temporarily understood for the sake of brevity, should inadvertently be left out of account altogether; and, on the other hand, lest factors which have been already taken into account should again be needlessly introduced when their services are no longer required. For these reasons I think the abbreviations cannot be employed with safety till some familiarity has been first acquired with the longer but easier method adopted in the text.

APPENDIX.

NOTE a.—This is obtained as follows:—Multiplying the last two indeterminate factors, namely,

$$x_1 y_{0'.1'} + x_2 y_{0.2} + x_3 y_{2'.1},$$

and
$$x_{4'.0} z_0 + x_0 z_{0'} + x_{4'} z_{0'},$$

we get for our product

$$x_{4'.1.0}\, y_{0'.1'}\, z_0 + x_{4'.2.0}\, y_{0.2}\, z_0 + \underline{x_{4'.3.0}\, y_{2'.1}\, z_0}$$
$$+ x_{1.0}\, y_{0'.1'}\, z_{0'} + x_{2.0}\, y_{0.2}\, z_{0'} + x_{3.0}\, y_{2'.1}\, z_{0'}$$
$$+ x_{4'.1}\, y_{0'.1'}\, z_{0'} + x_{4'.2}\, y_{0.2}\, z_{0'} + x_{4'.3}\, y_{2'.1}\, z_{0'}.$$

But the third term (the one underlined) is zero, since it contains the inconsistent compound factor $y_{2'.1} z_0$; for (by Rule 8)

$$y_{2'.1} = y_{2'.1}\, p(y_2 - y_1) = y_{2'.1}\, p(-z - \tfrac{1}{2}z) = y_{2'.1} z_{0'},$$

and $z_{0'}$ is inconsistent with z_0.

Omitting the underlined term therefore, and multiplying the terms left by $u_3 x_{0'} + u_4 x_0$, we get (omitting the terms which contain the inconsistent factor $x_{0'.0}$)

$$u_3 x_{0'} (\underline{x_{4'.1}\, y_{0'.1'}} + \underline{x_{4'.2}\, y_{0.2}} + \underline{x_{4'.3}\, y_{2'.1}})\, z_0$$
$$+ u_4 x_0 (x_{4'.1}\, y_{0'.1'}\, z_0 + x_{4'.2}\, y_{0.2}\, z_0 + x_1\, y_{0'.1'}\, z_{0'}$$
$$+ x_2\, y_{0.2}\, z_{0'} + x_3\, y_{2'.1}\, z_{0'} + x_{4'.1}\, y_{0'.1'}\, z_{0'}$$
$$+ \underline{x_{4'.2}\, y_{0.2}\, z_{0'}} + \underline{x_{4'.3}\, y_{2'.1}\, z_{0'}}).$$

But each of the compound factors $x_{0'.1}\, y_{0'}$, $x_{0'.3}\, y_0$, $x_{0'.3}\, z_{0'}$, $x_{4'.0}\, z_{0'}$ is zero, as may be seen by application of Rule 8 or mere inspection of the Table of Limits. Hence the terms underlined vanish. Substituting the terms left, we get

$$A = u_{2'.4}\, (x_{4'.1.0}\, y_{0'.1'}\, z_0 + x_{4'.2.0}\, y_{2.0}\, z_0 + x_{1.0}\, y_{0'.1'}\, z_{0'}$$
$$+ x_{2.0}\, y_{2.0}\, z_{0'} + x_{3.0}\, y_{2'.1}\, z_{0'}).$$

But the factors *dotted underneath* in the respective terms may be omitted, for they are implied in their co-factors *of the same variable*, as

1877.]   *Equivalent Statements and Integration Limits.*   19

may be seen by mere inspection of the table, or by application of Rules 6 and 7. Taking, for example, the factors of the first term, we have
$$x_{1.0} = x_1 a + x_0 \beta,$$
in which $a = p(x_1 - x_0) = p(-y) = y_{0'},$
and $\beta = p(x_0 - x_1) = y_0 = 0,$
because of the co-factor $y_{0'}$;

and $y_{0'.1'} = y_{0'} a + y_{1'} \beta,$
in which $a = p'(y_0 - y_1) = p'(-\tfrac{1}{2}z) = z_0,$
and $\beta = p'(y_1 - y_0) = p'(\tfrac{1}{2}z) = z_{0'} = 0.$

The cancelling in the other terms may be similarly verified by Rules 6 and 7.

NOTE β.—Substituting this value of $u_{7.4}$, we get
$$A = u_{7.4}(x_{7.4.1} y_{0'} + x_{7.4.2} y_0) z_0$$
$$+ u_{7.4}(x_{7.1} y_{1'} + x_{7.2} y_2 + x_{7.3} y_{3.1}) z_{0'}.$$

In the *first term* of the first bracket,
$$x_{7.4'} = x_7 a + x_{4'} \beta,$$
in which $a = p'(x_7 - x_4) = p'\left(4a - \dfrac{y^2}{2z}\right)$
$= p(y^2 - 8az),$

for $z_{0'}$ is inadmissible because of the co-factor z_0 outside the bracket,
thus, $a = p\{(y - \sqrt{8az})(y + \sqrt{8az})\}$
$= p'(y + \sqrt{8az}),$

for positive values of y are inadmissible because of the co-factor $y_{0'}$;
also $\beta = p'(x_4 - x_7) = p'(y^2 - 8az)$
$= p'\{(y - \sqrt{8az})(y + \sqrt{8az})\}$
$= p(y + \sqrt{8az}).$

Thus, $a = y_{4'}$ and $\beta = y_4.$

Again, in the *second term* of the first bracket, we have
$$x_{7.4'} = x_7 a + x_{4'} \beta,$$
in which (as before)
$$a = p(y^2 - 8az) = p\{(y - \sqrt{8az})(y + \sqrt{8az})\}$$
$$= p(y - \sqrt{8az}),$$

for *negative* values of y are inadmissible this time because of the co-factor y_0. Hence $a = y_3$.

Similarly, we get $\beta = y_{4'}.$

Substituting these values of $x_{7'.4}$ in the first and second terms respectively, we get

$$A = u_{2'.4}\{(x_{7'.1} y_{4'.0'} + x_{4'.1} y_{0'.4} + x_{7'.2} y_{8.0} + x_{4'.1} y_{8'.0}) z_0 \\ + (x_{7'.1} y_{1'} + x_{7'.2} y_2 + x_{7'.3} y_{2'.1}) z_{0'}\}.$$

Applying Rule 8 to the x-statements in each term, we get

in the first term, $\quad x_{7'.1} = x_{7'.1} y_0$;

in the second term, $\quad x_{4'.1} = x_{4'.1} y_{8'}$,

because of the co-factors z_0 and $y_{0'}$;

in the third term, $\quad x_{7'.2} = x_{7'.2} y_8$;

in the fourth term, $\quad x_{4'.2} = x_{4'.2} y_7$,

because of the co-factors z_0 and y_0.

Taking next the second bracket, we have

in the fifth term, $\quad x_{7'.1} = x_{7'.1} y_0$;

in the sixth term, $\quad x_{7'.2} = x_{7'.2} y_{8'}$;

in the seventh term, $\quad x_{7'.3} = x_{7'.3} z_1$.

Substituting in every term, we get

$$A = u_{2'.4}\{(x_{7'.1} y_{4'.0'.6} + x_{4'.1} y_{0'.8'.4} + x_{7'.2} y_{8.8.0} + x_{4'.2} y_{8'.0.7}) z_0 \\ + (x_{7'.1} y_{1'.6} + x_{7'.2} y_{8'.2} + x_{7'.3} y_{2'.1} z_1) z_{0'}\}.$$

But by inspection of the table, or by application of Rules 6 and 7, we have $y_{4'.0'} = y_{4'}$, $y_{0'.8'} = y_{8'}$, $y_{8.0} = y_8$, and $y_{0.7} = y_7$, so that the factors dotted underneath may be cancelled. Hence we get

$$A = u_{2'.4}\{(x_{7'.1} y_{4'.6} + x_{4'.1} y_{8'.4} + x_{7'.2} y_{8'.3} + x_{4'.2} y_{8'.7}) z_0 \\ + (x_{7'.1} y_{1'.6} + x_{7'.2} y_{8'.2} + x_{7'.3} y_{2'.1} z_1) z_{0'}\}.$$

Applying Rule 8 to the y-statements in each term, we get

in the *first* term, $\quad y_{4'.6} = y_{4'.6} z_{2'}$;

in the *second* term, $\quad y_{8'.4} = y_{8'.4} z_{2'}$;

in the *third* term, $\quad y_{8'.3} = y_{8'.3} z_{2'}$;

in the *fourth* term, $\quad y_{8'.7} = y_{8'.7} z_{2'}$;

in the *fifth* term, $\quad y_{1'.6} = y_{1'.6} z_1$;

in the *sixth* term, $\quad y_{8'.2} = y_{8'.2} z_1$;

in the *seventh* term, $\quad y_{2'.1} = y_{2'.1} z_{0'}$.

Substituting in every term, we get

$$A = \{z_{2'.0}(y_{4'.6} x_{7'.1} + y_{8'.4} x_{4'.1}) + z_{8'.0}(y_{8'.3} x_{7'.2} + y_{8'.7} x_{4'.2}) \\ + z_{0'.1}(y_{1'.6} x_{7'.1} + y_{8'.2} x_{7'.2} + y_{2'.1} x_{7'.3})\} u_{2'.4},$$

which, except in the arrangement of the terms, agrees with the result in the text.

[1877p]: The Calculus of Equivalent Statements (II). *Proceedings of the London Mathematical Society*, vol. 9, pp. 177-186.

The Calculus of Equivalent Statements (Second Paper).

By HUGH McCOLL, B.A.

[*Read June* 13*th*, 1878.]

The following additions to my former article on this subject (see Vol. IX., pp. 9—20), though perhaps belonging more strictly to the province of logic than to that of mathematics, will, I hope, be found interesting.

DEF. 12.—The symbol $A : B$ (which may be called an *implication*) asserts that the statement A implies B; or that whenever A is true B is also true.

Note.—It is evident that the implication $A : B$ and the equation $A = AB$ are equivalent statements. (See Rule 2.)

RULE 9.—As an extension of Rule 3 we may give the following:—

$$(ABCD\ldots)' = A' + B' + C' + D' + \ldots$$
$$= A' + AB' + ABC' + ABCD' + \ldots,$$

there being as many terms on the right-hand side of this equation as there are factors on the left-hand side.

Note.—It will be observed—(1) that the terms of the second equivalent for $(ABCD\ldots)$, are mutually exclusive; (2) that the rth term of this equivalent contains r factors; and (3) that the *last factor only* of each term is accented.

RULE 10.—As an extension of Rule 5 we may give the following:

$$A + B + C + D + \ldots = A + A'B + A'B'C + A'B'C'D + \ldots;$$

there being as many terms on the right-hand side as on the left.

Note.—It will be observed—(1) that the terms on the right-hand side are mutually exclusive; (2) that the rth term contains r factors; and (3) that in every term *all the factors except the last* are accented. It will also be observed that the equivalents (with mutually exclusive terms) for $(ABCD\ldots)'$ and $A + B + C + D + \ldots$ are so related that we can obtain either from the other by simply interchanging A and A', B and B', C and C', D and D', and so on throughout.

RULE 11.—If $A : B$, then $B' : A'$. Thus the implications $A : B$ and $B' : A'$ are equivalent, each following as a necessary consequence of the other. This is the logical principle of "contraposition."

Note.—The implication $A : BCD\ldots$ gives us as many simple implications as there are *factors* on the *right-hand side*, namely, the implications $A : B$, $A : C$, $A : D$, &c. The implication $A + B + C + \ldots : M$ gives us as many simple implications as there are *terms* on the *left-hand side*, namely, the implications $A : M$, $B : M$, $C : M$, &c. The equation $A = B$ gives us the two implications $A : B$ and $B : A$, from which again we get by contraposition $B' : A'$ and $A' : B'$. The implication $A : B$ is *not* of course equivalent to $B : A$, but to $B' : A'$; for, when we transpose the statements, we must change their signs, i.e., affix or remove the accent of negation. The implication $A : 0$ asserts that A is false; but $A : 1$ does not assert that A is true.

RULE 12.—If $A : B$, then $AC : BC$, whatever the statement C may be.

RULE 13.—If $A : \alpha$, $B : \beta$, $C : \gamma$, then $ABC : \alpha\beta\gamma$, and so on for any number of implications.

RULE 14.—If $AB = 0$, then $A : B'$ and $B : A'$. More generally, if $ABCD... = 0$; then $A : (BCD...)'$, $B : (ACD...)'$, and so on; also $AB : (CD)'$, $AC : (BD)'$, and so on. In other words, when a compound statement is false, if one or more of its factors be true, one at least of the remaining factors is false.

I quote the following from Bain's "Deductive Logic":—

"The author (i.e. Professor Boole) extends his analysis so as to comprise a more difficult order of examples, typified thus: Suppose the analysis of a particular class of substances has conducted us to the following general conclusions, namely:—

"First, Wherever the properties A and B are combined, either the property C or the property D is present also; but they are not present jointly.

"Secondly, Wherever B and C are combined, A and D are either both present or both absent.

"Thirdly, Wherever A and B are both absent, C and D are both absent also; and vice versâ, where C and D are both absent, A and B are both absent also.

"Let it then be required from these conditions to determine what may be concluded in any particular instance from the presence of the property A with respect to the presence or absence of the properties B and C, paying no regard to the property D.

"The working of the corresponding equations leads to the answer:—Wherever A is present, there either C is present and B absent, or C is absent. And, inversely, wherever C is present and B is absent, there A is present."

Professor Bain does not give Professor Boole's solution of this problem, and I have not yet seen the latter's celebrated works, "The Mathematical Analysis of Logic," and "The Laws of Thought," but the problem may be easily solved by the preceding rules as follows:—

Let A denote the statement that the property A is present, and let A' denote the statement that the property A is absent. Let similar interpretations be given to B, B', C, C', D, D'.

The first two given implications are

$$AB : (C+D)(CD)' \quad \ldots\ldots\ldots\ldots\ldots\ldots(1),$$
$$BC : AD + A'D' \quad \ldots\ldots\ldots\ldots\ldots\ldots(2);$$

and from these data alone we can find implications which contain neither D nor D', thus:

$$(C+D)(CD)' = (C+D)(C'+D') = CD' + C'D,$$

omitting the zero-terms CC' and DD'.

Hence (1) may be written $AB : CD' + C'D$.

Multiplying both members of this implication by C, we get (see rule 12), $$ABC : CD' \quad \dots\dots\dots\dots\dots\dots\dots\dots\dots\dots (4).$$

From (2) we get in the same way
$$ABC : AD \quad \dots\dots\dots\dots\dots\dots\dots\dots\dots\dots (5).$$

From (4) and (5) we get, by rule 13,
$$ABC : CD'AD : 0.$$

Hence $ABC = 0.$

That is, $A : (BC)'$ or $B'C + C'$ (see rules 14 and 3).

This is the first conclusion required, and it follows from the first two data alone, independently of the third.

The third premise is $$A'B' = C'D' \quad \dots\dots\dots\dots\dots\dots\dots\dots\dots\dots (3),$$

and from this we get $A'B'C = CC'D' = 0.$

Hence $B'C : A$ (from rule 14).

This is the second conclusion required, and it follows from the third premise alone, independently of the first two.

Note.—In my quotation from Prof. Bain's work (second edition), I have ventured to correct what I believe to be two misprints; I have substituted B for D in the conclusion of the paragraph beginning "Thirdly," and I have substituted B for A in the conclusion of the final paragraph.

With reference to the preceding note and solution, the referee who reported on this paper makes the following remarks :—

"These corrections agree with the problem in Boole's 'Laws of Thought,' p. 118.

"Boole is only giving some of the conclusions from the premises, not all, nor does he imply that the conclusions are all which are deducible from the premises; on page 129 he has three other conclusions:

"(1) Wherever the property C is found, either the property A or the property B will be found with it, but not both of them together.

"(2) If the property B is absent, either A and C will be jointly present, or C will be absent.

"(3) Conversely, if A and C are jointly present, B will be absent."

From the equations $ABC = 0$ and $A'B'C = 0$, which have been already deduced from the premises, it is evident that several conclusions may be drawn by mere inspection (see rule 14). Thus from the equation $ABC = 0$ we get

$$A : (BC)', \quad B : (AC)', \quad C : (AB)',$$
$$AB : C', \quad AC : B', \quad BC : A';$$

and from the equation $A'B'C = 0$ we get

$$A' : (B'C)', \quad B' : (A'C)', \quad C : (A'B'),$$
$$A'B' : C', \quad A'C : B, \quad B'C : A.$$

180   Mr. Hugh McColl *on*   [June 13,

From the two implications $C : (AB)'$ and $C : (A'B')'$, we get $C : (AB)'(A'B')'$, that is, $C : AB' + A'B$, as may be seen immediately by writing suitable equivalents for $(AB)'$ and $(A'B')'$ and multiplying. This is the *first* of the additional conclusions. The *second* is the implication $B : (A'C)'$, and the *third* is the implication $AC : B'$.

Note.—Prof. Boole's third premise is $A'B' = C'D'$, which is equivalent to two implications, namely, $A'B' : C'D'$ and $C'D' : A'B'$. The first of these implications is the only one I have used; possibly additional conclusions might be obtained by the help of the second implication, but at present I cannot see how.

This paper, as originally sent to the Mathematical Society, contained an application of the preceding principles and notation to the Syllogism. The gentleman (already referred to) who reported on my paper, very justly objected that, like Prof. Boole, I introduced a new and quite unnecessary term into the syllogism. I now propose, instead of the method thus objected to, to substitute the following modification of it, which is not open to the same objection.*

DEF. 13.—The symbol $A \div B$ asserts that A does *not* imply B; it is thus equivalent to the less convenient symbol $(A : B)'$.

Note.—The symbol $A \div B$ thus asserts that the truth of B is not a *necessary* consequence of the truth of A; in other words, it asserts that the statement A is consistent with B', but it makes no assertion as to whether A is consistent with B or not.

RULE 15.—If A implies B and B implies C, then A implies C.

RULE 16.—If A does not imply B, then B' does not imply A'; in other words, the non-implications $A \div B$ and $B' \div A'$ are equivalent.

Note.—This is easily proved as follows: $A \div B = (A : B)' = (B' : A')' = B' \div A'$. Thus in a non-implication, as in an implication, the rule is *Transpose and change signs* (see Rule 11).

RULE 17.—If A implies B but does not imply C, then B does not imply C; in other words, from the two premises $A : B$ and $A \div C$, we get the conclusion $B \div C$.

RULE 18.—If A (assuming it to be a consistent statement) implies B, then A does not imply B'; in other words, from the implication $A : B$ we deduce the non-implication $A \div B'$, which is equivalent to $B \div A'$.

Note.—The non-implication $AB \div 0$ is the same as $A \div B'$.

Rules 15, 17, 18 combined with the principle of contraposition, expressed in Rules 11 and 16, will be found to include all the valid syllogisms, to the examination of which we now proceed.

---

* See the geometrical illustrations on p. 182, 183.

1878.]   *the Calculus of Equivalent Statements.*   181

The statement *All X is Y* may be denoted by the implication $x : y$, in which x denotes the statement that a certain representative individual belongs to the class X, and y denotes the statement that he belongs to the class Y. Similar interpretations may be put upon z, u, v, α, β, &c. The number of individuals belonging to any class may be known or unknown, constant or varying, finite or infinite.

Note.—There is some analogy between this analytical kind of logic and analytical geometry; for, just as in the former the statements $x, y, z,$ &c. refer to *a representative individual of a class*, so in the latter the distances $x, y, z,$ &c. have reference to *a representative point in a locus.*

The statement *No X is Y* will thus be expressed by the implication $x : y'$, which, on the principle of contraposition, is equivalent to the implication $y : x'$ (see Rule 11).

The statement *Some X is Y* will be expressed by the non-implication $x \div y'$, which, by Rule 16, is equivalent to the implication $y \div x'$. *Some* means *one individual at least.*

The statement *Some X is not Y* will be expressed by the non-implication $x \div y$, which, by Rule 16, is equivalent to the implication $y' \div x'$.

These conventions being laid down, the ordinary nineteen syllogisms are as follows, the order for each syllogism (read downwards) being *Major Premise, Minor Premise, Conclusion :—*

First Figure.

$y : z$	$y : z$	$y : z$	$y : z$
$x : y$	$x : y$	$x \div y'$	$x \div y'$
$x : z$	$x : z'$	$x \div z'$	$x \div z$

Note.—The first two moods in this figure are direct applications of Rule 15, and the other two are applications of Rule 17 combined with the principle of contraposition. Thus, by contraposition, the second premise in the third mood may be written $y \div x'$, and from the two premises we then get the conclusion $z \div x'$, which, by contraposition, is equivalent to $x \div z'$.

Second Figure.

$z : y'$	$z : y$	$z : y'$	$z : y$
$x : y$	$x : y'$	$x \div y'$	$x \div y$
$x : z'$	$x : z'$	$x \div z$	$x \div z$

Note.—As in the preceding figure, each mood may be brought under Rule 15 or 17 by transposing one or more of the premises, and making the necessary alteration in the accent.

Third Figure.

$y : z$	$y \div z'$	$y : z$	$y : z'$	$y \div z$	$y : z'$
$y : x$	$y : x$	$y \div x'$	$y : x$	$y : x$	$y \div x'$
$x \div z'$	$x \div z'$	$x \div z'$	$x \div z$	$x \div z$	$x \div z$

Note.—The second premise in the first mood of this figure is $y : x$; from the first premise we deduce the non-implication $y \div z'$ by Rule 18, and then the conclusion follows from Rule 17. The conclusion in the second mood follows directly from the premises by Rule 17. In the third mood we get $z \div x'$ from the premises by Rule 17, and this by contraposition (see Rule 16) is equivalent to $x \div z'$. The remaining three moods in like manner come under Rule 17 either directly or by contraposition, or by the help of Rule 18.

Fourth Figure.

$z : y$	$z : y$	$z \div y'$	$z : y'$	$z : y'$
$y : x$	$y : x'$	$y : x$	$y : x$	$y \div x'$
$x \div z'$	$x : z'$	$x \div z'$	$x \div z$	$x \div z$

Note.—In the first mood we use Rules 15 and 18. In the second mood we use Rule 15 and contraposition. In the third mood we use contraposition and Rule 17. In the fourth mood we proceed as follows:—The first premise is equivalent to $y : z'$ by contraposition, the second premise gives the non-implication $y \div x'$ by Rule 18, the conclusion $z' \div x'$ follows from Rule 17, and this by contraposition is equivalent to $x \div z$. In the fifth mood $y : z'$ is an equivalent for the first premise; from this and the second premise we get the conclusion $z' \div x'$ by Rule 17, and this conclusion by contraposition is equivalent to $x \div z$.

Fresh syllogisms may be obtained by changing the sign of x throughout—*i. e.*, by affixing or removing the accent of negation; but these nineteen fresh syllogisms, like the nineteen ordinary ones already given, will all come under Rule 15 or 17, either directly or by the principle of contraposition, or by the aid of Rule 18.

The cardinal difference between my present method of treating the syllogism and my former one consists in the introduction of the symbol $\div$ as the denial of $:$, so that the statements $A : B$ and $A \div B$ stand in precisely the same relation to each other as a and a'. In fact, if we denote either statement by a, we must denote the other by a'. In my former method I expressed the statement *Some X is Y* by the implication $v : xy$, in which v denoted the statement that the representative individual spoken of belonged to some class V common both to the class X and the class Y. This was the point objected to in the report on my former paper.

Geometrical Illustrations.

The meanings of the symbols $x : y$, $x : y'$, $x \div y$, $x \div y'$, $x = y$ may be illustrated geometrically as follows:

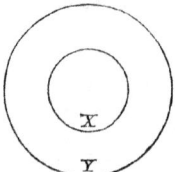

Fig. 1. $(x : y)(y \div x)$.

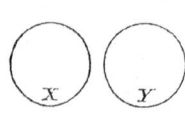

Fig. 2. $x : y'$.

1878.]   *the Calculus of Equivalent Statements.*   183

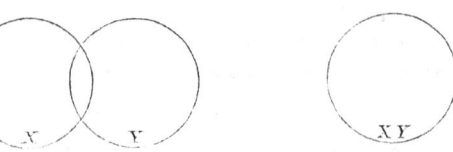

Fig. 3. $(x \div y)(x \div y')$.   Fig. 4. $x = y$.

Let x denote the statement that a representative point is within the boundary X, and let y denote the statement that it is within the boundary Y. Then Fig. 1 illustrates the compound statement $(x : y)(y \div x)$, the first factor of which is an implication, the second a non-implication; $x : y'$ is illustrated by Fig. 2; Fig. 3 illustrates the product of the two non-implications $x \div y$ and $x \div y'$; and Fig. 4 illustrates $x = y$, the boundary X coinciding with the boundary Y.

The various syllogisms may be similarly illustrated. For instance, the first syllogism may be illustrated by Fig. 5.

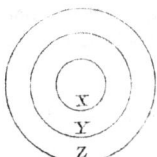

Fig. 5. $x : y : z$.

APPENDIX.

Since $\alpha : \beta$ asserts that β is a factor of α, the implication $\alpha : 0$ asserts that α is false, and the implication $1 : \alpha$ asserts that α is true; but the implications $0 : \alpha$ and $\alpha : 1$ make no assertion as to whether α is true or false. In other words, a statement which contains a false factor must be false, and a statement which is a factor of a true statement must be true; but a factor of a false statement may be true, and a statement which contains a true factor may nevertheless be false.

Every possible valid syllogism comes under one or other of the four following standard implications, either directly or by the substitution of equivalents:

$$(\alpha : \beta)(\beta : \gamma) : (\alpha : \gamma) \quad\quad\quad\quad (1),$$
$$(\alpha : \beta)(\beta : \gamma) : (\alpha \div \gamma') \quad\quad\quad\quad (2),$$
$$(\alpha : \beta)(\alpha \div \gamma) : (\beta \div \gamma) \quad\quad\quad\quad (3),$$
$$(\alpha : \beta)(\alpha : \gamma) : (\beta \div \gamma') \quad\quad\quad\quad (4).$$

The first of the above implications is the symbolical expression of Rule 15, and the third is the symbolical expression of Rule 17. The first may be considered an axiom; the second is only an application of the same axiom, since $(a : \gamma)$ implies $(a \div \gamma')$;* and the third and fourth may be deduced from the first as follows:

Multiplying both sides of the first implication by $(a \div \gamma)$, we get

$$(a : \beta)(\beta : \gamma)(a \div \gamma) : 0;$$

and the third implication follows from this by Rule 14. In the second implication, change γ into γ', and therefore γ' into γ, and it becomes

$$(a : \beta)(\beta : \gamma') : (a \div \gamma).$$

Multiplying both sides of this implication by $(a : \gamma)$, we get

$$(a : \beta)(\beta : \gamma')(a : \gamma) : 0;$$

and the fourth implication follows from this by Rule 14.

Let $f(a, \beta, \gamma)$, $\phi(a, \beta, \gamma)$, $\psi(a, \beta, \gamma)$, $\chi(a, \beta, \gamma)$ denote the foregoing four standard implications respectively. It is easily seen that $f(a, \beta, \gamma) = f(\gamma', \beta', a')$, and that $\chi(a, \beta, \gamma) = \chi(a, \gamma, \beta)$. Considering all those syllogisms equivalent which have equivalent premises and the same or equivalent conclusions, the ordinary nineteen syllogisms are the following:—

Fig. 1.
$$f(x, y, z), \quad f(x, y, z'), \quad \psi(y, z, x'), \quad \psi(y, z', x').$$

Fig. 2.
$$f(x, y, z'), \quad f(x, y', z'), \quad \psi(y, z', x'), \quad \psi(y', z', x').$$

Fig. 3.
$$\chi(y, x, z), \quad \psi(y, x, z'), \quad \psi(y, z, x'), \quad \chi(y, x, z'), \quad \psi(y, x, z), \quad \psi(y, z', x').$$

Fig. 4.
$$\phi(z, y, x), \quad f(x, y', z'), \quad \psi(y, x, z'), \quad \chi(y, x, z'), \quad \psi(y, z', x').$$

If we number these syllogisms 1, 2, 3, ... 18, 19, we may see that $2=5$, $3=11$, $4=7=14=19$, $6=16$, $10=17$, and $12=18$. Eight of the nineteen syllogisms may therefore be looked upon as redundant. On the other hand, several valid syllogisms which are not found among the nineteen might be introduced by admitting the implications and non-implications $y' : z$, $y' \div z$, $y' : x$, $y' \div x$ (or their equivalents $z' : y$, $z' \div y$, &c.) as factors in the premises, which would lead to similar expressions in x and z in the conclusion.

Note.—The implication $a : \beta'$ asserts that a and β are inconsistent with each other; the non-implication $a \div \beta'$ asserts that a and β are consistent with each other.

---

* This assumes that a is a *consistent* statement—*i. e.*, one which *may* be true.

Supplement.

Since the preceding article was put in type I have discovered the following method of elimination, under which Prof. Boole's problem may be brought as a particular case:

Let $f(x)$ denote any statement involving x; and let $f'(x)$ denote the denial of this statement, so that $f'(x)$ is equivalent to $\{f(x)\}'$. Let $\phi(x)$ and $\phi'(x)$ be similarly interpreted. Suppose now that the implication $f(x) : \phi(x)$ is given, and that we have to solve this implication with regard to x. The required solution is

$$f(0)\,\phi'(0) : x, \quad f(1)\,\phi'(1) : x'.$$

Let the antecedents of x and x' in this result be respectively denoted by a and β; that is, let a denote $f(0)\,\phi'(0)$, and let β denote $f(1)\,\phi'(1)$. Then we have

$$a : x, \quad \beta : x'.$$

Suppose next that, instead of the single given implication $f(x) : \phi(x)$, we have a *series* of such implications, and that from these given implications we get the solutions

$$a_1 : x, \quad a_2 : x, \quad a_3 : x, \quad \&c.;$$
$$\beta_1 : x', \quad \beta_2 : x', \quad \beta_3 : x', \quad \&c.;$$

that is,
$$a_1 + a_2 + a_3 + \ldots : x,$$
$$\beta_1 + \beta_2 + \beta_3 + \ldots : x'.$$

Then evidently

$$(a_1 + a_2 + a_3 + \ldots)(\beta_1 + \beta_2 + \beta_3 + \ldots) : 0.$$

When we have performed the multiplication, we shall have a series of implications, each of the form $a_m \beta_n : 0$, and none of these will involve x or x', so that we have eliminated x.

Note.—Zero terms may, of course, be omitted. Any implication of the form $0 : x$ conveys no information, and may therefore be omitted. If the final implication be of the form $0 : xx'$, that is, $0 : 0$, we may infer that the given implications are not sufficient for the elimination of x.

The *zero terms* here referred to are terms which are zero *independently of the given implications of the problem;* such, for instance, as $AA'B$, $BC \times 0$, &c. Thus an implication of the form $AA'B : x$ or $BC \times 0 : x'$ may be omitted; but we must not omit an implication of the form $AB : x$, unless, *independently of the given implications of the problem*, we know A and B to be inconsistent.

Since, as is evident from Rule 5, $a = a + a = a + a + a =$ &c., and, more generally, $a = a + a\beta = a + a\beta + a\gamma =$ &c., we may omit, in any indeterminate expression, any terms that are *multiples* of others. For example, in the expression $\alpha\beta+\gamma\delta+\gamma+\alpha\beta\delta+\alpha\beta$, we may omit the terms underlined, so that the expression reduces to $\alpha\beta+\gamma$.

In Prof. Boole's problem, putting x for D, and therefore x' for D', the given implications are,

$$AB : C'x+Cx',$$
$$BC : Ax+A'x',$$
$$A'B' : C'x',$$
$$Cx' : A'B'.$$

Putting, as before, a_1, a_2, a_3, a_4 for the four antecedents of x, and putting $\beta_1, \beta_2, \beta_3, \beta_4$ for the four antecedents of x', we get

$$a_1 = ABC', \quad a_2 = BCA, \quad a_3 = A'B'C, \quad a_4 = C'(A+B);$$
$$\beta_1 = ABC, \quad \beta_2 = BCA', \quad \beta_3 = A'B', \quad \beta_4 = 0;$$

and it will be found upon trial that

$$(a_1+a_2+a_3+a_4)(\beta_1+\beta_2+\beta_3) = (AB+A'B')C;$$

so that the complete solution of the problem is furnished by the equation

$$(AB+A'B')C = 0.$$

As another example, let it be required to eliminate C (instead of D) from the data of Prof. Boole's problem. Putting x for C, the given implications are

$$AB : x'D+xD',$$
$$Bx : AD+A'D',$$
$$A'B' : x'D',$$
$$x'D' : A'B'.$$

Interpreting a_1, a_2, &c., β_1, β_2, &c. as before, we get

$$a_1 = ABD', \quad a_2 = 0, \quad a_3 = A'B'D, \quad a_4 = D'(A+B)$$
$$\beta_1 = ABD, \quad \beta_2 = B(A'D+AD'), \quad \beta_3 = A'B', \quad \beta_4 = 0.$$

Hence $(a_1+a_2+a_3+a_4)(\beta_1+\beta_2+\beta_3+\beta_4) = ABD'+A'B'D;$

and the complete solution is

$$ABD'+A'B'D = 0.$$

Similarly, if we eliminate B, we get

$$AC'D'+A'CD = 0,$$

and if we eliminate A, we get

$$B(CD+C'D') = 0.$$

[1878d]: The Calculus of Equivalent Statements (III).
Proceedings of the London Mathematical Society, (1878-1879), vol. 10, pp. 16-28.

The Calculus of Equivalent Statements (Third Paper).
By HUGH McCOLL, B.A.
[*Read November* 14*th*, 1878.]

The following formulæ are all either self-evident or easily verified and some of them will be found useful in abbreviating the operations of the calculus :—

(1) $1' = 0$, $0' = 1$;
(2) $1 = 1+a = 1+a+b = 1+a+b+c$, &c.;
(3) $(ab+a'b')' = a'b+ab'$,
    $(a'b+ab')' = ab+a'b'$;
(4) $a : a+b : a+b+c$, &c.;
(5) $(a+A)(a+B)(a+C)\ldots = a+ABC\ldots$;
(6) $(a : b) : a'+b$;
(7) $(a = b) = (a : b)(b : a)$;
(8) $(a = b) : ab+a'b'$;
(9) $(A : a)(B : b)(C : c)\ldots : (ABC\ldots : abc\ldots)$;
(10) $(A : a)(B : b)(C : c)\ldots : (A+B+C+\ldots : a+b+c+\ldots)$;
(11) $(A : x)(B : x)(C : x)\ldots = (A+B+C+\ldots : x)$;
(12) $(x : A)(x : B)(x : C)\ldots = (x : ABC\ldots)$;
(13) $(A : x)+(B : x)+(C : x)+\ldots : (ABC\ldots : x)$;
(14) $(x : A)+(x : B)+(x : C)+\ldots : (x : A+B+C+\ldots)$.

RULE 19.—To test the equivalence of any two statements ; say
$$f(a, b, c, \ldots) \text{ and } \varphi(a, b, c, \ldots).$$

---

* It is easy to show that the only condition necessary for any octagonal frame, with 4 diagonal bars joining opposite vertices, to be capable of internal stress, is that the 4 intersections of each pair of opposite sides shall be *in directum*.

1878.]   *Calculus of Equivalent Statements.*   17

Let a denote the one, and β the other. If $a = \beta$, then $a\beta' + a'\beta = 0$; and, conversely, if $a\beta' + a'\beta = 0$, then $a = \beta$.

Symbolically, this rule may be briefly expressed thus:
$$(a = \beta) = (a\beta' + a'\beta = 0).$$

RULE 20.—To discover the redundant terms of any indeterminate statement.

These redundant terms are easily detected by mere inspection when they imply (or are multiples of) *single* co-terms, as in the expression
$$a'\beta\gamma + a'\gamma + a\beta\gamma' + \beta\gamma'.$$

But when they do not imply single co-terms, but the sum of two or more co-terms, they cannot generally be thus detected by inspection. They can always, however, be discovered by the following rule, which includes all cases:—

Any term of an indeterminate statement may be omitted as redundant when this term, multiplied by the denial of the sum of all its co-terms gives the product zero; and if the product is not zero, the term cannot be cancelled as redundant.

Take, for example, the indeterminate statement
$$CD' + C'D + B'C' + B'D'.$$

Beginning with the first term, we get
$$CD' (C'D + B'C' + B'D')' = BCD'.$$

Hence, the first term CD' cannot be cancelled as redundant

Taking next the second term, we get
$$C'D (CD' + B'C' + B'D')' = BC'D.$$

This shews that the second term $C'D$ cannot be cancelled as redundant.

We next take the third term, and get
$$B'C' (CD' + C'D + B'D')' = 0.$$

The third term $B'C'$ may therefore be omitted as redundant.

Omitting the third term as redundant, we finally try the fourth term $B'D'$, and get $\quad B'D' (CD' + C'D)' = B'C'D'.$

Thus the fourth term cannot be omitted as redundant, *if we omit the third term*. But, since $B'D'(CD' + C'D + B'C')'$

is also zero, the fourth term may be omitted as redundant, *if we retain the third term*.

We may therefore omit either the third or the fourth term, but not both.

Note.—The process for discovering redundant terms may be much abbreviated by means of Rule 22, which see further on.

Thus $\quad CD' (C'D + B'C' + B'D')' = CD' (0 + 0 + B')' = BCD'$;
and, again, $\quad B'D' (CD' + C'D + B'C')' = B'D' (C + 0 + C')' = 0$.

VOL. X.—NO. 142.   C

On the Reduction of Statements to their Primitive Form.

DEF. 14.—An indeterminate statement is said to be in its *primitive form* when it and all its parts are free from redundant terms and redundant factors.

For example, $a+ab+m+m'n$ is reduced to its primitive form when we omit the redundant term ab, and out of the last term strike out the unnecessary factor m'. For $a = a+ab$, and $m+n = m+m'n$, so that the primitive form of the expression is $a+m+n$.

RULE 21.—To reduce an indeterminate statement to its primitive form:—

Develope the denial of the statement by the formula
$$(A+B+C+\ldots)' = A'B'C'\ldots \quad \text{(see Rule 4)};$$
multiply together the indeterminate factors which the application of the formula
$$(ABC\ldots)' = A'+B'+C'+\ldots \quad \text{(see Rule 9)}$$
will then give rise to; and omit the redundant terms in the product (see Rule 20). Then develope the denial of this product, and go through the same process as before. The result will be the primitive form of the original given statement.

Take, for example, the indeterminate statement
$$A+BC+A'B'D+A'C'D.$$

Let a denote this statement.

Then $\quad a' = A'(BC)'(A'B'D)'(A'C'D)'$, by Rule 4.

Applying Rule 9 to the bracket factors, we get
$$a' = A'(B'+C')(\overline{A}+B+D')(\overline{A}+C+D').$$

The terms over-lined in the last two brackets may be cancelled since they are inconsistent with their co-factor A'. Hence
$$a' = A'(B'+C')(B+D')(C+D')$$
$$= A'(B'+C')(D'+BC) \quad \text{(see formula 5)}$$
$$= A'D'(B'+C').$$

Hence, by Rules 9 and 4, we get
$$a = A+D+BC.$$

The primitive form of the given statement is therefore
$$A+D+BC.$$

As another example, take the statement
$$AB'C'+ABD+A'B'D'+ABD'+A'B'D.$$

Let a denote this statement.

1878.]   *Calculus of Equivalent Statements.*   19

Then $\quad a' = A'B + AB'C$,

omitting redundant terms.

Hence $\quad a = (A+B')(A'+B+C') = AB + AC' + A'B' + B'C'$.

The second or fourth term (AC' or $B'C'$) may be omitted as redundant, but not both. The given statement has therefore *two* primitive forms, namely, $AB + A'B' + B'C'$ and $AB + AC' + A'B'$.

Note.—When the given indeterminate statement re-appears as the final result of the process, we may, of course, infer that it is already in its primitive form.

The Method of Unit and Zero Substitution.

DEF. 15.—Statements of the form
$$A, \; A+B, \; AB(C+D), \; AB+CD+E, \; \&c.,$$
will be called *unconditional statements*, to distinguish them from implications (or conditional statements) such as $A:B$, $A+B:CD$, &c.

RULE 22.—Let $f(x)$ denote any unconditional statement involving x.
Then $\quad xf(x) = xf(1)$, and $x'f(x) = x'f(0)$.

More generally, let $f(a, b, c, ..., a, \beta, \gamma, ...)$ be any unconditional statement involving a, b, c, &c. Then

$abc \ldots a'\beta'\gamma' \ldots f(a, b, c, \ldots a, \beta, \gamma, \ldots)$
$\quad = abc \ldots a'\beta'\gamma' \ldots f(1, 1, 1, \ldots, 0, 0, 0, \ldots)$,

always substituting 1 for the repetition of an outside factor, and 0 for its denial.

Proof:—Any unconditional statement $f(x)$ may evidently, by multiplication and the application of Rules 4 and 9, be reduced to an indeterminate statement of the form $ax + \beta x' + \gamma$. Let therefore
$$f(x) = ax + \beta x' + \gamma.$$
Then, however complicated may be the expressions denoted by a, β, γ, it is evident that $xf(x) = x(a+\gamma)$, and $x'f(x) = x'(\beta+\gamma)$;

that is, $\quad xf(x) = xf(1)$, and $x'f(x) = x'f(0)$.

By the same reasoning,
$$ab'f(a, b) = ab'f(1, b) = ab'f(1, 0);$$
and the same mode of proof may evidently be extended so as to prove the general formula

$abc \ldots a'\beta'\gamma' \ldots f(a, b, c, \ldots, a, \beta, \gamma, \ldots)$
$\quad = abc \ldots a'\beta'\gamma' \ldots f(1, 1, 1, \ldots, 0, 0, 0, \ldots)$.

Hence may be proved the formula (given in my second paper) for solving the implication $f(x) : \phi(x)$. Thus, from the given implication

we get $\qquad f(x)\, \phi'(x) : 0.$

Therefore $\qquad x'f(x)\, \phi'(x) : 0;$

that is, $\qquad x'f(0)\, \phi'(0) : 0.$

Hence $\qquad f(0)\, \phi'(0) : x,$ by Rule 14.

Also we have $\qquad xf(x)\, \phi'(x) : 0,$

that is $\qquad xf(1)\, \phi'(1) : 0.$

Hence $\qquad f(1)\, \phi'(1) : x',$ by Rule 14.

From these two formulæ we also get, by transposition and Rule 9,

$$x : f'(1) + \phi(1),$$
$$x' : f'(0) + \phi(0).$$

RULE 23.—Let $f(x)$ denote any complex statement (conditional or unconditional) involving x. Then

$$xf(x) : f(1), \text{ and } x'f(x) : f(0).$$

More generally, let $f(a, b, c, \ldots \alpha, \beta, \gamma, \ldots)$
denote any complex statement (conditional or unconditional) involving a, b, c, &c. Then

$$abc \ldots \alpha'\beta'\gamma' \ldots f(a\, b, c, \ldots, \alpha, \beta, \gamma, \ldots) : f(1, 1, 1, \ldots, 0, 0, 0, \ldots),$$

substituting 1 for the repetition of any statement, and 0 for its denial.

For when we know a statement to be true, we may replace it by 1 in any co-factor of it, and the denial of it by 0. In the antecedent of the above implication, if we know the factors $a, b, c, \ldots, \alpha', \beta', \gamma', \ldots$ to be true, it follows that we may replace each of the true statements $a, b, c, \ldots$ by 1, and the denials $\alpha, \beta, \gamma, \ldots$ of the true statements $\alpha', \beta', \gamma', \ldots$ by 0, in the co-factor $f(a, b, c, \ldots, \alpha, \beta, \gamma, \ldots)$.

As a simple illustration, let $f(x, y)$ denote

$$ax + b : cy + dy' + e.$$

Then $\qquad xy'f(x, y) : f(1, 0),$

that is, $\qquad xy'f(x, y) : (a+b : d+e).$

In other words, if we know x to be true, y to be false, and the implication $ax + b : cy + dy' + e$ to be true, we may infer that the implication $a+b : d+e$ is also true.

RULE 24.—When we have solved any implications with respect to any statement which they may involve and its denial (say x and x'), we may omit in the antecedents any terms which imply (or are multiples of) any other terms (or sum of terms), *whether the implying and implied statements belong to the same antecedent or not.*

Take, for example, the implications
$$\overline{m} + \underline{ab} + \overline{a'c} + b + d : x,$$
$$\overline{m} + \underline{a'b'} + a' + \overline{ad} + e : x',$$

which we may consider as the solution with respect to x and x' of some given implications. The terms under-lined and over-lined may both be omitted. The terms under-lined, ab and $a'b'$, may be omitted as *redundant*, because they are multiples respectively of their co-terms b and a'; while the terms over-lined, m, $a'c$, and ad, may be omitted as *zero*, because they are multiples respectively of m, a', and d (which are not their co-terms), and therefore contain the inconsistent compound factor xx'.

Note.—When our object is not the solution with respect to x and x' of any implication (or implications), but to *eliminate* x; then, though we may still omit redundant terms, such as ab and $a'b'$ in the preceding example, we must not omit zero terms such as m, $a'c$, and ad; for the discovery of zero terms, independent of x, is the very object of the elimination.

I will now illustrate the utility of the preceding rules and principles by one or two examples.

Let there be four given implications (or conditional statements), namely—

$$ax' : c + dy \quad \dots\dots\dots\dots\dots\dots\dots (1),$$
$$bx : c + dy + e \quad \dots\dots\dots\dots\dots\dots (2),$$
$$a'b' : x + de' + c \quad \dots\dots\dots\dots\dots\dots (3),$$
$$a + b + c : x + y \quad \dots\dots\dots\dots\dots\dots (4);$$

and from these four implications let it be required to deduce an implication containing neither x nor x', nor y nor y'.

Solving first with respect to x and x', we get
$$ac'(d' + y') + 0 + a'b'(d' + e)c' + (a + b + c)y' : x,$$
$$0 + bc'(d' + y')e' + 0 + 0 : x'.$$

Hence $\quad abc'(d' + y')e' + bc'(d' + y')e'(a + b)y' : 0.$

The factor $a + b$ in the second term may be omitted, since it is implied in its co-factor b (see Formula 4). Omitting this factor, we get
$$bc'(d' + y')e'(a + y') : 0.$$

We have now to solve this last implication with respect to y and y'. These solutions are * $\quad bc'e' : y, \quad bc'd'e'a : y'.$

Hence, finally, $\quad abc'd'e' : 0.$

This last implication constitutes the required solution; and from it all possible conclusions may be obtained by inspection (see Rule 14).

---

* Observe that, when $y = 0$, y' must $= 1$, so that the factors $d' + y'$ and $a + y'$, being each unity (by Formula 2), may be omitted; and when $y = 1$, $y' = 0$, so that $d' + y' = d'$, and $a + y' = a$.

The problem may also be solved as follows:—

Let $f(x, y)$ denote the product of the four given implications from which we have to eliminate x and y. Then, omitting implications of the form $0 : a$ and $a : 1$, because they convey no information, we get, by application of Rule 23 and Formulæ 6 and 2 of this paper,

$$xy\, f(x, y) : b' + c + d + e,$$
$$x'y\, f(x, y) : (a' + c + d)(a + b + de' + c),$$
$$xy'\, f(x, y) : b' + c + e,$$
$$x'y'\, f(x, y) : (a' + c)(a + b + de' + c)\, a'b'c'.$$

Hence, since $\quad xy + x'y + xy' + x'y' = 1,$

we get, by Formula 10 of this paper, adding antecedents to antecedents, and consequents to consequents, and omitting redundant terms,[*]

$$f(x, y) : b' + c + d + e + (a' + \underline{c} + \underline{d})(a + b + \underline{de'} + c).$$

The terms under-lined in the two brackets may be omitted, since they are multiples of c and d, the second and third terms before the brackets. Hence $f(x, y) : b' + c + d + e + a'(a + b)$;

that is, $\qquad f(x, y) : a' + b' + c + d + e,$

for $\qquad b' + a'b = a' + b'.$

Transposing, we get $\quad abc'd'e' : f''(x, y).$

But, since $f(x, y)$ denotes the product of the premises, if all the premises be correct, $f''(x, y) : 0$. Hence, assuming the correctness of the premises, $\qquad abc'd'e' : 0.$

Let us next take an example from Boole's "Laws of Thought" (p. 106), and reproduced with a different solution in Jevons's "Pure Logic," p. 69.

Following Prof. Jevons's notation (as far as my method will permit), but with a slight difference in the interpretation of his symbols, let A, B, C, D, E respectively denote the five *statements*, that a certain substance is *wealth, transferable, limited in supply, productive of pleasure, preventive of pain*. The premises of the problem will then be expressed by the equation $\qquad A = BC(D + E);$

and we are required to express C in terms of A, B, D, eliminating E.

Assuming the truth of the premises, and using Rule 23 and Formulæ 2 and 8 of this paper, we get

$$CE : (A = B) : AB + A'B',$$
$$CE' : (A = BD) : ABD + A'B' + A'D'.$$

---

[*] Observe that the fourth consequent is a multiple of the second, since a', c implies (or is a multiple of) $a' + c + d$ (see Formula 4).

1878.]   *Calculus of Equivalent Statements.*   23

Hence, adding antecedent to antecedent, and consequent to consequent (see Formula 10), we get (since $E+E'=1$)
$$C : AB+A'B'+A'D' \dots\dots\dots\dots\dots\dots\dots(1),$$
omitting two redundant terms ABD and $A'B'$.

Again, from the given premises we get
$$C'E : (A=0) : A',$$
$$C'E' : (A=0) : A'.$$
Hence $C' : A'$ .... (2).

From (1) and (2) we get, by transposition,
$$\overline{AB'}+A'BD : C',$$
$$A : C.$$

The term AB' over-lined in the first of these two implications may be cancelled as zero,* since it is a multiple of the term A in the second implication (see Rule 24). Cancelling this term and transposing again, we get, for our final result,
$$A : C : A+B'+D'.$$

Another problem from Boole (see "Laws of Thought," page 146), when translated into the language of this calculus, is the following :—

From the premises expressed by the complex statement
$$\{A'C' : E(B'D+BD')\}(ADE' : BC+B'C')\{A(B+E) = C'D+CD'\},$$
it is required, (1) to express A in terms of B, C, D, eliminating E; and (2) to express B in terms of A, C, D.

Using Rule 23 and Formulæ 6 and 8 of this paper, we get
$$AE : C'D+CD',$$
$$AE' : (D'+BC+B'C')\{B(C'D+CD')+B'(CD+C'D')\}.$$

Adding antecedent to antecedent, and consequent to consequent (see Formula 10), and omitting the term $B(C'D+CD')$ in the double bracket, because it is a multiple of the first consequent $C'D+CD'$, we get $A : C'D+CD'+B'(D'+\overline{BC}+B'C')(CD+C'D').$

We may cancel the term over-lined because it is inconsistent with the outside factor B'.

Hence $A : C'D+CD'+B'(C'D'+\underline{B'C'D'}),$
that is, $A : C'D+CD'+B'C'D'.$
Hence, by transposition, $CD+BC'D' : A'.$

---

* From this zero term we may deduce the two implications $A : B$ and $B' : A'$.

Again, from our premises we get
$$A'E : (C+B'D+BD')(CD+C'D'),$$
$$A'E': C(CD+C'D').$$

Now, since C implies $C+B'D+BD'$, by Formula 4, the consequent of the second implication implies (or is a multiple of) the consequent of the first implication, and may therefore be omitted as redundant when the two consequents are added together. Hence we get
$$A' : (C+B'D+BD')(CD+C'D'),$$
that is, $\qquad A' : CD+BC'D'.$

The consequent of A' in this last implication was before shewn to be the antecedent of A'. Hence
$$A' = CD+BC'D'.$$
Therefore $\qquad A = C'D+CD'+B'C' = C'D+CD'+B'D'.$

These two equivalents for A are the two primitive forms of
$$(CD+BC'D')',$$
the denial of the equivalent for A' (see Rule 21).

This will be found to agree with Boole's result when his expression for A is reduced to its primitive form (see Rule 21).

We are next required to express B in terms of A, C, D.

Now, in doing this, we might go back to the given premises of the problem and proceed as before; but it will be much shorter to avail ourselves of the results already obtained. If we had over-lined any *conditional zero terms* (*i.e.*, terms which are zero by virtue of, and not independently of, the given implications of the problem), we might, by equating these with zero, obtain expressions for B (*i.e.*, antecedents and consequents of B) in terms of C and D, independently of A, and then obtain the antecedents and consequents involving A by means of the equation $\qquad A' = CD+BC'D'.$

But since no conditional zero terms have been thus over-lined, all that remains to be done is to solve the last equation with respect to B and B', thus:—
$$B : A'(CD+C'D')+A(C'D+CD'),$$
$$B' : A'(CD)+A(CD)'.$$

Transposing, we get (by mere inspection, if we employ Formula 3)
$$A(CD+C'D')+A'(C'D+CD') : B',$$
$$A(CD)+A'(CD)' : B.$$

Writing the antecedents of B' and B in full (without brackets), we get
$$\overline{ACD}+AC'D'+\overline{A'C'D}+\overline{A'CD'} : B',$$
$$\overline{ACD}+A'C'+A'D' : B.$$

1878.] *Calculus of Equivalent Statements.* 25

Omitting the conditional zero terms over-lined (see Rule 24), and transposing in the first implication, we get
$$B : A' + C + D,$$
$$A'(C' + D') : B.$$
Boole's conclusions, when translated into this notation, are
$$B : A'(C'D' + CD) + A(C'D + CD'),$$
$$A'C'D' : B.$$
The first of these implications has been already obtained, and that its consequent is equivalent to the simpler consequent $A' + C + D$ may be proved by Rule 19; thus,—Putting a and β for the two consequents respectively, we get $a\beta' + a'\beta = A'C'D + A'CD' + ACD$. Now the three terms on the right-hand side of this equation have already been over-lined as zero. Hence $a\beta' + a'\beta$ vanishes, which shows that the statements denoted by a and β are equivalent. Similarly it may be shown that the two antecedents $A'(C' + D')$ and $A'C'D'$ are equivalent.

Let it be required, next, to express D in terms of A, B, C.

E having already been eliminated, we may, as before, avail ourselves of the equation
$$A' = CD + BC'D'$$
previously obtained. From this we readily get
$$D : A'C + AC',$$
$$D' : A'(BC') + A(BC')'.$$
Transposing, we get (see Formula 3)
$$AC + A'C' : D',$$
$$ABC' + A'B' + A'C : D.$$
This gives us Prof. Boole's conclusions $AC : D'$, $A'C : D'$, $A'C : D$ (which, being independent of B, might also have been obtained directly by inspection of the conditional zero terms already over-lined in the antecedent of B'), and also two conclusions $ABC' : D$ and $A'B' : D$, involving B and not given in Boole's solution.

If we were required to express C in terms of A, B, D, we might, as before, start from the equation
$$A' = CD + BC'D'.$$
Thus
$$C : A'D + AD',$$
$$C' : A'(BD') + A(BD')' ;$$
and, by transposition,
$$AD + A'D' : C',$$
$$A'B' + A'D + ABD' : C.$$
This solution with respect to C is not given in Boole's "Laws of Thought."

———

This calculus of statements and implications may, I believe, be employed with advantage in investigating the causes of natural phenomena. Let it be required, for example, to ascertain the cause of some phenomenon X, and let the data of the problem (data which may be the results of observation and experiment, or deductions from previously known laws) be expressed symbolically in one complex statement, say
$$\phi(x, a, b, c, \ldots, u, v, \ldots),$$
in which x asserts the existence (or occurrence) of the phenomenon X, while $a, b, c, \ldots, u, v, \ldots$ assert the existence of certain accompanying circumstances, $A, B, C, \ldots, U, V, \ldots$ If we know, or have strong reasons for suspecting, that certain of the monomial (or single-letter) statements, say $u, v, \ldots$, have reference to merely accidental and not necessary concomitants of X, our first step should be to eliminate these from the data. Let the implication left after this elimination be
$$\psi(x, a, b, c, \ldots) : 0.$$
We must now solve this implication with respect to x and x'. Let the solution (after omitting zero and redundant terms, and reducing the antecedents of x and x' to their primitive form) be
$$a_1 + a_2 + a_3 + \ldots : x,$$
$$\beta_1 + \beta_2 + \beta_3 + \ldots : x',$$
in which $a_1, a_2, a_3, \ldots, \beta_1, \beta_2, \beta_3, \ldots$ are some functions of $a, b, c, \ldots$

If the statements $a_1, a_2, a_3, \ldots$ have a common factor, the circumstance (or combination of circumstances), whose existence is asserted by this common factor, is probably the cause of X. This follows from the canon of inductive philosophy called the "Method of Agreement." When this common factor is an indeterminate statement, the circumstances (or combinations of circumstances), whose existence is asserted by the several terms of this indeterminate statement, may (provisionally, till a common factor of the terms is discovered) be regarded as separate and independent causes of X.

Let m, asserting the existence of the circumstance M, be a common factor of the statements $a_1, a_2, a_3, \ldots$. If $\beta_1, \beta_2, \beta_3, \ldots$ have also the common factor m', asserting the non-existence of M, the probability that M is the cause of X will be greatly increased. This is in accordance with the inductive canon called the "Joint Method."

If, finally, any term in the antecedent of x be of the form qm, while some term in the antecedent of x' is of the form qm', the sole difference being that m is accented in the one term and unaccented in the other; then the inference that M is the cause of X will be decisively confirmed by the inductive "Method of Difference."

1878.]   *Calculus of Equivalent Statements.*   27

In the abstract of this paper, read at the Meeting of the Mathematical Society, on the 14th of last November, mention was made of some criticism of mine on Prof. Jevons's method of solving logical problems.

This criticism, I wish to say now, was founded on an insufficient acquaintance with his method,—an acquaintance derived solely from his "Elementary Lessons in Logic,"—and for this reason I now ask permission to retract it. Prof. Jevons has since kindly sent me a copy of his "Pure Logic," dated 1864, from which I find that there is a much closer resemblance between his method and mine than I had any idea of, and that he had anticipated me in several things of which I believed myself to be the first discoverer.

I am also indebted to the kindness of the Rev. Robert Harley for the loan of Boole's "Laws of Thought," in which, as in Jevons's "Pure Logic," in spite of a very different notation and mode of treatment, I find many points of resemblance to my method. My method, however, differs both from Prof. Boole's and Prof. Jevons's in three cardinal points, which a perusal of this paper will show to be so important as to necessitate an essentially different treatment of the whole subject. These three points are:—

(1) With me every single letter, as well as every combination of letters, always denotes a *statement*.

(2) I use a symbol (the symbol :) to denote that the statement following it is true provided the statement preceding it be true.

(3) I use a special symbol—namely, an accent—to express denial; and this accent, like the minus sign in ordinary algebra, may be made to affect a multinomial statement of any complexity.

On these three points of difference are founded some symmetrical rules on which depends the whole working of the calculus.

It may interest some of the readers of this paper to be told that this method owes its origin to a question in probability (No. 3440), proposed by Mr. Stephen Watson in the *Educational Times*. My solution of this question, with an introductory article, entitled "Probability Notation," was published in the *Educational Times* for August, 1871; and in this introductory article may be seen the germs of the present method. Shortly after this I gave up all mathematical investigations, and my thoughts did not again revert to the subject till two or three months before the appearance of my article on "Symbolical Language" in the *Educational Times* for July, 1877. This article was in the editor's hands before I noticed that this "Symbolical Language," as I called it, might also be employed, without change or modification, and with unerring certainty, in tracing to their last hiding-place the limits which often escape so mysteriously from the mathematician's grasp when he ventures to change the order of integration or the variables in a multiple integral. This was a pleasant and encouraging surprise, and gave rise to hopes that the discovery of fresh fields of application would soon follow. I still looked upon the method, however, as an essentially mathematical one—grafted indeed on a logical stem, but destined to yield mathematical fruit, and mathematical fruit only. In the application of my method to examples, I was in the habit (when applying Rule 8) of using the symbol : for the word *implies*, writing

$$x_{m'n} : p(x_m - x_n) : p\{\phi(y, z)\} : \&c.$$

as a convenient abbreviation for

$$x_{m'n} = x_{m'n} p(x_m - x_n) = x_{m'n} p\{\phi(y, z)\} = \&c.;$$

but I did not use this symbol in anything that I sent for publication, as it did not seem to involve any important principle, and I was unwilling to complicate my method by the introduction of any symbol that might, without any serious inconvenience, be dispensed with. Afterwards, however, when I began to turn my attention to the more purely logical aspect of the method, I saw that very important advantages might be secured by employing this or some other convenient symbol to express logical inference, especially in the applications of the very useful principle of "contraposition"; and the result of my cogitations upon this point was my second paper printed in these "Proceedings."

[1878e]: **Symbolical or Abbreviated Language, with an Application to Mathematical Probability.** *The Educational Times (Reprint)*, vol. 28, pp. 20-23.

SYMBOLICAL OR ABBREVIATED LANGUAGE, WITH AN APPLICATION TO MATHEMATICAL PROBABILITY. *By* HUGH McCOLL, B.A.

DEF. 1.—The symbols A, B, C, &c. denote *statements* (or propositions) registered for convenience of reference in a table. The equation $A = 1$ asserts that the statement A is *true*; the equation $A = 0$ asserts that the statement A is *false*; and the equation $A = B$ asserts that A and B are equivalent statements.

DEF. 2.—The symbol $A \times B \times C$ or ABC denotes a *compound* statement, of which the statements A, B, C may be called the *factors*. The equation $ABC = 1$ asserts that *all three statements* are *true*; the equation $ABC = 0$ asserts that all three statements are *not* true, *i.e.*, that at least *one* of the three is false. Similarly a compound statement of any number of factors may be defined.

21

DEF. 3.—The symbol $A + B + C$ denotes an *indeterminate* statement, of which the statements A, B, C may be called the *terms*. The equation $A + B + C = 0$ asserts that *all the three statements* are *false*; the equation $A + B + C = 1$ asserts that all the three are *not* false, *i.e.*, that at least *one* of the three is true. Similarly $A + B$, $A + B + C + D$, &c. may be defined.

DEF. 4.—The symbol A' is the *denial* of the statement A. The two statements A and A' are so related that they satisfy the two equations $AA' = 0$ and $A + A' = 1$. The same symbol (*i.e.*, a *dash*) will convert any complex statement into its denial. For example, $(AB)'$ is the denial of the compound statement AB.

DEF. 5.—When an *indeterminate statement* is a factor of a compound statement, the former is enclosed in brackets. For example, $(A + B)(C + D)$ denotes a compound statement of two indeterminate factors. Exactly as in ordinary algebra, we have the identity

$$(A + B)(C + D) = AC + AD + BC + BD;$$

and the same rule may be extended to a compound statement of any number of indeterminate factors, and whatever be the number of terms in the respective factors.

The following examples will illustrate the preceding definitions:—

(1.) The equation $A = AB$ asserts either that the statement A implies the statement B, or else that B is true necessarily. As particular cases of this, we have $A = AA = AAA = $ &c., as repetition neither strengthens nor weakens the logical value of a statement. Again,

$$A = A(B + B') = A(B + B')(C + C') = \&c.;$$

for $\qquad B + B' = C + C' = 1$ (def. 4).

(2). $\qquad (AB)' = AB' + A'B + A'B'$
$\qquad\qquad = AB' + A'(B + B') = AB' + A'$, for $B + B' = 1$.

(3). $\qquad (AB)' = BA' + B'$, by exchanging A and B in (2).

(4). $\qquad (A + B)' = A'B'$; $(A + B + C)' = A'B'C'$, &c.

(5). $\{AB(C + D)\}' = AB(C + D)' + (AB)'$, from (2)
$\qquad\qquad = ABC'D' + (AB)'$, from (4)
$\qquad\qquad = ABC'D' + AB' + A'$, from (2).

RULE 1.—If there be any statements (simple or complex), say A, B, C, ..., of such a nature that they neither strengthen nor weaken each other (the truth or falsehood of one in no way affecting the truth or falsehood of the rest); then the chance that the compound statement ABC ... is true is the *product abc* ..., in which $a = $ the chance that A is true, b the chance that B is true, and so on.

RULE 2.—If there be any statements A, B, C, ... (simple or complex) of such a nature that only *one* of them can be true; then the chance that the *indeterminate statement* $A + B + C + ...$ is true is $a + b + c + ...$, in which a, b, c ... denote the same chances as in the preceding rule.

By an extension of the preceding principles and notation we may obtain a solution of the following general problem:—

Given that certain functions, P, Q, R, ... say, of any variables $x, y, z, ...$ are respectively positive (or negative), find the chance that certain other functions U, V, W, ... of the same variables are positive (or negative), the values of $x, y, z, ...$ being taken at random.

22

The method of effecting the solution may be briefly indicated as follows:—

DEF. 6.—The symbol p prefixed to any algebraical (or arithmetical) expression converts the expression into a statement, namely, that the expression is real and *positive*; the prefix p' in like manner asserts that the expression affected by it is real and *negative*; and the prefix q asserts that the expression affected by it is *impossible*. For example, $p\,x\,p'(y-x)\,q\,\{y-(m^2-x^2)^{\frac{1}{2}}\}$ is a compound statement of three factors: the first factor asserts that x is *positive*; the second asserts that $y-x$ is *negative*; and the third asserts that $y-(m^2-x^2)^{\frac{1}{2}}$ is *impossible*. Again, let it be given that x, y, a are all three real and positive, it is easy to perceive the identity

$$p(x^2+y^2-a^2) = p\{y-(a^2-x^2)^{\frac{1}{2}}\}\,p(a-x) + p'(a-x).$$

DEF. 7.—The symbols $x_1, x_2, x_3, \ldots$ denote the 1st, 2nd, 3rd, &c. limits of x, registered in any convenient order in a table of reference. The limits of the other variables of the expression (or expressions) under consideration are denoted similarly.

DEF. 8.—The symbol $x_{mn r's't'}$ is a compound statement of 5 factors, and asserts that x_m and x_n are *inferior* limits of x, while x_r, x_s, and x_t are *superior* limits of x. In fact, it is an abbreviation for the compound statement $p(x-x_m)\,p(x-x_n)\,p'(x-x_r)\,p'(x-x_s)\,p'(x-x_t)$. Compound statements of any number of factors, and having reference to the limits of x (or any other variable), may be similarly abbreviated.

RULE 3.—The compound statement $x_{m'n'r's'} = x_{m'}A + x_{n'}B + x_{r'}C + x_{s'}D$, in which $x_{m'}$, $x_{n'}$, &c. are abbreviations for the statements that the mth, n^{th}, &c. limits of x are *superior* limits; while the capitals A, B, C, &c. respectively denote the statements that x_m is the nearest superior limit of x, that x_n is the nearest limit, that x_r is the nearest limit, and so on. In other words, A is an abbreviation for the compound statement $p'(x_m-x_n)\,p'(x_m-x_r)\,p'(x_m-x_s)$. The value of B is obtained from this by simply exchanging m and n; the value of C is obtained from it by exchanging m and r; and so on.

RULE 4.—This is obtained from the preceding rule by simply copying all the words in it (except *superior*, for which we must write *inferior*), and *omitting all the accents*, both on the numbers m, n, r, s and on the symbol p.

RULE 5.—The statement $x_{mn'} = x_{mn'}A$, in which $A = p(x_n-x_m)$.

By repeated applications of Rules 3, 4, 5, any compound statement of the form $x_{1.2'.3'.4.5.6'\ldots}\,y_{1'.2.3'.4\ldots}\,z_{1'.2.3'.4.5\ldots}$, say, may be reduced either to a single elementary term of the form $x_{mn'}\,.\,y_{rs'}\,.\,z_{tu'}$, in which x_m and x_n are the *nearest limits* (inferior and superior) of x, and so on for y and z; or it may be reduced to an indeterminate statement consisting of several such terms. This effected, the integral calculus must do the rest. If all the values of x, y, and z between the assigned limits are equally probable, the chance that the elementary term given above is a true statement is proportional to the definite integral

$$\int_{x_m}^{x_n} dx \int_{y_r}^{y_s} dy \int_{z_t}^{z_u} dz.$$

In this notation it is evident that only one meaning is attached to the

23

symbol x_m, which is always an abbreviation for the statement $p'(x-x_m)$, and asserts that x_m (the mth limit of x) is a *superior* limit. With the symbol x_m the case is different; in the table of limits, and when effected by the symbol p or p', it simply denotes the mth limit of x, as registered in the table; but in all other cases the symbol x_m denotes a *statement*, namely, the statement $p(x-x_m)$, and asserts that x_m (the mth limit of x) is an *inferior* limit. It thus appears that x_m sometimes denotes a *limit* and sometimes a *statement* with reference to that limit; but no ambiguity or confusion can result from this double meaning, as the context will always show what meaning is to be attached to the symbol. For example, if one side of an equation denotes a *statement*, the other side also must denote a statement; if one factor of an expression is a statement, all the factors must be statements; if one term of an expression is a statement, all the other terms of the expression must be statements. There is, therefore, no danger of confounding the *statement* x_m with the *limit* x_m, just as in ordinary language there is no fear of confounding the noun *bear* with the verb *bear* in such sentences as "the *bear* will *bear* his share of the burden."

The symbols px, $p'x$, py, $p'y$, &c., occur so frequently that they may be conveniently denoted always by x_0, $x_{0'}$, y_0, $y_{0'}$, &c.

In a compound statement, therefore, which contains x_0 as a factor, x is always positive; and in a compound statement which contains $x_{0'}$ as a factor, x is always negative. The same remarks of course apply to y_0, $y_{0'}$, z_0, $z_{0'}$, &c.

[1878f]: Symbolical Language:--No. 2. *The Educational Times (Reprint)*, vol. 28, p.100.

SYMBOLICAL LANGUAGE :—No. 2. *By* HUGH MCCOLL, B.A.

The method briefly sketched in my former article on this subject [see pp. 20—23 of this volume of the *Reprint*] is not restricted to Probability. It will also enable us to ascertain, by a sure, simple, and almost mechanical process, the new limits of integration which arise when we change the order of integration or the variables in a multiple integral. Take, for example, a multiple integral of 4 variables of the form

$$\int_{\lambda_7}^{\lambda_8} du \int_{\lambda_5}^{\lambda_6} dx \int_{\lambda_3}^{\lambda_4} dy \int_{\lambda_1}^{\lambda_2} dz \; \phi(u, x, y, z),$$

in which the limits λ_1, λ_2 are functions of the variables y, x, u; the limits λ_3, λ_4 functions of x, u; and so on. From this integral we get a compound statement of 8 factors (2 for each variable), namely, $p(u-\lambda_7) \, p'(u-\lambda_8)$ for the variable u, $p(x-\lambda_5) \, p'(x-\lambda_6)$ for the variable x, and so on. Suppose we have to *reverse* the order of integration : we must from these given statements deduce others in which the limits of u are functions of x, y, z, the limits of x functions of y, z, and so on. These deduced statements will constitute one or more terms of some such form as the following :

$$u_{1 \cdot 2' \cdot 3' \dots 0 \cdot 4 \cdot 7 \dots} \, x_{6' \cdot 2' \cdot 3' \dots 5 \cdot 1 \cdot 0} \, y_{0' \cdot 2' \dots 3} \, z_{1' \cdot 0' \cdot 3},$$

in which $u_0, x_0, y_{0'}, z_{0'}$ are convenient equivalents for $pu, px, p'y, p'z$; while $u_{1'}, u_{2'}$, &c. are to be interpreted as in my former article.

We must now apply Rules 3, 4, 5 (see my former article) to the u-statements in every term. This will probably introduce fresh limits of x, y, z. We next apply Rules 3, 4, 5 to the x-statements in every term ; this may introduce fresh limits of y, z ; and so on. When we have thus applied Rules 3, 4, 5 successively to u, x, y, z, the required limits for the new order of integration will be obtained. The final result *may* consist of only one elementary term ; but generally it will consist of more, and in complicated cases it may consist of several such terms. In Question 5373, for instance, the transformed integral required will be found to consist of *seven* terms, each with its own limits of integration, though the limits of u will be the same for all the terms.

Much labour may generally be saved by dropping *zero terms* (*i.e.*, terms with inconsistent factors) as we go along, when mere inspection of the table (without having recourse to Rule 5) will suffice to detect them. But if we overlook these zero terms, they will eventually disappear of themselves in the subsequent evolutions of the process. We may also shorten the process by cancelling factors which mere inspection of the table (instead of having recourse to Rule 3 or 4) will show to be implied in their co-factors *of the same variable*.

[1879e]: The Calculus of Equivalent Statements (IV). *Proceedings of the London Mathematical Society*, vol. 11, pp. 113-21.

The Calculus of Equivalent Statements. (*Fourth Paper.*)
By HUGH McCOLL, B.A.

[*Read* Feb. 12*th*, 1880, *revised April* 22*nd*, 1880.]

Probability Notation.

In applying symbolical logic to probability, the following notation will, I think, be found useful.

DEF. 1.—The symbol x_a denotes the *chance* that the statement x is true *on the assumption that the statement a is true*.

As illustrations we may take $x_x = 1$, $x_{x'} = 0$, $x'_x = 0$, $(a+x)_{ax} = 1$, $(ax)_{a'+x'} = 0$.

Again, speaking of a point taken at random within the square $ABCD$, the sides of which are bisected at the points E, F, G, H, let a denote the statement, "It will fall within the triangle ABD," and let x denote the statement, "It will fall within the square $EFGH$." Then, evidently, $x_a = \frac{1}{2}$. Again, speaking of the same random point, let y denote the statement, "It will fall within the triangle EHD." Then 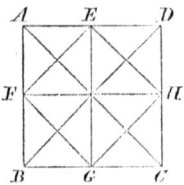 $y_a = \frac{1}{8}$, $y_x = 0$, $y_{x'} = \frac{1}{4}$, $x_y = 0$, $x_{y'} = \frac{4}{7}$, $y'_x = 1$, $(ax)_{y'} = \frac{2}{7}$.

DEF. 2.—The symbol ϵ is an equivalent for the symbol 1, and denotes any statement whose truth is taken for granted *throughout the whole of an investigation*.

Hence x_ϵ simply denotes *the chance that x is true*.

In the geometrical illustration to Def. 1, ϵ would denote the statement, "The random point will fall within the square $ABCD$;" and, attaching the same meanings as before to the other letters, we have $x_\epsilon = \frac{1}{2}$, $(ax)_\epsilon = \frac{1}{4}$, $a_\epsilon = \frac{1}{2}$, $(a+x)_\epsilon = \frac{3}{4}$, $(a'x)_\epsilon = \frac{1}{4}$, $(ax')_\epsilon = \frac{1}{4}$.

NOTE.—It is worth noticing that the implication $a : x$ is equivalent to the statement $x_a = 1$.

DEF. 3.—Any two statements a and x are said to be *independent*, when we have any one of the equivalent equations $x_a = x_\epsilon$, $a_x = a_\epsilon$, $x_a = x_{a'}$, $a_x = a_{x'}$. (See Appendix, Note 1.)

Independent statements may be illustrated geometrically thus. Suppose a point to be taken at random within the rectangle $ABCD$, which is made up of 12 equal squares, 4 in each row, and 3 in each column. Let a denote the statement, "The random point will fall within the rectangle DF;" and let x denote the statement, "The random point will fall within the rectangle BE." Then, evidently, a and x are independent, for

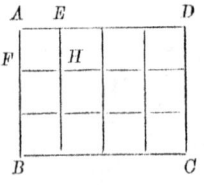

$$a_x = AH \div BE = \tfrac{1}{3}, \quad a_\epsilon = DF \div AC = \tfrac{1}{3}, \quad a_{x'} = DH \div EC = \tfrac{1}{3};$$
$$x_a = AH \div DF = \tfrac{1}{4}, \quad x_\epsilon = BE \div AC = \tfrac{1}{4}, \quad x_{a'} = BH \div FC = \tfrac{1}{4}.$$

Let $AB = m$, and $BC = n$; and let x, y be the coordinates of a random point in the rectangle AC, referred to AB and BC as axes. Then, supposing the rectangle AC to be divided into $\dfrac{mn}{dx\,dy}$ equal parts, each part being an infinitesimal rectangle whose area is $dx\,dy$, the chance that the random point will fall within some particular one of

1880.] *the Calculus of Equivalent Statements.* 115

these rectangles, is $\dfrac{dx\,dy}{mn}$; the chance that the infinitesimal rectangle within which it falls is at a distance x from AB is $\dfrac{dx}{n}$, and the chance that it is at a distance y from BC is $\dfrac{dy}{m}$. It is clear that the statement, the chance of whose truth is $\dfrac{dx}{n}$, in no way affects the chance of the truth of the statement whose chance is $\dfrac{dy}{m}$, so that the chance of the truth of these two independent statements combined is $\dfrac{dx\,dy}{mn}$. If these two independent statements be denoted by x and y respectively, it may be shown, by the same reasoning as before, that $x_y = x_\epsilon = x_y$, and that $y_x = y_\epsilon = y_{x'}$.

RULE 1.—$(ab)_x = a_x b_{ax}$, $(abc)_x = a_x b_{ax} c_{abx}$, and so on, for any compound statement. Putting ϵ for x in these equations, we get

$$(ab)_\epsilon = a_\epsilon b_a, \quad (abc)_\epsilon = a_\epsilon b_a c_{ab},$$

and so on. (See Note after Rule 2.)

NOTE.—From this rule, we see that $x_a = \dfrac{(ax)_\epsilon}{a_\epsilon}$, and therefore $x_a \div a_x = x_\epsilon \div a_\epsilon$.

RULE 2.— $$(a+b)_x = a_x + b_x - (ab)_x;$$
$$(a+b+c)_x = (a_x + b_x + c_x) - \{(ab)_x + (ac)_x + (bc)_x\} + (abc)_x;$$
and, generally,
$$(a_1 + a_2 + a_3 + \ldots + a_n)_x = s_1 - s_2 + s_3 - \ldots + (-1)^{n-1} s_n,$$

in which s_r denotes the sum of the $\dfrac{\lfloor n}{\lfloor r\,\lfloor n-r}$ chances (the statement x being taken for granted) of the $\dfrac{\lfloor n}{\lfloor r\,\lfloor n-r}$ combinations that can be made of the n statements $a_1, a_2, a_3 \ldots a_n$, taken r at a time.

NOTE.—The following geometrical illustrations will help to illustrate and verify Rules 1 and 2. Let ϵ, a, x, y respectively denote the four

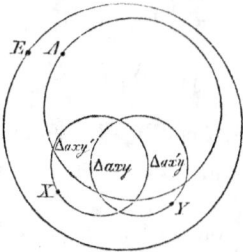

statements—(1) *A point is taken at random within the circle E*, (2) *It* *falls within the circle A*, (3) *It falls within the circle X*, (4) *It falls within the circle Y*. Let $\Delta\epsilon$, Δx, Δax, Δaxy, $\Delta ax'y$, $\Delta axy'$, $\Delta(x+y)$ respectively denote the *areas* with reference to which the *statements* ϵ, x, ax, axy, $ax'y$, axy', $x+y$ respectively, can possibly be true; and, similarly, let the symbol $\Delta f(\epsilon, a, x, y)$ generally denote the area with reference to which the statement $f(\epsilon, a, x, y)$ can possibly be true. Then

$$(xy)_a = \frac{\Delta axy}{\Delta a} = \frac{\Delta ax}{\Delta a} \times \frac{\Delta axy}{\Delta ax} = x_a y_{ax},$$

$$(x+y)_a = \frac{\Delta(ax+ay)}{\Delta a} = \frac{\Delta(axy' + axy + ax'y)}{\Delta a}$$

$$= \frac{\Delta axy' + \Delta axy + \Delta ax'y}{\Delta a}$$

$$= \frac{\Delta ax + \Delta ay - \Delta axy}{\Delta a}$$

$$= \frac{\Delta ax}{\Delta a} + \frac{\Delta ay}{\Delta a} - \frac{\Delta axy}{\Delta a}$$

$$= x_a + y_a - (xy)_a.$$

The above expressions are of course simplified, when for a we put ϵ. In this case, we have

$$(xy)_\epsilon = \frac{\Delta xy}{\Delta \epsilon} = \frac{\Delta x}{\Delta \epsilon} \times \frac{\Delta xy}{\Delta x} = x_\epsilon y_x,$$

$$(x+y)_\epsilon = \frac{\Delta(x+y)}{\Delta \epsilon} = \frac{\Delta(xy' + xy + x'y)}{\Delta \epsilon}$$

$$= \frac{\Delta xy' + \Delta xy + \Delta x'y}{\Delta \epsilon}$$

$$= \frac{\Delta x + \Delta y - \Delta xy}{\Delta \epsilon}$$

$$= \frac{\Delta x}{\Delta \epsilon} + \frac{\Delta y}{\Delta \epsilon} - \frac{\Delta xy}{\Delta \epsilon}$$

$$= x_\epsilon + y_\epsilon - (xy)_\epsilon.$$

RULE 3.— $\qquad (x : a) : \left(x_a = \dfrac{x_\epsilon}{a_\epsilon}\right).$

This follows readily from Rule 1. For, since $(ax)_\epsilon = a_\epsilon x_a$, we have $x_a = \dfrac{(ax)_\epsilon}{a_\epsilon}$. But $x = ax$ by supposition, so that $x_a = \dfrac{x_\epsilon}{a_\epsilon}$.

NOTE.—From this we see that $x_{x+y} = \dfrac{x_\epsilon}{(x+y)_\epsilon}$, and that $(xy)_x = y_x$, whatever be the statements x and y.

RULE 4.—Taking $(a-ab)_x$ or $\{a(1-b)\}_x$ as a convenient abbreviation for $a_x - (ab)_x$, and so on for similar expressions, we may clear of accents a compound chance of the form $(abc \ldots d'e'f'\ldots)_x$ by means of the equation

$$(abc \ldots d'e'f'\ldots)_x = \{(abc \ldots)(1-d)(1-e)(1-f)\ldots\}_x,$$

the factors on the right-hand side of the equation to be multiplied like ordinary algebraical factors, and then each term of the product to be affected by the suffix x.

Beginning with the simplest case, we have

$$(\alpha\beta')_x = \alpha_x \beta'_{ax} = \alpha_x(1-\beta_{ax})$$
$$= \alpha_x - \alpha_x \beta_{ax} = \alpha_x - (\alpha\beta)_x$$
$$= (\alpha - \alpha\beta)_x = \{\alpha(1-\beta)\}_x.$$

Similarly, we have

$$(\alpha\beta'\gamma')_x = \alpha_x(\beta'\gamma')_{ax} = \alpha_x\{1-(\beta+\gamma)_{ax}\}$$
$$= \alpha_x\{1-\beta_{ax}-\gamma_{ax}+(\beta\gamma)_{ax}\}$$
$$= \alpha_x - \alpha_x\beta_{ax} - \alpha_x\gamma_{ax} + \alpha_x(\beta\gamma)_{ax}$$
$$= \alpha_x - (\alpha\beta)_x - (\alpha\gamma)_x + (\alpha\beta\gamma)_x$$
$$= (\alpha - \alpha\beta - \alpha\gamma + \alpha\beta\gamma)_x$$
$$= \{\alpha(1-\beta)(1-\gamma)\}_x.$$

And the same method of proof may evidently be extended to a compound statement with any number of accented factors.

To illustrate the use of the preceding rules and notation, I will apply them to two problems taken from Boole's "Laws of Thought." In the first of these my result agrees with that of Boole, but in the second our results differ.

The chance that a witness A. speaks the truth is p, the chance that another witness B. speaks the truth is q, and the chance that they disagree in a statement is r. What is the chance that, if they agree, their statement is true? (See Boole's "Laws of Thought," p. 279.)

Let x denote the statement, "A. speaks the truth," and let y denote the statement, "B. speaks the truth."

The data are $x_i = p$, $y_i = q$, $(x'y + xy')_i = r$; and we are required to find $(xy)_{xy+x'y'}$.

Now $(xy)_{xy+x'y'} = \dfrac{(xy)_i}{(xy+x'y')_i}$ (See Rule 3, Note)

$$= \frac{(xy)_i}{1-r};$$

and, to find $(xy)_i$, we have

$$r = (x'y + xy')_i = (x'y)_i + (xy')_i$$
$$= \{(1-x)y\}_i + \{x(1-y)\}_i \quad \text{(from Rule 4)}$$
$$= x_i + y_i - 2(xy)_i = p + q - 2(xy)_i.$$

118 Mr. Hugh McColl *on* [April 22,

Therefore $(xy)_* = \frac{1}{2}(p+q-r)$,

and, substituting, we get $(xy)_{xy+x'y'} = \dfrac{p+q-r}{2(1-r)}$.

Note.—It is worth noticing that in this solution no assumption is made as to whether or not the statements x and y are independent.

The next problem is this :—

The chances of two causes A_1 and A_2 are c_1 and c_2 respectively. The chance that, if the cause A_1 present itself, an event E will accompany it, whether as a consequence of the cause A_1 or not, is p_1; and the chance that, if the cause A_2 present itself, the event E will accompany it, whether as a consequence of it or not, is p_2. Moreover, the event E cannot appear in the absence of both the causes A_1 and A_2. Required the chance of the event E. (See Boole's "Laws of Thought," p. 321.)

Let $x =$ the cause A_1 will present itself,

let $y =$ the cause A_2 will present itself,

and let $z =$ the event E will occur.

The data are $x_* = c_1$, $y_* = c_2$, $z_x = p_1$, $z_y = p_2$, and the implication $z : x+y$; and we are required to find z_*.

Now the given implication $z : x+y$ is equivalent to the equation $z = z(x+y)$. Hence

$z_* = (xz+yz)_* = (xz)_* + (yz)_* - (xyz)_*$ (by Rule 2)

$= x_* z_x + y_* z_y - (xy)_* z_{xy}$ (by Rule 1)

$= c_1 p_1 + c_2 p_2 - (xy)_* z_{xy}$.

If (guided by the concrete illustration in Boole's foot-note, see Appendix, Note 2) we assume A_1 and A_2 to be independent, and the event E to be more probable when both causes exist, than when only one of them exists, we shall have $(xy)_* = x_* y_* = c_1 c_2$, and z_{xy} greater than z_x and greater than z_y, that is, greater than p_1 and greater than p_2.

Hence $c_1 p_1 + c_2 p_2 - c_1 c_2 p_1$ and $c_1 p_1 + c_2 p_2 - c_1 c_2 p_2$ are superior limits to z_*, while (since 1 is a superior limit to z_{xy}) $c_1 p_1 + c_2 p_2 - c_1 c_2$ is evidently an inferior limit to z_*.

If we take $c_1 = \cdot 1$, $c_2 = \cdot 2$, $p_1 = \cdot 6$, and $p_2 = \cdot 7$, we shall find z_* to be within the limits $\cdot 18$ and $\cdot 186$.

Boole's equation* for finding the *exact* value of the required chance gives (on substituting these numerical values of c_1, c_2, p_1, p_2) $\cdot 19069$, &c., as the required result, which is too much by at least $\cdot 00469...$

Rule 5.—From the compound implication $(\alpha : x)(x : \beta)$ we may infer that α_* is an inferior limit of x_*, and β_* a superior limit.

This equation (in which u denotes the required chance) is

$$\frac{(u-c_1 p_1)(u-c_2 p_2)}{c_1 p_1 + c_2 p_2 - u} = \frac{\{1-c_1(1-p_1)-u\}\{1-c_2(1-p_2)-u\}}{1-u}.$$

1880.] *the Calculus of Equivalent Statements.* 119

This rule is useful in such cases as the following: Suppose we have, among other data, a series of implications, each of the form $f(x) : \phi(x)$, and that we are required to find the limits of x. Solving with respect to x, the given implications combined will be equivalent to the compound implication $(a : x)(x : \beta)(\gamma : 0)$, in which a denotes the weakest antecedent of x cleared of zero terms, β the strongest consequent of x cleared of zero terms, and γ the sum of the zero terms which do not involve x. The solution can hardly be called complete till a, β, γ (if complex disjunctive statements) are reduced to their primitive forms. (See Rule 21 of my Third Paper.)

Question 6258, proposed in the "Educational Times" for last March, may be taken as a simple illustration of the use of this rule. In this question the given compound implication

$$(ax + bx' : c)(bx + ax' : c')$$

is equivalent to $(ac + bc' : x)(x : a'c' + b'c)(ab : 0)$.

Appendix.

NOTE 1.—To prove that (as stated in Def. 3) the four equations $x_a = x_{,}$, $a_x = a_{,}$, $x_a = x_{a'}$, and $a_x = a_{x'}$ are equivalent statements, each implying the rest.

From Rule 1, we have $x_{,}a_x = a_{,}x_a$; consequently, if we assume $x_a = x_{,}$, we shall have $a_x = a_{,}$; and, conversely, if we assume $a_x = a_{,}$, we shall have $x_a = x_{,}$.

We will next assume the equation $a_x = a_{x'}$, and from it deduce the equation $x_a = x_{,}$.

From Rule 1, we get $a_x = \dfrac{a_{,}x_a}{x_{,}}$, and $a_{x'} = \dfrac{a_{,}x'_a}{x'_{,}}$. But, by supposition, $a_x = a_{x'}$; therefore $\dfrac{x_a}{x_{,}} = \dfrac{x'_a}{x'_{,}}$;

that is, $\dfrac{x_a}{x_{,}} = \dfrac{1 - x_a}{1 - x_{,}}$;

whence we get $x_a = x_{,}$.

By reversing this proof, we may from the equation $x_a = x_{,}$ deduce the equation $a_x = a_{x'}$.

We have thus proved the three equations $a_x = a_{,}$, $x_a = x_{,}$, $a_x = a_{x'}$ to be equivalent.

In the same way as it was shown that $a_x = a_{x'}$ was equivalent to $x_a = x_{,}$, we may show that $x_a = x_{a'}$ is equivalent to $a_x = a_{,}$. Hence we finally get $(x_a = x_{,}) = (a_x = a_{,}) = (x_a = x_{a'}) = (a_x = a_{x'})$.

NOTE 2.—The foot-note referred to, page 118, is the following:—

The mode in which such data as the above might be furnished by experience, is easily conceivable. Opposite the window of the room in

The Calculus of Equivalent Statements. [April 22,

which I write, is a field, liable to be over-flowed from two causes, distinct, but capable of being combined, viz., floods from the upper sources of the River Lee, and tides from the ocean. Suppose that observations made on N separate occasions have yielded the following results: on A occasions the river was swollen by freshets, and on P of those occasions it was inundated, whether from this cause or not. On B occasions the river was swollen by the tide, and on Q of those occasions it was inundated, whether from this cause or not. Supposing, then, that the field cannot be inundated in the absence of *both* the causes above mentioned, let it be required to determine the total probability of its inundation.

Here the elements a, b, p, q of the general problem represent the ratios $\frac{A}{N}$, $\frac{P}{A}$, $\frac{B}{N}$, $\frac{Q}{B}$, or rather the values to which those ratios approach, as the value of N is indefinitely increased. (Boole's "Laws of Thought," p. 321.)

[The following is a very simple proof of the fundamental rule in the Inverse Method of Probability.

Let x assert the occurrence of a certain event, and let $\alpha, \beta, \gamma, \ldots$ respectively assert the existence of several mutually exclusive circumstances, any one of which may be the cause of the event. We have to prove that $a_x = \frac{x_a}{x_a + x_\beta + x_\gamma + \ldots}$, on the assumption that $x : \alpha + \beta + \gamma + \ldots$, and that $a_\epsilon = \beta_\epsilon = \gamma_\epsilon = \&c.$

Since
$$(ax)_\epsilon = x_\epsilon a_x = a_\epsilon x_a,$$
therefore
$$a_x = \frac{a_\epsilon x_a}{x_\epsilon}.$$

But $x_\epsilon = (ax + \beta x + \gamma x + \ldots)_\epsilon$, since $x : \alpha + \beta + \gamma + \ldots$, that is,
$$x_\epsilon = (ax)_\epsilon + (\beta x)_\epsilon + (\gamma x)_\epsilon + \ldots$$
$$= a_\epsilon x_a + \beta_\epsilon x_\beta + \gamma_\epsilon x_\gamma + \ldots$$

Substituting this value of x_ϵ, we get
$$a_x = \frac{a_\epsilon x_a}{a_\epsilon x_a + \beta_\epsilon x_\beta + \gamma_\epsilon x_\gamma + \ldots} \quad \ldots \ldots \ldots (1).$$

But, by supposition, $a_\epsilon = \beta_\epsilon = \gamma_\epsilon = \&c.$; therefore
$$a_x = \frac{x_a}{x_a + x_\beta + x_\gamma + \ldots} \quad \ldots \ldots \ldots (2).$$

The same rule, of course, holds good when we interchange α and β, α and γ, and so on. The expression (1) is, of course, more general than (2), and should be used when we cannot legitimately assume that $a_\epsilon = \beta_\epsilon = \gamma_\epsilon = \&c.$]

In concluding this paper I wish to record my obligations to Mr. C. J. Monro, of Hadley, Barnet, for much kind and patient criticism of my method generally, as well as for some valuable suggestions as to the directions which my investigations should take. It was by his advice that I resumed the study of probability, in which my method first originated; and it was in working out a problem in probability, which he proposed to me as a test of the power of my method, that I hit upon the notation which I here propose.

[1880n]: On the Diagrammatic and Mechanical Representation of Propositions and Reasoning. The London, Edinburgh and Dublin philosophical Magazine and Journal of Science, vol. 10, pp. 168-171.

XXV. *On the Diagrammatic and Mechanical Representation of Propositions and Reasonings.*

To the Editors of the Philosophical Magazine and Journal.

GENTLEMEN,

MR. VENN has kindly sent me a copy of his very interesting paper in the Philosophical Magazine for July, in which he explains a method which he has invented for solving logical problems by means of diagrams. The method is certainly ingenious, and for verifying analytical solutions of easy and elementary problems it would, I think, be useful in the hands of a teacher; but I cannot agree with its inventor's estimate of its practical utility in other respects, much less with his opinion as to its superiority over rival methods. Speaking of his diagram for five-letter problems, Mr. Venn says:—

"It must be admitted that such a diagram is not quite so simple to draw as one might wish it to be; but then we must remember what are the alternatives before any one who wishes to grapple effectively with five terms and all the thirty-two possibilities which they yield. *He must either write down, or in some way or other have set before him, all those thirty-two compounds of which* X Y Z W V *is a sample; that is, he must contemplate the array produced by 160 letters.*"

Fom the words in italics it is evident that Mr. Venn does not yet appreciate the advantages of my own method, which assuredly lays one under no such onerous obligation as he mentions. It grapples effectively, not merely with problems

* "Vortex Atoms," Proc. Roy. Soc. Edinb. Feb. 18, 1867.

Representation of Propositions and Reasonings. 169

of five terms, but with problems of six, seven, eight, or even more terms; and it does so because it does *not* oblige one to take into separate consideration all those perplexing possibilities with which Mr. Venn's and similar methods are hampered. That the readers of this Magazine may be able to judge fairly as to the respective capabilities of Mr. Venn's method and mine, I will first solve one of his four-letter problems, and then a six-letter problem of my own, which though exceedingly easy by my method, would, if I am not greatly mistaken, subject his diagrammatic method to a severe strain.

"Every X is either Y or Z; every Y is either Z or W; every Z is either W or X; and every W is either X or Y: what further condition, if any, is needed to ensure that every XY shall be W?"

This is a special case of the following more general problem:—

Given a series of implications, $A : a$, $B : b$, $C : c$, &c.; what is the weakest implication that need be added to these data to justify the inference $m : n$?

The answer is $mn' : Aa' + Bb' + Cc' + \ldots$

When A, a, B, b, &c. are complex expressions involving m or n or both, great simplification may be effected by substituting in these expressions 1 for m and n', and therefore 0 for m' and n. In Mr. Venn's problem the data are (when expressed in my notation)

$$x : y+z, \quad y : z+w, \quad z : w+x, \quad w : x+y,$$

and the weakest addition to the premises to justify the inference $xy : w$ is therefore

$$xyw' : xy'z' + yz'w' + zw'x' + wx'y'.$$

Substituting 1 for every x, y, and w' (and therefore 0 for every x', y', and w) in the consequent of this implication, the implication becomes $xyw' : z'$, which is equivalent to $xyw'z : 0$, or $xyz : w$, the result required. In actual practical working these substitutions of unity and zero would be made mentally while writing down the consequent of the required implication, so that the result may fairly be said to follow directly from mere inspection of the data.

This and the other problems given by Mr. Venn are much too easy: the following problem, involving six letters, would be a fairer test of the power of his method; and I should much like to see his solution of it.

Taking $ax + by : cd'$ as the symbolical expression of the statement "whenever the event A happens with X, or B with Y, then C happens without D," and so on for similar state-

170   *Representation of Propositions and Reasonings.*

ments; when may we infer from the four implications $ax+by : cd'$, $bx+ay : c'd$, $cx'+dy' : ab'$, and $dx'+cy' : a'b$, (1) that either X or Y has happened but not both, (2) that both X and Y have happened or else neither of them? What combinations among the events A, B, C, D are impossible?

In other words, we are required to find in terms of a, b, c, d, (1) the weakest antecedent of $x'y + xy'$, (2) the weakest antecedent of $xy + x'y'$, and (3) what combinations among the statements a, b, c, d are inconsistent with the data.

From the data

$$(ax+by : cd')(bx+ay : c'd)(cx'+dy' : ab')(dx'+cy' : a'b)$$

we get by mere inspection

$$xy : (a'b'+cd')(a'b'+c'd) :: a'b'^{*},$$
$$x'y' : (c'd'+ab')(c'd'+a'b) :: c'd'.$$

Hence
$$xy + x'y' : a'b' + c'd';$$
that is,
$$(a+b)(c+d) : x'y + xy'. \quad \ldots \quad \ldots \quad (1)$$

Again,

$$x'y : (b'+cd')(a'+c'd)(c'+ab')(d'+a'b)$$
$$:: (a'b'+a'cd'+b'c'd)(c'd'+a'bc'+ab'd') :: a'b'c'd'.$$

From the symmetry of the conditions we thence get

$$xy' : a'b'c'd'.$$

Hence
$$x'y + xy' : a'b'c'd';$$
that is,
$$a+b+c+d : xy + x'y'. \quad \ldots \quad \ldots \quad (2)$$

Also, since
$$xy + x'y' + x'y + xy' = 1,$$

we get
$$1 : a'b' + c'd' + \underline{a'b'c'd'};$$

that is, omitting the redundant term underlined,

$$(a+b)(c+d) : 0. \quad \ldots \quad \ldots \quad (3)$$

This completes the solution. From (1) we learn that we may infer that one and one only of the events X and Y has happened, provided we know *that* A *or* B *has happened in conjunction with* C *or* D. But since (3) informs us that this conjunction is impossible, it follows that *the data are not suffi-*

---

* The symbol :: expresses *equivalence*; thus for $\alpha : \beta :: \gamma :: \delta$ read "α implies β, β is equivalent to γ, and γ is equivalent to δ."

cient to justify the inference. From (2) we learn that we may infer that both X and Y or neither of them have happened provided we know that any of the events A, B, C, D (one or more) has happened; while (3) informs us that the combinations AC, AD, BC, BD are impossible.

Other inferences, besides those required, may easily be drawn. For instance, since

$$(xy)' :: x' + y' :: x'y' + xy' + x'y' : c'd',$$

we get $c+d : xy$. Similarly we get $a+b : x'y'$. The product of these two inferences implies (2).

Where is the formidable array of 6×2^6 (or 384) letters which Mr. Venn, unless I misunderstood his words, supposes the logician obliged to face as a necessary preliminary to all inference in every problem requiring six letters? Whether Dr. Boole's or Prof. Jevons's method can fairly be charged with imposing this heavy labour I am not prepared to say; but my method certainly does not impose it.

Yours &c.
Hugh M'Coll.

73 Rue Siblequin, Boulogne-sur-Mer,
August 3, 1880.

V. *Implicational and Equational Logic.*
By HUGH McCOLL, *B.A., University of London*[*].

PROF. JEVONS, in his new work, 'Studies in Deductive Logic,' of which he has kindly sent me a copy, refers to my papers in 'Mind' and in the 'Proceedings of the London Mathematical Society' in terms which might give rise to some misapprehension as to the real nature of my symbolical method. He says that "I reject equations in favour of *implications*," and in so doing "ignore the necessity of the equation for the application of the Principle of Substitution."

Now, it is quite true that I reject equations in favour of implications in those classes of logical problems (and they are very numerous) in which implications lead to the simplest, shortest, and most elegant solutions; but there are other classes of problems, especially in mathematics, which necessitate the equational form of statement; and in these I do not hesitate to adopt it. The simple truth is, that my method admits of both forms; and, as a matter of fact, I employ both, sometimes even in the same problem. In my first paper in the Proceedings of the London Mathematical Society (which treats of the limits of multiple integrals) I adopt the equational form throughout; in my second and third papers, which relate entirely to questions of pure logic, I generally adopt the implicational form, as the simplest and most effective; while in my fourth paper, which treats of probability, I mainly adopt the equational form.

As to the statement that "I ignore the necessity of the equation in the application of the Principle of Substitution," I am not quite sure that I understand what it means. I certainly recognize the principle that if $\alpha = \beta$, then $f(\alpha) = f(\beta)$, or, as the rule may be expressed symbolically in my notation, $(\alpha = \beta) : \{f(\alpha) = f(\beta)\}$; but I cannot in the least understand what bearing this has upon the advantages or disadvantages of my system of implications.

The question whether the implication $\alpha : \beta$, or its equivalent, the equation $\alpha = \alpha\beta$, should be preferred in a symbolical system of logic, must be decided on the broad grounds of practical convenience. I believe it may be taken as a useful

---

clearly kept in view, viz. that this assumption or theory, by opening out an absolutely limitless field of speculative hypothesis, completely annihilates all *method* or rational system in physical inquiry, and therefore that all progress or insight into the physical processes underlying phenomena is absolutely brought to a standstill so long as this theory is adhered to.

[*] Communicated by the Author.

On Implicational and Equational Logic. 41

principle in symbolical reasoning generally, that conventional symbols of abbreviation should be adopted for all expressions *which have to be employed frequently*. On this principle a^3 was probably used as a convenient abbreviation for aaa, a^4 for $aaaa$, and so on, before the discovery of the important law expressed by the equation $a^m \times a^n = a^{m+n}$. The same necessity for symbolical abbreviation originated the useful symbols $f(x), f(x, y), f'(x)$, and many others. On this principle, since I find that such statements as "If α is true β is true," or "α implies β," *are extremely common in all reasoning*, I use the simple symbol $\alpha : \beta$ as a very convenient abbreviation*. Granted that the equation $\alpha = \alpha\beta$ will also accurately express the statement "α implies β," it is a much less simple and suggestive expression for it. Compare, again, the implication, $\alpha\beta + \gamma\delta : ab + cd$, with its equivalent, the equation $\alpha\beta + \gamma\delta = (\alpha\beta + \gamma\delta)(ab + cd)$, and the superior simplicity of the implication will be still more striking. But the abbreviating power of my symbol of implication becomes most conspicuous in what may be called *implications of the second order*, as in the syllogism

$$(\alpha : \beta)(\beta : \gamma) : (\alpha : \gamma).$$

May I ask Prof. Jevons how he would express this syllogism in his equational notation, *in pure symbols and entirely without words*. I can only see one way in which he could do this consistently with his views, namely by the very clumsy equation

$$(\alpha = \alpha\beta)(\beta = \beta\gamma) = (\alpha = \alpha\beta)(\beta = \beta\gamma)(\alpha = \alpha\gamma).$$

This looks so exceedingly like a *reductio ad absurdum*, that I cannot help hoping that it will lead Prof. Jevons to reconsider his opinion that *the equational form alone* should be employed in symbolical logic.

So far I have argued on the assumption that my $\alpha : \beta$ is equivalent to Prof. Jevons's $\alpha = \alpha\beta$; and both Prof. Jevons and I agree, I believe, in the opinion that practically this is the case. At the same time, it must be borne in mind that, for this assumption to be strictly true, the letters α and β must have the same meanings in the implication $\alpha : \beta$ as in the equation $\alpha = \alpha\beta$; and therefore either each letter must in both forms

---

* The equivalence of $\alpha : \beta$ and $\alpha = \alpha\beta$ may be proved formally in my notation as follows:—
From the formula
$$(\alpha = \beta) = (\alpha : \beta)(\beta : \alpha)$$
we get
$$(\alpha = \alpha\beta) = (\alpha : \alpha\beta)(\alpha\beta : \alpha) = (\alpha : \alpha)(\alpha : \beta)(\alpha\beta : \alpha) = (\alpha : \beta);$$
for the factors $\alpha : \alpha$ and $\alpha\beta : \alpha$ are each equal to unity—that is to say, *always true*, whatever the statements α and β may be. (See formula 3 of this paper further on.)

42   *On Implicational and Equational Logic.*

denote a *statement*, as in my system, or else each letter must in both forms denote a *quality or thing*, as in Prof. Jevons's system. On the supposition that each letter denotes a statement, my notation exhibits clearly the very remarkable fact that in the syllogism $(\alpha:\beta)(\beta:\gamma):(\alpha:\gamma)$, the very same relation which connects α with β, and β with γ, connects also the combined premises $(\alpha:\beta)(\beta:\gamma)$ with the conclusion $\alpha:\gamma$. On the assumption that each letter denotes a statement, Prof. Jevons's notation (as I have shown) could only show this coincidence of relation in a very clumsy and roundabout manner; while, on the assumption that each letter denotes a thing or quality (as in his system), his notation could scarcely be used in this extended way at all.

The same remarks apply to many other useful and symmetrical formulæ, which, so far as I can see, are altogether uninterpretable on Prof. Jevons's hypothesis that each letter should denote a thing or quality; while on my hypothesis, that each letter should denote a statement, every formula conveys a clear and precise meaning, which it is scarcely possible to misunderstand. Take, for example, the formulæ:—

(1) $(A:a)(B:b)(C:c)\ldots:(ABC\ldots:abc\ldots)$;
(2) $(A:a)(B:b)(C:c)\ldots:(A+B+C+\ldots:a+b+c+\ldots)$;
(3) $(x:a)(x:b)(x:c)\ldots=(x:abc\ldots)$;
(4) $(a:x)(b:x)(c:x)\ldots=(a+b+c+\ldots:x)$;
(5) $(a:x)+(b:x)+(c:x)+\ldots:(abc\ldots:x)$,
(6) $(x:a)+(x:b)+(x:c)+\ldots:(x:a+b+c+\ldots)$.

These formulæ express logical laws of undoubted truth, which Prof. Jevons could scarcely express in his notation without the help of words.

Prof. Jevons approves to some extent of my accent to express denial, and occasionally adopts this notation in his new work; but he finds it difficult, he says, to believe that there is any advantage in my innovations in other respects, and he is of opinion that "my proposals tend towards throwing Formal Logic back into its ante-Boolian confusion." To this general condemnatory opinion it is difficult to make any definite reply; I can only express my regret that Prof. Jevons has nowhere throughout his book given a single example of this tendency in my proposals "towards throwing Formal Logic back into its ante-Boolian confusion." Abundant materials were at his disposal for comparing my method with his own in the fairest and most decisive way possible, namely in the actual solution of problems. Out of the various problems of which I have published solutions he might surely have found *one* with which

43

to point and illustrate his criticism. Friendly contests are at present being waged in the 'Educational Times' among the supporters of rival logical methods; I hope Prof. Jevons will not take it amiss if I venture to invite him to enter the lists with me, and there make good the charge of "ante-Boolian confusion" which he brings against my method.

November 29, 1880.

[1880p]: Symbolical Reasoning (I). Mind, vol. 5, pp. 45-60.

III.—SYMBOLICAL REASONING.

SYMBOLICAL reasoning may be said to have pretty much the same relation to ordinary reasoning that machine-labour has to manual labour. In the case of machine-labour we see some ingeniously contrived arrangement of wheels, levers, &c., producing with speed and facility results which the hands of man without such aid could only accomplish slowly and with difficulty, or which they would be utterly powerless to accomplish at all. In the case of symbolical reasoning we find in an analogous manner some regular system of rules and formulæ, easy to retain in the memory from their general symmetry and interdependence, economising or superseding the labour of the brain, and enabling any ordinary mind to obtain by simple mechanical processes results which would be beyond the reach of the strongest intellect if left entirely to its own resources.

46   *Symbolical Reasoning..*

The most striking achievements of symbolical reasoning are to be found in the various branches of pure and applied mathematics, and in short in all subjects of human inquiry that admit of more or less exact *measurement*. In all these the symbols employed represent numbers or ratios, or the various relations of numbers and ratios. Till within very recent times symbolical reasoning was exclusively restricted in its application to questions of this nature—questions to which the final practical answer was always an arithmetical expression.

The first person to show that symbolical reasoning might also be employed with advantage in the investigation of matters usually considered altogether beyond the sphere of mathematics was the late Professor Boole. This he did first in his *Mathematical Analysis of Logic*, and afterwards more fully in his celebrated *Laws of Thought*, published in 1854.

These works excited much admiration in the mathematical world, and, it may be added, caused no small trepidation among logicians, who saw their hitherto inviolate territory now for the first time invaded by a foreign power, and with weapons which they had but too much reason to dread. With these potent mysterious symbols mathematicians had already extended their dominion far and wide, whilst they, the successors of the illustrious Aristotle, had not added a single acre to the very restricted possessions bequeathed to them by their great predecessor. And now their aggressive rivals threatened to wrest from them these very possessions and annex the sacred province of logic also to the already over-grown empire of mathematics.

If the attack led by Prof. Boole had been vigorously followed up by his fellow-mathematicians, it might have gone hard with the logicians; but it was not thus followed up, and the logicians had therefore time to recover from their consternation. Ten years after, in 1864, Professor Jevons published his able treatise on [1] *Pure Logic* and in a skilfully planned attack regained possession of the ground which the invader had occupied with his symbols. He showed that the whole of Prof. Boole's results might have been obtained more briefly, as well as much more simply, from purely logical considerations, and with no more symbolism than logicians were already in the habit of using. Other champions, and notably Mr. A. J. Ellis, though a mathematician himself, soon followed and espoused the cause of logic against mathematics, each planting his own system like

---

[1] It may be here remarked that Prof. Jevons uses the expression *Pure Logic* in a somewhat different sense from that in which I use it further on in this paper.

a strong fortress where he thought it would most effectually protect the ancient and venerable science from future invasion.

The writer of this paper would like to contribute his humble share as a peacemaker between the two sciences, both of which he profoundly respects and admires. He would deprecate all idea of aggression or conquest on either side, and yet it is quite plain to him that the two cannot henceforth remain distinct and independent as they have hitherto done. Union for the future there must be; this is written in clear and indelible letters in the book of fate. But can there be no union without conquest and annexation? Would England be happier or more prosperous now if she had conquered and annexed Scotland, as she very nearly succeeded in doing in the reign of Edward I.? Would Scotland be freer or more contented if she had stubbornly rejected and resisted the act of union in the reign of Queen Anne? Do not Englishmen and Scotchmen alike now both "glory," as George III. said he did, "in the name of *Briton*"? Why should not logicians and mathematicians unite in like manner under some common appellation?

That logic, when treated symbolically, is capable of rendering important services to mathematics was shown by Prof. Boole in the latter portion of his *Laws of Thought*, in which he applies his method to certain classes of questions in mathematical probability. Quite recently I have myself shown in my papers on "Symbolical Language" published in the *Educational Times*, and much more fully in the first of my three papers on the "Calculus of Equivalent Statements," published in the *Proceedings of the London Mathematical Society*, that by the help of logic (treated symbolically) we may clear away with the greatest ease a complete jungle of difficulties which had vexatiously arrested the progress of mathematical science in a direction in which its cultivators were most eager to advance it. This jungle of difficulties presented itself in that part of the Integral Calculus which treats of the limits of multiple integrals, a subject which had occupied the attention of some of the most eminent mathematicians for the last fifty years or more, and which they had found extremely perplexing. In the "Calculus of Equivalent Statements" (as I have called my symbolical invention) logic presents the mathematician with an instrument at whose touch all these difficulties vanish. It is an instrument too of so simple a construction and so easy of application that a mere school-boy may speedily learn to use it. To give a detailed description of this invention is not my purpose. But I may be allowed to explain as clearly as I can, and with as few technicalities as possible, the elementary logical principles upon which it is based.

48   *Symbolical Reasoning.*

Such an explanation, discussing, as it will do to some extent, the fundamental rules of our ordinary reasoning, should, if I properly perform my work, possess some interest, not for the professed logician or mathematician merely, but for all educated readers.

Logic may conveniently be divided into two kinds, namely,[1] *Pure Logic* and *Applied Logic*. In a strictly analogous sense mathematicians usually distinguish between [1] *Pure Mathematics*, including geometry, algebra, the differential calculus, &c., and *Applied Mathematics*, which includes mechanics, optics, astronomy, and in short any science whose subject-matter is capable of more or less exact measurement or numerical calculation. And it is only from this point of view of measurement and numerical calculation that such sciences are in any way dependent upon mathematics. In optics, for example, there are many remarkable phenomena which not only may be explained very clearly without any aid from mathematics, but even exclude all mathematical considerations as irrelevant. It will conduce much to clearness of thought if we in like manner draw a distinct boundary line between *Pure* Logic and *Applied* Logic.

Pure Logic may be defined as the general science of reasoning considered in its most abstract sense, that is to say, as far as possible independently of any special subject of investigation. In other words, Pure Logic is the Science of Reasoning considered with reference to those general rules and principles of thinking which hold good whatever be the matter of thought.

Applied Logic may be simply defined as the application of the general rules of pure logic to *special subjects*, such as mathematics, physics, medicine, politics, or in fact anything (down to the most ordinary concerns of life) that offers any workable material for the human reason.

Reasoning again, whose fundamental rules and principles it is the business of logic to investigate and explain, may also be conveniently divided into two kinds, *mental* and *symbolical*.

In *mental* reasoning we dispense entirely with symbols, and in the dark and silent recesses of our own minds endeavour to evolve fresh knowledge from the knowledge already existing there. In this kind of reasoning all our outward senses are quiescent and we might still be capable of it even if we were blind, deaf and dumb, and unable to move hand or foot.

In *symbolical* reasoning, on the other hand, we use symbols as artificial aids to our naturally more or less defective memories.

---

[1] The adjectives *abstract* and *concrete* would be preferable to *pure* and *applied*, were it not for the fact that the expressions *pure mathematics* and *applied mathematics* are already established by usage.

Symbolical Reasoning.

These symbols may be divided into two kinds, *permanent* and *temporary*. By a *permanent* symbol I mean a symbol whose meaning is permanent or always the same. By a *temporary* symbol I mean a symbol to which we attach only a temporary meaning. Here again the science of mathematics supplies us with a useful analogy. In common arithmetic, in which the same symbol or collection of symbols always denotes the same number or ratio, we have examples of *permanent* symbols; and in algebra, in which the same letter or symbol may denote sometimes one number or ratio and sometimes another, we have examples of *temporary* symbols. In our common ordinary reasoning, in like manner, the same word or collection of words has always the same meaning; whereas in symbolical reasoning, the same letter or symbol may sometimes have one meaning and sometimes another. The analogy between the algebra of mathematics and the algebra of logic may be carried a step further. In both, *permanent* as well as *temporary* symbols are employed when convenient, and in both, the symbols of *relation*, such as $+$ and $\times$, are always permanent.

In my system of symbolical reasoning I have found it convenient to make my *temporary* symbols denote *statements*, while my *permanent* symbols, such as $+$, $\times$, :, usually denote the various relations in which these statements stand with respect to each other. That each individual temporary symbol, as well as every combination of such symbols, always denotes a *statement*, is one of the leading characteristics of my logical system, to the fuller explanation of which I now proceed.

Definition 1.—When two or more statements are made, each is called a *factor* of the *compound statement* which they collectively make up; and this whole compound statement is called a *multiple* of each separate factor, and the *product* of all the factors combined.

Thus, let a denote the statement "He is tall," let b denote the statement "He is dark," and let c denote the statement "He is a German". Then abc will denote "He is a tall, dark German". This compound statement abc is the *product* of the three factors a, b, c, and a *multiple* of any single factor a.

In the foregoing example the three statement-factors a, b, c refer all to one common subject; but this need not always be the case. For example, if a denote "His father is German," and b denote "His mother is French," then ab will denote the compound statement "His father is German and his mother is French".

This combination of factors into multiples may also, as in common algebra, be expressed by the symbol $\times$. Thus, the symbols abc and $a \times b \times c$ are equivalent.

Symbolical Reasoning.

Def. 2.—When a *disjunctive* statement is made, the several statements of which it is formed are called the *terms* of this statement; and, on the other hand, the disjunctive statement is called the *sum* of the terms. These terms are connected, as in mathematics, by the sign $+$. Thus, "He will go to Paris, Vienna, or Berlin, this summer," may be expressed by the disjunctive statement $a + b + c$, if we agree that the first term a denotes the statement "He will go to Paris this summer," that the second term b denotes the statement "He will go to Vienna this summer," and that the third term c denotes the statement "He will go to Berlin this summer".

The terms of a disjunctive statement do not necessarily refer (as in this example) to the same subject. For example, if a, b, c, respectively denote the three statements, "Henry will go to Paris," "Richard will go to Vienna," and "Robert will go to Berlin," then the symbol $a + b + c$ will denote the disjunctive statement "Henry will go to Paris, or else Richard will go to Vienna, or else Robert will go to Berlin". The disjunctive symbol $a + b + c$ asserts that *one* of the three events named will take place, but it makes no assertion as to whether or not *more than one* will take place.

In my papers in the *Educational Times* and in the *Proceedings of the Mathematical Society* these disjunctive statements are called *indeterminate statements*.

It is evident that in this algebra of statements the words *factor, multiple, product, sum,* and *term*, have not by any means the same meaning as in ordinary algebra; but there are some remarkable analogies which render it desirable and convenient to borrow these mathematical terms, instead of inventing new ones. The most remarkable of these analogies is to be found in the rule of multiplication, which is precisely the same in logic as in mathematics. In logic as in mathematics the product of the factors $a + b$ and $c + d$ is $ac + ad + bc + bd$, and the same rule holds good whatever be the number of factors, and whatever be the number of terms in the respective factors. Thus, if a, b, c, d, respectively denote the four statements "He will go to Aberdeen," "He will go to Brighton," "He will go to Chester," "He will go to Dublin," the expression $(a + b)(c + d)$ may be read "He will go either to Aberdeen or to Brighton, and he will also go either to Chester or to Dublin"; while the equivalent expression $ac + ad + bc + bd$ may be read "He will go to Aberdeen and to Chester, or else he will go to Aberdeen and to Dublin, or else he will go to Brighton and to Chester, or else he will go to Brighton and to Dublin".

Def. 3.—The symbol $:$, which may be read "implies," asserts

Symbolical Reasoning. 51

that *the statement following it must be true, provided the statement preceding it be true.*

Thus, the expression $a:b$ may be read "a implies b," or "If a is true, b must be true," or "Whenever a is true, b is also true". As a simple verbal illustration, let a denote the statement "He received the letter yesterday," and let b denote the statement "The letter was posted more than a week ago"; then the symbol $a:b$ may be read "If he received the letter yesterday, it must have been posted more than a week ago". Again, let a denote the statement "No foreigners are eligible for that appointment, and this man is a foreigner," and let b denote the statement "This man is not eligible for that appointment". Then the symbol $a:b$ may be read "If no foreigners are eligible for that appointment, and this man is a foreigner, he cannot be eligible for that appointment". Here a denotes a compound statement which may be resolved into two factors. Let m denote the first factor, "No foreigners are eligible for that appointment," and let n denote the second factor, "This man is a foreigner"; then $a=mn$, and the symbols $a:b$ and $mn:b$ denote exactly the same statement, namely, "If no foreigners are eligible for that appointment, and this man is a foreigner, he cannot be eligible for that appointment".

Expressions of the form $a:b, mn:b, a+b:c+d$, &c. (involving the symbol :) are called *Implications* or *Conditional Statements*. The statement to the left of the sign : is called the *Antecedent*, and the statement to the right of the sign : is called the *Consequent*.

Def. 4.—The Symbol =, when placed between two statements, asserts that the two statements are equivalent, each implying the other. Thus the equation $a=b$ is equivalent to the compound implication $(a:b)(b:a)$; or, as it may be expressed symbolically $(a=b)=(a:b)(b:a)$.

It is easy to see that the implication $a:b$ is a brief and convenient equivalent for the equation $a=ab$. The economical advantages secured by adopting the former as an abbreviation for the latter do not seem so great when the antecedent is a simple expression as above. But let the antecedent be a complex expression, and the advantages secured by the symbol : become apparent at once. It requires no formal proof to show, for instance, that the implication $a+bc+de:x$ is a much simpler and more manageable expression than its equivalent, the equation $a+bc+de=(a+bc+de)x$. But mere economy of mechanical labour is not the sole advantage which results from the adoption of this symbol of inference or conditional statement. It is also a very simple and suggestive representation of a great and fundamental law which runs through all reasoning, and which, from want of a better name, I will call the *Law of Implication*. This law expresses the broad fact that the sole function of the reason is to evolve fresh knowledge from the antecedent knowledge already laid up in the store-house of the memory, and that unless we supply it with this material to work upon, it will not work at all. A mere sense-impression, too fleeting and transitory to enter into combination with some previous recollection, is a material of which the reason can make no use. This law of implication seems to have presided over the very birth and infancy of human speech, as well as over its subsequent growth and development. Every verbal statement, as we all know, may be divided into two distinct parts, which are technically called *subject* and *predicate*. But if we examine very closely the meanings of these terms, we shall find that the relation in which they stand to each other is strikingly analogous to that connecting the terms *antecedent* and *consequent* in any implication $a:b$. Take for example the statement "Man is mortal". Let a denote the statement "He is a man," and let b denote the statement "He is mortal". Then the implication $a:b$ is an exact equivalent for the statement "Man is mortal". But this subject will be considered more fully when we come to speak of the syllogism.

Since the implication $a:b$ is an equivalent for the equation $a = ab$, it follows that the antecedent a is a *multiple* of the consequent b.

Def. 5.—An *accent* (') is the symbol of *denial*, and simply negatives any statement (simple or complex) to which it is affixed. Thus a' is the *denial* (or negative) of a. It does not assert that a (if a compound statement) may not possibly contain *some* true factors; it only asserts that there is falsehood somewhere, that *one* factor at least of a is false. If a denotes the statement "He will go to Aberdeen," a' will denote the statement "He will *not* go to Aberdeen". So if ab denote the compound statement "He will go to Aberdeen, and she will go to Brighton," then $(ab)'$ will simply deny this, and may be read "*It is not true* that he will go to Aberdeen and that she will go to Brighton".

A little consideration will show that the symbol $(ab)'$ is equivalent to the symbol $a' + b'$. Attaching to the former the same meaning as before, namely, "It is not true that he will go to Aberdeen and that she will go to Brighton," the latter may be read "Either he will not go to Aberdeen or she will not go to Brighton". In like manner we get $(abc)' = a' + b' + c'$, and so on.

As another example take the equivalent symbols $(a + b)'$ and $a'b'$. If a and b respectively denote the same statements as

Symbolical Reasoning.

before, the symbol $(a+b)'$ may be read "*It is not true* that either he will go to Aberdeen or that she will go to Brighton"; and the equivalent symbol $a'b'$ may be read "He will not go to Aberdeen and she will not go to Brighton". Similarly we get $(a+b+c)' = a'b'c'$, and so on.

Def. 6.—Statements represented by letters or any other arbitrary symbols, which we adopt for the convenience of the moment and to which we attach only a *temporary* meaning, are usually statements whose truth or falsehood may be considered an open question, like the statements of witnesses in a court of justice. It is convenient therefore to have an invariable symbol which shall be applicable to any statement whose truth is admitted and unquestioned, and to such a statement only. The conventional symbol used for this purpose is the symbol 1. For a like reason it is convenient to have an invariable symbol to represent any statement whose *falsehood* is admitted and unquestioned. The symbol used for this purpose is 0.

These symbols, 1 and 0, I have borrowed from the mathematical theory of probability, which, I need hardly say, was the suggestive origin of my whole method, as it in all probability was the suggestive origin of the similar yet fundamentally different method of Professor Boole.

The following are a few among many useful and symmetrical formulæ that may be readily deduced as necessary consequences of the preceding definitions:—

(1.) $aa' = 0$
(2.) $a + a' = 1$
(3.) $(abc \ldots)' = a' + b' + c' + \ldots$
(4.) $(a+b+c+ \ldots)' = a'b'c' \ldots$
(5.) $(ab+a'b')' = ab' + a'b$.
(6.) $(a:b) = (b':a')$, and $(a:b)' = (b':a')'$.
(7.) $(a:b) : a' + b$.
(8.) $(a=b) : ab + a'b'$.
(9.) $(x: abc \ldots) = (x:a)(x:b)(x:c) \ldots$
(10.) $(a+b+c+ \ldots : x) = (a:x)(b:x)(c:x) \ldots$

The first of the above formulæ, namely, the formula $aa' = 0$, symbolically expresses the law of thought to which logicians have given the name of the *Law of Contradiction*. It asserts that a statement and its denial cannot both be true. By virtue of this formula any compound statement that contains any inconsistent combination like aa' vanishes as an impossibility. For example $abcc' = 0$.

The formula $a + a' = 1$ is the symbolical expression of the law of thought to which logicians have given the name of the *Law* of *Excluded Middle*. By virtue of this formula we have the equation $a = a(b+b')$, whatever the statement b may be.

The formula $(ab+a'b')' = ab'+a'b$ may be proved as follows:
$(ab+a'b')' = (ab)'(a'b')'$, by formula 4,
$= (a'+b')(a+b)$, by formula 3,
$= ab'+a'b$, by actual multiplication, omitting the two impossible or zero terms aa' and bb'.

The formula $(a:b) = (b':a')$ is the symbolical expression of the logical Principle of "Contraposition". It asserts that the statement "If a is true, b is true," is the exact logical equivalent of the statement "If b is false, a is false". The truth of either of these two conditional statements follows as a necessary consequence of the truth of the other. This principle of Contraposition is a very important one.

The formula $(a:b) : a'+b$ deserves some consideration. Let a denote the statement "He will persist in his extravagance," and let b denote the statement "He will be ruined". Then the implication $a:b$ may be read "If he persists in his extravagance he will be ruined," while the disjunctive statement $a'+b$ may be read "He will either discontinue his extravagance, or he will be ruined". To some readers these two statements may seem logically equivalent, so that they should be connected by $=$, the symbol of equivalence, and not by $:$, the symbol of inference or implication. We will therefore subject the statements to a closer analysis. If they are really equivalent their denials will also be equivalent. Let us see if this is the case. The denial of $a:b$ is $(a:b)'$, and this denial may be read "He may persist in his extravagance without necessarily being ruined". The denial of $a'+b$ is $(a'+b)'$ or ab' (see formula 4), which may be read "He *will* persist in his extravagance, *and* he will *not* be ruined". Now it is quite evident that the second denial is a much stronger and more positive statement than the first. The first only asserts the *possibility* of the combination ab'; the second asserts the *certainty* of the same combination. The *denials* of the statements $a:b$ and $a'+b$ having thus been proved to be not equivalent, it follows that the statements $a:b$ and $a'+b$ are themselves not equivalent, and that, though $a'+b$ is a necessary consequence of $a:b$, yet $a:b$ is not a necessary consequence of $a'+b$.

It is easy to see that the implications $a:1$ and $0:a$ give us no information whatever as to the truth or falsehood of a, but that the equations $a=1$, and $a=0$ are the exact equivalents of the implications $1:a$ and $a:0$ respectively, and that from the former we infer that a is true, and from the latter that a is false.

Consistency of notation in this algebra of logic requires that the implications $a:1$ and $0:a$ should each be equivalent to 1

Symbolical Reasoning.

whether the statement a be true or false. In a strictly analogous manner consistency of notation in the common algebra of mathematics requires that a^0 should be equal to 1 whatever be the value of the number or ratio a. In the one algebra we may thus have the anomalous looking equation $(0:0) = 1$, and in the other the anomalous looking equation $0^0 = 1$.

The symbol : for the word *implies* is not quite an equivalent for the symbol .·. commonly used for the word *therefore*. The difference in meaning between the two symbols may be seen from the equation $(a .\cdot. b) = a(a:b)$. The statement $a .\cdot. b$ is stronger than the conditional statement $a:b$ and implies the latter. The former asserts that b is true *because* a is true; the latter asserts that b is true *provided* a be true.

We will now examine the syllogisms of Aristotle in the light of this notation. The first thing that strikes anyone on reading these syllogisms for the first time is the constant recurrence of the words *all, some,* and *no*, as in the syllogism " *No* men are gods; *all* men are living beings; therefore *some* living beings are not gods" (Felapton). Moreover, in applying these syllogistic rules to examples of ordinary reasoning, expressed in the simple untechnical language of daily life, logicians usually subject this language to more or less distortion, resulting sometimes in extremely uncouth and scarcely comprehensible phrases, in order to introduce one or more of these quantitative words, without the express use of which they apparently think that no argument can be strictly and rigorously logical. For myself, so far am I from regarding those words as indispensable accessories to logic that I look upon them rather as the fatal cords that have for centuries held prisoner this noblest of the sciences and effectually prevented its flight beyond the limits of a meagre and barren circle of insipid truisms. By cutting these cords asunder we can, I believe, set logic free to soar on vigorous pinions into new and fruitful regions, where it will join the other sciences, and in the common pursuit of truth exercise over them all that sovereign authority which is its undoubted right, and which its Grecian fetters alone have hitherto prevented it from exercising. The precise mode in which I think this desirable result may be brought about will become apparent as I proceed.

It has already been pointed out that, by the principle of Contraposition, the implication $a:b$ is equivalent to the implication $b':a'$. By changing b into b', and therefore b' into b, these give us two other implications $a:b'$ and $b:a'$, which are also equivalent to each other. And when two implications are equivalent their *denials* are also equivalent. Thus $(a:b)'$ is equivalent to $(b':a')'$, and $(a:b')'$ is equivalent to $(b:a')'$.

These denials of implications I have in my second paper in

Symbolical Reasoning.

the *Proceedings of the Mathematical Society* called *non-implications*; and instead of the accent to express negation I have used as the denial of : the symbol $\div$. Thus $(a:b)'$ and $a \div b$ are symbols which have exactly the same meaning, each asserting that the statement a does *not* imply the statement b, or, in other words, that a may possibly be true without b being so. I shall use the latter symbol in the rest of this paper.

All possible valid syllogisms, including the 19 syllogisms usually given and many others besides, come either directly or by the substitution of equivalents under one or other of the following four standard-implications, which are however more general than the syllogisms, since they are not like the latter restricted to mere classification :—

$$(a:b)(b:c):(a:c) \quad \ldots \quad (1)$$
$$(a:b)(b:c):(a \div c') \quad \ldots \quad (2)\,[1]$$
$$(a:b)(a \div c):(b \div c) \quad \ldots \quad (3)$$
$$(a:b)(a:c):(b \div c') \quad \ldots \quad (4)\,[1]$$

This is shown in my second paper in the *Proceedings of the Math. Society*, to which the reader is referred for a full and formal proof. I shall content myself here with giving one or two illustrations of the way in which syllogisms may be thus converted into implications.

Take the syllogism called Barbara, "All Y is Z, and all X is Y; therefore all X is Z".

Now by the first premiss, "All Y is Z," is meant that every single individual that possesses the attribute Y (or belongs to the class Y) possesses also the attribute Z (or belongs to the class Z); and the other premiss and the conclusion may be similarly interpreted. No matter who or what any individual may be, we are told in the first premiss that if it possesses the attribute Y, it also possesses the attribute Z. In other words, speaking throughout of some originally [2] unclassed individual, we are told that the statement "It possesses the attribute Y" implies the state-

---

[1] The statement a is here understood to be a *consistent* statement, i.e., a statement which *may* be true. When this restriction is removed the second of the above implications should be written

$$(a:b)(b:c):(a \div c') + (a:0),$$

and the fourth should be written

$$(a:b)(a:c):(b \div c') + (a:0).$$

This note has been suggested by some friendly criticism of my logical system with which the Rev. J. Venn has kindly favoured me.

[2] The mathematical reader will notice the analogy between this representation of a whole class by a single individual possessing the distinguishing attribute of the class and the representation in analytical geometry of a whole series of points or locus by a single specimen-point belonging to this locus.

Symbolical Reasoning. 57

ment " It possesses the attribute Z"; so that if y denotes the first statement and z the second, we have the implication $y:z$, which may be read "If any individual possesses the attribute Y, it must also possess the attribute Z". In like manner " All X is Y " may be expressed by the implication $x:y$, and " All X is Z " by the implication $x:z$. Speaking throughout therefore of some originally unclassed individual, the syllogism becomes

$$(y:z)(x:y):(x:z).$$

If we change the order of the premisses, this becomes

$$(x:y)(y:z):(x:z),$$

which may be read " If any individual possesses the attribute X, it must also possess the attribute Y, and if it possesses the attribute Y, it must also possess the attribute Z. Consequently, if it possesses the attribute X, it must also possess the attribute Z." In this form the syllogism is an example of the first of the four standard-implications given above, for we only substitute x for a, y for b, and z for c.

Take next the syllogism called Fresison, " No Z is Y, and some Y is X; therefore some X is not Z ".

As before let x, y, z respectively denote the three statements that a certain individual belongs to the class X, that it belongs to the class Y, and that it belongs to the class Z. Then " No Z is Y " is equivalent to the implication $z:y'$, which may be read " The statement that an individual belongs to the class Z implies that it does *not* belong to the class Y". The second premiss " Some Y is X" is equivalent to the *non-implication* $y \div x'$ or $(y:x')'$, which may be read " The statement that any individual belongs to the class Y does not imply that it is excluded from the class X". From these two premisses, on the understanding that the same individual is spoken of throughout, follows the conclusion expressed by the implication $x \div z$ or $(x:z)'$, which may be read " The statement that an individual belongs to the class X does not imply that it belongs to the class Z ". The symbolical expression of this syllogism is therefore

$$(z:y')(y \div x'):(x \div z).$$

This may be brought under the third of the four standard-implications given above as follows. By the principle of Contraposition (or transposition) we get $(z:y') = (y:z')$. Substituting and reading y for a, z' for b, and x' for c in the third standard-implication we get

$$(y:z')(y \div x'):(z' \div x');$$

and the conclusion $z' \div x'$, by the principle of Contraposition, is equivalent to $x \div z$.

It is needless to give any more examples. The method of

conversion is the same throughout. Each premiss of a syllogism is expressed as a simple implication or non-implication, and the whole syllogism thus becomes a complex implication. The principle of logical Contraposition, which bears a remarkable analogy to the rule of "transposition" in algebraical equations, is appealed to frequently. By this principle we may transpose the antecedent and consequent of any implication, provided we at the same time *change the sign* of each. By "changing the sign" of a statement I mean affixing or removing the accent of negation. Thus, by Contraposition, the implication $x:y'$ is equivalent to the implication $y:x'$; and the non-implication $x \div y'$ to the non-implication $y \div x'$. In both cases of equivalence the unaccented x moves from the left to the right and *assumes an accent*, while the accented y moves from the right to the left and *drops its accent*.

No part of logic has received so much attention and given rise to so much discussion as the syllogisms of Aristotle. This is why I have selected them as illustrations of the application of my symbolical method. A far more important subject of application however is the great and ever recurring problem of physical science—how to discover the general laws which regulate the various phenomena of the material universe. How logical symbolism may be systematically and advantageously employed even in those difficult researches and in cases quite beyond the reach of the ordinary mathematical symbolism has been briefly indicated in my third paper, published in the *Proceedings of the Mathematical Society*.

The reader will observe that the whole of my symbolical system of logic rests upon very simple and easily grasped principles. Though this system does not necessarily exclude metaphysical considerations, it is both theoretically and practically independent of such considerations, and rests upon a surer and firmer basis because it is thus independent. It may be a branch and a very important branch of *applied* logic to investigate the primary source and origin of the knowledge which we find existing in our mind, but it certainly is no part of *pure* logic. *Pure* logic must take this knowledge for granted. We must reason from the known to the unknown whatever be the subject of investigation, and it is the proper and special function of pure logic to explain how this may be done most safely and most certainly. If teaches us (if I may be allowed the metaphor) how to use our intellectual oars and steer the boat of reason whatever be the direction in which we wish to travel, and without troubling ourselves as to the exact source of the river on which that boat is moving. A short distance up this river we may be able to go, in spite of the strong opposing current,

Symbolical Reasoning.

and we may even succeed in tracing to its source a little tributary here and another there; but up the river or down the river the motion of the oars is always the same, and it is by the help of the water within their sweep that the boat is ever propelled onwards into new positions. It is through and by means of the knowledge expressed by the *antecedent* that the reason reaches the knowledge expressed by the *consequent*. The latter becomes a means and medium of progression in its turn, and so the reason moves onward from knowledge to knowledge.

Just as my symbolical method, though not necessarily excluding metaphysical considerations, is yet independent of such considerations, so, though it does not necessarily exclude inquiries into grammatical distinctions, it is yet independent of all such distinctions. Grammar, like metaphysics, may be an important branch—indeed it *is* an important branch—of *applied* logic, but, like metaphysics and many other special subjects of investigation, it is no essential part of *pure* logic. The student of pure logic need know nothing of grammar, absolutely nothing. The grammatical structures of sentences are matters with which he has no special concern. His business is to investigate the logical relations in which *statements* stand to each other, and if he understands the exact meaning of each statement that enters into his argument, he need not trouble himself as to the exact form of words in which that statement is expressed. Nay more; the statements of his argument need not be expressed in words at all. If he understands their meaning singly and collectively, it is enough. Any sign or symbol that conveys any intelligible information to the mind may be regarded as a *statement* so far as the logician is concerned with it. The deaf and dumb make use of many signs among themselves (apart from their regular alphabet) which in this sense are real statements, and if sufficiently numerous might answer all the purposes of ordinary speech in any logical argument; and yet these are statements which cannot easily be resolved into *subject* and *predicate*. Much has been written in praise of, and a good deal has been written in disparagement of, Hamilton's logical scheme called the "Quantification of the Predicate". Prof. Jevons, in his *Elementary Lessons in Logic*, speaks of the scheme in terms of high approval. Mr. Ellis, on the other hand, in the *Educational Times* for August, 1872, calls it "an unfortunate conception" and "a barbarous abuse of language". My system, which adopts full and complete *statements* as the ultimate constituents into which any argument can be resolved, steers clear of the discussion altogether. At the same time, though it does not recognise the question as in any way an essential one, or one properly belonging to pure logic at all, it is not on that account debarred from discussing it if it so chooses. It certainly is worth remarking as a matter of some interest, though not as a matter of paramount importance, that Hamilton's "All X is all Y" is expressed by the equation $x = y$, which is equivalent to the compound statement $(x:y)(y:x)$; that his "All X is some Y" is expressed by the compound statement $(x:y)(y \div x)$; that his "Some X is some Y" is expressed by the compound statement $(x \div y)(x \div y')$; and that all his other "quantifications" may be similarly translated into the language of my calculus. In this translation the letters x, y, y' are to be understood in the sense which I attached to them when discussing the syllogism; that is to say, they all denote *classifying statements*, referring to some one originally unclassed individual as their common subject.

I think I have now sufficiently explained the fundamental principles on which my Symbolical System of Logic is constructed, though not the rules of symbolical operations which are founded upon these principles, nor yet the various practical applications of which the method is capable. A full and detailed account of these, such as I have given in the *Proceedings of the Mathematical Society*, would be altogether beyond the aim of the present paper. In explaining what I believe to be the advantages of my own system, I have carefully avoided drawing any comparison between it and other systems which are already before the public. With two of these, Prof. Boole's and Prof. Jevons's, it has much in common, but it has been conceived and developed quite independently of theirs, and the points of difference which distinguish it are fundamental and important.

HUGH MCCOLL.

[1897a] Symbolic Reasoning(II). Mind, Vol. 6, pp. 493-510.
III.—SYMBOLIC REASONING. (II.)

(For I. see MIND, January, 1880.)

BY HUGH MACCOLL.

SYMBOLIC Logic (including Mathematics) may be defined as 'the science of reasoning by the aid of representative symbols; these symbols being employed as *synonymous substitutes for longer expressions that are required frequently*'. The words in italics contain the pith and principle of the whole subject. When any expression, verbal or symbolic, of inconvenient length has to be written frequently in the course of an argument or investigation, we naturally cast about for some short and simple symbol to represent and replace it. This desire to economise time, space and labour is always, always has been, and always will be, the great motive power that sets going and keeps going the evolutionary progress of the science. What, for instance, was the primary object of the symbol × in such a case as 27365×7? Clearly to save the trouble of writing down an addition sum of seven rows of figures, each row being 27365. The same may be said of the symbol a^5 as a substitute for $aaaaa$, and of many others, including the remarkable and highly general symbol $\phi(x)$, which plays such an important part in the higher mathematics as a substitute for any expression whatever that contains x in any relation whatever as one of its constituents. As Symbolic Science advances and tackles more difficult problems these conventional abbreviations afterwards combine among themselves and produce fresh expressions of inconvenient length and frequent recurrence which give birth in *their* turn to fresh representative substitutes—to abbreviations of abbreviations—which in their powers of thought-condensation bear, on an average, the same ratio to the symbols they replace as these had done to *their* immediate progenitors. Thus it has been that the science of mathematics has slowly acquired its present marvellous power within the limits of its application; and thus it will be that the newer but more general science of symbolic logic, with a wider sweep and bolder aim, will before long develop into a still more powerful instrument of research.

The problems with which our reason has to deal are of various kinds and cannot be exhaustively classified. Sometimes from certain data or premises A, *known or admitted to be true*, we have to prove a conclusion Q; that is to say, we seek to convince ourselves or others that Q *as well as* A *is true*. Sometimes we have data or premises P which are *not* always certain or admitted to be true, and then we seek to prove—*not* the conclusion Q—that cannot be done from uncertain or unadmitted data —but the proposition that P *implies* Q, *i.e.*, that *if* P is true Q is true—which is quite another matter. Sometimes our reason has to deal with the *inverse* problem, namely, to find from what data or premises Q we can derive a conclusion P, as when we seek the possible roots of a given equation. These are only two or three of the kinds of problems of which it is the business of logic to find solutions, but as they are among the commonest we will take them first in the order of consideration.

Since the conditional proposition *If* P *is true* Q *is true* (or P *implies* Q) is one of frequent recurrence we want some symbol to represent it. What symbol should we adopt? Various logicians have adopted various symbols, each giving some reason founded on some mathematical analogy for his own special choice. Boole adopts $P = \frac{0}{0} Q$ or $P = v Q$; Pierce takes $P \prec Q$; Schröder uses $P = (= Q$; and there are many others; each writer, as I have said, justifying his choice on the ground of some real or fancied mathematical analogy. My own choice has been the symbol P : Q, *not* (as has been erroneously supposed) on the ground of any analogy to a ratio or division, but simply because a *colon* symbol is easily formed, occupies but little space—two important considerations—and— though this is less important—because it is not unpleasing to the eye. I hold that we may claim the same liberty of definition and interpretation for any of our symbols of relation (+, =, :, etc.) as we claim for any letter of the alphabet, x, when in one problem (some unit of reference being understood) we say "Let x denote his gain"; in another "Let x denote his loss"; and in another "Let x denote the distance of the planet Neptune". So long as it suits our purpose to attach the same meaning to any symbol, so long we should adhere to that meaning—so long,

SYMBOLIC REASONING. 495

and no longer. As a general rule we should be as conservative as circumstances will permit in the significations we give to our symbols of relation (+, =, :, etc.): these are the *constants*, the *fixed stars*, as it were, of our logical systems; while we may deal more freely with our planetary ever-varying symbols, x, y, a, b, etc., which generally denote numbers or ratios in mathematics, and classes, properties or statements in logic. But as even the so-called fixed stars are only found to be *relatively* fixed, so our so-called *constant* symbols of relation need only be *relatively* constant. In my former paper in MIND ("Symbolical Reasoning," MIND, Jan., 1880), and in my papers in the *Proceedings of the Mathematical Society* which had preceded it, I adhered throughout to the symbol $a : \beta$ as my representative of an implication; but when, after several years' abandonment, I recently returned to my logical studies and began to consider the complex relations of the higher orders of implications, as in the formula

$$(a : \beta) : \{(u : a) : (u : \beta),$$

I felt the necessity of further abbreviations and adopted a_β as a synonym of $a : \beta$, so that the above formula might appear as

$$a_\beta : (u_a : u_\beta).$$

I also tried how the symbol a^u would act as the converse of a_u and meaning that a *is implied* in u. This led immediately and of necessity to the discovery that the formulæ $a^u a^v = a^{u+v}$, $(a\beta)^u = a^u \beta^u$, and $a^0 = 1$ would on this interpretation hold good in logic as well as in mathematics; but I found that the series of analogies stopped when I tried $a^{uv} = (a^u)^v$. Then I tried a^u as a substitute for $a + u'$ (the alternative implied by u_a) and found that the preceding three analogies still held good, and also this fourth analogy $a^{uv} = (a^u)^v$, but that, on the other hand, this interpretation of a^u would lead to an additional logical formula $a^{uv} = a^u + a^v$, which does *not* generally hold good in mathematics.

If we take a^u as synonymous with u_a it is evident that every formula with *indices* may be converted at once into an equivalent formula with subscripts, and *vice versâ*. Thus

$$a^u a^v = a^{u+v} \text{ corresponds to } u_a v_a = (u + v)_a$$
$$(a\beta)^u = a^u \beta^u \text{ corresponds to } u_{a\beta} = u_a u_\beta$$
$$a_\beta : (u_a : u_\beta) \text{ corresponds to } \beta^a : (a^u : \beta^u),$$

and so on. The last formula may be read thus: "If β is a factor of a (that is, if the statement β is implied in the statement a); then, if a is a factor of u, β must also be a factor of u ". In spite however of the mathematical analogies obtained by the above interpretation of indices, I found it generally more convenient to employ the subscript form u_a as a synonym for $u : a$, and to reserve indices for other uses. What chiefly led me to this decision was the discovery that in dealing with implications of the higher degrees (*i.e.* implications of implications) a calculus of two dimensions (unity and zero) is too limited, and that for such cases we must adopt a *three*-divisional classification of our statements. We have often to consider not merely whether a statement is *true* or *false*, but whether it is a *certainty*, like $2 + 3 = 5$; an *impossibility*, like $2 + 3 = 8$; or a *variable* (neither always true nor always false), like $x = 4$. To illustrate the meaning of a *variable* statement, we may suppose x in the last statement to be taken at random out of three possible and equally probable values 2, 4, 6. If this experiment be repeated often enough, the statement ($x = 4$) will be sometimes true and sometimes false; its chance of being true will, in fact, be *one-third*. The formula

$$(a : \beta) : (u_a : u_\beta)$$

is another example of a *certainty*; for it holds good whether its elementary constituents (u, a, β) be *certainties*, *impossibilities*, or *variables*—separately or conjointly. But this does not apply to the converse implication

$$(u_a : u_\beta) : (a : \beta),$$

which fails for some values of its constituents, as, for example, when u is a *certainty*, a a *variable*, and β an *impossibility*.

This necessity for a[1] three-divisional classification of statements naturally suggested the adoption of some corresponding modification in notation; so I chose the symbol ϵ (as in my *fourth* paper in the *Proceedings of the Mathematical Society*) to replace *unity* as the symbol of *certainty*; η (instead of *zero*) as the symbol for an *impossibility*; and θ as a suitable symbol to denote a statement which is neither a *certainty* nor an *impossibility*, whose chance of being true is neither *unity* nor *zero*, and which, therefore, may fitly be called a *variable*. For distinguishing these three classes of statements the notation of indices is most convenient. The three equational symbols ($a = \epsilon$), ($\beta = \eta$), and

---

[1] For other divisions and a logic of 3^n dimensions, see note at the end of this paper.

SYMBOLIC REASONING.

($\gamma = \theta$) will very well express that a is a *certainty*, β an *impossibility*, and γ a *variable*; but in complicated cases (in order to economise space and dispense as far as possible with those necessary evils called *brackets*) still simpler symbols were desirable. I therefore chose indices to denote *classes* of statements, so that, for example, the symbol a^u asserts that *the statement a belongs to the class of statements denoted by the symbol u*. On this interpretation, the three symbols a^ϵ, β^η, γ^θ respectively assert that a is a *certainty*, β an *impossibility*, and γ a *variable*, and are therefore synonymous with the three longer symbols $(a = \epsilon)$, $(\beta = \eta)$, and $(\gamma = \theta)$. Again, putting τ for a *true* statement (not necessarily a *certainty*) and ι for a *false* statement (not necessarily an *impossibility*), a^τ will assert that a is true, and a^ι that a is false. Thus $a^\tau : \beta^\tau$ in my new notation becomes equivalent to $a : \beta$ in my former; $a^\tau : \beta^\iota$ becomes equivalent to $a : \beta'$; $a^\tau \beta^\iota$ to $a\beta$; $a^\tau + \beta^\iota$ to $a + \beta'$; and so on; while a^ϵ (*not* a^τ) becomes equivalent to the former $(a = 1)$, and a^η (*not* a^ι) becomes equivalent to the former $(a = 0)$. The symbols 1 and 0 being thus displaced by ϵ and η were *ipso facto* set at liberty for other purposes. I use the former in conjunction with its old comrades 2, 3, 4, etc., in such cases as u_1, u_2, u_3, etc., which respectively denote *particular* statements of the class u. Thus the equational statement $a\beta : a + \beta = \epsilon_1$ asserts that the implication $a\beta : a + \beta$ is a particular statement of the class called *certainties*, and is the *first certainty* that has entered into our present argument. Similarly the equational statement $\theta^\eta = \eta_3$ asserts that θ^η is a particular statement of the class called *impossibilities*, and is the *third impossibility* that has entered into our argument. And so on for other *particular statements* η_2, ϵ_5, θ_4, etc., of their respective classes. On the other hand, I use the symbols $a_{o\beta}$, a^{ou}, $a_{o(u+v)}$, etc. (chiefly to avoid cumbersome brackets) as convenient synonyms for the denials $(a_\beta)'$, $(a^u)'$, $(a_{u+v})'$, etc. For example, the formula $(a + \beta)_{ou} = a_{ou} + \beta_{ou}$ is much more convenient than its equivalent, the formula $\{(a + \beta)_u\}' = (a_u)' + (\beta_u)'$. Other symbols of abbreviation employed by me in my fifth paper (recently published) in the *Proceedings of the Mathematical Society* are the following:—

(1) a^{uw} is short for $(a^u)^w$, so that $a^{\eta\iota}$, $a^{o\eta}$, $(a^\eta)^\iota$, and $(a = \eta)^\iota$ are all four synonymous, each denying the truth of a^η and asserting that a is a *possibility*. Similarly, a^{uvw} means $(a^{uv})^w$; and so on.

(2) $a^{u,v}$ is short for $a^u + a^v$; $a_{u,v}$ is short for $a_u + a_v$; $(a, \beta)^u$ is short for $a^u + \beta^u$. Hence it is evident that $a^{\epsilon\iota}$, $a^{o\epsilon}$,

$a^{\eta,\theta}$, and $a^\eta + a^\theta$ are all four synonymous, each asserting that a is an *uncertainty*.

(3) $a \,!\, \beta$ is equivalent to $\beta : a$ and asserts that a *is implied in* β.

(4) $a : \beta : \gamma : \delta$ is synonymous with $(a : \beta)\,(\beta : \gamma)\,(\gamma : \delta)$; and $a\,!\,\beta\,!\,\gamma\,!\,\delta$ is synonymous with $(a\,!\,\beta)\,(\beta\,!\,\gamma)\,(\gamma\,!\,\delta)$. The former is a chain of *deductive*, and the latter a chain of *inductive, sorites*.

(5) The symbol :: is synonymous with =, but of *shorter reach*, so that the equation $a :: \beta = \gamma$ means $(a :: \beta) = \gamma$, and does *not* mean $a :: (\beta = \gamma)$, which would be denoted by $a = \beta :: \gamma$. The main object of the symbol ::, as a synonym of =, is to avoid a multiplicity of brackets. For example, the formula

$$(a + \beta)_u :: a_u + \beta_u = a_u :: \beta_u$$

asserts that the equational statement on the left side of the sign (=) is equivalent to the equational statement on its right side, each implying the other.

(6) The symbol $\frac{a}{\beta}$ means $a_\beta\, a'_{o\beta}$ or its equivalent $a_\beta\, \beta^u$. It asserts (like a_β) that whenever a is true β is true; and it also asserts (what a_β neither asserts nor denies) that β is *not* always true when a is *not* true. The equivalence of $a_\beta\, a'_{o\beta}$ and $a_\beta\, \beta^u$ is easily proved, so that either may be taken as a definition of $\frac{a}{\beta}$. The statement $\frac{a}{\beta}$ is called a *Causal* implication, as it indicates some causal connexion between a and β, whereas its factor, the *general* implication a_β, is synonymous with $(a\beta')^\eta$ and does not necessarily indicate any causal connexion. Thus a_β always holds good when β is a *certainty*, whatever a may be; and it also always holds good when a is an *impossibility*, whatever β may be; so that a_ϵ and η_a are always *certainties* even when $a = \eta$. On the other hand, the causal implication $\frac{a}{\beta}$ contradicts its definition and becomes an impossibility when β is a certainty.

(7) The symbol $a > \beta$ means $a_\beta\, \beta_{oa}$. It asserts that a implies β, but that β does not imply a. In this case a is said to be *stronger* than β. On the other hand, $a < \beta$ means $\beta > a$ and asserts that a is *weaker* than β. It is evident that, by this definition, $a > \eta$ is an impossibility, as it implies η_{oa}, which is easily proved to be inconsistent with our definitions. As a rule, the greater the number of factors in a statement (that is to say, the more it asserts) the *stronger* it is; but, on the other hand, the greater the chance as a rule that it contains an *inconsistency* some-

where; and a single inconsistency η (like a zero-factor in mathematics) makes the whole an inconsistency. Hence, no statement can be stronger than an *impossibility*. By parity of reasoning, no statement can be weaker than a *certainty*. A witness whose testimony consisted of a restatement of facts already admitted and unquestioned would not be very helpful in any serious and *bonâ fide* inquiry or investigation.

(8) Two more symbols remain, and they involve an important principle. These are WA and SA. The first denotes the *weakest premise from which we can infer* A; and the second denotes the *strongest conclusion which we can draw from* A. The symbol A is here understood to denote some function $\phi(a, \beta)$ of two or more constituents a, β, etc.; while WA and SA denote some other functions $\psi_1(a^u, \beta^v)$, $\psi_2(a^u, \beta^v)$, with constituents a^u, β^v, etc., in which u and v may each denote ϵ, or η, or θ, as the case may be. For example, the formula

$$W(a\beta)^\theta = a^\epsilon \beta^\theta + a^\theta \beta^\epsilon.$$

asserts that the *weakest premise* (with subject a or β, and predicate ϵ or η or θ) from which we can infer that $a\beta$ is a *variable* is the alternative that either a is a *certainty* and β a *variable*, or else a a *variable* and β a *certainty*; while the formula

$$S(a\beta)^\theta = a^\eta \beta^\theta + a^\theta \beta^\eta$$

asserts that the *strongest conclusion* we can draw from $(a\beta)^\theta$ alone (*i.e.* without further data) is the alternative that either a is a *possibility* and β a *variable*, or else a a *variable* and β a *possibility*.

It is evident that the formula WA : A : SA is a *certainty*; and a little consideration will show the validity also of the formulæ WA′ = S′A and SA′ = W′A, in which S′A and W′A denote the denials of SA and WA, and are therefore short for (SA)′ and (WA)′. In other words, the weakest premise from which we can infer the denial of A (or that A is false) is the denial of the strongest conclusion we can draw from A; and the strongest conclusion we can draw from the denial of A is the denial of the weakest premise (or data) from which we can infer A. For example, assuming the formula $W(a : \beta) = a^\eta + \beta^\epsilon$, we get

$$S(a : \beta)' = W'(a : \beta) = (a^\eta + \beta^\epsilon)' = a^\eta \beta^{\epsilon u};$$

so that the strongest inference we can draw from the *denial* of the implication $a : \beta$ is that a is a *possibility* and β an *uncertainty*.

I think I may predict that *synonyms* are destined to play an important part in the future development of symbolic logic, as they undoubtedly have done in the natural evolution of ordinary language. As new needs and new ideas arise with the growth of civilisation and the general advance of humanity, do we not often find that two words which were at first synonyms gradually differentiate and, while still remaining synonymous in some combinations, cease to be so in others?—just as a_β and $\dfrac{a}{\beta}$ are interchangeable in the equivalent statements $a^\theta \beta^\theta a_\beta$ and $a^\theta \beta^\theta \dfrac{a}{\beta}$, but not in the non-equivalent statements $a^\theta \beta^\epsilon a_\beta$ and $a^\theta \beta^\epsilon \dfrac{a}{\beta}$, of which the second, but not the first, is always an *impossibility*. And does not this principle of evolution powerfully contribute to the precision and utility of a language, both as an instrument of research and as a medium for communicating our ideas to others? In connexion with these remarks it will not be irrelevant to mention that before the idea of differentiating between the symbols a_β and $\dfrac{a}{\beta}$ occurred to me I was in the habit of using sometimes the one and sometimes the other as a synonym for the general implication $a : \beta$; so that when I afterwards found it convenient to give symbolic expression to the idea of a *causal* implication I at the same time found a suitable symbol for the purpose ready to my hand.

In ordinary speech not only do we find different words with the same meaning, but also different meanings to the same word; yet it seldom happens that this leads to real ambiguity: the context nearly always removes all danger of miscomprehension. The same rule holds good in symbolic logic, with this difference that here, from our complete liberty to define our symbols as we please, we can guard against the danger with absolute certainty. In my first paper in the *Proceedings of the Mathematical Society*—a paper which deals almost exclusively with the limits of multiple integrals and with probability—I employed each of the symbols x_1, x_2, x_3, etc., y_1, y_2, y_3, etc. (and so on for any number of variables x, y, z, etc.), in two different senses in the same argument; the symbol y_8, for example, denoting the eighth limit of the variable y registered in an accompanying table of reference, and also denoting the *statement* that this limit y_8 is positive; yet this double signification of the

SYMBOLIC REASONING.

same symbol never produces ambiguity, the context always showing what meaning attaches to it in each particular case. The following simple example of a double-meaning symbol is taken from my recent paper in the *Proceedings of the Mathematical Society*.

Let x^u, x^v, x^w respectively assert that the number (or ratio) x is real and positive, that x is real and negative, that x is imaginary, so that the disjunctive statement $x^u + x^v + x^w$ is a *certainty*; and let y^u, y^v, y^w, $(x+y)^u$, $(x+y)^v$, $(x+y)^w$ be interpreted in the same manner. What is the weakest premise in classifying x and y from which we can infer the conclusion $(x+y)^u$? And what is the strongest conclusion we can draw from $(x+y)^u$ as our only premise? The answers are obtained by an easy symbolic process and are

$$W\,(x+y)^u = x^u y^u;$$
$$S\,(x+y)^u = x^w y^w + x^{wi} y^{wi}(x^u + y^u).$$

The first answer may be read: "The *least* that we *must* know, and the *most* that we *need* know, as to the classification of x and y into positive, negative or imaginary, in order to infer that their sum is real and positive, is that each is real and positive. We cannot infer it from less (*i.e.*, weaker) data." The second answer may be read: "The most that we can infer about x and y when we only know that their sum is real and positive is the alternative that either both or neither are imaginary, and that in the latter case one at least is real and positive".

Observe that the sign $+$ is here used in two different senses. In $(x+y)^u$ it connects two *numbers* (or ratios), so that the sum of those numbers, must also be a number; while in $x^u + y^u$, it connects two *statements*, so that the disjunctive $x^u + y^u$ must also be a *statement*. It may be objected that the employment of x^u, x^v, x^w in this way as *statements* might interfere with the free use of the same symbols when u, v, w represent mathematical quantities, such as 2, 3, $\frac{1}{2}$, etc. The reply to this is that we have only to agree or define that the letters u, v, w (or any others we choose) shall, *during the same argument or investigation*, be restricted as indices to the meanings we have assigned to them; while other indices may retain their usual mathematical signification. For example, x^{3v}, as an abbreviation for $(x^3)^v$, would assert that x^3 is real and negative; while x^{v3}, if accompanied by no definition, would be meaningless.

To show the working of this logical calculus of three dimensions I may take the following problem, which Dr.

Venn in his *Symbolic Logic* (second edition, p. 442) calls "Alice's Problem," and which I understand has (under another form) been already discussed by logicians, but with varying conclusions.

Given the statement $A_B (C : A_B')$, can C be true?

My answer is (1) that the data are not sufficient to justify the conclusion that C is possible; and (2) that the data are not sufficient to justify the conclusion that C is impossible. This I prove as follows:—

Putting ϕ for $A_B (C : A_B')$, we get, by a process explained in my fifth paper in the *Proceedings of the Mathematical Society*,

$$W\phi = A^\eta + B^\epsilon C^\eta$$
$$S\phi = W\phi + A^\theta B^\theta C^\eta = A^\eta + B^\epsilon C^\eta + A^\theta B^\theta C^\eta.$$

Now, if $C^\eta S\phi$ could be proved $= \eta$, we should have $S\phi : C^\eta$, and then C^η would be a legitimate conclusion from ϕ. But

$$C^\eta S\phi = C^\eta(A^\eta + B^\epsilon + A^\theta B^\theta) = C^\eta S(A_B),$$

which, without further data, can *not* be proved $= \eta$. Hence, *we cannot from ϕ draw the conclusion that C is possible*.

Again, if $C^{\eta\iota} S\phi$ could be proved $= \eta$, we should have $S\phi : C^\eta$, and C^η would then be a legitimate conclusion from ϕ. But $C^{\eta\iota} S\phi = A^\eta C^{\eta\iota}$, which, again, without further data, cannot be proved $= \eta$. Hence, *we cannot from ϕ conclude that C is impossible*. Thus we have not sufficient data from which to infer either C^η or its denial $C^{\eta\iota}$. In other words, though C must either be possible or impossible, we cannot from the given statement $A_B(C : A_B')$, without some additional premise, ascertain which alternative is the true one.

In obtaining the results $W\phi$ and $S\phi$ I assumed a proposition which is not quite self-evident, namely, that the data $A_B(C : A_B')$ constitute an *impossibility* when A, B, C are all three *variables*. This proposition I now proceed to prove.

No implication $a : \beta$ can be a *variable*[1] when (as in the data of this problem and throughout the preceding argument) its antecedent a and consequent β are both *singulars;* that is to say, when each letter denotes only *one statement, and always the same statement*, be it of the class ϵ or η or

---

[1] This is a point which I did not make sufficiently clear in my recent paper in the *Proceedings of the Mathematical Society* when discussing $W(a : \beta)^\theta$ and $S(a : \beta)^\theta$. The implication $a : \beta$ would be a variable *if a* and β were taken at random *repeatedly out of statements belonging some to the class ϵ, some to the class η, and some to the class θ*. For another case in which $a : \beta$ would be a variable, see the concluding paragraph of this paper, preceding the note on a logic of 3^n dimensions.

SYMBOLIC REASONING.   503

θ. Hence, $a : \beta$ being synonymous with $(a\beta')^\eta$ must be either $= \epsilon$ or else $= \eta$; it cannot be $= \theta$. This being assumed, we have to prove that

$$A^\theta B^\theta C^\theta A_B(C : A_B') = \eta.$$

Since, by hypothesis, the premise A_B is an implication with *singular constituents*, it must be either $= \epsilon$ or else $= \eta$. If $A_B = \eta$, the data contain an impossible factor A_B, and therefore must $= \eta$, which was the proposition to be proved. On the other hand, if A_B does *not* $= \eta$, then A_B must $= \epsilon$. Hence, A_B' must $= \eta$; otherwise we should both have $A_B = \epsilon$ and also $A_B' = \epsilon$, which would imply A^η, which would contradict our assumption A^θ. Thus we have $A_B = \epsilon$ and $A_B' = \eta$, which reduces $A^\theta B^\theta C^\theta A_B(C : A_B')$ to the form $A^\theta B^\theta C^\theta(C : \eta)$, which implies C^η, which contradicts our assumption C^θ. Hence, whether we take $A_B = \epsilon$ or $A_B = \eta$ (the assumption $A_B = \theta$ being here inadmissible) the statement $A^\theta B^\theta C^\theta A_B (C : A_B')$ reduces to η; *quod erat demonstrandum*.

A problem of somewhat more complexity is the following: Let ϕ denote the implication

$$u_{a,\beta}\, (a,\beta)_v : u_v,$$

which, by definition, is equivalent to

$$(u_a + u_\beta)(a_v + \beta_v) : u_v,$$

and may easily be proved equivalent to

$$u_a \beta_v + u_\beta a_v : u_v.$$

For what values of u, a, β, v (expressed in terms of ϵ, η, θ) is ϕ true? For what values false?

The answers which I find are (1) that ϕ is true for every term in the disjunctive statement

$$u^\eta + v^\epsilon + a^\epsilon \beta^\epsilon + a^\eta \beta^\eta + u^\epsilon v^\eta (a,\beta)^\theta + u^\epsilon v^\theta (a^\theta \beta^{\epsilon\iota} + a^{\epsilon\iota} \beta^\theta)$$
$$+ u^\theta v^\eta (a^\theta \beta^{\eta\iota} + a^{\eta\iota} \beta^\theta),$$

and (2) that ϕ is false for every term in the product

$$(u^\epsilon v^{\epsilon\iota} + u^\theta v^\eta)(a^\epsilon \beta^\eta + a^\eta \beta^\epsilon).$$

These two results, I believe, include all cases; that is to say, the first result is the value of $W\phi$, and the *denial* of the second result is the value of $S\phi$. To show how any case may be verified let us take the term $u^\theta v^\eta a^\eta \beta^\epsilon$ in the second result. If this term be correct ϕ should reduce to η when we put $u = \theta$, $v = \eta$, $a = \eta$, and $\beta = \epsilon$. Substituting these values in the third form of ϕ, we get

$$\theta_\eta \epsilon_\eta + \theta_\epsilon \eta_\eta : \theta_\eta \text{ which } = \eta\eta + \epsilon\epsilon : \eta = \epsilon : \eta = \eta,$$

as we should have.

Many persons who are ready enough to admit the utility of mathematics profess scepticism as to the advantages of symbolic logic. To these I would remark, firstly, that the latter is the more general science and includes the former; and, secondly, that, apart from its aid to accurate thinking in general, symbolic logic has already (as shown in my first paper in the *Proceedings of the Mathematical Society*) rendered important assistance in one of the most difficult and perplexing parts of mathematics. It was a true instinct that led Boole to attempt the construction of what he called a "General Method in Probabilities," founded on symbolic logic; but unfortunately Boole allowed an essentially false principle to vitiate his whole reasoning, so that his elaborate "General Method in Probabilities," into whose service he presses the highest branches of mathematics, becomes, alas! from this one flaw, a gigantic and imposing fallacy. His solution of the "challenge problem" which he proposed as a test of the power and efficacy of his method I have proved to be wrong in my fourth paper in the *Proceedings of the Mathematical Society*; and, in my controversy with Dr. MacAlister (see vol. xxxvii. of *Mathematical Questions with their Solutions from the* Educational Times), I think I have succeeded in laying bare the subjective fallacy over which Boole and many others after him have stumbled into error. "*When the probabilities of events are given,*" says Boole, "*but all information respecting their dependence withheld, the mind regards them as independent*" (*Laws of Thought*, p. 256). And further on he says: "*We must regard the events as independent of any connexion beside that of which we have information*". In other words, when we have no information as to any connexion between A and B, but know (from observation or otherwise) that the chance of A happening is a and that of B happening is b, we may infer that the chance of *both* happening is ab—an utterly fallacious principle. Cases in which, from these data, the chance of the concurrence is ab, and cases in which it is *not ab*, may be exhibited to the eye by a simple geometrical construction, fixed and unvarying, as I have shown in my fourth paper in the *Proceedings of the Mathematical Society*, so that the chance remains always the same whether or not "the mind regards the events as independent".

No one can admire Boole's *Laws of Thought* more than I do. As a philosophical and speculative work it is brimful of profound thought and original suggestions, while its style is charmingly lucid and attractive; but none the less must

SYMBOLIC REASONING.

I express my opinion that the work is weakest where it is generally supposed to be strongest, namely, in the power and originality of its logical calculus, and that what Boole himself, and others after him, considered his greatest achievement and "the crowning triumph of his method," namely, the application of his calculus to probability, is precisely that portion of his work in which his failure has been complete and absolute. Boole may not inaptly be compared to Shakespeare. Both authors possessed a remarkable *analytical* insight into the workings of the human mind; the one of its secret motives and passions; the other of the subtle laws of its intellectual operations; yet both—the one judged by his plays, the other by his *Laws of Thought*—showed but little *constructive* ability. Just as Shakespeare limited himself to skilful adaptations to his purpose of the dramatic plots of preceding playwrights or of the accepted facts of history, so Boole limited himself to skilful adaptations to new uses of the rules and formulæ which he found ready to his hand in mathematics. Boole undoubtedly showed great originality and ability in his *application* of these rules and formulæ, and an unfortunate thing I believe it has been for symbolic logic that he did show this originality and ability. I cannot help thinking that the seeming success which attended his efforts to squeeze all reasoning into the old cast-iron formulæ constructed specially for numbers and quantities has tempted many other able logicians to waste their energies in the like futile endeavours; when those energies might have been employed with far greater chances of success in inventing new and independent formulæ, more elastic and more suitable for the highly general and widely varying kinds of problems which are destined to enter more and more largely into the ever-expanding subject of Symbolic Logic. As regards the introduction of absolutely new symbols, such as Schröder's $\in$ and Pierce's $-<$, I think it should be avoided as much as possible. Generally speaking, it is better to put to fresh uses the familiar symbols of old acquaintance than have recourse to strangers. This may be a conservative prejudice on my part, but it is a fact that I have myself introduced no new symbol, though I have freely exercised my right of definition and interpretation as regards some of the symbols and combinations of symbols already in common use among mathematicians. The symbols $+$, $=$, $:$, $::$, with subscripts, indices, and commas, are all old friends; while the symbol !, though a more recent arrival on mathematical territory, has been there long enough to have acquired the rights of naturalisation.

Diagrammatic Illustrations. The accompanying figures will help us to understand, and fix in our mind, the meanings of the symbols $a : \beta$, $\dfrac{a}{\beta}$, $a' + \beta$, $a\beta'$, and others which we have been discussing in the preceding pages.

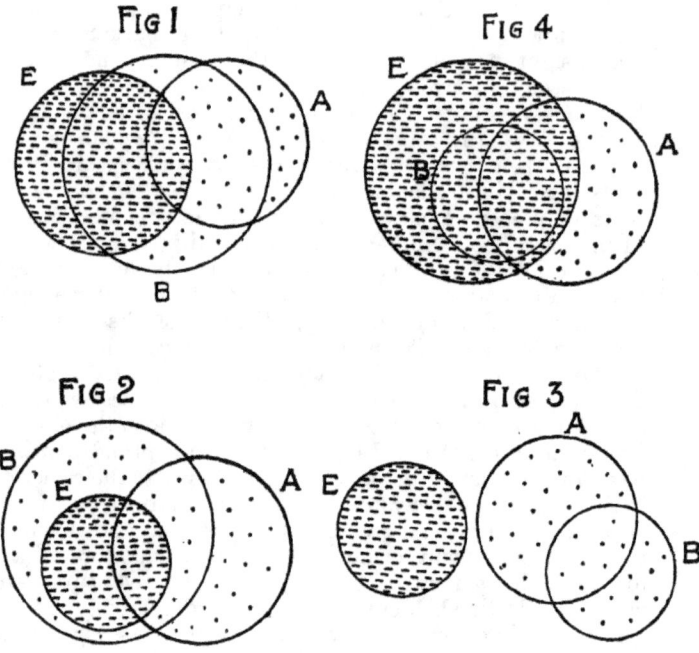

Consider first Fig. 1. Let the collection of points in the circle E be our *universe of discourse*, all points not included in E being left out of account. Let a point P be taken at random in the collection E, and let the symbols ϵ, a, β, respectively assert that P will belong to the collection E, that it will belong to the collection A, that it will belong to the collection B. From the disposition of the circles in the case we are considering (Fig. 1), this random point P may or may not turn out to be in A, and it may or may not be in B, so that the statements a and β may be true or false; and if the experiment be repeated often enough, each of the two statements will be sometimes true and sometimes

SYMBOLIC REASONING.

false. Hence, in this case a and β are *variables*, and so are their denials a' and β'. For Fig. 1 therefore we have $a^\theta \beta^\theta$.

Next consider Fig. 2; and as before let P be taken at random in the collection E; while, as before, ϵ, a, β, respectively assert that P will belong to the collection E, that it will belong to the collection A, that it will belong to the collection B. As before, a may turn out true or false, but this time β *cannot be false*, since the point P is restricted, by hypothesis, to the collection E, which is wholly included in the collection B. In this case therefore a is a *variable*, but β is a *constant* of the class ϵ or *certainties*; so that for this figure we have $a^\theta \beta^\epsilon$.

Next consider Fig. 3. Here neither a nor β can possibly be true, since the collections A and B are both excluded from our universe of discourse; so that we have $a^\eta \beta^\eta$.

Now let us consider the implication $a : \beta$ and its implied alternative $a' + \beta$. The alternative $a' + \beta$ is a *certainty* in Fig. 1, for here its denial $a\beta'$ is an *impossibility*, as no point *taken at random in the collection* E can, at the same time, belong to the class A and *not* belong to the class B. In Fig. 1 therefore we have $(a\beta')^\eta$, which is equivalent to $(a' + \beta)^\epsilon$ and also to $a : \beta$.

In Fig. 4 the alternative $a' + \beta$, and therefore also its denial $a\beta'$, are *variables*; for in Fig. 4 either statement (*i.e.* the alternative or its denial) may be true or false; and if the experiment of the random point P be repeated often enough, each will be sometimes true and sometimes false. The statement $a' + \beta$ is true, and its denial $a\beta'$ false, every time P happens *not* to be in A, and also every time it happens to be in B; and *vice versâ*.

In Fig. 1 not only have we $a : \beta$ true (with its synonym a_β) but also the *causal* implication $\frac{a}{\beta}$, which implies both a_β and β^{ϵ_1} by definition. But this is not the case in Fig. 2; for in Fig. 2 we have β^ϵ, as already shown, which contradicts β^{ϵ_1}; so that in Fig. 2 the *causal* implication $\frac{a}{\beta}$ is false.

In Fig. 4 the *general* implication $a : \beta$, and therefore also the *causal* implication $\frac{a}{\beta}$, are false. For $a : \beta$ means $(a\beta')^\eta$ and asserts that $a\beta'$ is an *impossibility*; whereas in Fig. 4 we have the denial of this, namely $(a\beta')^n$, which asserts that $a\beta'$ is a *possibility*.

In Fig. 3 the *causal* implication $\frac{a}{\beta}$, and therefore also the general implication $a : \beta$, are true; for in Fig. 3 we have $(a\beta')^\eta \beta^u$; the compound statement $a\beta'$ being an *impossibility* because of its impossible factor a, and β being an *uncertainty* because it is here an *impossibility*—the stronger statement implying the weaker.

Throughout the preceding discussion, in considering the varying positions of the point P, we assumed the collections of points E, A, B to be fixed and always the same; so that while the statements $a, \beta, a' + \beta$, and $a\beta'$ might be *variables*, the implication $a : \beta$ was always a *constant* belonging, according to the positions of the *fixed circles*, either to the class ϵ or to the class η. But if we abandon this assumption of *fixedness* and suppose the circles E, A, B to be formed *randomly* under some limiting conditions—such, for example, as taking two random points in some given area and considering the straight line joining them as the diameter of a random circle—the case will be altered. If this experiment be repeated often enough, the implication $a : \beta$ will now be sometimes true and sometimes false; that is to say, it will be a *variable*, and in some cases it will be a variable whose chance of being true admits of accurate calculation.

Note on a Logic of 3^n Dimensions. The preceding three-divisional scheme of logic is more especially suited for problems in probability, the statements $a^\epsilon, a^\eta, a^\theta$ respectively asserting that the chance of a being true is *unity*, that it is *zero*, that it is *less than unity and greater than zero*. On this understanding, the symbol $a : \beta$, being synonymous with $(a\beta')^\eta$, asserts that the chance of the truth of the compound statement that affirms a while denying β is zero. But this is not always the meaning of the word 'implies' in mathematics. When we say that a formula a (such as Taylor's or Maclaurin's Theorem in the Differential Calculus) *implies* another formula β (such as the expansion of sin x or cos x in Trigonometry) there is no question of *chance* or *probability*: both formulæ are always true, and what we really mean is that β is a particular case of a. Similarly, when we say that the equational proposition or formula $a^2 - b^2 = (a - b)(a + b)$ *implies* the equational proposition

$$763^2 - 761^2 = (763 - 761)(763 + 761)$$

we mean that the numerical statement is a particular case of the algebraic one. So in logic we have

$$\phi^\epsilon(a, \beta) : \phi^\epsilon(a_1, \beta_1) \phi^\epsilon(a_2, \beta_2);$$

that is to say, if the formula $\phi(a, \beta)$ be valid for *all* values

SYMBOLIC REASONING.   509

of a and β, it must be true for the particular statements a_1, β_1 and also for the particular statements a_2, β_2, whatever these may be. In other words, $\phi(a_1, \beta_1)$ and $\phi(a_2, \beta_2)$ are particular cases of $\phi(a, \beta)$.

Again, let us suppose that we wish to establish some proposition x in mathematics or physics about whose truth we do not feel quite certain, and that we find its validity to depend upon the truth of another proposition a which seems easier to investigate, but whose truth is also uncertain. Here, as before, we may write $a : x$, and here also we have the element of uncertainty, as in problems of chance; but the uncertainty in this case is purely *subjective*, and can hardly be expressed by a numerical ratio;[1] for the chance of a being true is in this case always the same, *unity* or *zero*, whether we choose to consider it so or not, and our *uncertainty* is as to which of the two values we ought to assign to the unknown chance—if indeed the term *chance* is not here altogether a misnomer. To meet all these cases and bring them within the sweep of one logical scheme, we must have a logic not of three dimensions only, but of 3^n dimensions. Thus, let κ denote every statement *known to be true*, λ every statement *known to be false*, and μ every *doubtful* statement, *neither known to be true nor known to be false*; then, any formula $\phi(\epsilon, \eta, \theta)$ of the scheme described in the preceding pages may be converted at once,[2] by simple substitution of letters, into a formula $\phi(\kappa, \lambda, \mu)$ of this new and more subjective scheme. And these two corresponding three-dimensional schemes may be united into a *nine-dimensional* scheme. For example, $a^{\epsilon\kappa}$ would express $(a^\epsilon)^\kappa$ and assert that a^ϵ *is known to be true*, or, in other words, that a *is known to be always true;* while $a^{\kappa\epsilon}$ would mean $(a^\kappa)^\epsilon$ and assert that a^κ *is always true*, or, in other words, that a *is always known to be true;* which is quite a different statement from $a^{\epsilon\kappa}$. The statement $a^{\epsilon\kappa}$ (or *a is known to be always true*) might apply to a difficult mathematical proposition whose truth I had just discovered, but which I might afterwards forget; and to such a proposition $a^{\kappa\epsilon}$ (or *a is always known to be true*) would *not* apply. To any well-known truism, or any simple formula, like $(a + b) x = ax + bx$, which I could hardly, with a healthy brain, ever forget, both statements $a^{\kappa\epsilon}$ and $a^{\epsilon\kappa}$ would apply.

[1] Dr. Venn, in his *Logic of Chance*, holds substantially the same view; while Boole and De Morgan maintain the *subjective fallacy*. As a clear exposition of first principles Dr. Venn's work is unsurpassable.

[2] For example, the formula $(a + \beta)^\epsilon : a^\epsilon + \beta^\epsilon + a^\theta \beta^\theta$ in the one system becomes $(a + \beta)^\kappa : a^\kappa + \beta^\kappa + a^\mu \beta^\mu$ in the other.

Another three-divisional scheme would be the following: We might agree to make $\phi^a(x, y)$ assert that the formula $\phi(x, y)$ is true for *all* values of x and y; $\phi^n(x, y)$ that it is true for *no* values of x and y; and $\phi^s(x, y)$ that it is true for *some* value or values of x or y, or both, but not for all. These three schemes might, in like manner, be united into a twenty-seven-divisional scheme of symbolic logic; and so on *ad libitum*. It is evident that a logic of 3^n dimensions, constructed on these lines, though complicated for high values of n, would be in no way transcendental, like the geometry of a four-dimensional or n-dimensional space, but would be founded upon, and give results in accordance with, the daily and ordinary facts of our consciousness and experience.

[The problem discussed on page 502 has been discussed by its proposer in MIND for January, 1895.—EDITOR.]

[1901f] La Logique Symbolique et ses Applications. Bibliothèque du 1° Congrès International de Philosophie. Logique et Histoire des Sciences, pp. 135-183.

BIBLIOTHÈQUE DU CONGRÈS INTERNATIONAL DE PHILOSOPHIE

III

Logique

et

Histoire des Sciences

PARIS

Librairie Armand Colin

5, rue de Mézières, 5

1901

Tous droits réservés.

LA LOGIQUE SYMBOLIQUE ET SES APPLICATIONS

Par Hugh Mac Coll,
de l'Université de Londres.

I. — *Logique pure*.

La logique *pure*, ou logique *abstraite*, s'occupe des règles générales du raisonnement, c'est-à-dire des règles qui restent valides quel que soit le sujet de recherche. Or, quel que soit le sujet de recherche, tout raisonnement, pour pouvoir s'exprimer, demande des *propositions*. Donc, pour rendre notre raisonnement parfaitement général, et nos formules universellement applicables, nous devons prendre la classification des différentes espèces de propositions et les rapports entre elles comme le premier but de notre recherche, et appeler ce travail la Logique *pure*. Par « proposition », je veux dire tout son ou symbole (ou combinaison de sons ou de symboles) qu'on emploie pour donner quelque information, ou pour se rappeler quelque information déjà reçue. Le « Cröa » poussé par un corbeau en sentinelle, pour avertir ses camarades qu'il vient de voir un homme avec un fusil, est, dans ce sens, une proposition.

Le pavillon qui indique la nationalité d'un vaisseau qui passe, est aussi une proposition. De telles propositions simples et indivisibles, sans sujet et sans prédicat, furent probablement les sources primitives des langues humaines (voir § 26). Mais, quoi qu'il en soit, il est certain que les propositions, qu'elles soient simples et indivisibles, ou complexes et divisibles, sont les *unités* sur lesquelles nous basons, et avec lesquelles nous exprimons, tous nos raisonnements. La conclusion naturelle est donc que, dans la Logique la plus générale, la Logique *abstraite*, ces unités que nous appelons *propositions* devraient être représentées par les signes les plus élémentaires; et ces signes, ce sont les lettres de l'alphabet. Pour la classification des propositions, et pour exprimer leurs liens de rapport, nous pouvons emprunter aux Mathématiques, non seulement des signes spéciaux, comme $+$, $:$, etc., mais aussi des combinaisons, telles que A^B, A_B, $\frac{A}{B}$, etc., sans trop nous occuper des significations primitives de ces signes ou de ces combinaisons. Il est même permis, au courant du même argument, de *varier par une nouvelle définition* le sens que nous donnons à tel signe ou à telle combinaison. Certes, c'est une liberté dont il ne faut pas abuser; nous ne devrons changer le sens de nos symboles que dans les cas (qui sont toutefois assez nombreux) où les avantages du changement en dépassent les inconvénients; mais il n'y a à cela aucune objection de principe. Exactement comme on peut dire en Mathématiques : « soit un tel nombre représenté

LA LOGIQUE SYMBOLIQUE ET SES APPLICATIONS 137

par x », on peut dire en Logique : « soit une telle proposition complexe représentée par telle ou telle combinaison ». La seule chose à considérer, c'est l'utilité ou l'inutilité de la nouvelle convention (voir § 26).

Cette question de notation est très importante. En général, nos yeux s'habituent difficilement aux symboles absolument nouveaux ou inusités; c'est pourquoi, dans mon système de notations, je m'en passe entièrement, préférant employer en des sens variables les symboles, simples ou complexes, que nous trouvons déjà emmagasinés et à notre disposition dans le riche arsenal des Mathématiques. Ai-je tort ou raison? C'est à ceux qui connaissent mieux que moi les travaux de mes confrères à en décider[1]; et pour leur fournir les moyens d'arriver à une conclusion, je vais maintenant expliquer mon système et donner quelques exemples de son application.

1. Le symbole A^B affirme[2] que la proposition A appartient à la classe B. Le symbole A^{BC} est un abrégé de $(A^B)^C$; il affirme que la proposition A^B appartient à la classe C. Par exemple, si P veut dire *probable*, et C veut dire *certain*, alors A^P veut dire A *est probable*, et A^{PC} veut dire *il est certain que* A *est probable*; tandis que A^{CP} veut dire *il est probable que* A *est certain* (voir § 23). Le symbole A^{BCD} veut dire $(A^{BC})^D$, et ainsi de suite.

---

1. Parmi ces logiciens, je crois devoir nommer spécialement M. Louis Couturat, dont les critiques justes et bien raisonnées m'ont donné récemment l'occasion d'éclaircir certains points que mes premières recherches avaient laissés un peu obscurs.
2. Dans la Logique concrète ou appliquée, A peut représenter une chose (longueur, surface, etc.) qui n'est pas une proposition (voir § 27).

H. MAC COLL

2. Quand un symbole A représente une *classe*, les symboles A_1, A_2, A_3, etc., représentent des *individus* appartenant à cette classe (voir § 26).

3. Les lettres grecques τ, ι, ε, η, θ représentent les termes *vrai, faux, certain, impossible, variable*. Toutes les propositions intelligibles[1] peuvent être divisées en deux classes, les *vraies* (τ) et les *fausses* (ι). Toutes les propositions intelligibles peuvent être aussi divisées en trois classes : les *certaines* ($\varepsilon_1, \varepsilon_2, \varepsilon_3$, etc.), les *impossibles* (η_1, η_2, η_3, etc.) et les *variables* ($\theta_1, \theta_2, \theta_3$, etc.). Une proposition est *certaine* quand sa probabilité est *un*, c'est-à-dire quand elle résulte nécessairement de nos données ou de nos définitions. Une proposition est *impossible* quand sa probabilité est *zéro*, c'est-à-dire quand elle contredit quelque donnée ou définition. Une proposition est *variable*, quand sa probabilité n'est ni *un* ni *zéro*, mais quelque fraction entre les deux. Par exemple, en parlant d'un point pris au hasard parmi les cinq points marqués dans le cercle E (Fig. 1), supposons que les lettres A, B, C (comme propositions) affirment respectivement que le point qui se présentera sera dans le cercle A, qu'il sera dans le cercle B, qu'il sera dans le cercle C. Il est évident que nous aurons $A^\varepsilon B^\theta C^\eta$; c'est-à-dire : A est *certain*, B *variable*, C *impossible*.

Une proposition peut être *vraie* sans que sa probabilité soit *un*, et *fausse* sans que sa probabilité soit *zéro*. Par exemple, dans le cas actuel, la proposition B (dont la pro-

---

[1]. Toute proposition *non-intelligible* (qui *ne dit rien* dans notre univers du discours) peut être représentée par le symbole 0. (Voir p. 147, note 1.)

LA LOGIQUE SYMBOLIQUE ET SES APPLICATIONS

babilité est $\frac{2}{5}$) peut être *vraie*, mais elle n'est jamais *certaine*; et elle peut être *fausse*, mais elle n'est jamais *impossible*.

4. Le symbole $A^\alpha B^\beta$, comme proposition, affirme deux choses, que A appartient à la classe α, *et* que B appartient à

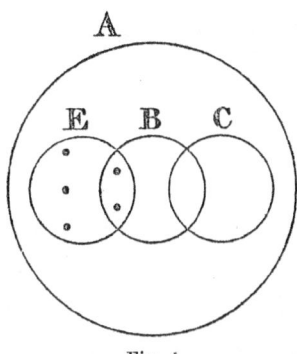

Fig. 1.

la classe β; tandis que le symbole $A^\alpha + B^\beta$ n'affirme qu'une chose, que A appartient à la classe α, *ou* B à la classe β; de sorte que, dans ce cas, on peut considérer le symbole $+$ comme le *synonyme* de la conjonction *ou*. Le symbole $A^\alpha + B^\beta$ affirme qu'une *au moins* des deux propositions A^α et B^β est vraie. On peut appeler A^α et B^β les *facteurs* du *produit* $A^\alpha B^\beta$, et les *termes* de la *somme* $A^\alpha + B^\beta$. Des propositions comme $A^\alpha B^\beta C^\gamma$ ou $A^\alpha + B^\beta + C^\gamma$ doivent être interprétées de la même manière. Il est évident que le signe $+$ n'exprime nullement une *addition*, de sorte que *somme* en Logique n'a pas le même sens que *somme* en Mathématiques. Le symbole $A^{\alpha, \beta}$ est synonyme de $A^\alpha + A^\beta$; et $(A, B)^\alpha$ est synonyme de $A^\alpha + B^\alpha$.

H. MAC COLL

Soit A un nom *générique* (proposition ou chose concrète) appartenant à chaque individu (A_1, A_2, A_3, etc.) de la classe. Alors le symbole A^B affirme[1] qu'un individu (sans dire lequel) de la classe A appartient aussi à la classe B.

Le symbole A^{B_1} affirme non seulement que A est un B, mais aussi que A est le B spécial de la série B_1, B_2, B_3, etc., que nous avons nommé B_1. Donc nous avons $A^B = A^{B_1} + A^{B_2} + A^{B_3} +$ etc. Les formules suivantes[2] sont évidentes :

$$A^{\iota\eta} = A^{\varepsilon}, \quad A^{\eta\iota} = A^{\varepsilon} + A^{\theta},$$
$$A^{\iota\varepsilon} = A^{\eta}, \quad A^{\varepsilon\iota} = A^{\eta} + A^{\theta},$$
$$A^{\iota\theta} = A^{\theta}, \quad A^{\theta\iota} = A^{\varepsilon} + A^{\eta},$$
$$\varepsilon^{\eta} = \eta, \quad \varepsilon^{\theta} = \eta, \quad \theta^{\varepsilon} = \eta, \quad \theta^{\eta} = \eta,$$
$$\eta^{\varepsilon} = \eta, \quad \eta^{\eta} = \varepsilon, \quad \varepsilon^{\iota} = \eta, \quad \eta^{\iota} = \varepsilon, \quad \varepsilon^{\tau} = \varepsilon,$$
$$(A^{\tau}B^{\tau})^{\iota} = A^{\iota} + B^{\iota}, \quad (A^{\tau} + B^{\tau})^{\iota} = A^{\iota}B^{\iota}.$$

Toutes ces équivalences, comme toutes les formules valides, sont des *certitudes formelles* (voir § 21).

5. L'exposant τ (qui affirme qu'une proposition est *vraie*) est souvent sous-entendu[3]; et l'exposant ι (qui affirme qu'une proposition est *fausse*) est souvent remplacé par un accent de négation. Les symboles $A, A^{\tau}, A', A^{\iota}$, par exemple, peuvent exprimer respectivement *Il ira en Amérique, Il*

---

1. Dans un de mes mémoires dans les *Proceedings of the London Mathematical Society*, j'ai posé comme définition que le symbole A^B (quand A indique une *classe*) affirme que *tout* individu de la classe A appartient à la classe B; mais, après expérience, je trouve cette convention incommode.
2. Il est sous-entendu dans ces formules que les propositions A et B n'appartiennent pas à la classe 0; mais elles peuvent appartenir à la classe η sans détruire la validité des formules. (Voir § 3, note 1. Voir aussi p. 147, note 1.)
3. Comme *termes* et comme *facteurs*, τ et ι sont équivalents respectivement à ε et η; mais pas toujours comme *exposants*. Car comme *terme* ou *facteur* (puisque A veut dire A^{τ}), $\tau = \tau^{\tau} = \varepsilon$, et $\iota = \iota^{\tau} = \eta$.

LA LOGIQUE SYMBOLIQUE ET SES APPLICATIONS

est vrai qu'il ira en Amérique, Il n'ira pas en Amérique, Il est faux qu'il ira en Amérique. Avec cette convention, les trois formules :

$$(AB)' = A' + B', \quad (A + B)' = A'B', \quad A(B + C) = AB + AC$$

sont synonymes respectivement de :

$$(A^\tau B^\tau)^\iota = A^\iota + B^\iota, (A^\tau + B^\tau)^\iota = A^\iota B^\iota, A^\tau(B^\tau + C^\tau) = A^\tau B^\tau + A^\tau C^\tau.$$

(Mais voir p. 147, note 1.)

Des cas cependant pourraient se présenter où la suppression de l'exposant τ changerait le sens. Par exemple, puisque A^A (quelle que soit la classe A) est une certitude (voir § 4), il résulte que τ^τ est une certitude. Donc $\tau^{\tau\varepsilon}$, qui veut dire $(\tau^\tau)^\varepsilon$, est aussi une certitude. Mais si nous supprimons l'exposant τ dans la certitude $\tau^{\tau\varepsilon}$, elle devient τ^ε, qui n'est pas nécessairement une certitude ; car une proposition τ, vraie dans le cas actuel C_1, n'est pas nécessairement vraie dans tous les cas possibles C_1, C_2, C_3, etc. Supposons[1] (voir la fig. 2) que les propositions *vraies* $\tau_1, \tau_2,$ etc. (vraies dans le cas actuel C_1) soient représentées par des points dans les deux carrés supérieurs ; les propositions *fausses* $\iota_1, \iota_2,$ etc. (fausses dans le cas actuel C_1) par des points dans les deux carrés inférieurs ; les *variables* $\theta_1, \theta_2,$ etc. (vraies dans

$\tau\varepsilon$	$\tau 0$
$\iota\eta$	$\iota 0$

Fig. 2.

---

[1]. Ces quatre carrés représentent les quatre modalités de la Logique traditionnelle. Les propositions dans le carré $\tau\varepsilon$ sont *vraies maintenant et toujours*; celles dans le carré $\iota\eta$ sont *fausses maintenant et toujours*; celles dans le carré $\tau 0$ sont *vraies maintenant mais pas toujours*; celles dans le carré $\iota 0$ sont *fausses maintenant mais pas toujours*. (Voir § 17.) Le symbole τ' (sans l'exposant τ sous-entendu) n'est pas synonyme de $\tau^\iota (= \eta)$, mais de ι; et ι' n'est pas synonyme de $\iota^\iota (= \varepsilon)$, mais de τ. L'accent de négation n'est pas un exposant.

certains cas, fausses dans d'autres) par des points dans les deux carrés de droite; et les *certitudes* ε_1, ε_2, etc., et les *impossibilités* η_1, η_2, etc., respectivement par des points dans les deux carrés de gauche. Il est évident qu'un point τ pris au hasard dans les deux carrés $\tau\varepsilon$ et $\tau\theta$ ne sera pas nécessairement un des points dans le carré $\tau\varepsilon$; c'est-à-dire τ^ε n'est pas une certitude. Sans exposants (exprimés ou sous-entendus) les symboles τ, τ', ι, ι' (comme A et A') sont des *sujets sans attributs*; mais avec l'exposant ι sous-entendu, nous avons :

$$\tau = \tau^\tau = \varepsilon,\ \tau' = (\tau')^\tau = \iota^\tau = \eta,\ \iota = \iota^\tau = \eta,\ \iota' = (\iota')^\tau = \tau^\tau = \varepsilon.$$

6. Les symboles $A:B$ et $B!A$ sont synonymes[1] de $(AB')^\eta$, et par conséquent de $(A^\iota + B)^\varepsilon$. Le symbole $A:B$ affirme que *si A est vrai B est vrai*; $B!A$ affirme que *B est vrai si A est vrai*; $(AB')^\eta$ affirme qu'*il est impossible que A soit vrai et (en même temps) B faux*; et $(A' + B)^\varepsilon$ affirme qu'*il est certain que A est faux ou B vrai*. Si l'antécédent A représente les données d'un argument, le conséquent B représente la conclusion. Puisqu'il ne faut pas confondre A^ε avec A^τ, ni A^η avec A^ι, il ne faut pas non plus confondre $A:B$, c'est-à-dire $(A'+B)^\varepsilon$, avec $A'+B$. La proposition $A:B$, comme son synonyme $(A'+B)^\varepsilon$, affirme que $A'+B$ est vrai (et, par conséquent, AB' faux) *dans tous les cas admissibles*; tandis que la proposition $A'+B$, comme son synonyme $(AB')^\iota$, affirme seulement que $A'+B$ est vrai (et

---

[1]. A noter que $A:B$ est l'abrégé de $A^\tau:B^\tau$, qui veut dire $(A^\tau B^\iota)^\eta$, dont l'abrégé est $(AB')^\eta$.

LA LOGIQUE SYMBOLIQUE ET SES APPLICATIONS

par conséquent AB' faux) *dans un certain cas*. Nous appelons la proposition A : B une *implication*, parce qu'elle affirme que A *implique* B.

7. Le symbole A : B : C veut dire (A : B)(B : C), et A!B!C veut dire (A!B)(B!C). Une série d'*implications déductives* A : B : C : D, etc., ou d'*implications inductives* A!B!C!D, etc., doit être interprétée de la même manière.

8. Le symbole (A = B), ou son synonyme (A :: B), veut dire (A : B) (B : A). On appelle B : A (comme son synonyme A!B) l'*inverse* de A : B. Le symbole A ≡ B affirme que A est *synonyme* de B.

9. Pour éviter autant que possible l'emploi des parenthèses, il est commode de poser certaines conventions pour fixer la *portée* des signes de relation, +, :, !, ::, =. Convenons donc que les trois signes, :, !, :: ont tous la même portée, et que chacun a plus de portée que le signe +, et moins de portée que le signe =.

Ainsi, l'implication A + B : C veut dire (A + B) : C, et non pas A + (B : C); et la double implication A = B :: C veut dire A = (B :: C), et non pas (A = B) :: C, qu'on peut exprimer sans parenthèse par A :: B = C. Les formules suivantes sont évidentes ou faciles à prouver :

(1) $\quad \varepsilon : A = A^\varepsilon,$
(2) $\quad A : \eta = A^\eta,$
(3) $\quad A : \varepsilon = \varepsilon,$
(4) $\quad \eta : A = \varepsilon,$
(5) $\quad \varepsilon + A = \varepsilon,$
(6) $\quad \eta + A = A,$

$$
\begin{align}
&(7) & A : B &= B' : A', \\
&(8) & A : BC &= (A : B)(A : C), \\
&(9) & A + B : C &= (A : C)(B : C), \\
&(10) & (A = \varepsilon) &= A^\varepsilon, \\
&(11) & (A = \eta) &= A^\eta, \\
&(12) & \varepsilon : \eta &= \eta, \\
&(13) & \varepsilon : \theta &= \eta, \\
&(14) & \tau : A &= A^\varepsilon, \\
&(15) & A : \iota &= A^\eta, \\
&(16) & A : \tau &= \varepsilon, \\
&(17) & \iota : A &= \varepsilon, \\
&(18) & \tau + A &= \varepsilon, \\
&(19) & \iota + A &= A, \\
&(20) & (A = \tau) &= A^\varepsilon, \quad \text{et non } A^\tau, \\
&(21) & (A = \iota) &= A^\eta, \quad \text{et non } A^\iota, \\
&(22) & \tau : \iota &= \eta, \\
&(23) & \tau : \theta &= \eta, \\
&(24) & \varepsilon A &= A, \\
&(25) & \eta A &= \eta.
\end{align}
$$

Nous prouvons (4), (5), (9), (13), (14), (17), (20), (21) comme suit :

(4) $\quad \eta : A = (\eta A')^\eta = \eta^\eta = \varepsilon.$

(5) $\quad \varepsilon + A = (\varepsilon' A')' = (\eta A')^\iota = \eta^\iota = \varepsilon.$

(9) $\quad A + B : C = C' : (A + B)' = C' : A'B' = (C' : A')(C' : B')$
$= (A : C)(B : C),$ en vertu des formules 7 et 8.

(13) $\quad \varepsilon : \theta = (\varepsilon \theta')^\eta = (\theta')^\eta = \theta^\varepsilon = \eta.$

(14) $\quad \tau : A = \tau^\tau : A^\tau = \varepsilon : A = A^\varepsilon$ (voir p. 140, note 3).

(17) $\quad \iota : A = \iota^\tau : A^\tau = \eta : A = \varepsilon$ (voir p. 140, note 3).

(20) $\quad A :: \tau = A^\tau :: \tau^\tau = A :: \varepsilon = (A : \varepsilon)(\varepsilon : A) = \varepsilon(\varepsilon : A) = \varepsilon : A = A^\varepsilon$

(21) $\quad A :: \iota = A^\tau :: \iota^\tau = A :: \eta = (A : \eta)(\eta : A) = A^\eta \varepsilon = A^\eta.$

LA LOGIQUE SYMBOLIQUE ET SES APPLICATIONS

La proposition $(A = B)$ affirme que A est *équivalent* à B, mais non nécessairement que A et B sont *synonymes* (voir § 8); car nous avons les équivalences $(\varepsilon_1 = \varepsilon_2)$ et $(\eta_1 = \eta_2)$, quelles que soient les certitudes ε_1 et ε_2, ou les impossibilités η_1 et η_2. Nous prouvons la seconde comme suit :

$$(\eta_1 = \eta_2) = (\eta_1 : \eta_2)(\eta_2 : \eta_1) = (\eta_1 \eta_2')^\eta (\eta_2 \eta_1')^\eta = (\eta_1 \varepsilon_1)^\eta (\eta_2 \varepsilon_2)^\eta$$
$$= \eta_1^\eta \eta_2^\eta = \varepsilon$$

car la négation d'une impossibilité est une certitude. Évidemment $(A \equiv B) : (A = B)$.

10. Le symbole A_B est synonyme de $A:B$, et A_{0B} est synonyme de $(A:B)^\iota$ (mais voir § 26). Ces abréviations sont souvent commodes. Ainsi :

(1) $\qquad (A : B) : (x_A : x_B)$,
(2) $\qquad (A : B) : (B_x : A_x)$,
(3) $\qquad (A + B)_x : A_x + B_x : (AB)_x$,
(4) $\qquad x_{AB} : x_A + x_B : x_{A+B}$,
(5) $\qquad (A + B)_{0x} = A_{0x} + B_{0x}$,
(6) $\qquad (A + B)_x :: A_x + B_x = A_x :: B_x$.

11. Pour montrer l'utilité de cette notation, je prouve la dernière formule comme suit :

$$(A + B)_x :: A_x + B_x =$$
$$A_x B_x :: A_x + B_x = (A_x B_x : A_x + B_x)(A_x + B_x : A_x B_x) = \varepsilon(A_x + B_x : A_x B_x)$$
$$= A_x + B_x : A_x B_x = (A_x : A_x B_x)(B_x : A_x B_x) \qquad \text{(voir § 9, formule 9)}$$
$$= (A_x : A_x)(A_x : B_x)(B_x : A_x)(B_x : B_x) \qquad \text{(voir § 9, formule 8)}$$
$$= \varepsilon (A_x : B_x)(B_x : A_x) \varepsilon = (A_x : B_x)(B_x : A_x) = A_x :: B_x$$

12. Soient $A_1, A_2, A_3, \ldots A_n$ des propositions qui sont

H. MAC COLL

toutes possibles, mais dont *une seulement* est vraie. Sur ces n propositions, supposons que $A_1, A_2, ..., A_r$ impliquent (chacune séparément) une conclusion φ; que $A_{r+1}, A_{r+2}, ..., A_s$ impliquent φ'; et que les autres $A_{s+1}, A_{s+2}, ..., A_n$ n'impliquent ni φ ni φ'. Alors, nous posons comme définitions (voir § 8) :

$$W\varphi \equiv A_1 + A_2 + A_3 + ... + A_r,$$
$$W\varphi' \equiv A_{r+1} + A_{r+2} + ... + A_s,$$
$$V\varphi \equiv V\varphi' \equiv A_{s+1} + A_{s+2} + ... + A_n,$$
$$S\varphi \equiv W\varphi + V\varphi,$$
$$S\varphi' \equiv W\varphi' + V\varphi.$$

De plus, $W'\varphi$ veut dire $(W\varphi)'$; $S'\varphi$ veut dire $(S\varphi)'$.

Puisque A est plus fort[1] que $A + B$, que $A + B$ est plus fort que $A + B + C$, etc., nous pouvons appeler $W\varphi$ la *plus faible donnée dont on puisse déduire la proposition* φ, tandis que $S\varphi$ indique *la plus forte conclusion que l'on puisse déduire de* φ; de sorte que nous aurons $W\varphi:\varphi:S\varphi$. De ces définitions on déduit facilement les formules :

$$W\varphi' = S'\varphi, \quad S\varphi' = W'\varphi,$$
$$V^0\varphi = (W\varphi = S\varphi = \varphi) = (V\varphi = 0).$$

Le symbole $V^0\varphi$, ainsi que son synonyme $V\varphi = 0$, affirme, *non* que $V\varphi$ est impossible, mais que $V\varphi$ *n'existe pas dans notre univers actuel*; c'est-à-dire, que toute proposition de la série $A_1, A_2, ..., A_n$, ou implique φ ou implique φ', de sorte que (dans cette série) il n'y a pas de proposition $V\varphi$ dont on puisse dire qu'elle n'implique ni φ ni φ'.

---

[1]. Car A implique $A + B$; $A + B$ implique $A + B + C$, etc. Symboliquement : $A : A + B : A + B + C$, etc.

LA LOGIQUE SYMBOLIQUE ET SES APPLICATIONS

Le symbole A^0 affirme que A est une proposition *muette*[1], qui *n'affirme rien* et qui *ne nie rien*, et qui, par conséquent, n'appartient à aucune des classes τ, ι, ε, η, θ.

13. Supposons que notre totalité (ou « univers ») d'hypothèses possibles se compose des neuf termes de la multiplication $(A^\varepsilon + A^\eta + A^\theta)(B^\varepsilon + B^\eta + B^\theta)$, et de leurs combinaisons conjonctives ou disjonctives ; nous obtenons de nombreuses formules, telles que :

(1) $\qquad W(AB)^\theta = A^\varepsilon B^\theta + A^\theta B^\varepsilon$

(2) $\qquad S(AB)^\theta = A^{\eta\iota} B^\theta + A^\theta B^{\eta\iota}$

(3) $\qquad W(AB)^{\theta\iota} = S'(AB)^\theta = A^\eta + B^\eta + A^\varepsilon B^\varepsilon$

(4) $\quad S(AB)^{\theta\iota} = W'(AB)^\theta = A^\eta + B^\eta + A^\varepsilon B^\varepsilon + A^\theta B^\theta$

La formule (1) affirme que la plus faible donnée dont on peut conclure que AB est une proposition *variable* est la proposition disjonctive $A^\varepsilon B^\theta + A^\theta B^\varepsilon$, qui affirme que *ou* A *est certain et* B *variable, ou* A *variable et* B *certain*. La formule (2) affirme que la plus forte conclusion qu'on peut tirer de la proposition $(AB)^\theta$ est la proposition $A^{\eta\iota}B^\theta + A^\theta B^{\eta\iota}$, qui affirme que *ou* A *est possible et* B *variable, ou* A *variable et* B *possible*. D'autres formules évidentes ou faciles à

---

[1]. D'après notre définition, ni $W\varphi$ ni $V\varphi$ ni $S\varphi$ ne peuvent appartenir à la classe η: mais un ou deux des trois peuvent être absents, et, par conséquent, appartenir à la classe 0. Dans nos formules $\varphi(A)$, $\varphi(A, B)$, etc., il est sous-entendu que A, B, etc., ne sont pas des *propositions muettes*, autrement quelques-unes des formules ne seraient pas valides, comme par exemple la formule $A^{\tau\iota} = A^{\iota\tau}$. Car $0^{\tau\iota} = (0^\tau)^\iota = \eta^\iota = \varepsilon$; tandis que $0^{\iota\tau} = (0^\iota)^\tau = \eta^\tau = \eta$. Mais la formule est vraie dans le cas A^η, comme dans tous les autres; car $\eta^{\tau\iota} = (\eta^\tau)^\iota = \eta^\iota = \varepsilon$, et aussi $\eta^{\iota\tau} = (\eta^\iota)^\tau = \varepsilon^\tau = \varepsilon$. Si nous admettons le cas $A^0 B^0$, la formule $(AB)^\iota = A^\iota + B^\iota$ devrait être $(A^\tau B^\tau)^\iota = A^{\tau\iota} + B^{\tau\iota}$. La formule $(A^\alpha B^\beta)^\iota = A^{\alpha\iota} + B^{\beta\iota}$ est vraie toujours, même dans le cas $A^0 B^0$. Une proposition η contredit quelque donnée; tandis que 0 ne contredit rien.

prouver sont :

(5)	$W\varphi : \varphi : S\varphi$
(6)	$W\varphi :: S\varphi = W\varphi :: \varphi :: S\varphi$
(7)	$W(AB)^\varepsilon = S(AB)^\varepsilon = A^\varepsilon B^\varepsilon$
(8)	$W(A+B)^\varepsilon = A^\varepsilon + B^\varepsilon$
(9)	$S(A+B)^\varepsilon = A^\varepsilon + B^\varepsilon + A^\theta B^\theta$
(10)	$W(A+B)^\eta = S(A+B)^\eta = A^\eta B^\eta$
(11)	$W(A+B)^\theta = A^\eta B^\theta + A^\theta B^\eta$
(12)	$S(A+B)^\theta = A^{\varepsilon\iota} B^\theta + A^\theta B^{\varepsilon\iota}$
(13)	$W(AB)^\eta = A^\eta + B^\eta$
(14)	$S(AB)^\eta = A^\eta + B^\eta + A^\theta B^\theta$
(15)	$W(A:B) = W(AB')^\eta = A^\eta + B^\varepsilon$
(16)	$S(A:B) = S(AB')^\eta = A^\eta + B^\varepsilon + A^\theta B^\theta$
(17)	$W(A:B)' = S'(A:B) = A^\varepsilon B^{\varepsilon\iota} + A^{\eta\iota} B^\eta$
(18)	$S(A:B)' = W'(A:B) = A^{\eta\iota} B^{\varepsilon\iota}$

On peut déduire les formules (15) et (16) de (13) et (14), en changeant B en B'.

La formule (17) affirme que la plus faible donnée dont on peut déduire que A n'implique pas B est la proposition que A *est certain et* B *incertain ou* A *possible et* B *impossible*. On prouve cette formule de la manière suivante :

$W(A:B)' = S'(A:B) = (A^\eta + B^\varepsilon + A^\theta B^\theta)'$, (voir formule 16)

$= A^{\eta\iota} B^{\varepsilon\iota}(A^{\theta\iota} + B^{\theta\iota})$; car $(\alpha+\beta+\gamma)' = \alpha'\beta'\gamma'$, et $(\alpha\beta)' = \alpha' + \beta'$,

$= A^\varepsilon B^{\varepsilon\iota} + A^{\eta\iota} B^\eta$; car $A^{\eta\iota} A^{\theta\iota} = A^\varepsilon$, et $B^{\varepsilon\iota} B^{\theta\iota} = B^\eta$.

14. Ces formules nous aident souvent à trouver des cas où une implication n'est pas valide. Par exemple, $(x_A : x_B) : A_B$, l'inverse de la formule 1 du § 10, n'est pas tou-

LA LOGIQUE SYMBOLIQUE ET SES APPLICATIONS 149

jours vraie. Pour trouver des cas où cette inverse est fausse, représentons-la par φ. Alors, par une série d'implications inductives, nous aurons (voir les formules 15 et 17) :

$$\varphi^\iota \,!\, (x_\text{A}:x_\text{B})\text{A}_{\text{OB}} \,!\, \text{W}(x_\text{A}:x_\text{B})\, \text{WA}_{\text{OB}} \,!\, \{(x_\text{A})^\eta + (x_\text{B})^\varepsilon\} (\text{A}^\varepsilon \text{B}^{\varepsilon\iota} + \text{A}^{\eta\iota} \text{B}^\eta)$$
$$!\, \{(x^\varepsilon \text{A}^{\varepsilon\iota} + x^{\eta\iota}\text{A}^\eta) + (x^\eta + \text{B}^\varepsilon)\} (\text{A}^\varepsilon \text{B}^{\varepsilon\iota} + \text{A}^{\eta\iota} \text{B}^\eta)$$
$$!\, x^\eta \text{A}^\varepsilon \text{B}^{\varepsilon\iota} + x^\varepsilon \text{A}^\theta \text{B}^\eta + x^\eta \text{A}^{\eta\iota} \text{B}^\eta.$$

Ainsi nous trouvons qu'il y a *au moins* trois cas où l'implication du troisième degré φ n'est pas vraie : 1° quand x est *impossible* (η), A *certain* (ε), B *incertain* ($\varepsilon\iota$); 2° quand x est *certain* (ε), A *variable* (θ), B *impossible* (η); 3° quand x est *impossible* (η), A *possible* ($\eta\iota$), B *impossible* (η). Si nous excluons toute proposition variable de notre univers, de sorte que toute proposition *vraie* soit *certaine*, et toute proposition *fausse impossible*; alors, le second terme $x^\varepsilon \text{A}^\theta \text{B}^\eta$ disparaît, à cause du facteur A^θ, et le premier et le troisième deviennent chacun synonyme de $x'\text{AB}'$. Ces résultats sont faciles à vérifier. Prenons le cas $x^\varepsilon \text{A}^\theta \text{B}^\eta$. En remplaçant x, A, B dans l'implication φ par leurs exposants ε, θ, η, nous aurons :

$$\varphi = (\varepsilon_\theta : \varepsilon_\eta) : \theta_\eta = (\eta : \eta) : \eta = \varepsilon : \eta = \eta,$$

c'est-à-dire que l'implication φ est fausse dans le cas $x^\varepsilon \text{A}^\theta \text{B}^\eta$.

15. L'implication A : B est synonyme de l'équivalence (A = AB); car (voir § 8, et § 9, formule 8) :

$$(\text{A} = \text{AB}) = (\text{A} : \text{AB})(\text{AB} : \text{A}) = (\text{A} : \text{A})(\text{A} : \text{B})(\text{AB} : \text{A})$$
$$= \varepsilon (\text{A} : \text{B}) \varepsilon = \text{A} : \text{B}.$$

H. MAC COLL

Donc nous pouvons appeler l'antécédent A un *multiple* du conséquent B, et le conséquent B un *facteur* de l'antécédent A. L'implication $A : \eta$ affirme que la proposition A contient un facteur *impossible*, sans dire quelle impossibilité de la série η_1, η_2, η_3, etc., de sorte que $A : \eta$ est synonyme de A^η.

16. L'implication $(A : B) : A' + B$ est toujours vraie, mais pas toujours son *inverse*, l'implication $A' + B : (A : B)$. Soit φ cette *inverse*. En prenant les mêmes données qu'au § 13, nous trouverons facilement : $W\varphi' = A^\varepsilon B^\theta + A^\theta B^\eta$. Donc il y a deux cas, $A^\varepsilon B^\theta$ et $A^\theta B^\eta$, dans lesquels φ est nécessairement faux; de sorte que $A : B$ et $A' + B$ ne sont pas synonymes. On peut vérifier comme dans § 14. Prenons le second cas $A^\theta B^\eta$. En remplaçant A et B par leurs exposants θ et η, nous aurons :

$$\varphi = A' + B : (A : B) = \theta' + \eta : (\theta : \eta) = \theta' : \eta = \eta;$$

car (puisque $A^\theta = A^{\iota\theta}$, voir § 4, formule 3) la négation d'une variable est aussi une variable, et, par conséquent, ne peut pas avoir une impossibilité comme facteur.

17. Les quatre modalités (voir p. 141, note 1) de la Logique traditionnelle méritent quelques mots. Le produit des deux certitudes $(A^\varepsilon + A^\theta + A^\eta)$ et $(A + A^\iota)$ est :

$$A^\varepsilon + AA^\theta + A^\iota A^\theta + A^\eta \qquad (1)$$

qui est synonyme de

$$A^\varepsilon + AA^{\varepsilon\iota} + A^\iota A^{\eta\iota} + A^\eta \qquad (2)$$

car

$$AA^{\varepsilon\iota} = A(A^\eta + A^\theta) = AA^\theta, \quad \text{et} \quad A^\iota A^{\eta\iota} = A^\iota(A^\varepsilon + A^\theta) = A^\iota A^\theta.$$

LA LOGIQUE SYMBOLIQUE ET SES APPLICATIONS

Les quatre termes de (2) sont les quatre modalités; car:

A^e affirme que A est *nécessairement vrai*; c'est-à-dire la supposition que A soit faux *contredit quelque donnée ou définition*.

A A^{ei} affirme que A est vrai *dans un cas particulier*, mais qu'il est incertain *comme loi générale*; c'est-à-dire la supposition qu'il soit faux *ne contredit aucune donnée ou définition*.

A^i A^{ni} affirme que A est faux *dans un cas particulier*, mais qu'il est possible *comme loi générale*; c'est-à-dire, la supposition qu'il soit vrai *ne contredit aucune donnée ou définition*.

A^n affirme que A est *nécessairement faux*; c'est-à-dire, la supposition qu'il soit vrai *contredit quelque donnée ou définition*.

18. Quelques mots sont aussi dus au syllogisme. Tout syllogisme valide peut être déduit de la formule

$$(A : B)(B : C) : (A : C),$$

dont l'abrégé est $A_B B_C : A_C$, et que nous désignerons par le symbole $\varphi(A, B, C)$.

Sur les individus P_1, P_2, P_3, etc. (qui peuvent être des propositions ou des choses concrètes) formant l'univers de notre discours, prenons un individu P au hasard. Supposons que les symboles A, B, C affirment respectivement que P appartient à la classe A, que P appartient à la classe B, que P appartient à la classe C. Ces conventions posées, la formule $\varphi(A,B,C)$ exprimera le syllogisme appelé *Barbara*.

Cette formule (comme toute formule valide; voir § 21) est vraie *quelles que soient les propositions* A, B, C *qui la composent*, même, par exemple, si elles sont des *impossibilités*. Supposons que $A = \eta_1$, $B = \eta_2$, $C = \eta_3$; nous aurons (voir § 15) :

$$\varphi(A, B, C) = \varphi(\eta_1, \eta_2, \eta_3) = (\eta_1 : \eta_2)(\eta_2 : \eta_3) : (\eta_1 : \eta_3) = \varepsilon_1 \varepsilon_2 : \varepsilon_3 = \varepsilon_4.$$

Pour montrer comment les autres syllogismes valides peuvent être déduits de la formule $\varphi(A, B, C)$, prenons le syllogisme appelé *Fresison*. Nous aurons (voir §§ 10 et 14) :

$$Fresison = C_{B'} B_{0A'} : A_{0C} = C_{B'} B_{0A'} A_C : \eta$$
$$= A_C C_{B'} : B_{A'} = A_C C_{B'} : A_{B'} = \varphi(A, C, B').$$

C'est-à-dire, si dans $\varphi(A, B, C)$ nous changeons B en C, et C en B', nous aurons une formule qui est équivalente à *Fresison*.

Le syllogisme appelé *Darapti* n'est pas valide sous sa forme habituelle; pour le rendre valide, il faut que nous ajoutions à nos deux prémisses une troisième prémisse B^{η_1}, car, sans cette troisième prémisse, nous aurions :

$$Darapti = B_C B_A : A_{0C'} = B_{CA} A_{C'} : \eta = B_{CA} (AC)^\eta : \eta.$$

Or cette dernière implication est évidemment fausse dans le cas $B^\eta (CA)^\eta$; car alors elle devient $\eta_\eta \eta^\eta : \eta$, qui est impossible, puisque η_η et η^η sont tous les deux des certitudes (voir § 15), et une certitude ne peut pas impliquer une impossibilité.

Mais *Darapti* devient une formule valide si nous ajoutons B^{η_1} à nos prémisses; et alors on peut le déduire, comme les

LA LOGIQUE SYMBOLIQUE ET SES APPLICATIONS   153

autres syllogismes, de la formule $\varphi(A, B, C)$. Car, avec cette troisième prémisse B^η, nous aurons :

$Darapti = B^\eta B_C B_A : A_{0C'} = B^\eta B_C B_A A_{C'} : \eta = B^\eta B_{CA}(AC)^\eta : \eta$
$= B_{0\eta} B_{AC}(AC)_\eta : \eta = B_{AC}(AC)_\eta : B_\eta = \varphi(B, AC, \eta).$

C'est-à-dire que *Darapti*, augmenté d'une troisième prémisse B^η, devient un cas particulier de la formule $\varphi(A, B, C)$, en changeant A en B, B en AC, et C en η.

II. — *Probabilités*.

19. Soient A et B deux propositions. Dans ce cas le symbole $\frac{A}{B}$ exprime la probabilité que A soit vraie *en supposant que B soit vraie*. Le symbole $\frac{A}{\varepsilon}$ exprime la probabilité que A soit vraie, *en ne supposant rien que les données du problème à résoudre*. Le symbole $\delta\frac{A}{B}$, ou $\delta(A, B)$, est synonyme de $\frac{A}{B} - \frac{A}{\varepsilon}$, et représente *la dépendance de A par rapport à B*; c'est-à-dire, il représente l'augmentation, ou (s'il est négatif) la diminution, subie par la probabilité $\frac{A}{\varepsilon}$, quand on suppose B ajoutée aux données du problème.

Si $\delta\frac{A}{B}$ est zéro, on dit que A est indépendante de B; ce qui implique (comme nous allons voir) que B est indépendante de A. Le symbole $\delta^0(A, B)$ veut dire $(\delta\frac{A}{B} = 0)$.

Sur la totalité des points marqués dans le cercle E, en fixant notre attention sur une seule des trois figures 3, 4, 5, prenons un point P au hasard; et supposons que le symbole A affirme que P sera un des points marqués dans le cercle

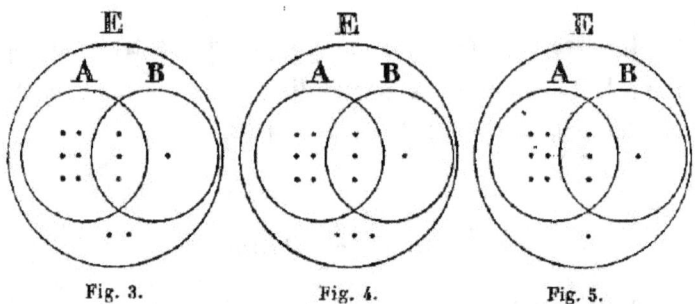

Fig. 3.   Fig. 4.   Fig. 5.

A, et que B affirme que P sera un des points marqués dans le cercle B. Nous aurons :

Fig. 3 : $\quad \delta \dfrac{A}{B} = \dfrac{A}{B} - \dfrac{A}{\varepsilon} = \dfrac{AB}{B} - \dfrac{A}{\varepsilon} = \dfrac{3}{4} - \dfrac{9}{12} = 0;$

Fig. 4 : $\quad \delta \dfrac{A}{B} = \dfrac{A}{B} - \dfrac{A}{\varepsilon} = \dfrac{AB}{B} - \dfrac{A}{\varepsilon} = \dfrac{3}{4} - \dfrac{9}{13} = \dfrac{3}{52};$

Fig. 5 : $\quad \delta \dfrac{A}{B} = \dfrac{A}{B} - \dfrac{A}{\varepsilon} = \dfrac{AB}{B} - \dfrac{A}{\varepsilon} = \dfrac{3}{4} - \dfrac{9}{11} = -\dfrac{3}{44}.$

Ainsi, nous voyons que, dans la Fig. 3, A est indépendante de B ; que, dans la Fig. 4, la dépendance de A par rapport à B est positive ; et que, dans la Fig. 3, la dépendance est négative.

20. Il est souvent commode, en discutant des questions de probabilité, de représenter les propositions par des lettres majuscules, A, B, C, etc., et leurs probabilités respectives par les minuscules a, b, c, etc.; de sorte que

nous aurons $\frac{A}{\varepsilon}=a$, $\frac{B}{\varepsilon}=b$, $\frac{C}{\varepsilon}=c$, etc. Si nous supposons aussi $a'=1-a$, $b'=1-b$, $c'=1-c$, etc., nous aurons $\frac{A'}{\varepsilon}=a'$, $\frac{B'}{\varepsilon}=b'$, $\frac{C'}{\varepsilon}=c'$. Ces conventions posées, nous trouvons plusieurs formules, telles que :

(1) $\quad \frac{AB}{\varepsilon}=\frac{A}{\varepsilon}\cdot\frac{B}{A}=\frac{B}{\varepsilon}\cdot\frac{A}{B}$, $\quad \frac{AB}{x}=\frac{A}{x}\cdot\frac{B}{Ax}=\frac{B}{x}\cdot\frac{A}{Bx}$

(2) $\quad \frac{A+B}{\varepsilon}=\frac{A}{\varepsilon}+\frac{B}{\varepsilon}-\frac{AB}{\varepsilon}$, $\quad \frac{A+B}{x}=\frac{A}{x}+\frac{B}{x}-\frac{AB}{x}$

(3) $\quad \frac{A}{B}=\frac{a}{b}\cdot\frac{B}{A}$, $\quad \delta\frac{A}{B}=\frac{a}{b}\delta\frac{B}{A}$

(4) $\quad \frac{A'}{B}=1-\frac{A}{B}$, $\quad \delta\frac{A'}{B}=-\delta\frac{A}{B}$

(5) $\quad \frac{A}{B'}=\frac{a}{b'}-\frac{b}{b'}\frac{A}{B}$, $\quad \delta\frac{A}{B'}=-\frac{b}{b'}\delta\frac{A}{B}$

(6) $\quad \frac{A'}{B'}=1-\frac{A}{B'}$, $\quad \delta\frac{A'}{B'}=-\delta\frac{A}{B'}=\frac{b}{b'}\delta\frac{A}{B}$.

Par la formule : $\delta\frac{A}{B}=\frac{a}{b}\delta\frac{B}{A}$, on voit que $\delta\frac{A}{B}$ doit avoir le même signe (positif, négatif, ou zéro) que $\delta\frac{B}{A}$, puisque les probabilités a et b sont nécessairement positives. Donc, quand A est indépendant de B, B doit être indépendant de A ; théorème exprimé symboliquement par $\delta^0(A, B)=\delta^0(B, A)$ (voir § 19).

21. Il est quelquefois commode de diviser nos propositions en *certitudes formelles* (x_1, x_2, etc.), *impossibilités formelles* (λ_1, λ_2, etc.), et *variables formelles* (μ_1, μ_2, etc.).

Les *certitudes formelles* résultent des définitions mêmes

de nos symboles, sans autres données; les *impossibilités formelles* contredisent chacune quelque définition, ou combinaison de définitions; les *variables formelles* ne sont ni formellement certaines ni formellement impossibles. Par exemple,

$$(AB:A)^\varkappa, \quad (\eta:A)^\varkappa, \quad (A^\varepsilon A^\theta)^\lambda, \quad (\theta:\eta)^\lambda, \quad (A:B)^\mu, \quad (A+B)^\mu.$$

Ces définitions de $\varkappa$, λ, μ nous donnent les formules :

(1) $\quad A^\varkappa : A^\varepsilon,$ (2) $\quad A^\lambda : A^\eta,$ (3) $\quad A^\theta : A^\mu.$

La troisième peut être déduite des deux autres ainsi :

$$\varepsilon : (A^\varkappa : A^\varepsilon)\,(A^\lambda : A^\eta) : (A^{\varepsilon\iota} : A^{\varkappa\iota})\,(A^{\eta\iota} : A^{\lambda\iota}) : (A^{\varepsilon\iota}A^{\eta\iota} : A^{\varkappa\iota}A^{\lambda\iota}) : (A^\theta : A^\mu)$$

en faisant appel aux formules :

$$\alpha : \beta = \beta' : \alpha' \quad \text{et} \quad (\alpha : \beta)(\gamma : \delta) : (\alpha\gamma : \beta\delta).$$

Toute *variable formelle*, comme A : B, peut être une *certitude* (pas une certitude *formelle*), une *impossibilité*, ou une *variable*, d'après nos données (voir § 23). Prenons, par exemple, l'implication A : B, où A et B peuvent représenter des propositions simples ou des implications compliquées. Quelles sont les probabilités respectives que l'implication A : B soit une *certitude*, une *impossibilité*, une *variable*? A cette question, ainsi posée, on ne peut pas donner de réponse; les données nécessaires nous manquent. Mais si nous prenons comme données que les 9 termes du produit $(A^\varepsilon + A^\eta + A^\theta)(B^\varepsilon + B^\eta + B^\theta)$ sont également probables, nous pouvons trouver des limites plus proches

LA LOGIQUE SYMBOLIQUE ET SES APPLICATIONS

que 1 et 0 pour les probabilités demandées. Le produit est :

$$A^\varepsilon(B^\varepsilon + \underline{B^\eta} + \underline{B^\theta}) + A^\eta (\underset{..}{B^\varepsilon} + \underset{..}{B^\eta} + \underline{B^\theta}) + A^\theta (\underset{..}{B^\varepsilon} + \underline{B^\eta} + B^\theta).$$

Trois des 9 termes (les trois soulignés, $A^\varepsilon B^\eta, A^\varepsilon B^\theta, A^\theta B^\eta$) impliquent $(A:B)^\eta$; cinq termes (ceux qui ont des points au-dessous) impliquent $(A:B)^\varepsilon$; et un seul terme, $A^\theta B^\theta$, n'implique (sans autres données) ni $(A:B)^\varepsilon$, ni $(A:B)^\eta$, ni $(A:B)^\theta$, ni aucune de leurs négations $(A:B)^{\varepsilon\iota}$, $(A:B)^{\eta\iota}$, $(A:B)^{\theta\iota}$.

Donc, si nous posons comme définition que le symbole $\varphi^x(a, b)$ affirme que la probabilité de φ^x est entre les limites a et b, nous aurons [1] (en posant $\varphi = A:B$, et x successivement $= \varepsilon, \eta, \theta$) :

$$(A:B)^\varepsilon \left(\frac{5}{9}, \frac{6}{9}\right), \qquad (A:B)^\eta \left(\frac{3}{9}, \frac{4}{9}\right), \qquad (A:B)^\theta \left(0, \frac{1}{9}\right).$$

Si nous excluons A^θ et B^θ de notre univers de possibilités et si nous prenons comme données que les quatre termes du produit $A^{\varepsilon,\eta} B^{\varepsilon,\eta}$ sont également probables, nous aurons :

$$\frac{(A:B)^\varepsilon}{\varepsilon} = \frac{(A^\varepsilon B^\eta)^\iota}{\varepsilon} = 1 - \frac{A^\varepsilon B^\eta}{\varepsilon} = 1 - \frac{1}{4} = \frac{3}{4},$$

$$\frac{(A:B)^\eta}{\varepsilon} = \frac{A^\varepsilon B^\eta}{\varepsilon} = \frac{1}{4}.$$

Il est évident qu'ici notre donnée $A^{\theta\iota} B^{\theta\iota}$ implique $(A:B)^{\theta\iota}$.

---

[1]. La probabilité $\frac{\varphi^x}{\varepsilon}$ est toujours entre les limites $\frac{W\varphi^x}{\varepsilon}$ et $\frac{S\varphi^x}{\varepsilon}$; car $W\varphi^x : \varphi^x : S\varphi^x$, et $\frac{\varphi^x}{\varepsilon} = \frac{W\varphi^x}{\varepsilon} + \frac{V\varphi^x}{\varepsilon} \cdot \frac{\varphi^x}{V\varphi^x}$. Ici $V\varphi^x = A^\theta B^\theta$, pour chaque valeur de x (voir §12 et 13); mais $W\varphi^\varepsilon = W\varphi = A^\eta + B^\varepsilon$; $W\varphi^\eta = W\varphi^\iota = A^\varepsilon B^{\theta\iota} + A^{\eta\iota} B^\eta$; $W\varphi^\theta$ est ici une proposition *muette*, puisqu'aucune des 9 propositions qui constituent notre univers n'implique (sans plus de données) la conclusion φ^θ.

H. MAC COLL

22. Bon nombre de logiciens confondent l'implication complexe A : (B : C) (*Si* A *est vrai, alors, si* B *est vrai* C *est vrai*), avec l'implication simple AB : C (*Si* A *et* B *sont vrais*, C *est vrai*); mais d'après mon interprétation de la conjonction *si*, ces deux implications ne sont pas synonymes. (Voir § 6.)

Désignons la première par φ_1 et la seconde par φ_2; alors nous aurons :

$$\varphi_1 = A : (B : C) = A : (BC')^\eta$$
$$\varphi_2 = AB : C = (ABC')^\eta = A : (BC')^\iota$$

Or il est clair, puisque le conséquent $(BC')^\eta$ implique le conséquent $(BC')^\iota$, que φ_1 implique φ_2 (voir la première formule de § 10). Mais φ_2 n'implique pas nécessairement φ_1, car il y a des cas où φ_2 est vrai et φ_1 faux. Supposons, par exemple, que BC' soit une *variable* θ_1; il suit que sa négation θ'_1 est aussi une variable. Supposons aussi (comme c'est permis) que A soit cette variable θ'_1. Alors nous aurons :

$$\varphi_2 = \theta'_1 : \theta'_1 = \varepsilon_1 \text{ (une } \textit{certitude}\text{)}$$
$$\varphi_1 = \theta'_1 : \theta_1^\eta = \theta'_1 : \eta_1 = \eta_2 \text{ (une } \textit{impossibilité}\text{)}$$

car une impossibilité η_1 ne peut pas être un facteur d'une variable θ'_1.

Une autre preuve est la suivante. Sur les 15 points dans le cercle E, qu'un point P soit pris au hasard (Fig. 6). Supposons que les quatre symboles E, A, B, C, comme propositions, affirment respectivement que P sera un des 15 points dans le cercle E, que P sera un des 5 points dans le cercle A, que P sera un des 5 points dans le cercle B, que P sera un des

LA LOGIQUE SYMBOLIQUE ET SES APPLICATIONS

6 points dans le cercle C. La proposition E est une *certitude* ; les propositions A, B, C sont des *variables* dont les probabilités respectives sont $\frac{5}{15}$, $\frac{5}{15}$, $\frac{6}{15}$; et la proposition BC' est aussi une variable dont la probabilité est $\frac{3}{15}$.

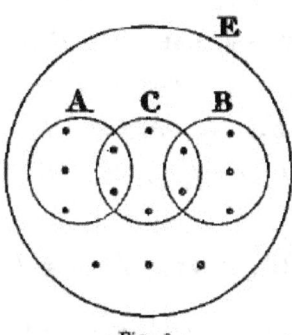

Fig. 6.

Comme ces quatre dernières conclusions résultent nécessairement de nos données, toute proposition qui contredit n'importe laquelle des quatre doit être une *impossibilité*. Or (BC')$^\eta$, qui affirme que la probabilité que BC' soit vrai est zéro, contredit notre dernière conclusion, qui affirme que cette probabilité est $\frac{3}{15}$. Donc (BC')$^\eta$ n'est pas simplement une proposition (ou prédiction) que le hasard rend fausse dans un cas particulier et à propos d'un certain point ; c'est une proposition qui *contredit nos données*, et qui est, par conséquent (dans les limites de nos données), une *impossibilité*. Que η_1 désigne cette impossibilité (BC')$^\eta$, et que θ_1 désigne la proposition variable A. Alors nous aurons :

$$\varphi_1 = A : (BC')^\eta = \theta_1 : \eta_1 = \eta_2,$$

car l'implication $\theta_1 : \eta_1$, qui affirme qu'une impossibilité η_1 est un facteur d'une variable θ_1, est une seconde impossibilité η_2. Donc, dans les conditions données, φ_1 est une impossibilité.

Maintenant prenons φ_2, qui veut dire $A : (BC')^\iota$. D'après notre définition d'une implication, φ_2 est synonyme de $(ABC')^\eta$, et affirme seulement que ABC' est *impossible*, affirmation qui (voir la figure) est évidemment une *certitude*, puisque AB est impossible. Appelons cette certitude ε_1. Ainsi nous avons, *dans les limites des mêmes données*,

$$\varphi_1 = A : (BC')^\eta = \eta_2, \quad \text{et} \quad \varphi_2 = A : (BC')^\iota = \varepsilon_1,$$

de sorte que, dans ce cas, comme dans le cas précédent, les deux propositions φ_1 et φ_2 ne sont pas synonymes.

23. Quand nous avons une proposition de la forme A^x, où x indique la classe à laquelle appartient la proposition A, nous appelons A une *proposition primaire* à l'égard de A^x, et A^x une *proposition secondaire* (ou une proposition du *second degré*) à l'égard de A.

Quand nous avons une proposition de la forme A^{xy}, qui est synonyme de $(A^x)^y$, et affirme que la proposition A^x appartient à la classe y, nous appelons A^{xy} une proposition du *second degré* à l'égard de A^x, mais une proposition du *troisième degré* à l'égard de A. Une proposition de la forme A^{xyz} ou A^{xyzu}, etc., doit être interprétée et nommée d'après le même principe.

Maintenant, supposons que A représente une proposition

LA LOGIQUE SYMBOLIQUE ET SES APPLICATIONS 161

simple ou compliquée[1], comment expliquer des propositions du troisième ou quatrième degré, telles que A^{xz}, $A^{\epsilon\eta\theta}$, etc.? Traduite en langage ordinaire, la proposition $A^{\epsilon\eta\theta}$, par exemple, veut dire : « Il est *possible mais incertain* (θ) qu'il est *impossible* (η) que A soit *certain* (ε) ». L'explication générale a déjà été donnée (voir § 21); mais allons aux

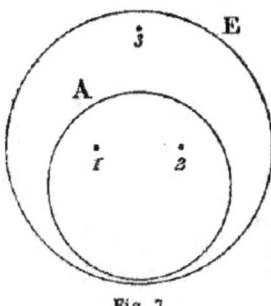

Fig. 7.

principes élémentaires, et examinons quelques exemples concrets.

Sur les 3 points marqués (Fig. 7) dans le cercle E, prenons un point P au hasard. Quelle est la probabilité que P sera un des deux points marqués dans le cercle A? La réponse est évidemment $\frac{2}{3}$; mais, pour faire mieux comprendre la méthode générale, je vais l'appliquer même à ce cas simple.

Que les symboles P_1, P_2, P_3 affirment respectivement comme propositions que P sera le point 1, 2, 3; tandis que

---

[1]. Par exemple, A peut représenter une implication quelconque de n'importe quel degré, comme α : β ou (α : β) : (α^x : β^x), qui formellement peut être vraie ou fausse.

le symbole A, comme proposition, affirme que P sera dans le cercle A. La proposition disjonctive $(P_1 + P_2 + P_3)$ est une *certitude*; donc :

$$A = (P_1 + P_2 + P_3) A = P_1 A + P_2 A + P_3 A$$

$$\frac{A}{\varepsilon} = \frac{P_1 A}{\varepsilon} + \frac{P_2 A}{\varepsilon} + \frac{P_3 A}{\varepsilon} \quad \text{(puisque } P_1 P_2, P_1 P_3, P_2 P_3 = \eta_1, \eta_2, \eta_3\text{)}$$

$$= \frac{P_1}{\varepsilon} \cdot \frac{A}{P_1} + \frac{P_2}{\varepsilon} \cdot \frac{A}{P_2} + \frac{P_3}{\varepsilon} \cdot \frac{A}{P_3} \qquad \left(\text{car } \frac{\alpha\beta}{\varepsilon} = \frac{\alpha}{\varepsilon} \cdot \frac{\beta}{\alpha}\right)$$

Mais :
$$\frac{P_1}{\varepsilon} = \frac{P_2}{\varepsilon} = \frac{P_3}{\varepsilon} = \frac{1}{3}.$$

Donc :
$$\frac{A}{\varepsilon} = \frac{1}{3}\left(\frac{A}{P_1} + \frac{A}{P_2} + \frac{A}{P_3}\right) = \frac{1}{3}(1 + 1 + 0) = \frac{2}{3}.$$

Mais si nous avions pris la Fig. 8 au lieu de la Fig. 7, nous aurions eu :

$$\frac{A}{\varepsilon} = \frac{1}{3}\left(\frac{A}{P_1} + \frac{A}{P_2} + \frac{A}{P_3}\right) = \frac{1}{3}(1 + 1 + 1) = 1.$$

Ainsi, quand les conditions de la Fig. 7 sont données, nous avons A^θ; mais quand les conditions de la Fig. 8 sont données, nous avons A^ε.

Maintenant supposons que nous ayons *cinq* figures : deux semblables à la Fig. 7, et trois à la Fig. 8; et que, sur ces cinq figures, nous prenions une figure F au hasard. Sur les trois points du cercle E, dans la figure que le hasard présentera, prenons un point P au hasard. La proposition A ayant le même sens qu'avant, quelles sont les probabilités respectives de A et de A^ε?

LA LOGIQUE SYMBOLIQUE ET SES APPLICATIONS

Soit F_1 le symbole de la proposition : « F *sera semblable à la Fig.* 7 » et soit F_2 le symbole de la proposition : « F *sera*

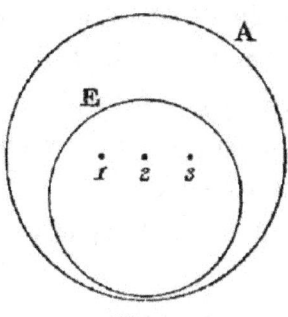

Fig. 8.

semblable à la Fig. 8 »; de sorte que $(F_1 + F_2)$ est une *certitude*. Donc, $A = (F_1 + F_2) A$, et

$$\frac{A}{\varepsilon} = \frac{F_1 A}{\varepsilon} + \frac{F_2 A}{\varepsilon} = \frac{F_1}{\varepsilon} \cdot \frac{A}{F_1} + \frac{F_2}{\varepsilon} \cdot \frac{A}{F_2} = \frac{2}{5} \cdot \frac{A}{F_1} + \frac{3}{5} \cdot \frac{A}{F_2} = \frac{2}{5} \cdot \frac{2}{3} + \frac{3}{5} \cdot \frac{3}{3} = \frac{13}{15}.$$

$$\frac{A^\varepsilon}{\varepsilon} = \frac{F_1 A^\varepsilon}{\varepsilon} + \frac{F_2 A^\varepsilon}{\varepsilon} = \frac{F_1}{\varepsilon} \cdot \frac{A^\varepsilon}{F_1} + \frac{F_2}{\varepsilon} \cdot \frac{A^\varepsilon}{F_2} = \frac{2}{5} \cdot \frac{A^\varepsilon}{F_1} + \frac{3}{5} \cdot \frac{A_\varepsilon}{F_2} = \frac{2}{5} \cdot 0 + \frac{3}{5} \cdot 1 = \frac{3}{5}.$$

La dernière conclusion, $\frac{A\varepsilon}{\varepsilon} = \frac{3}{5}$, implique $A^{\varepsilon\theta}$. Mais si nous prenons comme données que toutes les cinq figures sont semblables à la Fig. 8, la proposition F_2 sera une *certitude*, et nous aurons

$$\frac{A^\varepsilon}{\varepsilon} = \frac{F_2 A^\varepsilon}{\varepsilon} = \frac{F_2}{\varepsilon} \cdot \frac{A^\varepsilon}{F_2} = 1 \times 1 = 1;$$

c'est-à-dire, nous aurons la conclusion $A^{\varepsilon\varepsilon}$, « *Il est certain que* A *est certain* ».

Pour expliquer le sens de la proposition $A^{\varepsilon\eta\theta}$, supposons qu'il y ait deux *collections* de figures : la collection C_1, qui contient 3 figures, toutes semblables à la Fig. 7 ; et la collection C_2, qui contient 2 figures semblables à la Fig. 7, et 3 semblables à la Fig. 8. Des deux collections, C_1 et C_2, prenons une collection C au hasard. Dans la collection qui se présentera, prenons une figure F au hasard ; et dans la figure qui se présentera, prenons un point P au hasard. Quelle est la probabilité de $A^{\varepsilon\eta}$?

Les propositions A, F_1, F_2 ayant les mêmes significations qu'avant, soit C_1 le représentant de la proposition : « La collection C_1 se présentera », et C_2 le représentant de la proposition : « La collection C_2 se présentera ». Puisque $(C_1 + C_2)$ est une *certitude*, nous avons : $A^{\varepsilon\eta} = (C_1 + C_2) A^{\varepsilon\eta}$, et par conséquent (voir la première formule du § 20) :

$$\frac{A^{\varepsilon\eta}}{\varepsilon} = \frac{C_1 A^{\varepsilon\eta}}{\varepsilon} + \frac{C_2 A^{\varepsilon\eta}}{\varepsilon} = \frac{C_1}{\varepsilon} \cdot \frac{A^{\varepsilon\eta}}{C_1} + \frac{C_2}{\varepsilon} \cdot \frac{A^{\varepsilon\eta}}{C_2} = \frac{1}{2} \cdot \frac{A^{\varepsilon\eta}}{C_1} + \frac{1}{2} \cdot \frac{A^{\varepsilon\eta}}{C_2}.$$

Mais, puisque $(F_1 + F_2)$ est une certitude, $A^{\varepsilon\eta} = (F_1 + F_2) A^{\varepsilon\eta}$; donc (par la formule $\frac{\alpha\beta}{x} = \frac{\alpha}{x} \cdot \frac{x\alpha}{\beta}$; voir § 20) nous avons :

$$\frac{A^{\varepsilon\eta}}{C_1} = \frac{F_1}{C_1} \cdot \frac{A^{\varepsilon\eta}}{F_1 C_1} + \frac{F_2}{C_1} \cdot \frac{A^{\varepsilon\eta}}{F_2 C_1} = (1 \times 1) + \left(0 \times \frac{A^{\varepsilon\eta}}{F_2 C_1}\right) = 1$$

$$\frac{A^{\varepsilon\eta}}{C_2} = \frac{F_1}{C_2} \cdot \frac{A^{\varepsilon\eta}}{F_1 C_2} + \frac{F_2}{C_2} \cdot \frac{A^{\varepsilon\eta}}{F_2 C_2} = \left(\frac{2}{5} \times 0\right) + \left(\frac{3}{5} \times 0\right) = 0$$

Donc, $\frac{A^{\varepsilon\eta}}{\varepsilon} = \frac{1}{2}(1 + 0) = \frac{1}{2}$; conclusion qui implique $A^{\varepsilon\eta\theta}$.

24. Quand A est une proposition, et x une *fraction numérique*, le symbole A^x est un abrégé de l'équation $\frac{A}{\varepsilon} = x$, et affirme que la probabilité que A soit vrai est x. Cette défi-

LA LOGIQUE SYMBOLIQUE ET SES APPLICATIONS

nition nous conduit à plusieurs formules, telles que :

(1) $A^x B^y (AB)^z : (A+B)^{x+y-z}$; (2) $A^x B^y : \left(\dfrac{A}{B} = \dfrac{x}{y} \cdot \dfrac{B}{A}\right)$.

25. Soit $p = $ *probable* (plus souvent vrai que faux), et soit $q = $ *improbable* (plus souvent faux que vrai), nous aurons :

(3) $(A^p + A^q + A^{\frac{1}{2}})^c$ (4) $A^{\prime p} = A^q$ (5) $A^{p\prime} = A^q + A^{\frac{1}{2}}$
(6) $A^{\prime q} = A^p$ (7) $A^{q\prime} = A^p + A^{\frac{1}{2}}$ (8) $(AB)^p : A^p B^p$
(9) $(A+B)^q : A^q B^q$ (10) $(A:B) : (A^p : B^p)(B^q : A^q)$.

Par (4), (5), (6), (7) on voit que $A^{\prime p}$ n'est pas synonyme de $A^{p\prime}$, ni $A^{\prime q}$ synonyme de $A^{q\prime}$.

Toutes ces formules sont ou évidentes ou faciles à prouver. Il suffit de prouver la dernière, la formule 10. Cette formule est le produit de deux implications, $(A:B) : (A^p : B^p)$ et $(A:B) : (B^q : A^q)$. Appelons-les respectivement $\varphi(A, B)$ et $\psi(A, B)$. La première est évidente : si, toutes les fois que A est vrai, B est vrai ; alors, si A est plus souvent vrai que faux, B doit aussi être plus souvent vrai que faux. Il ne reste donc qu'à prouver $\psi(A, B)$. Puisque $\varphi(A, B)$ est une certitude *formelle*, $\varphi(B', A')$ aussi doit être vrai ; or :

$\varphi(B', A') = (B' : A') : (B^{\prime p} : A^{\prime p}) = (A : B) : (B^q : A^q) = \psi(A, B)$;
car $(B' : A') = (A : B)$, $B^{\prime p} = B^q$, $A^{\prime p} = A^q$ (voir formule 4).

26. Dans toutes les discussions précédentes, le symbole A_B était synonyme (voir § 10) de l'implication $(A : B)$; mais il est quelquefois commode de l'employer en d'autres sens. Par exemple, au lieu de désigner les individus (choses concrètes ou propositions) qui forment une classe A par A_1, A_2, A_3, etc., nous pouvons les désigner par A_α, A_β, etc., où A_α indique l'individu de la série A_1, A_2, A_3, etc., pour lequel la proposition α est vraie ; A_β celui pour lequel la proposition β est vraie ; etc. Ainsi A_B^C, qui veut dire $(A_B)^C$, est une proposition dont le sujet est A_B, et le prédicat C. Par exemple, soit A = *animal*, P = il est *petit*, T = il est *tué*. Alors, le symbole A_P^T = le *petit* animal est *tué* ; tandis que A_T^P = l'*animal tué* est *petit*.

Comme je l'ai déjà dit, il est probable que dans le langage primitif (ou les langages primitifs) de l'humanité, chaque mot ou son distinctif était une *proposition*. Supposons, en parlant d'une chose indéfinie et sans nom (appelons-la x), que C veut dire « x est un *cerf* », que B veut dire « x est *brun* », et que T veut dire « x a été *tué* par moi », ou « j'ai *tué* x ». Alors, C_B^T veut dire « Le *cerf brun* a été *tué* par moi », ou J'ai *tué* le *cerf brun* » ; tandis que C_T^B veut dire « Le *cerf* que j'ai *tué* est *brun* », et B_T^C veut dire « L'animal *brun* que j'ai *tué* est un *cerf* ». Ainsi nous voyons comment des propositions primitives, A, B, C, sans sujet et sans prédicat, auraient pu se combiner pour former une proposition complexe A_C^B où la proposition A devient un substantif concret, B un adjectif, A_B le sujet complet, et C le prédicat.

III. — *Calcul des limites.*

27. Le symbole A^x affirme que le nombre ou fraction A appartient à la classe x, où l'exposant x peut représenter un mot tel que *positif, négatif, zéro, imaginaire*, etc. Le symbole $A^x B^y$ affirme que A appartient à la classe x, *et* que B appartient à la classe y; tandis que $A^x + B^y$ affirme que A appartient à la classe x, *ou* B à la classe y. Dans les exemples qui se présenteront ici nous supposons $u = positif$, $v = négatif$, $z = zéro$, $\omega = imaginaire$. Par exemple, nous avons $(5-3)^u$, $(3-5)^v$, $(6-6)^z$, $\left(\frac{1}{2}+\frac{1}{3}\sqrt{-1}\right)^\omega$. Si nous n'admettons que des valeurs *positives* (u) et des valeurs *négatives* (v), excluant les *zéros* et les *imaginaires*, nous aurons les certitudes suivantes :

(1) $(AB)^u = A^u B^u + A^v B^v$; (2) $(AB)^v = A^v B^u + A^u B^v$.

(3) $(Ax - B)^u = \left\{A\left(x - \frac{B}{A}\right)\right\}^u = A^u \left(x - \frac{B}{A}\right)^u + A^v \left(x - \frac{B}{A}\right)^v$

(4) $(Ax - B)^v = \left\{A\left(x - \frac{B}{A}\right)\right\}^v = A^v \left(x - \frac{B}{A}\right)^u + A^u \left(x - \frac{B}{A}\right)^v$

(5) $\qquad (6x^2 - 13x + 6)^u =$

$\left\{6\left(x^2 - \frac{13}{6}x + 1\right)\right\}^u = \left\{6\left(x - \frac{3}{2}\right)\left(x - \frac{2}{3}\right)\right\}^u =$

$\left(x - \frac{3}{2}\right)^u \left(x - \frac{2}{3}\right)^u + \left(x - \frac{3}{2}\right)^v \left(x - \frac{2}{3}\right)^v = \left(x - \frac{3}{2}\right)^u + \left(x - \frac{2}{3}\right)^v$

car $\left(x - \frac{3}{2}\right)^u$ implique $\left(x - \frac{2}{3}\right)^u$, et $\left(x - \frac{2}{3}\right)^v$ implique $\left(x - \frac{3}{2}\right)^v$.

168   H. MAC COLL

(6) $(6x^2-13x+6)^v = \left(x-\dfrac{3}{2}\right)^v\left(x-\dfrac{2}{3}\right)^u + \left(x-\dfrac{3}{2}\right)^u\left(x-\dfrac{2}{3}\right)^v$

$\qquad = \left(x-\dfrac{3}{2}\right)^v\left(x-\dfrac{2}{3}\right)^u + \eta$ (une impossibilité)

Ainsi, dans la certitude (5), nous trouvons que x doit être *ou au-dessus* de $\dfrac{3}{2}$ *ou au-dessous* de $\dfrac{2}{3}$; et dans la certitude (6) nous trouvons que x doit être entre les limites $\dfrac{3}{2}$ et $\dfrac{2}{3}$.

Si nous admettons les valeurs *positives*, les valeurs *négatives* et les *zéros*, excluant toutes les autres, nous aurons[1] :

(7) $\quad (AB)^z = A^z + B^z = A^z B^{z\iota} + A^{z\iota} B^z + A^z B^z$

(8) $\quad (Ax-B)^z = A^z B^z + A^{z\iota}\left(x-\dfrac{B}{A}\right)^z$

(9) $\quad (Ax-B)^u = A^u\left(x-\dfrac{B}{A}\right)^u + A^v\left(x-\dfrac{B}{A}\right)^v + A^z B^v$

(10) $\quad (Ax-B)^v = A^v\left(x-\dfrac{B}{A}\right)^u + A^u\left(x-\dfrac{B}{A}\right)^v + A^z B^u$.

Quand nous avons $(x-a)^v$, ou son synonyme $x<a$, nous appelons a une *limite supérieure* de x; et quand nous avons $(x-a)^u$, ou son synonyme $x>a$, nous appelons a une *limite inférieure* de x.

28. Quand nous avons à parler souvent de plusieurs limites, x_1, x_2, x_3, etc., d'une variable x, il est commode de les enregistrer dans une *table de référence*, l'une après

---

[1]. Le symbole $A^{z\iota}$ affirme que A^z est faux; c'est-à-dire il affirme que A n'est pas *zéro*.

LA LOGIQUE SYMBOLIQUE ET SES APPLICATIONS

l'autre, dans l'ordre où elles se présentent (voir §§34, 35, 36). Le symbole $x_{m'.n}$, ou son synonyme x_n^m, affirme que x_m est une limite supérieure, et x_n une limite inférieure de x. Le symbole $x_{m'.n'.r.s}$, ou son synonyme $x_{r.s}^{m.n}$, affirme que x_m et x_n sont des limites supérieures, et x_r et x_s des limites inférieures de x.

29. Le symbole $x_{m'}$ (avec l'accent sur l'm) est toujours une proposition, synonyme de $(x-x_m)^v$, qui est synonyme de $x < x_m$; il affirme que la $m^{\text{ième}}$ limite sur notre liste ou table de référence, est une limite *supérieure* de x. Le symbole x_m (sans accent sur l'm) affirme comme proposition que la $m^{\text{ième}}$ limite de x sur notre liste est une limite *inférieure*. Donc $x_m = (x - x_m)^u = (x > x_m)$.

Ainsi [1] (voir § 30) :

$$x_{m'} = x^m = (x - x_m)^v = (x < x_m)$$
$$x_m = (x - x_m)^u = (x > x_m)$$
$$x_{m'.n} = x_n^m = (x - x_m)^v (x - x_n)^u = (x < x_m)(x > x_n)$$
$$x_{m'.n'.r.s} = x_{r.s}^{m.n} = (x < x_m)(x < x_n)(x > x_r)(x > x_s).$$

30. L'emploi des symboles x^m et x_m tantôt comme propositions, tantôt comme nombres ou quantités, n'amène aucune confusion d'idées; car le *contexte* indique toujours la vraie signification. Par exemple, si nous avons $x_6 = \frac{3}{2}$, $x_7 = \frac{2}{3}$, nous déduisons (voir § 27, exemple 5) :

$$(6x^2 - 13x + 6)^u = x_6 + x^7.$$

---

[1]. Les symboles x^m, x_n^m, $x_{r.s}^{m.n}$, etc., comme synonymes respectifs des propositions $x_{m'}$, $x_{m'.n}$, $x_{m'.n'.r.s}$, m'ont été proposés par M. L. Couturat. Ils occupent moins de place, et sont souvent commodes pour d'autres raisons.

Ici nous avons trois propositions, dont chacune emploie le signe d'égalité ou équivalence. Dans les deux premières, puisque $\frac{3}{2}$ et $\frac{2}{3}$ sont des nombres, les symboles x_6 et x_7 représentent aussi des nombres; mais dans la troisième, puisque $(6x^2 - 13x + 6)^u$ est une proposition, les deux termes de la disjonction équivalente, $x_6 + x^7$, sont aussi des propositions; et cette disjonction affirme que x est *ou* au-dessus de $\frac{3}{2}$ *ou* au-dessous de $\frac{2}{3}$.

31. Les opérations de ce calcul sont fondées principalement sur les trois formules suivantes :

(1) $\qquad x^{m.n} = x^m \alpha + x^n \beta$
(2) $\qquad x_{m.n} = x_m \alpha + x_n \beta$
(3) $\qquad x_n^m = x_n^m (x_m - x_n)^u.$

Dans les deux premières formules, le symbole α affirme que x_m est *la limite la plus proche*, et le symbole β que x_n est la limite la plus proche (voir §§ 33, 35, 36). C'est-à-dire, dans la première formule, $\alpha = (x_m - x_n)^v$, $\beta = (x_n - x_m)^v$; tandis que dans la seconde, $\alpha = (x_m - x_n)^u$, $\beta = (x_n - x_m)^u$. De la même manière on trouve [1] :

(4) $\qquad x^{m.n.r} = x^m \alpha + x^n \beta + x^r \gamma$
(5) $\qquad x_{m.n.r} = x_m \alpha + x_n \beta + x_r \gamma$

Dans la formule (4) :

$$\alpha = (x_m - x_n)^v (x_m - x_r)^v, \ \beta = (x_n - x_m)^v (x_n - x_r)^v, \text{ etc.}$$

---

[1]. Il ne faut pas oublier que, dans toutes ces formules, les symboles x^m, x_m, x^n, x_n, etc., représentent des *propositions* et *non* des nombres.

LA LOGIQUE SYMBOLIQUE ET SES APPLICATIONS

Dans la formule (5) :

$$\alpha = (x_m - x_n)^u (x_m - x_r)^u,\ \beta = (x_n - x_m)^u (x_n - x_r)^u,\ \text{etc.}$$

et ainsi de suite pour n'importe combien de limites, supérieures ou inférieures.

32. La limite *zéro* se présente si souvent, qu'au lieu de la représenter par x_1, ou x_2, ou x_3, etc., comme d'autres limites, il est commode de la représenter par un symbole x_0 ou y_0 ou z_0, etc., selon la variable en question. Ainsi le symbole x_0 comme *limite* est synonyme de *zéro*; mais, comme *proposition*, x_0 est synonyme de x^u, et affirme que x est *positif*. Pareillement, $x_{0'}$, ou son synonyme x^0, affirme que *zéro* est *une limite supérieure* de x; c'est-à-dire, que x est *négatif*. Il est à noter que l'exposant 0 dans ce cas n'a pas du tout le même sens que dans les définitions précédentes (voir §§ 12 et 19). C'est seulement dans le calcul des *limites*, et quand il y a une *table de référence*, que x^0 est synonyme de x^v, et affirme que x est négatif. Dans les autres cas, x^0 affirme que x est *zéro*, c'est-à-dire *non-existant*.

33. Comme exemple géométrique des formules du § 31, prenons le diagramme (*Fig. 9*). Supposons que les équations $y = \varphi_1(x)$, $y = \varphi_2(x)$, $y = \varphi_3(x)$ soient vraies pour chaque point sur les courbes y_1, y_2, y_3 respectivement; que $y = 0$ soit vraie pour les points sur la ligne droite y_0; et que $x = 0$, $x = x_1$, $x = x_2$, $x = x_3$, $x = x_4$ soient vraies pour les lignes droites x_1, x_2, x_3, x_4 respectivement. Il est évident que (avec les conventions habituelles à l'égard des directions positives ou négatives) la proposition $y_0^{1,2,3} x_0^4$ est

applicable à tout point qui se trouve sur la surface ombrée, et à aucun point hors de cette surface (voir § 32). Il est évident aussi que :

$$y'^{1.2.3}_0 x^4_0 = y^4_0 x^4_0 + y^2_0 x^3_1 + y^3_0 x^4_3$$

car la proposition $y^4_0 x^4_0$ est vraie pour tout point de la surface limitée par les lignes y_1, y_0, x_1, x_0; la proposition $y^2_0 x^3_1$

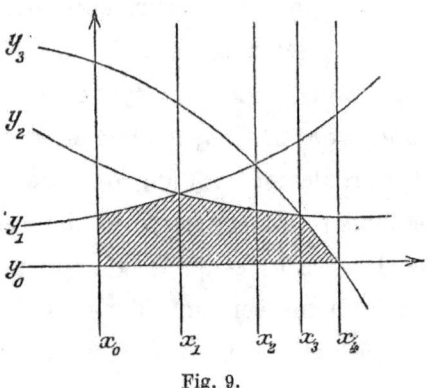

Fig. 9.

est vraie pour tout point de la surface limitée par les lignes y_2, y_0, x_3, x_1; et la proposition $y^3_0 x_3^4$ est vraie pour tout point de la surface limitée par les lignes y_3, y_0, x_4, x_3; et ces trois surfaces ensemble forment la surface totale contenant tous les points pour lesquels séparément la proposition $y^{1.2.3}_0 x^4_0$ est vraie. Il est à remarquer que $x^3_2 + x^2_1 = x^3_1$, de sorte que la limite x_2 disparaît.

34. Avant de donner des exemples de l'application de cette méthode à des problèmes qui relèvent du Calcul intégral, je vais expliquer certains symboles d'abréviations que j'ai employés pour la première fois, il y a seize ans,

LA LOGIQUE SYMBOLIQUE ET SES APPLICATIONS

dans mon mémoire sur *The Limits of Multiple Integrals*[1].

Les symboles $\varphi(x)_{m'.n}$ et $x_{m'.n}\,\varphi(x)$ diffèrent en signification. Le premier est synonyme de l'intégrale $\int_{x_n}^{x_m}\varphi(x)\,dx$; tandis que le second est synonyme de $\varphi(x_m) - \varphi(x_n)$. Par exemple, soit $\int \varphi(x)\,dx = \psi(x)$. D'après la convention proposée, nous aurons :

$$\varphi(x)\,x_{m'.n} = x_{m'.n}\,\psi(x) = \psi(x_m) - \psi(x_n).$$

Supposons, par exemple, que nous ayons à évaluer (voir § 32) l'intégrale $\int_{z_2}^{z_1} dz \int_{y_2}^{y_1} dy \int_{x_0}^{x_1} dx$, les limites étant

TABLE DE RÉFÉRENCE

$z_1 = y$	$y_1 = x$	$x_1 = a$
$z_2 = c$	$y_2 = b$	$x_0 = 0$

données comme dans la table ci-jointe. Le procédé, très détaillé, est le suivant (voir § 32)[2].

Intégrale $z_{1'.2}\,y_{1'.2}\,x_{1'.0} = (z_1 - z_2)\,y_{1'.2}\,x_{1'.0} = (y - c)\,y_{1'.2}\,x_{1'.0}$

$= y_{1'.2}\left(\dfrac{1}{2}y^2 - cy\right)x_{1'.0} = \left\{\left(\dfrac{1}{2}y_1^2 - cy_1\right) - \left(\dfrac{1}{2}y_2^2 - cy_2\right)\right\}x_{1'.0}$

$= \left\{\left(\dfrac{1}{2}x^2 - cx\right) - \left(\dfrac{1}{2}b^2 - cb\right)\right\}x_{1'.0} = \left(\dfrac{1}{2}x^2 - cx - \dfrac{1}{2}b^2 + bc\right)x_{1'.0}$

$= x_{1'.0}\left(\dfrac{1}{6}x^3 - \dfrac{1}{2}cx^2 - \dfrac{1}{2}b^2 x + bcx\right) = \dfrac{1}{6}a^3 - \dfrac{1}{2}ca^2 - \dfrac{1}{2}b^2 a + bca.$

---

1. Voir les *Proceedings of the London Mathematical Society* (13 nov. 1884).
2. Quand le symbole $x_{m'.n}$ exprime (comme ici) une *intégrale* à évaluer, et non une *proposition*, on peut laisser l'accent sous-entendu; mais dans ce cas, $x_{m.n}$ ne sera pas synonyme de $x_{n.m}$, mais de $-x_{n.m}$. (Voir § 38).

H. MAC COLL

35. Comme exemple de ce calcul, prenons le problème suivant : Chacune des trois variables x, y, z est prise au hasard entre 1 et 0; trouver la probabilité que la fraction $\dfrac{z(1-x-y)}{1-y-yz}$ soit aussi entre 1 et 0.

Solution : Supposons que le symbole Q affirme que l'événement en question arrivera, et que A affirme nos données, à savoir que x, y, z sont chacun entre 1 et 0.

TABLE DE RÉFÉRENCE

$x_1 = 1$	$y_1 = 1$	$z_1 = 1$
$x_2 = 1 - y$	$y_2 = \dfrac{1}{1+z}$	
$x_3 = \dfrac{y+z-1}{z}$	$y_3 = 1 - z$	

Nous avons à trouver la probabilité $\dfrac{Q}{A}$. Cette probabilité est évidemment une fraction dont le numérateur est l'intégrale $\int dx \int dy \int dz$ limitée par la condition QA, et le dénominateur la même intégrale limitée seulement par la condition A. Or, $A = x_0^1 \, y_0^1 \, z_0^1$, de sorte que l'intégrale $A \int dx \int dy \int dz = \int_0^1 dx \int_0^1 dy \int_0^1 d = 1$.

Donc $\dfrac{Q}{A} = AQ \int dx \int dy \int dz$, et nous n'avons qu'à trouver les limites compatibles avec la condition AQ.

Soit N le numérateur $z(1-x-y)$, et soit D le dénominateur $1-y-yz$. Alors (voir § 27) :

$$Q = N^u D^u (N-D)^v + N^v D^v (N-D)^u.$$

LA LOGIQUE SYMBOLIQUE ET SES APPLICATIONS

Nous pouvons prendre l'ordre de variation que nous voudrons. Prenons l'ordre x, y, z; et exprimons d'abord les limites de x en termes de y et de z; puis les limites de y en termes de z; et finalement les limites de z numériquement. Chaque limite doit être enregistrée dans la *table de référence* sitôt qu'elle sera trouvée, de sorte que la table se développe au courant de l'opération. Puisque z est positif[1] :

$$N^u = (1-x-y)^u = \{x-(1-y)\}^v = (x-x_2)^v = x^2, N^v = x_2$$
$$D^u = (1-y-yz)^u = \{y(1+z)-1\}^v = (y-y_2)^v = y^2, D^v = y_2$$
$$(N-D)^v = (z-zx+y-1)^v = (zx-y-z+1)^u = \left(x - \frac{y+z-1}{z}\right)^u = x_3$$
$$(N-D)^u = (z-zx+y-1)^u = x^3.$$

En substituant ces résultats dans notre expression pour Q, nous aurons :

$$Q = x^2 y^2 x_3 + x_2 y_2 x^3 = x_3^2 y^2 + x_2^3 y_2.$$

En multipliant par la certitude (ou donnée) x_0^1, nous aurons :

$$Q x_0^1 = x_{3.0}^{2.1} y^2 + x_{2.0}^{3.1} y_2$$

En appliquant les deux formules du § 31, nous trouvons (voir §§ 3 et 32)[2] :

$$x^{2.1} = x^2 (x_2 - x_1)^v + x^1 (x_1 - x_2)^v = x^2 \varepsilon + x^1 \eta = x^2$$
$$x_{3.0} = x_3 (x_3 - x_0)^u + x_0 (x_0 - x_3)^u = x_3 y_3 + x_0 y^3$$

---

1. Il ne faut pas oublier que x^n et x_n comme *propositions* veulent dire respectivement $x_n > x$ et $x_n < x$. (Voir § 30.) Le symbole ε désigne une *certitude*, et η une *impossibilité*.

2. Mais il est évident (sans avoir recours au § 31) que $x^{2.1} = x^2$, et que $x_{2.0} = x_2$, puisque x^2 implique x^1 et que x_2 implique x_0, par simple inspection de la table de référence.

H. MAC COLL

$$x^{3.4} = x^3(x_3-x_1)^v + x^1(x_1-x_3)^v = x^3\varepsilon + x^1\eta = x_3$$
$$x_{2.0} = x_2(x_2-x_0)^u + x_0(x_0-x_2)^u = x_2\varepsilon + x_0\eta = x_2.$$

En substituant ces résultats dans l'expression pour Qx_0^1, nous aurons :

$$Qx_0^1 = x^2(x_3y_3 + x_0y^3)y^2 + x_2^3y_2 = x_3^3y_3^2 + x_0^3y^{32} + x_2^3y_2.$$

Nous devons maintenant appliquer la formule 3 du § 31 aux propositions x_3^2, x_0^2, x_2^2. Mais $(x_3-x_1)^u = y^2$, $(x_1-x_0)^u = \varepsilon$ (une *certitude*), et $(x_3-x_2)^u = y_2$; de sorte que l'application de cette formule n'introduit aucune nouvelle limite en y, et ne supprime aucun terme en le prouvant impossible. Nous avons donc trouvé toutes les limites de x, et nous devons maintenant trouver les limites de y. En multipliant par la donnée y_0^1, nous aurons :

$$Qx_0^1y_0^1 = x_3^2y_{3.0}^{2.1} + x_0^2y_0^{3.2.1} + x_2^3y_{2.0}^1.$$

En appliquant les formules du § 31, ou par simple inspection de la table de référence, nous trouvons :

$$y^{2.1} = y^2, \qquad y_{3.0} = y_3, \qquad y^{3.2.1} = y^3, \qquad y_{2.0} = y_2$$

et, en substituant ces résultats dans le second membre nous aurons :

$$Qx_0^1y_0^1 = x_3^2y_3^2 + x_0^2y_0^3 + x_2^3y_2^1.$$

L'application de la formule $y_n^m = y_n^m(y_m - y_n)^u$ n'introduit aucune nouvelle limite en z, et ne supprime aucun terme en le prouvant impossible. Nous avons donc trouvé toutes les

LA LOGIQUE SYMBOLIQUE ET SES APPLICATIONS

limites de y, et il ne reste qu'à trouver les limites de z. En multipliant par la donnée z_0^1, nous aurons :

$$QA = Q x_0^1 y_0^1 z_0^1 = (x_3^2 y_3^2 + x_0^2 y_0^3 + x_2^3 y_2^1) z_0^1.$$

L'application de la formule 3 du § 31 au facteur z_0^1 ne produit aucun changement, puisque $(z_1 - z_0)^u$ est une certitude formelle. Le procédé pour trouver les limites est donc fini. Il reste à évaluer les trois intégrales simples dont la somme est $QA \int dx \int dy \int dz$ (voir § 34). Le résultat est $\frac{5}{4} - log\, nép.\, 2$, qui est un peu supérieur à $\frac{5}{9}$.

36. Le problème suivant[1] est un peu plus compliqué : Étant donné que a est positif, que n est un nombre entier et positif, et que les variables x et y sont prises chacune au hasard entre a et $-a$, quelle est la probabilité que $(x+y)^n - a$ soit négatif et $(x+y)^{n+1} - a$ positif ?

Solution : Soit A notre donnée $y_2^1 x_2^1$ (voir la table de référence) qui affirme que y et x sont entre a et $-a$; soit Q la proposition $\{(x+y)^n - a\}^v$, et R la proposition $\{(x+y)^{n+1} - a\}^u$, les exposants u et v ayant les mêmes significations que dans les conventions précédentes. Nous avons à trouver la probabilité $\frac{QR}{A}$. Puisque toutes les valeurs, positives ou négatives, de y et de x entre a et $-a$ sont également probables, la probabilité demandée est $\frac{QRA \int dy \int dx}{A \int dy \int dx}$; et nous n'avons

---

[1] C'est encore un problème de probabilité; mais il est évident que la méthode est aussi applicable à d'autres espèces de problèmes.

qu'à trouver les limites d'intégration. Cette fois-ci nous prendrons l'ordre de variation y, x.

TABLE DE RÉFÉRENCE

$y_1 = a$	$x_1 = a$	$a_1 = 1$
$y_2 = -a$	$x_2 = -a$	$a_2 = \dfrac{1}{2^{\frac{n}{n-1}}}$
$y_3 = a^{\frac{1}{n}} - x$	$x_3 = a^{\frac{1}{n}} - a$	$a_3 = \dfrac{1}{2^{\frac{n+1}{n}}}$
$y_4 = -a^{\frac{1}{n}} - x$	$x_4 = a - a^{\frac{1}{n}} = -x_3$	
$y_5 = a^{\frac{1}{n+1}} - x$	$x_5 = a + a^{\frac{1}{n+1}}$	
$y_6 = -a^{\frac{1}{n+1}} - x$	$x_6 = a + a^{\frac{1}{n}}$	
	$x_7 = a^{\frac{1}{n+1}} - a$	
	$x_8 = -a - a^{\frac{1}{n+1}} = -x_5$	
	$x_9 = a - a^{\frac{1}{n+1}} = -x_7$	

Supposons d'abord que n soit un nombre pair. Alors,

$$Q = \left\{(x+y) - a^{\frac{1}{n}}\right\}^v \left\{(x+y) + a^{\frac{1}{n}}\right\}^u =$$
$$\left\{y - (a^{\frac{1}{n}} - x)\right\}^v \left\{y + (a^{\frac{1}{n}} + x)\right\}^u = y_4^3,$$
$$R = \left\{(x+y) - a^{\frac{1}{n+1}}\right\}^u = \left\{y - (a^{\frac{1}{n+1}} - x)\right\}^u = y_5.$$

Donc $QR = y_{4,5}^3$; et en multipliant par la donnée y_2^1 (car $A = y_2^1 x_1^1$) nous aurons :

$$QRy_2^1 = y_{2,4,5}^{3,4} = (y^3 x_3 + y^1 x)(y_2 x_{4,5} + y_5 x^5)$$
$$= (y^3 x_3 + y^1 x^3)(y_2 x_5 + y_5 x^5), \qquad (\text{car } x_{4,5} = x_5)$$
$$= y_2^3 x_5 + y_5^3 x_3^5 + y_5^1 x^3 \ (\text{car } x_{5,3} = x_5, x_5^3 = \eta, x^{3,5} = x^3).$$

Maintenant nous devons appliquer la formule 3 du § 31 aux

LA LOGIQUE SYMBOLIQUE ET SES APPLICATIONS 179

expressions en y. Ainsi, $y_2^2 = y_2^1 x^5$, $y_5^2 = y_5^1 a_1$, $y_5^1 = y_5^1 x_7$.
En substituant ces résultats, nous aurons :

$$QR y_2^1 = y_2^1 x_5^6 + y_5^3 x_3^5 a_1 + y_5^4 x_7^3.$$

Nous avons maintenant trouvé les limites de la variable y; et nous devons appliquer les formules 1, 2, 3 du § 31 aux expressions en x. En multipliant par la donnée x_2^1, nous trouvons :

$$QR y_2^1 x_2^1 = y_2^3 x_{5.2}^{1.6} + y_5^3 x_{3.2}^{1.5} a_1 + y_5^1 x_{7.2}^{3.1} = y_5^3 x_3^1 a_1 + y_5^1 x_7^{3.1}$$

Car $x_5^1 = \eta$ (une impossibilité); $x^{1.5} = x^1$, $x_{3.2} = x_3$, $x_{7.2} = x_7$. On arrive à ces résultats tout de suite par simple inspection de la table de référence, sans être obligé d'employer les formules du § 31. En appliquant ces formules aux expressions en x qui restent, nous trouvons :

$$x^{3.1} = x^3 a_1 + x^1 a^3, \quad x_3^1 = x_3^1 a_2, \quad x_7^3 = x_7^3 a_1, \quad x_7^1 = x_7^1 a_3.$$

En substituant ces valeurs, il vient :

$$QR y_2^1 x_2^1 = QRA = y_5^3 x_3^1 a_{1.2} + y_5^1 (x_7^3 a_{2.1} + x_7^1 a_3^2)$$
$$= y_5^3 x_3^1 a_1 + y_5^1 x_7^3 a_1 (y_5^3 x_3^1 + y_5^1 x_7^3) a_1;$$

$$a_{1.2} = a_1 = a_{2.1}, \quad \text{et} \quad a_3^2 = \eta \text{ (une impossibilité)}$$

C'est là la démarche finale pour trouver les limites; et le résultat nous apprend que, *quand n est un nombre pair*, QRA n'est possible que dans le cas où a_1 est une limite inférieure de a. Autrement dit, quand n est un nombre pair, et que a n'est pas > 1, la probabilité de QR est *zéro*. Pour trouver la probabilité quand n est un nombre pair et $a > 1$, on n'a qu'à évaluer les intégrales, en employant la notation et la méthode du § 31 :

$$\text{Int A} = \text{int } y_{1'.2}\, x_{1'.2} = (y_1 - y_2)\, x_{1'.2} = (2a)\, x_{1'.2} = x_{1'.2}(2a)x = 4a^2$$

$$\text{int QRA} = y_{3'.5}\, x_{1'.3} + y_{1'.5}\, x_{3'.7} = (y_3 - y_5)\, x_{1'.3} + (y_1 - y_5)\, x_{3'.7}$$

$$= (a^{\frac{1}{n}} - a^{\frac{1}{n+1}})\, x_{1'.3} + (a - a^{\frac{1}{n+1}} + x)\, x_{3'.7}$$

$$= x_{1'.3}(a^{\frac{1}{n}} - a^{\frac{1}{n+1}})\, x + x_{3.7}\left(ax - a^{\frac{1}{n+1}}x + \frac{1}{2}x^2\right)$$

$$= (a^{\frac{1}{n}} - a^{\frac{1}{n+1}})(x_1 - x_3) + (a - a^{\frac{1}{n+1}})(x_3 - x_7) + \frac{1}{2}(x_3^2 - x_7^2)$$

$$= (a^{\frac{1}{n}} - a^{\frac{1}{n+1}})\left(2a - \frac{1}{2}a^{\frac{1}{n}} - \frac{1}{2}a^{\frac{1}{n+1}}\right)$$

$$\frac{\text{int QRA}}{\text{int A}} = \frac{\text{int QRA}}{4a^2} = \frac{1}{8a^2}(a^{\frac{1}{n}} - a^{\frac{1}{n+1}})(4a - a^{\frac{1}{n}} - a^{\frac{1}{n+1}}) = \frac{\text{QR}}{\text{A}}$$

Il reste à trouver la probabilité quand n est un nombre *impair*. Par le même procédé qu'avant nous trouvons :

$$\text{QRA} = (y_5^3\, x_3^1 + y_5^1\, x_7^3)\, a_1 + y_2^6\, x_2^0\, a_3.$$

Ici nous avons *deux* limites inférieures de a, de sorte que le procédé n'est pas encore fini. Pour séparer les différents cas possibles, il faut multiplier le résultat obtenu par la certitude[1] $(a_1 + a^1)(a_3 + a^3) = a_1 + a_3^1 + a^3$ (puisque $a_1 > a_3$). En représentant le coefficient de a_1 par M_1, et celui de a_3 par M_3, nous aurons :

$$\text{QRA} = (M_1 a_1 + M_3 a_3)(a_1 + a_3^1 + a^3) = (M_1 + M_3)a_1 + M_3 a_3^1;$$

car $a_{1.3} = a_1$, et $a_1^3 = \eta$ (une impossibilité). Donc il n'y a que

---

1. Je dis « certitude », parce que la probabilité que x ou y soit *exactement* a_1 ou a_3 (au lieu d'être au-dessus ou au-dessous) est infinitésimale.

LA LOGIQUE SYMBOLIQUE ET SES APPLICATIONS

deux cas possibles quand n est un nombre *impair* : le cas a_1 (c'est-à-dire $a > 1$), et le cas a_3^1. Pour le dernier, a_3^1 :

$$\frac{QR}{A} = \frac{\operatorname{int} M_3}{\operatorname{int} A} = \frac{1}{8a^2}\left(2a - a^{\frac{1}{n+1}}\right)^2.$$

Pour le premier cas, où $a > 1$:

$$\frac{QR}{A} = \frac{\operatorname{int}(M_1 + M_3)}{\operatorname{int} A}$$
$$= \frac{1}{4a^2}(a^{\frac{1}{n}} - a^{\frac{1}{n+1}})(2a - a^{\frac{1}{n}}) + \frac{1}{8a^2}\left(a^{\frac{1}{n}} - a^{\frac{1}{n+1}}\right)^2 + \frac{1}{8a^2}\left(2a - a^{\frac{1}{n+1}}\right)^2.$$

Pour le cas $a < a_3$, la probabilité est *zéro*.

37. L'avantage principal de cette méthode de trouver les limites est qu'elle est indépendante de tout diagramme, et que, par conséquent, on peut l'appliquer avec un succès certain, non seulement à des expressions de deux ou trois variables, mais aussi à des expressions de quatre ou plusieurs variables. Quand il n'y a que deux variables, une méthode graphique est souvent plus expéditive, pourvu qu'on n'ait affaire qu'à des lignes droites, ou à des courbes bien connues et faciles à tracer, comme le cercle, la parabole, etc. Mais une méthode graphique de trouver les limites est, en général, difficile quand il y a trois variables, et impossible quand il y en a quatre.

38. Dans mon mémoire sur *The Limites of Multiple Integrals*[1], on trouvera plusieurs méthodes pour abréger et simplifier ce calcul de limites. Par exemple, *quelles que*

---

1. Publié dans les *Proceedings of the London Mathematical Society*, t. XVI (13 nov. 1884).

soient les limites désignées par les symboles, l'intégrale à trois termes :

$$z_{3'.0}\, y_{3'.0}\, x_{1'.3} + z_{1'.0}\, y_{1'.3}\, x_{1'.3} + z_{3'.0}\, y_{1'.0}\, x_{3'.0}$$

équivaut à l'intégrale

$$z_{3'.0}\, y_{1'.0}\, x_1 - z_{3'.1}\, y_{3'.1}\, x_{3'.1},$$

qui ne contient que deux termes, et qui sera, en général, plus facile à évaluer. Ces simplifications sont fondées sur les formules :

(1) $x_{m'.n} = -x_{n'.m}$, (2) $x_{m'.n} = x_{m'.r} + x_{r'.n}$,

(3) $x_{m'.n} + x_{r'.s} = x_{m'.s} + x_{r'.n}$, (4) $x_{m'.n}\, x_{r'.s} = x_{n'.m}\, x_{s'.r}$,

mais, en général, elles peuvent être effectuées plus facilement par une méthode graphique, dont on trouvera l'explication dans le mémoire précité.

Conclusion.

Si je me suis abstenu de tout commentaire sur les recherches de mes confrères, ce n'est pas parce que je n'en reconnais pas l'importance, mais simplement parce que je n'ai pas encore pu leur consacrer l'attention et l'étude qu'elles méritent. Celles qui ne sont écrites ni en français ni en anglais me sont inaccessibles sans l'aide d'un traducteur; je n'en suis pas moins reconnaissant aux auteurs étrangers qui ont eu l'amabilité de m'envoyer des copies de leurs travaux[1]. En comparant les avantages de différentes

---

1. Si ces mémoires étaient écrits dans une langue internationale possédant la merveilleuse simplicité de l'*Esperanto* (dont j'ai tout récemment appris les principes), je n'hésiterais pas à consacrer à cette langue les quelques semaines nécessaires pour l'apprendre.

LA LOGIQUE SYMBOLIQUE ET SES APPLICATIONS 183

méthodes, on trouve souvent qu'une méthode A résout facilement certains problèmes qu'une autre méthode B ne résout pas, ou résout difficilement; tandis que pour d'autres problèmes c'est la méthode B qui a l'avantage. Mon but a été surtout d'adapter mes méthodes aux problèmes de probabilités et de limites, à cause de l'importance de ces problèmes dans beaucoup de recherches scientifiques. C'est, en effet, à un problème de ce genre qu'est due l'origine de mon système symbolique; et c'est encore dans son application à ces espèces de problèmes qu'on peut le mieux en voir l'utilité [1].

---

[1]. On en trouvera des exemples assez nombreux dans les articles que j'ai publiés à différentes époques depuis 1877 dans les *Proceedings of the London Mathematical Society*, dans le *Mind*, et dans les *Mathematical Questions and Solutions from the Educational Times*. Voir l'histoire du développement dè ces méthodes dans les *Memoirs of the Manchester Literary and Philosophical Society*, 3ᵉ série, t. VII.

[1902i]: Symbolic[al] Reasoning (IV). *Mind,* Vol. 11, pp. 352-368.

III.—SYMBOLIC REASONING (IV.).[1]

By Hugh MacColl.

Pure, Abstract, or General Logic.

1. THE simplest, the most general, and the most easily applicable kind of logic is the logic of *statements* or *propositions*. To this, and to this alone, can we correctly give the name of *pure logic*. Unlike all other kinds, it has the immense advantage of being independent of the accidental conventions of language. How dependent other systems are on linguistic conventions is shown by the importance they attach to the grammatical distinction between subject and predicate (see §§ 3, 4, 11). In pure logic (as I understand it) "A struck B" and "B was struck by A" are exact equivalents, and any symbol we choose to represent the one may also be employed to represent the other. So in mathematics. The statements "A is greater than B," "B is less than A," "$A - B$ is positive," "$B - A$ is negative," are all four equivalent; and any symbol, $A > B$, or $B < A$, or $(A - B)^P$, or $(B - A)^N$, used to express one of them, will also express any of the others.

2. Statements or propositions are the *indispensable units* of every argument. If one of these units be ambiguous or wanting in clearness, the validity of the argument becomes doubtful. We then discuss the meaning of this faulty unit, taking for our data the grammatical and other linguistic conventions of the tongue employed; and this discussion again must be carried on *by means of propositions*.

3. It is generally assumed that a proposition must consist of a subject and a predicate. That, however, is a matter of convention or definition. If I accept it, I must in my system make a distinction between the two words *statement* and *proposition*. Let me therefore define a *statement* as *any sound or symbol* (or collection of sounds or symbols) *employed to give information*. In this sense the warning "Caw" of a

[1] For III. see MIND, January, 1900.

SYMBOLIC REASONING. 353

sentinel rook, and the Union Jack floating from the mast of a passing ship, are statements. The former is equivalent to "Beware; I see a man coming with a gun"; the latter is equivalent to "This is a British ship". These are *elementary* statements—statements that cannot be separated into subject and predicate. In the evolution of human language, that division came later (see § 30).

4. A *proposition* I define as a statement of the form A^B, in which A is the subject, and B the predicate. Thus, every proposition is a statement; but every statement is not a proposition. Let A = Alexander, and B = baker. The proposition A^B asserts that *Alexander is a baker*. If we represent a proposition A^B by a single letter a, we may then (considering the *form* alone) say that a is a statement but not a proposition; whereas A^B, by our definition, is both (see §§ 3, 25). Let B_1, B_2, B_3, etc., be the separate individuals that constitute the class B. Then

$$A^B = A^{B_1} + A^{B_2} + A^{B_3} + \text{etc.}$$

That is to say (giving the same meanings to A and B as before), the statement that *Alexander is a baker* is equivalent to the statement that *Alexander is either Baker No. 1, or Baker No. 2, or Baker No. 3*, etc.

5. Let A = animal, and let B = brown; also let n be the total number of animals under consideration. Then the symbol $A_1^B A_2^B A_3^B \ldots A_n^B$ asserts that A_1 is brown, that A_2 is brown, etc.; that is to say, it asserts that *All the animals of our limited universe are brown*. The symbol $A_1^B + A_2^B + A_3^B + \ldots + A_n^B$, on the other hand, asserts that *one at least* of the animals (either A_1 or A_2 or A_3, etc.) is brown.

6. Let A_1, A_2, A_3, etc., be the individuals forming a class A; and let B_1, B_2, B_3, etc., be the individuals forming a class B. Out of the series A_1, A_2, etc., let an individual A be taken at random. The symbol A^B, on this hypothesis, asserts that A is also one of the individuals in the series B_1, B_2, etc. Hence, $A^{B\epsilon}$, which is an abbreviation for $(A^B)^\epsilon$, asserts that the statement A^B is a *certainty* (ϵ). Thus $A^{B\epsilon}$ may be considered as synonymous with the traditional "All A is B," or "Every A is a B". Similarly, $A^{B\eta}$, which asserts that A^B is impossible (η), is equivalent to the "No A is B" of the traditional logic; while $A^{B\eta\iota}$ denies this, and asserts that "*Some* A is B". In like manner, $A^{B\epsilon\iota}$ denies $A^{B\epsilon}$ (that every A is B), and asserts that "Some A is not B". The symbol $A^{B\theta}$ is equivalent to the combination $A^{B\eta\iota} A^{B\epsilon\iota}$, and asserts that A^B is possible but uncertain; that is, it asserts that

23

one A at least is B, but that every A is not B. Thus $A^{B\epsilon} = A_1{}^B A_2{}^B A_3{}^B \ldots A_n{}^B$, the number n being the number of individuals in the universe A_1, A_2, A_3, etc. Similarly we get $A^{B\eta} = A_1{}^{B\iota} A_2{}^{B\iota} A_3{}^{B\iota} \ldots A_n{}^{B\iota}$; that is, $A^{B\eta}$ asserts that A_1 is not B, that A_2 is not B, and so on till the last A_n. Hence

$$A^{B\theta} = A^{B\eta\iota} A^{B\epsilon\iota} = (A_1{}^{B\iota} A_2{}^{B\iota} A_3{}^{B\iota} \ldots A_n{}^{B\iota})^\iota (A_1{}^B A_2{}^B A_3{}^B \ldots A_n{}^B)^\iota$$
$$= (A_1{}^B + A_2{}^B + A_3{}^B + \ldots + A_n{}^B)(A_1{}^{B\iota} + A_2{}^{B\iota} + A_3{}^{B\iota} + \ldots + A_n{}^{B\iota}).$$

That is to say, $A^{B\theta}$ asserts, firstly, that one at least of the series A_1, A_2, A_3, etc., is B, and, secondly, that one at least is not B. Out of the n^2 terms in the product of the last two bracket-statements, n terms, namely, $A_1{}^B A_1{}^{B\iota}$, $A_2{}^B A_2{}^{B\iota}$, etc., may be omitted as self-contradictory; for $A_x{}^B$ (which is an abbreviation for $A_x{}^{B\tau}$) asserts that $A_x{}^B$ is *true*, and $A_x{}^{B\iota}$ asserts that $A_x{}^B$ is *false*. Thus the syllogisms *Barbara* and *Frissison* may be expressed respectively by

$$A^{B\epsilon} B^{C\epsilon} : A^{C\epsilon} \text{ and } C^{B\eta} B^{A\eta\iota} : A^{C\epsilon\iota}.$$

7. But a far simpler, more symmetrical, and more general way of treating the syllogism is to regard it from the point of view of pure or abstract logic. From this point of view all valid syllogisms are but particular cases of the general formula, or formal certainty (see §§ 31, 32)

$$(x : y)(y : z) : (x : z),$$

which I will represent by $\phi(x, y, z)$, or briefly ϕ, and which may be read "If whenever the statement x is true, y is true, and whenever y is true, z is true; then whenever x is true, z is true". It may also be read as "If x implies y, and y implies z, it follows that x implies z". The symbol ϕ (A, B, C) will then denote what ϕ becomes when any statement A is put for x, B for y and C for z (see § 12). Out of our universe of discourse, consisting say of the individuals P_1, P_2, P_3, etc., let an individual P be taken at random; and let the symbols A, B, C, as *statements*, assert respectively that P will belong to the class A, that P will belong to the class B, that P will belong to the class C; while A′, B′, C′ will be the respective *denials* of these statements. It is evident that, assuming[1] the existence of the classes A, B, C in our universe P_1, P_2, P_3, etc., and considering those syllogisms equivalent which have equivalent premisses and the same or equivalent conclusions, we shall have

---

[1] This assumption of existence is not necessary except in the case of Darapti, Felapton, Fesapo and Bramantip.

SYMBOLIC REASONING. 355

Barbara $= \phi(A, B, C)$
Celarent $=$ Cesare $= \phi(A, B, C')$
Darii $=$ Datisi $= \phi(B, C, A')$
Ferio $=$ Festino $=$ Ferison $=$ Fresison $= \phi(A, C, B')$
Camestres $=$ Camenes $= \phi(A, B', C')$
Disamis $=$ Dismaris $= \phi(B, A, C')$
Baroko $= \phi(A, C, B)$
Bokardo $= \phi(B, A, C)$
Darapti $= \phi(B, AC, \eta)$
Felapton $=$ Fesapo $= \phi(B, AC', \eta)$
Bramantip $= \phi(C, BA', \eta)$.

8. All these can be easily proved; but to show the method of bringing all within the sweep of the general formula $\phi(x, y, z)$, it will be enough to prove three, namely, *Fresison*, *Darapti* and *Bramantip*.

Fresison $= (C : B')(B : A')' : (A : C)'$
$= (C : B')(B : A')'(A : C) : \eta$
$= (A : C)(C : B') : (B : A')$
$= (A : C)(C : B') : (A : B') = \phi(A, C, B')$
Darapti $= (B : C)(B : A) : (A : C')'$.

But, since the classes A, B, C are understood throughout to exist in our universe P_1, P_2, P_3, etc., we have $\epsilon = A^\eta = B^\eta = C^\eta$. Hence

Darapti $= B^\eta(B : C)(B : A) : (A : C')'$
$= B^\eta(B : CA) : (A : C')'$
$= B^\eta(B : CA)(A : C') : \eta$
$= B^\eta(B : CA)(AC : \eta) : \eta$
$= (B : AC)(AC : \eta) : B^\eta$
$= (B : AC)(AC : \eta) : (B : \eta) = \phi(B, AC, \eta)$
Bramantip $= (C : B)(B : A) : (A : C')'$
$= C^\eta(C : B)(B : A) : (A : C')'$, since $C^\eta = \epsilon$
$= C^\eta(C : B)(B : A) : (C : A')'$
$= C^\eta(C : B)(C : A')(B : A) : \eta$
$= C^\eta(C : BA')(BA' : \eta) : \eta$
$= (C : BA')(BA' : \eta) : (C : \eta) = \phi(C, BA', \eta)$.

9. It is evident, since $x : y = y' : x'$, that $\phi(x, y, z) = \phi(z', y', x')$; so that all the syllogisms remain valid if we reverse the order of their constituents, provided we at the same time change their signs. For example, Camestres and Camenes may each be expressed, not only in the form $\phi(A, B', C')$, but also in the form $\phi(C, B, A')$.

10. In the syllogisms, the statements A, A', B, B', etc., are understood to be abbreviations for the propositions P^A, $P^{A'}$, P^B, $P^{B'}$, etc., all of which have the same subject P, an

individual taken at random out of the universe of discourse P_1, P_2, P_3, etc. But in the general formula $\phi(x, y, z)$, of which all valid syllogisms are but particular cases, the statements x, y, z need not be understood to refer to the same subject. The formula $\phi(x, y, z)$ holds good whatever be its constituent statements x, y, z, which may, one and all, be *certainties, impossibilities* or *variables*. For example, take the case $x^\eta y^\eta z^\eta$, and suppose $x = \eta_1$, $y = \eta_2$, $z = \eta_3$, we get

$$\phi(\eta_1, \eta_2, \eta_3) = (\eta_1 : \eta_2)(\eta_2 : \eta_3) : (\eta_1 : \eta_3) = \epsilon_1 \epsilon_2 : \epsilon_3 = \epsilon_4;$$

for $\eta : a = (\eta a')^\eta = \eta^\eta = \epsilon$, whatever be the particular impossibility represented by the symbol η out of the series η_1, η_2, η_3, etc., and whatever be the statement a. Next, take the case $x^\epsilon y^\epsilon z^\eta$. Assuming x, y, z to be respectively ϵ_1, ϵ_2, η_1, we get

$$\phi(\epsilon_1, \epsilon_2, \eta_1) = (\epsilon_1 : \epsilon_2)(\epsilon_2 : \eta_1) : (\epsilon_1 : \eta_1) = \epsilon_3 \eta_2 : \eta_3 = \eta_4 : \eta_3 = \epsilon_4;$$

for, as before, $\eta : a = \epsilon$, whatever impossibility η may be out of the series η_1, η_2, η_3, etc., and whatever the statement a (see § 32).

11. There has been much discussion among logicians as to the "existential import of propositions," especially as to whether the proposition "All A is B" implies the existence of the subject A. The question does not appear to me to belong to the province of *pure logic*, which should treat of the relations connecting different classes of propositions, and not of the relations connecting the words of which a proposition is built up. The latter question is one properly of grammar and philology, and not of general or abstract logic. The answer depends upon the meaning we agree to give to the word *exist*. Take, for example, the proposition "*Non-existences* are *non-existent*". This is a self-evident truism; can we affirm that it implies the *existence* of its subject *non-existences?* In pure logic we have $\eta^\eta = \epsilon$, or more briefly $\eta^{\eta\epsilon}$, which asserts that it is certain that an impossibility is an impossibility. In pure logic the subject, being always a *statement, must* exist—that is, it must exist as a *statement*. It may be a certainty, an impossibility, or a variable—it may even (in the circumstances) be unmeaning; yet as a statement it always *exists*. But in pure logic we sometimes have to symbolise statements to which (in the circumstances considered) we can attach no meaning. Such statements belong, *not* to the class η, but to the class o. For example, A^η asserts that A is *impossible*—that is it *contradicts some datum or definition;* whereas A^o asserts that A, in the case considered, is a *meaningless* statement that affirms nothing

SYMBOLIC REASONING. 357

and contradicts nothing (see my recent memoir on "La Logique Symbolique et ses Applications" in the *Bibliothèque du Congrès International de Philosophie*: Librairie Armand Colin).

The Logic of Functions or Relations.

12. A symbol of the form $\phi(x)$ or $\psi(x)$ or $f(x)$, etc., is called a *function of x*. It denotes[1] any statement, or part of a statement, *containing the symbol x*. Similarly, $\phi(x, y)$ or $\psi(x, y)$, etc., is called a *function of x and y*. It denotes any statement, or part of a statement, *containing the symbols x and y*. The symbols $\phi(x, y, z)$, $\psi(x, y, z)$, etc., are to be interpreted in the same manner. The symbol ϕ_x or simply ϕ may be used as an abbreviation for $\phi(x)$. Similarly, $\phi_{x,y}$, or simply ϕ, may be used as an abbreviation for $\phi(x, y)$; and so on. The constituents x, y, z, etc., may each denote a word or collection of words or of other symbols; and they may (as generally with me), or may not (as in mathematics), separately represent complete propositions. When we have any function $\phi(x, y)$, then the symbol $\phi(a, \beta)$ denotes what $\phi(x, y)$ becomes when a is substituted for x, and β for y, *the other words or symbols remaining unchanged*. Similarly, $\phi(x)$ and $\phi(a)$, $\phi(x, y, z)$ and $\phi(a, \beta, \gamma)$, etc., are to be interpreted.

13. For example, let w = whale, h = herring, v = virtue; and let $\phi(w, h)$ denote the proposition "A small *whale* can swallow a large *herring*". Then $\phi(h, w)$ will denote "A small *herring* can swallow a large *whale*," the symbols w and h interchanging places, while the rest of the proposition remains unchanged. It is evident that this convention leads to the conclusion $\phi^\epsilon(w, h)$, $\phi^\eta(h, w)$. That is to say $\phi(w, h)$ is a *certainty* and $\phi(h, w)$ an *impossibility*. We also get $\phi^0_\cap(w, v)$; that is to say, the statement that "A small *whale* can swallow *virtue*" is *meaningless*.

14. The symbols $\phi(x)$, $\phi(x, y)$, etc., may thus be regarded as *blank forms* to be filled up, the blanks being represented by x, y, etc., and the words or other symbols to be substituted by a, β, etc. A statement of the form $\phi(x, y, z)$ may be represented by $\phi(x)$ or ϕ_x when the substitutions for x only have to be considered; by $\phi(x, y)$ or $\phi_{x, y}$ when we have to

[1] The definition of a *function* given here is more general than that given of a function in mathematics; and it includes the mathematical definition. I employed the functional symbol $f(x, y, z)$ to denote the complex implication $(x:y)(y:z):(x:z)$ in my second paper on the "Calculus of Equivalent Statements," published in the *Proceedings of the London Mathematical Society* in 1878.

358   HUGH MACCOLL :

consider the substitutions for x and y; and so on. When we speak of the *form alone*, without referring to any particular substitutions, we may denote the function simply by ϕ.

15. To show how dependent other systems of logic are upon mere linguistic conventions, which differ more or less in different countries, let us take the proposition " If A is the cousin of B, then B is the cousin of A," and denote it by ϕ (A, B). Translating this into French, let ψ (A, B) denote the proposition " Si A est le cousin de B, alors B est le cousin de A"; and suppose A to be a boy, and B to be a girl. We get the paradox ϕ^ϵ (A, B), ψ^η (A, B), which asserts that the English statement ϕ (A, B) is *certainly true* (ϵ), while its French translation ψ (A, B) is *certainly false* (η). For in French, " B est *le cousin* de A " implies that B is of the male sex, which is contrary to our data; whereas in English, " B is *the cousin* of A " implies nothing as to the sex of B.

16. Let ϕ_x, as an abbreviation for $\phi(x)$, denote the implication " If A is x of B, and B is x of C, then A is x of C"; and let a = an ancestor, s = a son, c = a cousin, h = the hat. We get ϕ_a^ϵ ϕ_s^η ϕ_c^θ ϕ_h°. That is ϕ_a is certain, ϕ_s impossible, ϕ_c variable (neither certain nor impossible), and ϕ_h meaningless.

17. According to writers on the *Logic of Relations*, a relation is said to be *transitive*, when the combination of the two propositions " A has the relation R to B, and B has the relation R to C," implies the conclusion that " A has the relation R to C ". Accepting this definition, and denoting the word *transitive* by T, and the proposition " R is transitive " by R^T, while the symbol $\phi_{x,y}$ or its equivalent $\phi(x, y)$, asserts that " x has the relation R to y," we may express the definition symbolically thus

$$R^T = \phi_{A,B} \phi_{B,C} : \phi_{A,C}.$$

But I think it would be simpler, as well as more general, to call, not the *relation* R but the variable *statement* or *function* ϕ, transitive, and to write the definition thus—

$$\phi^T = \phi_{A,B} \phi_{B,C} : \phi_{A,C} \text{ (see § 26)}.$$

18. Let ϕ_x denote the statement that " a has the ratio x to β," in which a and β (with all substitutes for them) are understood to be real positive magnitudes, neither infinite nor zero; in other words, let ϕ_x denote the equational statement ($a = x\beta$), we get by definition of T (see § 17)

$$\phi_x^T : (\phi_{a,\beta} \phi_{\beta,\gamma} : \phi_{a,\gamma}) : (x = 1).$$

That is to say, if the statement which asserts that a has a ratio x to β be *transitive*, the ratio x must be *unity* (*i.e.*, a

SYMBOLIC REASONING. 359

ratio of *equality*), and a must be equal to β. When a is not equal to β, the statement is not transitive (see § 19).

19. The preceding may be proved as follows—

$$(a = x\beta)(\beta = x\gamma) : (a = x^2\gamma).$$

But since $(a = x\beta)$ is, by hypothesis, transitive, we have also

$$(a = x\beta)(\beta = x\gamma) : (a = x\gamma).$$

Hence, from the logical formula $(A : B)(A : C) = (A : BC)$ we get

$$(a=x\beta)(\beta=x\gamma):(a=x^2\gamma)(a=x\gamma):(x^2\gamma=x\gamma):(x^2=x):(x=1)$$

for the supposition $(x = o)$ would contradict our hypothesis that neither a nor β nor γ is zero.

20. Considering the various ratios, $\frac{1}{2}$, $\frac{3}{5}$, $\frac{7}{5}$, etc., as forming a special class of *relations*, the implication

$$(a = x\beta)(\beta = x\gamma) : (a = x^2\gamma)$$

may be read: "If a has the relation x to β, and β the relation x to γ, then a has the relation x^2 to γ". We have proved that when this relation (or the statement asserting it) is transitive, then $x = x^2$, and $a = \beta = \gamma$, it being understood throughout that a, β, γ are real and positive.

21. *Ratio* is thus seen to be a particular species of *relation*. Before I attempt to give a *general* definition of the somewhat vague concept *relation*, let us examine another special case. Suppose ϕ_x to denote the statement "AxB" in which the variable x is to be replaced by some word or words, such as *strikes, was struck by, will speak to*, etc. Let $s =$ struck, $w =$ was struck by, $a =$ will strike, $\beta =$ will be struck by. Also, let ψ_x denote "BxA," what ϕ_x becomes when A and B interchange places. We get

$$(\phi_s = \psi_w)(\phi_w = \psi_s)(\phi_a = \psi_\beta)(\phi_\beta = \psi_a).$$

These four statement-factors are respectively synonyms of of (1) $(AsB = BwA)$; (2) $(AwB = BsA)$; (3) $(AaB = B\beta A)$; (4) $(A\beta B = BaA)$. In the four statements ϕ_s, ψ_w, ϕ_a, ψ_β, A stands in the active relation of *striker* towards B, and B in the passive relation of *being struck* towards A; whereas in the four others these relations are reversed. The relations s and w are therefore reciprocal, and so are a and β. The four relations s, w, a, β, taken in pairs, have also relations to each other. The relations s and a are active, w and β passive; s and w are past, a and β future; s is the *active* of w, w is *passive* of s; a is *future* in regard to s, s is *past* in regard to a.

22. These, however, are but discussions on *particular* rela-

tions; whereas what we want is a definition of the word *relation* in its widest and most general sense (see § 26). Such a definition is not easy. To meet the requirements of logic, especially of symbolic logic, I propose the following: Let $\phi\,(x, a, \beta, \pi)$ and $\psi\,(y, \beta, a, \pi)$, or their abbreviations ϕ and ψ, denote two equivalent [1] statements which nevertheless differ in three things: (1) that (in *position*) x in the former corresponds to y in the latter; (2) that a in the former corresponds to β in the latter; and (3) that β in the former corresponds to a in the latter — the remaining constant portion π occupying the same position in both. These conditions being satisfied, x (or more strictly x, π) is called the *relation of a to β*; and y (or more strictly y, π) is called the *relation of β to a*. Also the relation x is called the *reciprocal* of the relation y, and the relation y is called the *reciprocal* of the relation x. We may express this reciprocity by $(x=ry)\,(y=rx)$, or by any other symbol suited to the particular investigation upon which we happen to be engaged. When $x=y$, the relation connecting a and β is said to be *symmetrical*.

23. A few concrete examples will help to explain this definition and afford some test of its accuracy. Let ϕ assert that A *has lent money to* B, *which* B *has not yet paid;* and let ψ assert that B *has borrowed money from* A, *which* B *has not yet paid.* Here the relation of A to B is that of *creditor*, and that of B to A *debtor;* but we must proceed as if these words had not yet been invented. Numberless relations exist for which single words cannot be found in any language, symbolic or natural, and our definition would be very inadequate if it left these, as yet uncondensed relations, out of account. Let, therefore, $x=$ has-lent-money-to; let $y=$ has-borrowed-money-from; and let $\pi=$ which-B-has-not-yet-paid. Also let $a=$ A, and let $\beta=$ B. It will be seen at once that the statements ϕ and ψ with their constituents x, a, β, π, satisfy the definition of § 22. The reciprocal relations are x and y, — or rather x, π and y, π; for the fact that A has lent money to B (or that B has borrowed money from A) does not necessarily, and without the accompanying constituent π, imply that A *is now* B's creditor, though it implies that he *has been* so. If the words *creditor* and *debtor* did not exist in our language, we might compound the words of our

---

[1] "Equivalent" in the sense that each implies the other. The statements are supposed to be expressed in some non-inflectional language, symbolic or other, in which the value, effect, or meaning of a word or symbol generally varies with its position. Algebra and Chinese are good examples.

SYMBOLIC REASONING.

statements ϕ and ψ, and, putting the compound word *has-lent-money-to* for *creditor*, and *has-borrowed-money-from* for *debtor*, say that the statement ϕ asserts that A is the *has-lent-money-to* of B, and that the statement ψ asserts that B is the *has-borrowed-money-from* of A. In this form of the statements ϕ and ψ, we have $x = creditor$ (or its longer equivalent), $y = debtor$ (or its longer equivalent), and $\pi = is$. The following are self-evident cases of the defining formula of § 22 :—

(1) Let $\phi = (A > B)$, $\psi = (\beta < A)$.
Here $a = A$, $\beta = B$, $x = (>) = $ (greater than), $y = (<) = $ (less than), and π is non-existent.

(2) Let $\phi = (A : B) = $ (A implies B), $\psi = (B \,!\, A) = $ (B is implied by A).
Here $a = A$, $\beta = B$, $x = (:) = $ (implies), $y = (!) = $ (is implied by), and π is non-existent.

(3) Let $\phi = $ (A is now the teacher of B), and $\psi = $ (B is now the pupil of A).
Here $a = A$, $\beta = B$, $x = $ teacher, $y = $ pupil, and $\pi = $ (is now the).

(4) Let $\phi = $ (A was formerly the teacher of B), and let $\psi = $ (B was formerly the pupil of A).
Here $a = A$, $\beta = B$, $x = $ teacher, $y = $ pupil, and $\pi = $ (was formerly).

24. The last two examples (3) and (4) will show why I said that the stricter or more accurate relations were not x and y but x, π and y, π. Instead of saying "A *was* formerly the *teacher* of B," we may put the verb in the present tense, and say "A *is* the ex-teacher (or former teacher) of B"; and just as the words, *ex-king, queen-dowager*, etc., do not express the same relations as *king, queen*, etc., so x and y do not generally express the same relations as x, π and y, π.

25. This possibility of converting relations of the past or future into relations of the present is one of the many advantages of pure logic or the logic of statements. Let the symbol A denote the statement "The event *a did* happen," or let it denote the statement "The event *a will* happen". In either case we write $A^\tau, A^\iota, A^\epsilon$, etc.; that is, A *is* true, A *is* false, A *is* certain, etc. If A = "The event *a did* happen," then A^τ asserts that "It *is* true that *a did* happen"; and if A = "The event *a will* happen," then A^τ asserts that "It *is* true that *a will* happen." Whether A refers to the past, present, or future, A^τ (which replaces A in symbolic reasoning) always refers to the *present*; and the same may be said of $A^\epsilon, A^\eta, A^\theta$, and of A^x generally, whatever class of statements x may represent. Thus A^τ and A^ι are not exactly synonymous with A and A'.

26. The preceding discussion seems to me to make it clear that the so-called *logic of relations* bears pretty much the same relation to *pure logic* (the logic of statements) as the *theory of functions* bears to *pure mathematics* (see § 12); that is to say, in each case, the former is a special development in a particular direction of the latter. For this reason, in order to mark the analogy, the *logic of relations* should rather be called the *logic of functions*. The questions which it discusses are closely connected with philology and the theory of language is general. In mathematics the words *function* and *relation* are so closely allied that they may almost be considered synonymous. The statement $(A = f_B)$ may be read either as "A is the *function f* of B," or as "A has the *relation f* to B". The mathematical functions (or relations) f and F are *reciprocal* when we have

$$(A = f_B) = (B = F_A).$$

Applying to this case the defining formula of § 22, we find that ϕ here denotes $(A = f_B)$, that ψ denotes $(B = F_A)$, that x denotes f, that y denotes F, and that π denotes the sign $=$, common to the two leading[1] functional statements ϕ and ψ. A mathematical function (or relation) f is *symmetric* when we have

$$(A = f_B) = (B = f_A).$$

A mathematical function (or relation) f is *transitive* when we have

$$(A = f_B)(B = f_C) : (A = f_C).$$

27. Perhaps the most important principle underlying my system of notation is the principle that we may vary the meaning of any symbol or arrangement of symbols, provided, firstly, we accompany the change of signification by a new explanatory definition; and provided, secondly, the nature of our argument be such that we run no risk of confounding the old meaning with the new. Of course this variation of sense should not be resorted to wantonly and without cause; but the cases are numerous in which it leads both to clearness of expression and to an enormous economy in symbolic operations. This is especially the case when the nature of our researches requires the frequent repetition of a lengthy symbolic expression. Then, three courses are open to us. Firstly, we we may accept this repetition with all its inconvenience; or,

---

[1] Here we have *functions of functions*. The *statements* ϕ and ψ are functions of the mathematical functions f and F, *which are not statements*.

SYMBOLIC REASONING.

secondly, we may invent a wholly new symbol of unwonted form and unsuggestive of any analogy; or, thirdly, we may (as I usually do) borrow some familiar and, if possible, suggestive symbol (or combination of symbols), divest it of its old meaning, and, by the aid of a fresh definition, supply it with a new. This last course unquestionably requires much thought and deliberation in the choice of the symbol (or combination of symbols) thus to be entrusted with new duties. The great danger to be guarded against is, of course, the danger of ambiguity. The symbol (or combination of symbols) chosen should be such that the context and the general nature of the research must render its meaning unmistakable Does not the context usually prevent ambiguity in the ordinary language of daily life? Can we, for instance, ever confound a *verb* with a *noun*, because they now and then happen to be identical in form? What prisoner attempting to escape could misunderstand the stern warning of the sentinel, "If you move another step, I shall fire," and imagine that the latter was speaking of a *fire in a grate?* Suppose, when I enter upon some investigation in probability, I lay it down as a preliminary that capital letters must be understood throughout to denote statements; that small italics denote the numerical values of chances; and that the symbol A^x is an abbreviation for the proposition, "The chance that A is true is x". If at the end of my investigation I arrive at the conclusion $A^{\frac{1}{2}}$, what can this mean but that the chance required as to the truth of A is $\frac{1}{2}$? In ordinary algebra, when A is understood to denote a number or ratio, the symbol $A^{\frac{1}{2}}$ denotes the *square root* of A; but, in the sphere of pure logic, what meaning can we attach to the square root of a *statement?* No other logician or mathematician, so far as I know, has as yet insisted upon, and acted upon, this principle of absolute liberty to vary not only the meanings of our separate symbols, but also of their combinations or collocations, whenever clearness, brevity, or other convenience demands it. Prof. Peano (who may be regarded as the leader of the Italian school in symbolic logic) appears to go on the very opposite principle. He holds (if I rightly understand him) that each separate idea should be represented by its own special symbol, which we should never, if we can by any possibility avoid it, employ in any other sense. Now, I am not prepared to say that this is necessarily a wrong principle as regards *his* scientific explorations—some people make discoveries by travelling eastwards, others (like Columbus) by travelling westwards; but I feel quite sure that the principle would never succeed in *my* researches. For these Prof. Peano's notation is much too complicated. Should any one doubt this let him try his notation instead of mine in the solution of one of the complicated problems which I worked out in my recently published memoir on "La Logique Symbolique et ses Applications," in the third volume of the *Bibliothèque Internationale du Congrès de Philosophie*. It seems to me that our notation should always be shaped and suited to the nature of the investigation and to the kind of problems we encounter. Symbolic conventions that may be admirably adapted for one class of problems may be altogether unsuited for another. Even in dealing with the same class, synonymous symbols for the same thing, idea, or proposition, and variations of meaning for the same symbol, are often convenient. The symbols $A:B$ and $(AB')^\eta$ are synonyms; the latter being the definition or explanation of the former, and, therefore, by implication, the clearer of the two. But take the two synonymous complex statements :—

$$(A:B)(B:C):(A:C) \text{ and } \{(AB')^\eta (BC')^\eta (AC')^{n}\}^\eta,$$

the former of which dispenses entirely with the symbol η, and the latter entirely with the symbol :. The former is transparently evident, which is far from being the case with the latter. Two photographs or drawings of the same landscape may both be accurate from their respective points of view; yet one may appeal instantaneously to the memory, while the other is with difficulty recognised.

28. As an example of the same symbol used in different senses take the symbol A_B. In certain cases I use this symbol as a convenient representative of the implication $A:B$; but I also use it in other senses when convenient (of course after due warning), and entrust the expression of implication to $A:B$ alone (See *Bibliothèque du Congrès Internationale de Philosophie*, vol. iii. p. 166). One of these uses I define as follows: When we have a series of concrete things or abstract statements forming the class A_1, A_2, A_3, etc.; then A_B denotes the individual (or any one of the individuals) of the series for which A^B is true; A_C denotes the one (or any one) for which A^C is true; and so on. Thus in A_B the subscriptum B is *adjectival*; whereas in A^B the exponent B is *predicative*. For example, let S = *stag*, let B = *brown*, and let K = "*has been killed by me*" or "*I have killed*". Then S_B^K will mean "The *brown stag* (or a brown stag) has been *killed* by me," or "I have *killed* the (or a) *brown stag*"; whereas S_K^B would mean "The *stag* which I have *killed* (or which has been killed by me) is brown". Or again, supposing our universe of brown things to be restricted to animals,

SYMBOLIC REASONING.

we may have B_K^S, which would mean "The *brown* animal which I have *killed* is a *stag*. These examples bring us to the border of another class of questions which will be discussed in what follows.

BRUTE REASONING AND HUMAN REASONING.

29. It is probable that the primitive language, or primitive languages, of our remote ancestors, like the languages of the animals around them, consisted of mere elementary statements, such as, in our own day, the warning "Caw" of a sentinel rook, or the "Cluck" of a hen when she calls her chickens. These animal statements (like the more or less complex propositions of ordinary human speech) are simply *data*—unconsciously perhaps supplied in the case of some of the lower animals, and without foreknowledge of their effects, but purposely and with foresight in the case of the higher—*data* purposely offered in order that others may therefrom draw correct and useful conclusions. What does the "Caw" of the sentinel rook perched on the branch of a commanding tree say to the others on the ground busily feeding on the farmer's property? To one of these it may say, "A man is coming with a gun"; to another it may say, "A boy is coming with a catapult"; to all it says, "Danger approaches," though their respective ideas as to the precise danger may be vague and varied.

30. Let us now fly far back into the past and try to picture to our minds the origin of human language as we now know it—the language of *propositions*. When was it, and how was it, that primitive man—the desiderated "missing link" of anthropology—escaped from his chrysalis and passed from the brute condition into the human? I do not say *became* human: human he must have been before, or that barrier would have for ever remained impassable. It is the same germ that develops first into a caterpillar and then into a butterfly. To the first question—the question as to *when?*—we can give no answer. Geologists may fix within more or less exact epochs the structural variations that have taken place in animal bodies; they can hardly fix the dates corresponding to the changes in the delicate organ called the brain; still less those corresponding to the changes in that mysterious entity which works through the brain, which no microscope can detect, and which, in animals as in man, we may for the present agree to call the *mind*. To the *How* question we cannot give a precise and definite answer either. In the chain of mental evolution it would be idle to seek

the "missing link"; here the missing links are not one but many. The first of our far-off pre-historic ancestors that barked a tree or raised a heap of stones in order afterwards to remember where he had hidden some object which he prized, performed therein an act which ranks him at once as human. We may even go further and honour that great pre-historic unknown as the *first inventor of symbolic logic*. His arbitrary mark, whatever its nature, represented not one proposition merely but a whole train of reasoning, which we may translate freely as follows: "When I see this mark, it will remind me of the exact position of that spot yonder, where I am now going to hide this provision of nuts". Symbolic concentration of language could hardly be carried further. Yet it does not follow that the language of this 'missing link' and of his tribe had as yet attained the propositional stage. Probably it had not. A higher place must be assigned to that other, and probably later, 'missing link' who first grasped the idea of varying the order or collocation of the elementary sounds or symbols that individually represented *statements* in his (or her) language, so as thereby to form new and more precise statements (or *data*) suggestively allied to the old in their sound or in their form, yet differing from the old and from each other in their signification. Take the examples of § 28, namely, S_B^K, S_K^B, B_K^S. We may suppose that S, B, K were originally separate and complete, but not always clear and definite, statements. In the current language of the tribe the word or symbol S (or its equivalent) might have meant "I see a *stag*," or "I hear a *stag*," or "A *stag* is coming," or "It is a *stag*," etc. Otherwise expressed, the simple sound or symbol S might originally have done duty, sometimes for a proposition $\phi(s)$, sometimes for a proposition $\psi(s)$, and so on (see § 12). The same may be said of its co-symbols B and K. Let us suppose that S, B, K were respectively understood to mean "It is a *stag*," "It is *brown*," "I have *killed* it". Then S_B^K (or some other order or collocation $SB \cdot K$ or $BS \cdot K$ or $K \cdot BS$) would mean "The *brown stag* has been *killed* by me," or "I have *killed* the *brown stag*". Now, let it be observed that in this combination of the elementary statements S, B, K into the complex statement S_B^K the statements S and B *are taken for granted as already known*, while the statement K is *asserted as fresh knowledge*. Thus the categorical statement S_B^K is analogous but not equivalent to the implication $SB : K$, which *does not vouch for the truth of either* S *or* B *or* K. The statement S_B^K in fact means the same as the simple statement K, "I have *killed* it "; the only difference being that the *it* in the latter is replaced by the more definite symbol S_B (the *brown stag*) in the former.

31. This power of inventing and slowly developing a language suited to his needs distinguishes man from the brutes. The languages of the brutes appear to be inherited with their instincts, and to remain always the same; while that of man varies continually. What savage tribe or civilised community at the present time could understand the language spoken by their forefathers 3,000 years ago? Yet to-day the rook caws, and the dog barks, and the horse neighs, just as they did in the days of the ancient Chaldeans or Egyptians. To this difference between man and brute as regards *language* corresponds an analogous difference as regards intelligence. Brute and man alike are capable of *concrete* reasoning; man alone is capable of *abstract* reasoning. To explain my meaning I must have recourse to symbols. The brute as well as man is capable of the concrete inductive reasoning $AB:C$; that is to say, from experience—often painful experience—the brute as well as man can learn that the combination of events A and B is invariably followed by the event C. The higher order of brutes may also be able to communicate to others of their species a knowledge of each event A, B, C separately, or even collectively; but *no brute can communicate to another* a knowledge of the general inductive law $AB:C$, the equivalent of $(ABC')^n$, which it has learnt itself by experience. But this is rather a difference between the brute and the human in their respective powers of communicating their knowledge to others of their kind than a difference in their powers of reasoning, and thereby obtaining fresh knowledge for themselves. I will now show (again using symbols) that man possesses a higher reasoning faculty which no brute appears to possess even in the most rudimentary form. We have seen that from two *elementary* premisses A and B, brutes as well as men can, by inductive reasoning, draw a conclusion C. But no brute can, from the two *implicational* premisses $A:B$ and $B:C$ draw the *implicational* conclusion $A:C$. That is to say, the brute is capable of the concrete inductive reasoning

$$AB:C$$

but not of the abstract, deductive and formal reasoning [1]

$$(A:B)(B:C):(A:C).$$

[1] This formula was, I believe, introduced into logic for the first time, about twenty-four years ago, in my second paper on the "Calculus of Equivalent Statements," published in the *Proceedings of the London Mathematical Society.*

HUGH MACCOLL : SYMBOLIC REASONING.

It is evident that the latter is not only more difficult, but also that it is on a higher and totally different plane. In the former, the two premisses and the conclusion are all three *elementary statements* (see § 3), while the whole reasoning constitutes a *simple implication*. In the latter, the two premisses and the conclusion are all three *implications*, while the whole reasoning is an *implication of the second order*. The premisses A and B of the former are *percepts* supplied *directly by the senses ;* the premisses A : B and B : C of the latter are *hypothetical concepts of the mind*—concepts which may be true or false (as may also the conclusion), without in the least invalidating the formula (see §§ 10, 32).

32. Some writers[1] have supposed that certain of the inferior animals are capable of syllogistic reasoning. This error arises, I think, from a mistaken idea as to the real nature of a syllogism, and one for which the ordinary text-books on logic are in great measure responsible. These usually express Barbara somewhat as follows : " All A is B, All B is C ; *therefore* All A is C ". The syllogism, or any other argument, thus worded *is not a formal certainty ;* it is false whenever either of the premisses is false, whatever the conclusion may be ; and it is also false when the conclusion is false, whatever the premisses may be. Barbara should be worded as follows : " *If* all A is B, and all B is C : *then* All A is C ". In this form the syllogism is true whether premisses or conclusion be true or false (see § 10), and must, therefore, be classed amongst the *formal certainties*. Now, a statement is called a formal certainty when it follows necessarily from our formally stated conventions as to the meanings of the words or symbols which express it ; and until a language has entered upon the propositional stage those conventions (or definitions) cannot be formally expressed and classified. No language but the human has as yet reached this propositional stage ; and, therefore, no terrestrial animal except man is capable of syllogistic or other abstract reasoning.

---

[1] The late Prof. Max Müller, in his *Science of Language* (1861), speaking of a parrot that drops a light nut without attempting to crack it, supposes it to reason thus : " All light nuts are hollow ; this is a light nut ; therefore this nut is hollow ". But the parrot's reasoning is much more elementary. It is only the simple implication, " *Light nut* implies *no kernel*" ; an induction founded on perceptive experience, and *not necessarily* (or *formally*) *true.*

V.—SYMBOLIC REASONING (V.).[1]

By Hugh MacColl.

A RECENT controversy with a certain foreign logician has led me to examine with more care than I had hitherto done the points in which my symbolic logic resembles other modern systems, as well as the points in which it differs from them all. The result has been the discovery that the former are slight and superficial, while the latter are serious and fundamental. So much is this the case that it is hardly an exaggeration to say that no single formula in my system has exactly the same meaning as the formula which is supposed to be its equivalent in other systems. When both are valid, I usually find that mine is the more general and implies the other; when they are not both valid, I invariably find that the valid formula is mine, and the defective formula that of other systems. Examples of this will be given presently; meanwhile let me state the main points of difference.

1. Other logicians generally divide logic into two parts: the logic of *class inclusion* and the logic of *propositions*. Mine is *one simple homogeneous system* which comprises (either directly or as easy deductions), all the valid formulæ of their two divisions, as well as many other valid formulæ which their systems cannot even express.

2. My symbol of implication : they replace by some other, such as $\prec$, or $\in$, or $<$, etc. I shall adopt the first of these three throughout as their general representative, it being more easily formed than the second, and less likely to lead to ambiguity than the third. Now, this adoption of different symbols among logicians to express the same idea is a mere matter of taste or convenience, and if their symbol $\prec$ (or its equivalent) really expressed the same idea as my symbol :, I should not mention this circumstance as one of the points of difference. But their symbol $\prec$ *never does express the same idea as my symbol* :.

[1] For IV. see MIND, July, 1902.

3. They use their symbol $\prec$ in one sense in their logic of *class inclusion*, and in quite a different sense in their logic of *propositions*. I always use my symbol : in one and the same sense throughout, and a sense different from each of the meanings which they attach to their symbol $\prec$.

4. Even to the symbol of equivalence $=$ they attach two different meanings; and neither meaning corresponds exactly with that which the same symbol bears in my system.

5. They divide propositions into two classes, and *two only*, the *true* and the *false*. I divide propositions not only into true and false, but into various other classes according to the necessities of the problem treated; as, for example, into *certain, impossible, variable* ; or into *known to be true, known to be false, neither known to be true nor known to be false* ; or into *formal certainties, formal impossibilities, formal variables* (*i.e.*, those which are *neither*); or into *probable, improbable, even* (*i.e.*, with *chance even*); and so on *ad libitum*.

6. They make no distinction between the *true* and the *certain*, between the *false* and the *impossible*; so that, in their system, every *uncertain* proposition is *false*, and every *possible* proposition *true*. In other words, *variable* propositions— propositions that are possible but uncertain, propositions whose chance of being true is some proper fraction between 0 and 1—are excluded entirely from their universe. Many of their formulæ are therefore *not formal certainties*; they are only valid conditionally, and this defect, if it does not wholly destroy their utility, restricts within comparatively narrow limits their ranges of application.

7. Implications and other propositions of different *orders* or *degrees*,[1] such as $(A : B) : (C : D)$, $(A : B)^{\epsilon\iota}$, $A^{\theta\theta}$, $A^{\alpha\beta\gamma}$, etc., are not recognised (at least in my sense of the words) in other systems; so that the whole world of new ideas opened up by this exponential or predicative system of notation is a world with which they are utterly unable to deal; the bare attempt on the part of logicians would lead to a general break-up of all the systems now taught and a recasting of the whole of logic on different principles. This would be tantamount to the universal adoption of my system in all its essentials. Human nature being what it is, and professional prejudices being what they are, and what they can hardly help being, such a general recognition of the superiority of my system is hardly to be expected just yet; but I think it will come in

---

[1] For example my $(A : B) : (C : D)$ means $\{(AB')^{\eta\iota} + (CD')\eta\}^{\epsilon}$, whereas their $(A \prec B) \prec (C \prec D)$ means simply $AB'' + (CD')'$, and is therefore *only a statement of the first degree.*

SYMBOLIC REASONING. 357

time—after I have dropped into my place among the silent people of the past.

Let me now descend from generalities into particulars.

First with regard to point No. 3. In their logic of *class inclusion* they use the symbol $A \prec B$ to assert that *every individual of the class A belongs also to the class B*. In their logic of *propositions* they abandon this definition and use the same symbol to assert that *either A is false or B true*. I use the symbol $A : B$ in always one and the same sense, namely, to assert that *it is certain that either A is false or B true*. Hence, when A and B denote each a proposition, we get the following comparisons and definitions:—

$$A \prec B = A' + B = (AB')'$$
$$A : B = (A' + B)^\epsilon = (AB')^\eta$$
$$A : B = (A \prec B)^\epsilon;$$

So that my symbol $A : B$ is formally stronger than and implies their symbol $A \prec B$, just as A^ϵ is formally stronger than and implies A^τ. Thus, my symbol $A : B$ never coincides in meaning with their symbol $A \prec B$, when A and B are propositions.

They use the symbols 1 and 0 to denote *true* and *false* propositions respectively; so that 1 and 0 denote two mutually exclusive *classes* of propositions. Hence, consistency of notation requires that the symbol $0 \prec 1$ should assert that *every false proposition is a true proposition*, which is absurd. But, as a matter of fact, the statement $0 \prec 1$ is supposed in their systems, on the contrary, to be *always true*; and if we give its second meaning to the symbol $\prec$ and suppose 0 and 1 to be *single propositions* instead of *classes*, the statement $A \prec B$ *is* always true, as it then asserts that either 0 is false or 1 true, which is self-evident.

My symbol $\iota : \tau$, which is erroneously supposed to be equivalent to their $0 \prec 1$, does not lead to this inconsistency; for $A : B$, by its very definition, means simply $(A^\tau B^\iota)^\eta$. Hence

$$\iota : \tau = (\iota^\tau \tau^\iota)^\eta = (\eta \eta)^\eta = \epsilon.$$

Similarly, we get $\eta : \epsilon = (\eta^\tau \epsilon^\iota)^\eta = (\eta \eta)^\eta = \epsilon$.

Though the symbols $\tau, \iota, \epsilon, \eta, \theta$, as *exponents* (or *predicates*), denote *classes*, each denotes a *single statement* when it is the *subject* of a proposition. Thus η^τ asserts that *the impossible proposition η is true*, which is absurd. When it is necessary or convenient to distinguish between different propositions of the same class I use *subscripts*. Thus, in the propositions A^B, A_1^B, A_2^B, the subject A_1 or A_2 differs from the subject A pretty much as a proper noun differs from a common noun (see MIND, N.S., No. 43). In one or two places in my Sixth

Paper in the *Proceedings of the London Mathematical Society*, I employed the symbol A^B to assert, not (as here) that a certain unnamed individual of the class A belongs also to the class B, but that *every* A belongs to the class B. Subsequent experience however taught me that this convention was inconvenient; so I abandoned it.

Let us now consider point No. 4. In their logic of class inclusion their symbol $(A = B)$ asserts that *every individual of the class A is included in the class B, and every individual of the class B in the class A*. In their logic of propositions this same symbol $(A = B)$ asserts that *the propositions A and B are either both true or both false*, which is quite a different definition. In my system the symbol $(A = B)$ has neither of those meanings; it always asserts that *it is certain that either A and B are both true or both false*. Thus, when A and B denote each a single proposition, if we put $(A = B)_a$ for the symbol $(A = B)$ when the latter has *their* interpretation, and $(A = B)_\beta$ for the same symbol when it has *my* interpretation, we get the following comparison and definitions:—

$$(A = B)_a = AB + A'B'$$
$$(A = B)_\beta = (AB + A'B')^\epsilon$$
$$(A = B)_\beta = (A = B)_a^\epsilon;$$

so that my symbol $(A = B)$ is formally stronger than their symbol $(A = B)$, just as A^ϵ is formally stronger than A^τ. The symbol A^τ asserts that A *is true* (true at least in the case considered); whereas A^ϵ asserts that A *is certain* (that is to say, true in all circumstances consistent with our data and definitions).

In their logic of *class inclusion* they use the symbol AB (or its synonym $A \times B$) to denote *the class of individuals common to the classes A and B*. With irrefutable logic they then infer that their proposition $A \prec B$ is equivalent to their proposition $A = AB$. But consistency of notation demands that this convention as to the meaning of AB should hold good also as regards the classes 0 and 1, which (with them) denote false and true propositions respectively. Now, with this interpretation of their symbols, the class 0 we know, and the class 1 we know, but *what is the class 0×1 common to both?* Where can we find an intelligible and unambiguous proposition that can be described as *both true and false?* False propositions are numerous enough, as we often learn to our cost, and they are usually quite clear and unambiguous; but I have never yet come across an intelligible proposition that could be classed as *both true and false*. Such propositions I denote in my system, not by the symbol ι, which denotes a false but intelligible proposition, nor by the symbol η, which

SYMBOLIC REASONING.

denotes an intelligible proposition that contradicts our data, but by the symbol 0, which (with me) denotes a *meaningless proposition*. Thus, consistency of notation requires that the formula $(0 = 0 \times 1)$ should assert that *every false proposition is meaningless*, an assertion which we know to be untrue. But with their other interpretation of the symbol $=$, and supposing 0 and 1 to denote each a single proposition instead of a whole class, their formula $(0 = 0 \times 1)$ is true; for, on this convention, 0×1 will then denote *not* a class of propositions but a single compound proposition which is necessarily false because it contains a false factor 0. If I say "*Henry will go to Paris and Richard will go to Berlin*," and it turn out that Henry does *not* go to Paris, though Richard *does* go to Berlin, I make a *false* statement, though it is perfectly clear and unambiguous. We can neither call it *both true and false* nor *meaningless*. For, by our linguistic conventions, a compound statement is called *false*, if it contains a single false factor.

No inconsistency of this kind, or of any other, will be found in either of my statements $(\iota = \iota\tau)$ and $(\eta = \eta\epsilon)$, as I always use the symbol $=$ in one and the same sense. With me both statements are formal certainties, for

$$(\iota = \iota\tau) = \{\iota^\tau = (\iota\tau)^\tau\} = (\eta = \eta) = \epsilon,$$
$$\text{and } (\eta = \eta\epsilon) = \{\eta^\tau = (\eta\tau)^\tau\} = (\eta = \eta) = \epsilon;$$

the exponent or predicate τ being always understood when not expressed.

In most systems I find the formula
$$(A = 1) + (A = 0) = 1,$$
which, like my formula $(A^\tau + A^\iota)^\epsilon$, is meant to assert that the proposition A is necessarily either true or false. Considering 1 and 0 as single propositions, and adopting the second of their two interpretations of the symbol $=$, the formula is valid. But with my interpretation of the symbol $=$, the formula is *not* valid, whether the symbols 1 and 0 correspond to τ and ι or to ϵ and η. For (putting $::$ for $=$, to avoid brackets)

$$(A = \tau) + (A = \iota) :: \tau = (A^\tau = \tau^\tau) + (A^\tau = \iota^\tau) :: \tau^\tau$$
$$= (A = \epsilon) + (A = \eta) :: \epsilon$$
$$= (A^\epsilon + A^\eta)^\epsilon.$$

This asserts that *it is certain that the statement* A *is either certain or impossible*. Now, this may be true of some particular statement A; but it is not true of *every* statement A, for there are numberless statements (those I call *variables*) that are neither certain nor impossible. In other words, the statement $(A^\epsilon + A^\eta)^\epsilon$ is not a *formal certainty;* so that the formula of which it has been shown to be the simplification is not valid, or is only valid conditionally and within very narrow limits. If 1 and 0 be represented by ϵ and η respectively, we get the same result.

Now let me deal with points No. 1 and No. 7 and show that, as regards their valid formulæ, other systems are implied in mine; while mine, on the other hand, can work out problems and evolve new and fruitful ideas which their systems are unable even to express. First, as regards their logic of propositions. In my sixth paper on the "Calculus of Equivalent Statements," in the *Proceedings of the Mathematical Society*, I use a symbol ∂x in the following sense. When x denotes a statement A^ϵ, then ∂x denotes A. Hence, when x denotes A^η, ∂x must denote A', for A^η is synonymous with $(A')^\epsilon$. Also, when x denotes $A:B$, ∂x must denote their statement $A \prec B$; for $A:B$ means $(A'+B)^\epsilon$, and $A \prec B$ means $A' + B$. Thus my symbol $\partial\,(A:B)$ corresponds to their symbol $A \prec B$; and my symbol $A:B$ *would* correspond to their symbol $(A \prec B)^\epsilon$ if they adopted my notation of exponents with my signification of the symbol ϵ. On this understanding all the valid formulæ of their logic of propositions could be transferred from their systems into mine. Also, on the understanding that all *variable* propositions should be left out of account, my A^ϵ would be equivalent to my (and to their) A; my A^η to my A' and to the corresponding symbol in their notation; and my symbol $A:B$ to their symbol $A \prec B$; while my interpretation of the symbol $=$ would then be the same as theirs. But this arbitrary and unnecessary restriction of our universe of admissible statements would rob logic of nearly all its utility, whether as a practical instrument of scientific research (as in my *Calculus of Limits*), or as an educational instrument of mental training and culture.

The inability of other systems to express the new ideas represented by my symbols A^{xy}, A^{xyz}, etc., may be shown by a single example. Take the statement $A^{\theta\theta}$. This (unlike *formal certainties* such as ϵ^τ and $AB:A$, and unlike *formal impossibilities* such as θ^ϵ and $\theta:\eta$) may, in my system, be a *certainty*, an *impossibility*, or a *variable* according to the special data of our problem or investigation. But how could it be expressed in other systems? Not at all, for its recognition would involve an abandonment of their erroneous convention (assumed throughout) that *true* is synonymous with *certain*, and *false* with *impossible*. If they ceased to consider A as equivalent to $(A=1)$, and A' (or their corresponding symbol) as equivalent to $(A=0)$, and employed their $(A=1)$ as equivalent to my A^ϵ, and their $(A=0)$ as equivalent to my A^η, they *might* then express my statement $A^{\theta\theta}$ in their

SYMBOLIC REASONING. 361

notation; but the expression would be extremely long and intricate. Using $A \neq B$ as the denial of $(A = B)$, as is customary, A^θ would then be expressed by $(A \neq 0)(A \neq 1)$, and $A^{\theta\theta}$ by
$$\{(A \neq 0)(A \neq 1) \neq 0\}\{(A \neq 0)(A \neq 1) \neq 1\}.$$
This example of translation speaks for itself and renders all formal argument superfluous. Let any one try to express in this notation the formal certainty
$$A^{\theta\theta\epsilon} + A^{\theta\theta\eta} + A^{\theta\theta\theta}.$$
The expression needed would take up several lines, and it would be scarcely possible to extract the intended meaning from the bewildering jungle of symbols in which it would be enveloped.

It remains to show that my system also includes all valid formulæ of their logic of *class inclusion*. Their symbol $A \succ B$ asserts that *every individual of the class A belongs also to the class* B. This may be expressed by my symbol $A : B$ on the understanding that the two statements A and B *have the same subject* P, an individual taken at random out of our universe, P_1, P_2, P_3, etc. Thus $A : B$ becomes a mere abbreviation for $P^A : P^B$, which asserts that P cannot belong to the class A without also belonging to the class B, an assertion equivalent to the traditional *All A is B* and to their statement $A \prec B$. Thus, as I showed in MIND, January, 1880, and in MIND, July, 1902, the syllogism *Barbara* will become a particular case of my formula
$$(A : B)(B : C) : (A : C);$$
in which, let it be observed, the symbol : *has the same meaning throughout*, and A, B, C, as well as $(A : B)$, $(B : C)$, $(A : C)$, are propositions. But as this formula is a *formal certainty*, it holds good whether the statements A, B, C, have the same subject or not, so that it is more general than the syllogism. *Barbara* may also be expressed by
$$(A \prec B)(B \prec C) \prec (A \prec C),$$
but only on the condition that the symbol $\prec$ (unlike my symbol :) has *not* the same meaning throughout. For, though we may say that the *class* A is contained in the *class* B, the *class* B, in the *class* C, and the *class* A in the *class* C, we cannot logically speak of the *premisses* $(A \prec B)(B \prec C)$ as a *class* contained in the *conclusion* $A \prec C$. It is just the other way; if the word *contain* is to be used at all in this case, it is the conclusion that is contained in the premisses, and not the premisses in the conclusion.

If, in the last formula, the letters A, B, C denote propositions instead of classes, and we give $A \prec B$ its second meaning $A' + B$, the symbol $\prec$ will then (like my symbol :) have the same meaning throughout; but then the formula (unlike mine) will no longer represent *Barbara*. For $(A' + B)^e$, being synonymous with $A : B$, asserts that *every* P in the class A is also in the class B; whereas $A' + B$ (or its equivalent $A \prec B$) only asserts that a certain P of the series P_1, P_2, P_3, etc., is either excluded from A or included in B; it makes no assertion as to the other individuals of the series.

This comparison of the formulæ
$$(A : B)(B : C) : (A : C)$$
$$(A \prec B)(B \prec C) \prec (A \prec C),$$
which are erroneously supposed to be equivalent, is typical of many others. Another formula of mine that has led to misunderstandings is the formula (*Proceedings of the Mathematical Society*, Third Paper)
$$(A : x) + (B : x) : (AB : x).$$
Not that the validity of this formula has been called in question; it is indeed almost self-evident; but logicians have asserted that the symbol $=$ might with advantage replace the symbol $:$ before the conclusion $AB : x$, as (in their opinion) the converse implication is also true. Now, if my symbol $A : B$ (like their symbol $A \prec B$) meant $A' + B$, this converse implication *would* be true, and $=$ might replace $:$ before the conclusion $AB : x$. But this, as already explained, is *not* the signification of my symbol $A : B$, so that the substitution of $=$ for $:$ before the conclusion (or consequent) would destroy the validity of the formula. A geometrical illustration will

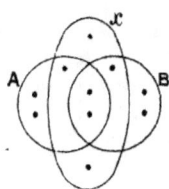

make this clear. Out of the total ten points marked in the ellipse x and the two circles A, B of the accompanying figure, take a point P at random, and let A, B, x assert respectively (as propositions) that P will be in A, that P will be in B, that P will be in x. It is evident that the respective chances of the four propositions A, B, x, AB are $\frac{5}{10}, \frac{5}{10}, \frac{6}{10}, \frac{2}{10}$; so that they are all *variables*. The implication $AB : x$ asserts that the point P cannot be in both the circles A and B without being also in the ellipse x, which is true. The implication $A : x$ asserts that P cannot be in A without being in x, which is false; and $B : x$ asserts that P cannot be in B without being in x, which is false also. Thus, the alternative $(A : x) + (B : x)$ is false while $AB : x$ is true, so that in this case the substitution of the symbol $=$ for $:$ before $AB : x$ in my formula would be wrong. But my formula is right in this case as in all others; for
$$\iota + \iota : \tau = \iota^\tau + \iota^\tau : . \tau^\tau = \eta + \eta : \epsilon = \eta : \epsilon$$
$$= (\eta^\tau \epsilon^\iota)^\eta = (\eta\eta)^\eta = \epsilon.$$

SYMBOLIC REASONING.

The same diagram will illustrate two other propositions which by most logicians are considered equivalent, but which, according to my interpretation of the conjunction *if*, are *not* equivalent. They are the complex conditional, *If A is true, then if B is true x is true*, and the simple conditional *If A and B are both true x is true*. Expressed in my notation, and with my interpretation of the conjunction *if*, these conditionals are respectively

$$A : (B : x) \text{ and } AB : x.$$

Giving to the propositions A, B, x, AB the same meanings as before (all having reference to the same subject, the random point P) it is evident that B : x, which asserts that the random point P cannot be in B without being also in x, *contradicts our data*, and is therefore *impossible*. The statement A, on the other hand, does *not* contradict our data, neither does its denial A', for both in the given conditions are possible though uncertain. Hence, A is *a variable*, and B : x being impossible, the complex conditional A : (B : x) becomes $\theta : \eta$, which is synonymous with θ^η and therefore an *impossibility*. But the simple conditional AB : x, instead of being impossible, is, in the given conditions, a *certainty*, for it is clear that P cannot be in both A and B without being also in x. Hence, though A : (B : x) always implies AB : x, the latter does not always imply the former, so that the two are not in all cases equivalent. In other words,

$$\{A : (B : x)\} : (AB : x)$$

is a formal certainty; but its converse

$$(AB : x) : \{A : (B : x)\}$$

is *not*.

Whether my interpretation of this troublesome little conjunction *if* is the most natural and the most in accordance with ordinary usage, I do not undertake to say; it certainly is the most convenient for the purposes of symbolic logic, and this alone is reason sufficient for its adoption. At the same time I may point out, as I did long ago (see MIND, Jan., 1880), that the usual *denial* of the conditional *If A is true B is true* is the categorical proposition A *may be true without B being true;* that is to say (A : B)' is equivalent to $(AB')^\pi$, which asserts that AB' is *possible*. From this equivalence necessarily follows the equivalence $A : B = (AB')^\eta$, which is my definition of the symbol A : B. The implication A : B expresses a *general law* and asserts that it has no exception. Its denial (A : B)' asserts that the law is *not* in all cases valid; it asserts $(AB')^\pi$, that an *exception* AB' is *possible*. The statement AB' (the denial of A' + B) asserts not merely the *possibility* of AB', but *an instance of its actual occurrence*.

Just as $A : B$, or its synonym $(AB')^\eta$, implies $A' + B$, so AB', the denial of the latter, implies $(AB')^{n}$, the denial of the former.

I did not enter upon the preceding discussion from any wish to provoke a controversy, but in order to remove misunderstandings. I find that several logicians are in error as to the precise meanings of my symbols and the relation in which my system stands to others that symbolically more or less resemble it. My main object has been to show that these resemblances of mere form hide important differences in matter, method, and limits of application. To effect this object without, at the same time, pointing out what, rightly or wrongly, I regard as serious defects in all the other symbolic systems of which I have any knowledge was impossible. But I have attacked no particular system; the faults that I have indicated are faults which they appear to have in common, and from which indeed my own earlier researches were not wholly free, though the central principle of these was sound and forms an important factor of the better and far more comprehensive system into which they have since developed. Modern symbolic logic, unlike the venerable logic of the schools, is a progressive science; it can lay claim to no finality or perfection. But, in the form which I have given it, it has now one great merit which it never possessed before; it has become a *practical* science; it can actually be applied as an instrument of research. As regards utility, logic used to be contrasted, much to its disadvantage, with mathematics; but now that the mathematician is obliged to hand over to the logician the disentanglement of some of his most difficult problems, he can no longer with justice or consistency look down upon the science of the latter and call it useless and inapplicable.

[1905o]: Symbolic[al] Reasoning (VI). *Mind,* Vol. 14, pp. 74-81.

VI.—SYMBOLIC REASONING (VI.).[1]

By Hugh MacColl.

1. There is no question on which logicians are so divided as that of the "Existential Import of Propositions". When we make any affirmation A^B, or any [2] denial A^{-B}, do we, at the same time, implicitly assert that the subject A really exists? Do we assert that the predicate B really exists? Do the four technical propositions of the traditional formal logic, namely, "Every (or all) A is B," "No A is B," "Some A is B," "Some A is not B," necessarily imply the actual existence of the class A? Do they necessarily imply the actual existence of the class B? These are questions upon which logicians have expended a great deal of thought and no small amount of ink; yet they appear to be as far from coming to an agreement upon them as ever. A simple theory of the subject, a theory to which they could all subscribe, should therefore be welcomed as a real boon. Such a theory I hope to be able to offer in what follows.

2. Let e_1, e_2, e_3, etc. (up to any number of individuals mentioned in our argument or investigation) denote our universe of *real existences*. Let 0_1, 0_2, 0_3, etc., denote our universe of *non-existences*, that is to say, of unrealities, such as *centaurs, nectar, ambrosia, fairies,* with self-contradictions, such as *round squares, square circles, flat spheres,* etc., including, I fear, the non-Euclidean geometry of four dimensions and other hyper-spatial geometries. Finally, let S_1, S_2, S_3, etc., denote our *Symbolic Universe,* or "Universe of Discourse," composed of all things real or unreal that are named or expressed by words or other symbols in our argument or investigation. By this definition we assume our Symbolic Universe (or "Universe of discourse") to consist of our universe of realities, e_1, e_2, e_3, etc., together with our universe of unrealities, 0_1, 0_2, 0_3, etc., *when both these enter into our argument.*

---

[1] For V., see Mind, July, 1903.
[2] The symbol A^{-B} is here used as a convenient symbol for $(A^B)'$, the denial of the proposition A^B (see the *Athenæum,* 5th September, 1903),

But when our argument deals only with *realities*, then our Symbolic Universe S_1, S_2, S_3, etc., and our Universe of realities, e_1, e_2, e_3, etc., will be the same; there will be no universe of unrealities 0_1, 0_2, 0_3, etc. Similarly, our Symbolic Universe may conceivably, but hardly ever in reality, coincide with our universe of unrealities.

3. Now, suppose we have a class A. The individuals, A_1, A_2, A_3, etc., compassing it must necessarily all belong to the *Symbolic* Universe S; but whether they all belong to the universe of *realities* e, or all to the universe of *unrealities* 0, or some to the universe e and the rest to the universe 0, depends upon the particular circumstances of our argument or investigation. When a class A belongs *wholly* to the universe e, or *wholly* to the universe 0, we may call it a *pure* class; when it belongs *partly* to the class (or universe) e and *partly* to the class 0, we may call it a *mixed* class. The *negative* class 'A (with a grave accent) consisting of the individuals 'A, 'A_2, 'A_3, etc., contains all the individuals of our symbolic universe which do not belong to the positive class A. Hence, the class 'e is synonymous with the class 0; and the class '0 with the class e. The class 'A may be called the *complement* of the class A, because both together make up the Symbolic Universe.

4. The subject A of any affirmative proposition A^B, or of any denial A^{-B}, is always understood to denote *a single individual*. If A happens to be the name of a class, then, in any proposition A^B or A^{-B}, the subject A is understood to denote a certain known, or previously indicated, individual of the series A_1, A_2, A_3, etc., whose special name or number it is unnecessary to state. For example, suppose A denotes *American*, and B *barrister*, the proposition A^B will then assert that "the *American* is a *barrister*". It does not say *which* American out of the whole series A_1, A_2, A_3, etc., is referred to; that is supposed to be known. When it is necessary to state which, then, instead of A^B, we must write A_1^B, or A_2^B, or A_3^B, as the case may be.

5. Let S be any individual taken at random out of our Symbolic Universe S, and let a, a', b, b', etc., be synonymous respectively with S^A, S^{-A}, S^B, S^{-B}, etc. We then get the following self-evident or easily proved formulæ, which we will name F_1, F_2, F_3, etc.

(1) A^s; (2) $(S^A)^{-\eta}$; (3) $a^{-\eta}$; (4) a^η; (5) $('A)^s$; (6) $('S)^s$; (7) $('A)^{-A} + (A \equiv S)$.

The formula F_6 looks somewhat paradoxical; but it can be easily proved. By our definitions (see §§ 2, 3), the symbol 'S denotes an individual that does not belong to the class S.

But, by definition, the class S denotes the *whole Symbolic Universe* (or "universe of discourse") to which every individual (real, unreal, or self-contradictory) named in our argument *must* belong. Hence, 'S is a self-contradiction. But, by our definition, all self-contradictions belong to the class 0. Hence 'S must belong to the class 0. But, by definition, the Symbolic Universe S contains all the individuals 0_1, 0_2, 0_3, etc., of the class 0, as well as all the individuals e_1, e_2, e_3, etc., which belong to the class of realities e. Hence, 'S must belong to the class S. In other words, the formula $('S)^s$, denoted by F_6, is always true. The preceding reasoning is a syllogism of the Barbara type, which may be expressed briefly as follows: "Every 'S is 0, and every 0 is S; therefore every 'S is S". The last formula F_7 asserts that an individual of the negative class 'A does not belong to the positive class A, except when A is synonymous with S and therefore denotes the whole Symbolic Universe. This Symbolic Universe, or "Universe of Discourse," may enlarge as the argument proceeds, seizing, appropriating, and firmly retaining every new entity (not excepting self-contradictory entities like 'S) which we designate by a symbol. Suppose for example, that in the course of our argument we have had to speak of several classes, *pure* or *mixed* (see § 3), and that all the individuals of all these classes amount to 82, of which 80 belong to the class e, and 2 to the class 0. Then, up till now, our Symbolic Universe S contains 82 individuals, so that we have

$$S = (S_1, S_2, \ldots S_{82}) = (e_1, e_2, \ldots e_{80}, 0_1, 0_2).$$

A fresh arrival 'S enters our Symbolic Universe, which immediately widens to make room for it; but the question has to be decided whether the stranger is to enter the class e or the class 0, just as parents have to decide the sex of a fresh addition to their family before they know whether to call it *Eva* or *Oscar*. The question presents as little difficulty in the one case as in the other; the new comer 'S (or S_{83}) is immediately recognised as belonging to the class 0, so that now we have

$$S = (S_1, S_2, \ldots S_{83}) = (e_1, e_2, \ldots e_{80}, 0_1, 0_2, 0_3),$$

the new comer 'S (or S_{83}) being synonymous with the new comer 0_3.

6. If every individual of a class A (whether a *pure* or a *mixed* class) belongs also to another class B; then, and then only, we say that "Every (or all) A is B". If this is not the case—if even a single A is excluded from the class B; then we say that "Some A is not B". If every individual in the class A be excluded from the class B; then, and then

only, we say that "No A is B". If this is not the case—
if a single individual of the class A belongs also to the class
B; then we say that "Some A is B". For example, let
the class A consist of the five individuals S_1, S_2, S_3, S_4, S_5;
let the class B consist of the eight individuals made up of
the preceding five individuals together with the three S_6, S_7, S_8; and let the class C consist of the three S_7, S_8, S_9. More
briefly expressed, let $A = (S_1, S_2, \ldots S_5)$; let $B = (S_1, S_2, \ldots S_8)$; and let $C = (S_7, S_8, S_9)$. Then, whether any of
these three classes, A, B, C, be *pure* or *mixed* (see § 3), the
following propositions follow necessarily from our data:—

(1) Every (or all) A is B, (2) Some A is B,
(3) Some B is A, (4) Some B is not A,
(5) No A is C, (6) Some A is not C,
(7) No C is A, (8) Some C is not A,
(9) Some B is C, (10) Some B is not C.

Any of the three classes A, B, C may consist wholly of
realities, or wholly of unrealities, or it may be a mixed class
containing both; whatever hypothesis we take in that way,
the preceding ten propositions are true. (See § 11.)

7. We may sum up briefly as follows: Firstly, when any
symbol A denotes an *individual*; then, any intelligible statement ϕ (A), containing the symbol A, implies that the individual represented by A has a *symbolic* existence; but
whether the statement ϕ (A) implies that the individual
represented by A has a *real* existence depends upon the
context. Secondly, when any symbol A denotes a *class*,
then, any intelligible statement ϕ (A) containing the symbol
A implies that the whole class A has a *symbolic* existence;
but whether the statement ϕ (A) implies that the class A is
wholly real, or *wholly unreal*, or *partly real and partly unreal*,
depends upon the context.

8. For example, let M denote "the *man* whom you see in
the garden"; let U denote "my *uncle*"; and let ϕ (M, U)
denote the statement "The *man* whom you see in the garden
is my *uncle*". In this case we generally have

$$\phi (M, U) : M^e \, U^e.$$

That is to say, the statement ϕ (M, U) would generally
imply that both M and U really exist. Next, let B denote
"a *bear*," and let ϕ (M, B) denote the statement "The *man*
whom you see in the garden is really a *bear*". Here we
should generally have

$$\phi (M, B) : M^0 \, B^e.$$

That is to say, the statement ϕ (M, B) would generally
imply that B really exists, but that the individual M is
imaginary—a mere optical illusion. Now take the state-

ment ϕ' (M, B), which denies ϕ (M, B) and asserts that "The *man* whom you see in the garden is *not* a *bear*". Here we should generally have

$$\phi' (M, B) : M^e\, B^0.$$

That is to say, the denying statement ϕ' (M, B) would usually be understood to imply that M (the *man* seen in the garden) really exists, but that the particular *bear* spoken of is imaginary and non-existent. Lastly, take ϕ' (M, U) which denies ϕ (M, U) and asserts that "The *man* whom you see in the garden is *not* my *uncle*". Here we should generally say

$$\phi' (M, U) : M^e;$$

but not necessarily ϕ' (M, U) : U^e. That is to say, the denying statement ϕ' (M, U) would usually imply the real existence of the *man* M, but not necessarily the real existence of "my *uncle*"; for the negative statement ϕ' (M, U) might be true even on the supposition that neither my father nor mother ever had a brother, so that the supposed uncle had never existed. Similarly, we may give examples of the implied existence or non-existence of *classes*, and show that as regards *real* and not mere *symbolic* existence, no absolute rule can be laid down; that, in each case, the conclusion depends upon the particular nature of the statement and upon the general context.

9. The preceding discussion seems to me to point to a serious and fundamental error in the commonly accepted systems of symbolic logic, founded on the Boolian principle of class-inclusion. These usually denote the class of individuals common to the classes A and B by the symbol AB, and they employ the symbol (A = AB) to assert that the class A and the class AB are the same, every individual in either being also found in the other. Thus interpreted, they say, and say truly, that the statement of equivalence (A = AB) is equivalent to the traditional "All A is B," or "Every A is B". So far I agree with them. But when they define 0 (or any other symbol) as indicating non-existence, and then assert that the equivalence (0 = 0A) is always true, whatever the class A may be, they appear to me to make an assertion which cannot easily be reconciled with their data or definitions. For suppose the class 0 to consist of the three unrealities 0_1, 0_2, 0_3, and the class A to consist of 0_3, e_1, e_2, e_3 (one unreality and three realities), the class 0A common to both contains but one individual, the unreality 0_3. We cannot here say that the class 0, which contains three individuals, is the same as the class 0A, which contains but one; neither can we say that every one of the three in-

dividuals 0_1, 0_2, 0_3, which form the class 0, is contained in the class A, which only contains one of them, namely, 0_3. And, à fortiori, an *infinite* class (0_1, 0_2, 0_3, etc.), cannot be contained in a *finite* class 0A, where $A=(A_1, A_2, \ldots A_m)$.

10. If in my system of logic my formula $(\eta=\eta A)$ asserted that the *class* η and the *class* ηA contained exactly the same individuals, this formula would be open to exactly the same objections as the formula $(0=0A)$ just criticised. But my formula $(\eta=\eta A)$ does *not* assert this; it only asserts the truism expressed by the double implication $(\eta : \eta A)(\eta A : \eta)$, namely, that it never happens that either of the two *statements* η and ηA is true while the other is false. The formula is equally valid in the form $(\eta_1 = \eta_2 A)$, whatever be the impossibilities η_1 and η_2, and whatever be the statement A. For, by a linguistic convention which I believe all logicians accept, any compound statement, say ABC, is considered *true* when, and only when, *all* its factors, A, B, and C, are true; but it is considered *false* if it has *a single false factor* A. Consequently, it must be *impossible* (or *always false*) if it has a single impossible factor η. Hence, $\eta_2 A$ is impossible because of the factor η_2. We may therefore denote $\eta_2 A$ by η_3 (impossibility No. 3), so that the formula $(\eta_1 = \eta_2 A)$ will then be equivalent to $(\eta_1 = \eta_3)$. Now, by definition,

$$(\eta_1 = \eta_3) = (\eta_1 : \eta_3)(\eta_3 : \eta_1) = (\eta_1 \eta'_3)^\eta (\eta_3 \eta'_1)^\eta.$$

But $\eta'_3 = \epsilon_1$, and $\eta'_1 = \epsilon_2$; for [1] the denial of any impossibility η_x is some certainty ϵ_y, so that the denial of η_3 is a certainty which we register as ϵ_1, and the denial of η_1 is another certainty which we register as ϵ_2. Hence, by substituting ϵ_1 for η'_3, and ϵ_2 for η'_1, we get

$$(\eta_1 = \eta_2 A) = (\eta_1 = \eta_3) = (\eta_1 \epsilon_1)^\eta (\eta_3 \epsilon_2)^\eta = \epsilon_3 \epsilon_4 = \epsilon_5.$$

11. Another disputed question which the preceding theory of the "Existential Import of Propositions" appears to decide is the validity or non-validity of the four traditional syllogisms, Darapti, Felapton, Fesapo and Bramantip. Now, as I pointed out in MIND, July, 1902, § 32, not one syllogism out of the whole nineteen is valid in its traditional form PQ ∴ R, as in this form it asserts without warrant that the two premisses P and Q are both true. In this form therefore any syllogism is false whenever either P or Q is false. To

---

[1] The denial of a certainty is an impossibility, the denial of an impossibility is a certainty, and the denial of a variable is a variable. If the chance of A is a, the chance of A' is $1-a$. When $a=1$, then A is a certainty and A' an impossibility. When $a=0$, then A is an impossibility and A' a certainty. When a is some fraction between 1 and 0, then $1-a$ is also a fraction between 1 and 0, so that, in this case, A and A' are both variables.

render the syllogism valid, it should be written in the form PQ : R ("*If* P and Q are true, *then* R is true"). Thus written, if in any syllogism we substitute for P, Q, R its special premisses and conclusion, we shall find that PQ : R, which means PQR' : η, is a formal certainty, whatever syllogism out of the nineteen we take as an example. Take Darapti, one of the four considered doubtful. Darapti, in its corrected or conditional form, says this, "*If* every B is C, and every B is A, *then* some A is C". This is supposed to fail when B is non-existent while A and C are existent but mutually exclusive. Let us see. Suppose
$$B = (0_1, 0_2, 0_3), C = (e_1, e_2, e_3), A = (e_4, e_5, e_6).$$
Here we have
$$P = \text{Every B is C} = \eta_1$$
$$Q = \text{Every B is A} = \eta_2$$
$$R = \text{Some A is C} = \eta_3,$$
three statements each of which contradicts our data, since, by our data in this case, the three classes A, B, C are mutually exclusive (see § 6). Hence, in this case, we have
$$(PQ : R) = (\eta_1 \eta_2 : \eta_3) = (\eta_4 : \eta_3) = (\eta_4 \eta'_3)^\eta = \epsilon_1;$$
so that Darapti, in its corrected form PQ : R, does *not* fail in the case supposed.

12. The fallacious reasoning by which the Boolian logicians have arrived at the conclusion that Darapti, even in its corrected form PQ : R, is not valid, is founded on the assumption that their definitions of their symbols lead to the conclusion that the statement $(0 = 0A)$ is a formal certainty; whereas, consistently with their definitions, this statement may be either true or false. For example, in the case $B^0 A^e$, given in § 11, the statement $(0 = 0A)$ is false.

13. It is curious that, by fallacious reasoning of a totally different kind, I formerly arrived at the same erroneous conclusion as the Boolian logicians about Darapti and the other three doubtful syllogisms. Finding, firstly, that the implication of the second degree
$$(b : c)(b : a) : (a : c')'$$
which, I may denote by F (a, b, c), expresses Darapti on the assumption that the propositions a, b, c have all three the same *subject*, namely, *an individual taken at random out of our "universe of discourse"*; and finding, secondly, that this formula, considered as a *general* formula, with no necessary reference to Darapti or any other syllogism, fails in the case $b^\eta (ac)^\eta$, I concluded, a little too hastily, that Darapti must also fail in this case. In this I overlooked the fact that, though the case of failure $b^\eta (ac)^\eta$ may arise in the *general* formula F (a, b, c), when a, b, c are understood to be wholly *unrestricted*,

the case need not arise, and, as a matter of fact, cannot arise when the propositions a, b, c are subject to the restrictions which render F (a, b, c) equivalent to Darapti. For these restrictions necessarily imply $a^{-\eta}$ $b^{-\eta}$ $c^{-\eta}$ (see § 5, Formulæ 2, 3).

14. Bearing upon this and similar pitfalls which waylay the too hasty investigator, in whatever branch of science, when he ventures to stretch his formulæ beyond their proper limits, the following rules and cautions may be found useful. Let ϕ_u denote any formula ϕ $(x, y, z,$ etc.) when the variables, x, y, z, etc., have an *unrestricted* (or very wide) range of values; and let ϕ_r denote the same formula when the variables have a *restricted* range (or a narrower range within the same limits). Then, employing the symbol ϕ^ϵ to assert that ϕ is true for *all* admissible values of its constituents, x, y, z, etc., we have the true formula $\phi_u^\epsilon : \phi_r^\epsilon$; but *we have no right to assume the converse formula* $\phi_r^\epsilon : \phi_u^\epsilon$, *nor its equivalent, the formula* $\phi_u^{-\epsilon} : \phi_r^{-\epsilon}$. The assumption $\phi_r^\epsilon : \phi_u^\epsilon$ is the common fallacy in scientific researches of a too hasty induction, which erroneously supposes that the validity[1] of ϕ^r implies the validity of ϕ_u. The assumption $\phi_u^{-\epsilon} : \phi_r^{-\epsilon}$ which erroneously supposes that the non-validity of ϕ_u implies the non-validity of ϕ_r, is the fallacy which formerly led me into the error referred to in § 13.

[1] Any formula ϕ $(x, y, z,$ etc.) is called *valid* when we have ϕ^ϵ; that is to say, when the formula is true whatever values, within the limits of our data, we assign to the variables x, y, z, etc. The statement $\phi^{-\epsilon}$, that ϕ is *not* valid, does not imply ϕ^η, that it is *never* true.

[1905p]: Existential Import. Mind, Vol. 14, pp. 295-296.

EXISTENTIAL IMPORT.

May I ask the Boolian logicians who still maintain that their formula $(0A = 0)$ is necessarily true, whatever the class A may be, to point out the error (if error they find) in the following reasoning?

According to their symbolic conventions, the statement $(XA = X)$ asserts that "Every X is A," whatever X and A may represent. By their conventions also the symbol 0 represents *non-existence*. Let A represent *existent*. It follows that the statement $(0A = 0)$ asserts that "Every *non-existence* is *existent*," an assertion which is self-contradictory. Hence, the statement $(0A = 0)$ is *not* always true for all values (*i.e.*, *meanings*) of A.

Of course, the formula $(0A = 0)$ holds good in mathematics for every number or ratio A; as, for example, $(0 \times 2 = 0)$. But then, in mathematics, $(0 \times 2 = 0)$ does *not* assert that "Every 0 is 2".

H. MacColl.

V.—SYMBOLIC REASONING (VII.).[1]

By Hugh MacColl.

Syllogistic Validity.

1. The validity tests of the traditional logic turn mainly upon the question whether or not a syllogistic "term" (*i.e.*, *class* X, Y or Z) is "distributed" or "undistributed". In ordinary language these words rarely, if ever, lead to any ambiguity or confusion of thought; but logicians have somehow managed to work them into a perplexing tangle. In the proposition "All X is Y," the class X is said to be "distributed," and the class Y "undistributed". In the proposition "No X is Y," the class X and the class Y are said to be both "distributed". In the proposition "Some X is Y," the class X and the class Y are said to be both "undistributed". Finally, in the proposition "Some X is not Y," the class X is said to be "undistributed" and the class Y "distributed".

2. Let us examine some consequences of this tangle of technicalities. Take the leading syllogism Barbara, the validity of which no one will question, provided it be expressed in its *conditional* form, namely, "*If* all Y is Z and all X is Y, then all X is Z". Being admittedly valid, this syllogism must hold good whatever values (or meanings) we give to its constituents X, Y, Z. It must therefore hold good (as every logician will surely admit) when X, Y and Z are synonyms, and therefore all denote *the same class*. In this case also the two premisses and the conclusion will be three truisms which no one would dream of denying. Consider now one of these truisms, say "All X is Y". Here, by the usual logical convention, the class X is said to be "distributed," and the class Y "undistributed". But when X and Y are synonyms, they denote the *same class*, so that the same class may, at the same time and in the same proposition, be both "*distributed*" and "*un*distributed". Does not this sound like a contradiction? Speaking of a certain

---

[1] For VI., see Mind, January, 1905.

SYMBOLIC REASONING.

concrete collection of apples in a certain concrete basket, can we consistently and in the same breath assert that "All the apples are already *distributed*" and that "All the apples are still *undistributed*"? Do we get out of the dilemma and secure consistency if on every apple in the basket we stick a ticket X and also a ticket Y? Can we then consistently assert that all the X apples are *distributed*, but that all the Y apples are *undistributed*? Clearly not, for every X apple is also a Y apple, and every Y apple an X apple. In ordinary language the classes which we can respectively qualify as *distributed* and *undistributed* are mutually exclusive; in the logic of our text-books this is evidently not the case. Students of the traditional logic should therefore disabuse their minds of the idea that the words "distributed" and "undistributed" of their text-books necessarily refer to classes mutually exclusive, as they do in everyday speech; or that there is anything but a forced and fanciful connexion between the "distributed" and "undistributed" of current English and the technical "distributed" and "undistributed" of logicians.

3. To make the traditional logic symmetrical, as well as more widely applicable, we must extend our *Symbolic Universe*, or 'Universe of Discourse,' so as to include not only the three syllogistic classes X, Y, Z, but also their *complementary* classes 'X, 'Y, 'Z (see my preceding paper); these being so related to the former that if we take any class X and its complement 'X, the two are, on the one hand, mutually exclusive, and, on the other, make up together the whole symbolic universe S. That is to say, all the m individuals X_1, X_2, X_3, etc., of the class X, together with the n individuals $'X_1, 'X_2, 'X_3$, etc., of the class 'X, make up the $m + n$ individuals S_1, S_2, S_3, etc., which constitute the whole symbolic universe S. The same thing may, of course, be said of any other two complementary classes Y and 'Y. It is evident that if any two positive classes X and Y are mutually exclusive, the complementary negative classes 'X and 'Y overlap; and, *vice versâ*, if 'X and 'Y are mutually exclusive, X and Y overlap. With this convention it follows that if S be any individual taken at random out of our Symbolic Universe (or Universe of Discourse) S_1, S_2, S_3, etc., neither the statement S^x nor the statement $S^{'x}$ is impossible, and neither of them is a certainty; that is to say, we shall always have x^θ and $(x')^\theta$, and never x^η nor x^ϵ, where x means S^x and x' means $S^{'x}$ or its synonym S^{-x}. In other words, simplicity, logical consistency, and unrestricted generality require the convention that the recognition of any class A in our *Symbolic Universe* necessitates also the recognition in

it of the complementary class 'A; each forming a portion only, and both constituting the whole.

4. With these conventions we shall always have

(1) All X is Y $= (xy')^\eta$; (2) No X is Y $= (xy)^\eta$;
(3) Some X is Y $= (xy)^{-\eta}$; (4) Some X is not Y $= (xy')^{-\eta}$.

When, in any of these four propositions, a letter x or y is affected by *one* negation (only) or by *three* negations, the class (X or Y) to which it refers is said (in text-book language) to be "distributed"; but if it is affected by *two* negations (only), it is said to be "undistributed". Let the symbol X^d assert that X is (in the text-book sense) "distributed," while X^u asserts that X is "undistributed". In (1) we have $X^d Y^u$; for here x is affected by *one* negation only, namely, the exponent or predicate η outside the bracket, while y is affected by *two* negations, namely, the exponent η outside the bracket and the accent of denial inside the bracket. In (2) we have evidently $X^d Y^d$. In (3) we have $X^u Y^u$; for here each letter is affected by *two* negations (only), namely, the negation η and the *minus* sign preceding it. In (4) we have $X^u Y^d$; for here x is affected by *two* negations (only), namely, the negation η and the *minus* sign preceding it; while y is affected by *three* negations, namely, the negation η, the *minus* sign, and the accent inside the bracket. Thus *one* and *three* negations indicate a class "distributed"; while *two* indicates a class "*undistributed*". Evidently X^d implies $('X)^u$, and X^u implies $('X)^d$.

5. If we change y into x in proposition (1) of § 4, we get

All X is X $= (xx')^\eta$.

Here we have $X^d X^u$. This shows that there is no necessary antagonism between X^d and X^u; that (in the text-book sense) the same class may be both "distributed" and "undistributed" at the same time.

6. Instead of the six customary canons of the traditional logic, some of which are not quite reliable, and others not quite self-consistent, I propose the following methods of testing the validity of syllogisms :—

Let *unaccented* capitals denote *implications*, and let *accented* capitals denote *non-implications* (or the denials of implications). Thus, if A denote the implication $x:y$, then A' will denote its denial, the non-implication $(x:y)'$. Let A, B, C denote any syllogistic implications, while A', B', C' denote their respective denials. Every valid syllogism in *general* logic (with *unrestricted* values of x, y, z), and every valid syllogism of the *traditional* logic, except Darapti, Felapton, Fesapo and Bramantip, must have one or other of the two forms

(1) AB : C; (2) AB' : C'.

SYMBOLIC REASONING. 393

That is to say, either the two premisses and the conclusion are all three implications (or "universals"), as in (1); or else one premiss only and the conclusion are both non-implications. If, in *general* logic, any syllogism fails to comply with one or other of these two forms, it can only be valid conditionally; that is to say, it will not be a formal certainty for all values ϵ or η or θ of x, y, z. The second form may be reduced to the first form by transposing the premiss B' and the conclusion C', and changing their signs; for AB' : C' is equivalent to AC : B. When thus transformed, the validity of AB' : C', that is of AC : B, may be tested in the same way as the validity of AB : C. The test is easy. Suppose the conclusion C to be $x : z$, in which z may be affirmative or negative. If, for example, $z = $ *He is a soldier;* then $z' = $ *He is not a soldier.* But if $z = $ *He is not a soldier;* then $z' = $ *He is a soldier.* The conclusion C being, by hypothesis, $x : z$, the syllogism AB : C, if valid, becomes either

$$(x : y : z) : (x : z)$$

or else

$$(x : y' : z) : (x : z),$$

in which y refers to the class Y (or "middle term") not mentioned in the conclusion $x : z$.

7. To take a concrete example, let it be required to test the validity of

$$(y : z') (y : x')' : (x : z)'.$$

Let Q denote this syllogism. Transposing the non-implications, we get

$$\begin{aligned} Q &= (y : z')(x : z) : (y : x'), \\ &= (y : z')(z' : x') : (y : x'), \\ &= (y : z' : x') : (y : x'). \end{aligned}$$

Thus, Q satisfies the necessary condition of validity as laid down in § 6. It is therefore valid both in *general* logic and in the *traditional* logic.

8. As an instance of a *non-valid* syllogism of the form AB : C, we may give

$$(x : y')(y : z') : (x : z');$$

for since the y's in the two premisses have *different signs*, the one being negative and the other affirmative, the combined premisses can neither take the form $x : y : z'$ nor the form $x : y' : z'$, which are respective abbreviations for $(x : y)(y : z')$ and $(x : y')(y' : z')$. The syllogism is therefore not valid.

9. Syllogisms of the form AB : C' include *Darapti, Felapton, Fesapo* and *Bramantip*. In *general* logic these are not formal certainties, and are therefore only valid conditionally. With the conventions of § 3, however, the required conditions sometimes hold good in the traditional logic. The following is a

valid syllogism, though the traditional logic would not recognise it as valid because it violates the canon that "the middle term must be distributed at least once in the premisses":—

> If all who approve of *Protection* are *Conservatives*, and all who approve of fiscal *Retaliation* are also *Conservatives*; then somebody (one person at least) who does not approve of fiscal *Retaliation* does not approve of *Protection*.

Speaking of an individual taken at random from our 'Universe of Discourse,' let P = "He approves of *Protection*"; let C = "He is a *Conservative*"; and let R = "He approves of fiscal *Retaliation*". Then, for our two premisses we have (P : C)(R : C), and for our conclusion (R' : P)'. Thus, putting ϕ for the whole syllogism, we have
$$\phi = (P : C)(R : C) : (R' : P)',$$
which may be read "If P implies C, and R also implies C; then R' does not imply P". In other words, the conclusion asserts that "One may disapprove of fiscal *Retaliation* without approving of *Protection*".

Now, let us first consider the formula ϕ from the standpoint of *general* logic, in which the symbols P, C, R may denote any statements whatever, possible or impossible. By a general method (which it would take too long here to explain) for testing the validity of formulæ, it may be shown that the formula ϕ fails in the case $C^\epsilon(R + P)^\epsilon$. The failure may be easily verified as follows. We have
$$\phi = (P + R : C) : (R'P')^{-\eta}$$
$$= (R + P : C) : (R + P)^{-\epsilon}.$$
Now, suppose $C = \epsilon_1$ and $R + P = \epsilon_2$. We get (see preceding paper)
$$\phi = (\epsilon_2 : \epsilon_1) : \epsilon_2^{-\epsilon} = \epsilon_3 : \epsilon_2^{-\epsilon} = \epsilon_3 : \eta_1 = \eta_2.$$
This shows that in *general* logic ϕ fails in the case $C^\epsilon(R+P)^\epsilon$. Let ϕ_1 now denote the general syllogism ϕ when restricted to the *concrete* example about "conservatives," "retaliation," "protection". The failure of the general formula ϕ, with *unrestricted* values of C, R, P, does not necessarily involve the failure of ϕ_1, in which the values (or meanings) of the statements C, R, P are *restricted* by the condition $C^\theta R^\theta P^\theta$ (see § 3). The *restriction* C^θ of the *traditional* logic is inconsistent with the factor C^ϵ in the failure case of *general* logic. Thus we have $\phi^{-\epsilon}\phi_1^\epsilon$, which asserts that the particular formula ϕ_1 is valid though the general formula ϕ (with unrestricted values of C, R, P) is not. To assert that the concrete syllogism ϕ_1 fails in the case $C^\epsilon(R + P)^\epsilon$ would be to assert that it fails on the supposition that *everybody* is a *Conservative*, and that also *everybody* is either a *Retaliationist*

or a *Protectionist*. This supposition is not only contrary to fact, but, by the convention of § 3, it cannot even *arise* in the traditional logic, since the existence of C (conservatives) in our 'Universe of Discourse' necessarily implies the existence (the *symbolic* existence) also of 'C (non-conservatives) in the same universe, even if the latter should be mere "men of straw".

10. The syllogism discussed in § 9 may also be expressed in the form
$$(PC')^\eta (RC')^\eta : (R'P')^{-\eta}.$$
Let $C' = y$, let $P = z'$, and let $R = x'$. It then becomes
$$(yz')^\eta (yx')^\eta : (xz)^{-\eta}$$
which is equivalent to
$$(y:z)(y:x):(x:z')',$$
and may be read "If all *non-conservatives* (Y) are *non-protectionists* (Z), and all *non-conservatives* (Y) are also *non-retaliationists* (X); then some *non-retaliationist* (one at least) is a *non-protectionist*. In this form it becomes a valid syllogism of the type Darapti, with the "middle term" no longer "undistributed". Yet the two syllogisms are necessarily equivalent since $a : \beta$ is always equivalent to $\beta' : a'$; so that the canon about "middle term" distribution refuses admittance to a syllogism when it presents itself under one form, but lets it pass as valid when it disguises itself under another.

11. Consider the syllogism "If all *non-existences* are *fictitious*, and all *non-existences* are represented by the symbol *zero*; then some *fictitious* things (or thing) are represented by the symbol *zero*".

Speaking of something S taken at random out of our Symbolic Universe, or Universe of Discourse, let the symbol 0 denote the statement "It is *non-existent*"; let f denote "It is *fictitious*"; and let z denote "It is represented by the symbol *zero*". Putting ϕ for the syllogism, we have
$$\phi = (0:f)(0:z):(f:z')'$$
$$= (0:fz):(fz:\eta)'.$$
Now, since the Symbolic Universe contains both real existences (e) and non-existences (0), the statement 0 (which is short for S^0) is not impossible, so that $0^{-\eta}$, or its synonym $(0:\eta)'$, is always understood among our data, and may be expressed whenever convenient. Hence, we get
$$\phi = (0:fz)(0:\eta)':(fz:\eta)'$$
$$= (0:fz)(fz:\eta):(0:\eta), \text{ by transposing.}$$
Thus, the syllogism ϕ, like all valid syllogisms, comes ultimately under the formula $(a:\beta)(\beta:\gamma):(a:\gamma)$, and is therefore a formal certainty.

12. Next, take the syllogism "If no *centaurs* are really

existent, and no *fairies* are really *existent* ; then some things (or thing) that are not *centaurs* are not *fairies* ".

This syllogism is perfectly valid, though, in the above form, it violates the traditional canon that no conclusion can be drawn from negative premisses.

Speaking of an entity S taken at random out of our Symbolic Universe, let $c =$ " It is a *centaur* "; let $e =$ " It *exists* really "; and let $f =$ " It is a *fairy* ". Also let ϕ denote the syllogism. We have

$$\phi = (c : e') (f : e') : (c' : f)' = (e : c') (e : f') : (c' : f)'$$
$$= (e : c'f') : (c'f' : \eta)' = (e : c'f') (e : \eta)' : (c'f' : \eta)'$$
$$= (e : c'f') (c'f' : \eta) : (e : \eta) \text{ by transposition ;}$$

so that here also ϕ is a formal certainty, as it is a particular case of the formula $(a : \beta) (\beta : \gamma) : (a : \gamma)$. The factor $(e : \eta)'$, which we introduced in the last complex implication but one, is, of course, understood throughout; as, by our convention of a 'Symbolic Universe,' the statement e (which is short for S^e) is not impossible, just as its denial, the statement e', or its synonym 0 or S^0, is not impossible. For our convention (see § 3) implies both e^θ and 0^θ, which respectively imply $e^{-\eta}$ and $0^{-\eta}$, these last being synonymous with $(e : \eta)'$ and $(0 : \eta)'$.

13. The three common-sense canons of the traditional logic, (1) that " All X is Y " implies " Some X is Y," (2) that " No X is Y " implies " Some X is not Y," and (3) that " All X is Y " and " No X is Y " are incompatible have been (somewhat paradoxically) called in question by some logicians; but on the assumption of a Symbolic Universe, including, when needed, both the real existences e_1, e_2, e_3, etc., and the non-existences or unrealities $0_1, 0_2, 0_3$, etc., they can be formally proved as follows :—

Speaking of an entity S, taken at random out of our Symbolic Universe, S_1, S_2, S_3, etc., let x and y respectively denote the two statements, " It belongs to the class X " and " It belongs to the class Y ". Let ϕ_1, ϕ_2, ϕ_3 denote the three canons respectively. We have

$$\phi_1 = (x : y) : (x : y')' = (xy)^\eta : (xy)^{-\eta} = (xy')^\eta (xy)^\eta : \eta$$
$$= (xy' : \eta) (xy : \eta) = (xy' + xy : \eta) : \eta$$
$$= \{x(y' + y) : \eta\} : \eta = (x\epsilon : \eta) : \eta = (x : \eta) : \eta$$
$$= (\theta : \eta) : \eta = \eta : \eta = \epsilon.$$

This proof is, of course, far fuller than is necessary, and I only give it thus in all its details to show how the same statement may sometimes be presented under different forms. The canon ϕ_2 may be proved similarly by merely interchanging y and y'. To prove ϕ_3, we have

$$\phi_3 = (x : y) (x : y') : \eta = (xy')^\eta (xy)^\eta : \eta,$$

which (as already shown) is equivalent to ϕ_1.

SYMBOLIC REASONING.

14. As already explained, my symbolic system assumes that any class A may be divided into individuals, or mutually exclusive divisions, A_1, A_2, A_3, etc. This convention may, when convenient or necessary, be carried farther. Any of these divisions, say A_8, may again be subdivided into $(A_8)_1$, $(A_8)_2$, $(A_8)_3$, etc.; which, to avoid brackets, may be denoted respectively by $A_{8\cdot 1}$, $A_{8\cdot 2}$, $A_{8\cdot 3}$, etc. And these again may be subdivided, so that, taking any subdivision $A_{8\cdot 3}$, we may have $A_{8\cdot 3\cdot 1}$, $A_{8\cdot 3\cdot 2}$, $A_{8\cdot 3\cdot 3}$, etc., *ad libitum*.

15. The convention that the symbol of non-existence 0 should be treated as a class symbol just like others, and that this supposed indivisible atom of symbolic logic may after all be broken up into individuals or elements 0_1, 0_2, 0_3, etc., like any other class symbol, appears to have caused as much astonishment as the modern discovery (or convention ?) that the physical atom may be broken up into electrons. But the convention will help us to get rid of some too hastily accepted paradoxes, not only in logic but also in mathematics, in metaphysics, and even in physics. We are all more or less subject to a certain mental disorder which, for want of a better name, I may call *symbolatry*. We mistake the symbol for the reality. We worship the formulæ of our own invention as if they were living oracles whose infallibility it would be impious to call in question. Yet all formulæ, being founded more or less on arbitrary conventions or definitions, must necessarily have their limits of applicability. When we force them beyond those limits, as we too often do, they evolve strange paradoxes, which some eminent logicians and mathematicians accept with surprising readiness, born of over-confidence in symbolic reasoning, but which the plain man of common sense stubbornly refuses to believe. Having myself been a victim more than once to symbolic hallucination, I have now become thoroughly sceptical. When rigorous symbolic reasoning brings me face to face with a startling paradox, I carefully scrutinise the fundamental assumptions, including definitions and conventions, in search of some lurking ambiguity; and, in nine cases out of ten, the search is successful. That is why I cannot accept some of the paradoxes of the non-Euclidean geometry, such as that "two straight lines may enclose a space," and that "a point moving always in a straight line and in the same direction may, finally, after an infinitely long journey, find itself at the point of starting". The path of the moving point may be "straight" *symbolically*, but it can hardly be so *really*.

[1905r]: The Existential Import of Propositions [A Reply to Bertrand Russell]. Mind, Vol. 14, pp. 401-402.

VI.—DISCUSSION.

THE EXISTENTIAL IMPORT OF PROPOSITIONS.

MR. MACCOLL'S interesting paper in the January number of MIND, together with his note in the April number, raises certain points which call for an answer from those who (like myself) adhere to the usual standpoint of symbolic logicians on the subject of the existential import of propositions.

The first point in regard to which clearness is essential concerns the meaning of the word "existence". There are two meanings of this word, as distinct as stocks in a flower-garden and stocks on the Stock Exchange, which yet are continually being confused, or at least supposed somehow connected. Of these meanings, only one occurs in philosophy or in common parlance, and only the other occurs in mathematics or in symbolic logic. Until it is realised that they have absolutely nothing to do with each other, it is quite impossible to have clear ideas on our present topic.

(a) The meaning of *existence* which occurs in philosophy and in daily life is the meaning which can be predicated of an individual, the meaning in which we inquire whether God exists, in which we affirm that Socrates existed, and deny that Hamlet existed. The entities dealt with in mathematics do not exist in this sense: the number 2, or the principle of the syllogism, or multiplication, are objects which mathematics considers, but which certainly form no part of the world of existent things. This sense of existence lies wholly outside Symbolic Logic, which does not care a pin whether its entities exist in this sense or not.

(b) The sense in which existence is used in symbolic logic is a definable and purely technical sense, namely this: To say that A exists means that A is a class which has at least one member. Thus whatever is not a class (*e.g.*, Socrates) does not exist in this sense; and among classes there is just one which does not exist, namely, the class having no members, which is called the null-class. In this sense, the class of numbers (*e.g.*) exists, because 1, 2, 3, etc., are members of it; but in sense (a) the class and its members alike do not exist: they do not stand out in a part of space and time, nor do they have that kind of super-sensible existence which is attributed to the Deity.

It may be asked: How come two such diverse notions to be confounded? It is easy to see how the confusion arises, by considering classes which, if they have members at all, must have

THE EXISTENTIAL IMPORT OF PROPOSITIONS.   399

members that exist in sense (*a*). Suppose we say: " No chimeras exist". We may mean that the class of chimeras has no members, *i.e.*, does not exist in sense (*b*), or that nothing that exists in sense (*a*) is a chimera. These two are equivalent in the present instance, because if there were chimeras, they would be entities of the kind that exist in sense (*a*). But if we say "no numbers exist," our statement is true in sense (*a*) and false in sense (*b*). It is true that nothing that exists in sense (*a*) is a number; it is false that the class of numbers has no members. Thus the confusion arises from undue preoccupation with the things that exist in sense (*a*), which is a bad habit engendered by practical interests.

Mr. MacColl assumes (p. 74) two universes, the one composed of existences, the other of non-existences. It will be seen that, if the above discrimination is accepted, these two universes are not to be distinguished in symbolic logic. All entities, whether they exist or whether they do not (in sense (*a*)), are alike real to symbolic logic and mathematics. In sense (*b*), which is alone relevant, there is among classes not a multitude of non-existences, but just one, namely, the null-class. All the members of every class are among realities,[1] in the only sense in which symbolic logic is concerned with realities.

But it is natural to inquire what we are going to say about Mr. MacColl's classes of unrealities, centaurs, round squares, etc. Concerning all these we shall say simply that they are classes which have no members, so that each of them is identical with the null-class. There are no Centaurs; 'x is a Centaur' is false whatever value we give to x, even when we include values which do not exist in sense (*a*), such as numbers, propositions, etc. Similarly, there are no round squares. The case of nectar and ambrosia is more difficult, since these seem to be individuals, not classes. But here we must presuppose definitions of nectar and ambrosia: they are substances having such and such properties, which, as a matter of fact, no substances do have. We have thus merely a defining concept for each, without any entity to which the concept applies. In this case, the concept is an entity, but it does not denote anything. To take a simpler case: "The present King of England" is a complex concept denoting an individual; "the present King of France" is a similar complex concept denoting nothing. The phrase intends to point out an individual, but fails to do so: it does not point out an unreal individual, but no individual at all. The same explanation applies to mythical personages, Apollo, Priam, etc. These words have a *meaning*, which can be found by looking them up in a classical dictionary; but they have not a *denotation*: there is no entity, real or imaginary, which they point out.

[1] This holds even of the null-class. Of all the members of the null-class, *every* statement holds, since the null-class has no members of which it does not hold. See below, on the interpretation of the universal affirmative A.

It will now be plain, I hope, that the ordinary view of symbolic logicians as to existential import does not require Mr. MacColl's modifications. This view is, that A and E do not imply the existence, in sense (b), of their subjects, but that I and O do imply the existence, in sense (b), of their subjects. No one of the four implies the existence, in sense (a), either of its subject or of any of the members of its subject. We have, adopting Peano's interpretation :—

A. All S is P = For all values of x, 'x is an S' implies 'x is a P'.

E. No S is P = For all values of x, 'x is an S' implies 'x is not a P'.

I. Some S is P = For at least one value of x, 'x is an S' and 'x is a P' are both true.

O. Some S is not P = For at least one value of x, 'x is an S' and 'x is not a P' are both true.

Thus I and O require that there should be at least one value of x for which x is an S, i.e., that S should exist in sense (b). I also requires that P should exist, and O requires that not-P should exist. But A and E do not require the existence of either S or P; for a hypothetical is true whenever its hypothesis is false,[1] so that if 'x is an S' is always false, 'All S is P' and 'No S is P' will both be true whatever P may be.

The above remarks serve to answer the objection raised by Mr. MacColl in the April number of MIND (p. 295) to the equation 0A = 0. To begin with, 0 does not represent the class of non-existences, but the non-existent class, i.e., the class which has no members. Thus, if "XA = X" means "every X is an A,"[2] then "0A = 0" means "every member of the class which has no members is an A," or "for every value of x, 'x is a member of the class which has no members' implies 'x is an A'". This hypothetical is true for all values of x, because its hypothesis is false for all values of x, and a hypothetical with a false hypothesis is true. Thus Mr. MacColl's objection rests upon his taking 0 to be the class of non existences, presumably in sense (a), since only so would 0 be a class with many members, all of them unreal, as he supposes it to be. The true interpretation of 0, as the non-existent class, in sense (b), at once disposes of the difficulty.

The same principles solve Lewis Carroll's paradox, noticed by "W" in the April number of MIND (p. 293). I cannot agree with "W" in regarding the paradox as merely verbal; on the contrary, I consider it a good illustration of the principle that a false proposition implies every proposition. Putting p for 'Carr is out,' q for 'Allen is out,' and r for 'Brown is out,' Lewis Carroll's two hypotheticals are :—

(1) q implies r.
(2) p implies that q implies not-r.

---

[1] See my *Principles of Mathematics*, vol. i., p. 18.
[2] Not "every X is A," as Mr. MacColl says, and as most logicians say.

THE EXISTENTIAL IMPORT OF PROPOSITIONS. 401

Lewis Carroll supposes that 'q implies r' and 'q implies not-r' are inconsistent, and hence infers that p must be false. But as a matter of fact, 'q implies r' and 'q implies not-r' must both be true if q is false, and are by no means inconsistent. The contradictory of 'q implies r' is 'q does not imply r,' which is not a consequence of 'q implies not-r'. Thus the only inference from Lewis Carroll's premisses (1) and (2) is that if p is true, q is false, *i.e.*, if Carr is out, Allen is in. This is the complete solution of the paradox.

B. RUSSELL.

MR. RUSSELL has very kindly and courteously sent me a proof of the above, so that logicians might, in the same number of MIND, have an opportunity of reading his criticism side by side with any comments I might desire to make. My comments shall be brief. With much of what Mr. Russell says in his able and interesting dissection of the question at issue I agree; but not with all. That the word *existence*, like many others, has various meanings is quite true; but I cannot admit that any of these "lies wholly outside Symbolic Logic". Symbolic Logic has a right to occupy itself with any question whatever on which it can throw any light. As regards Existential Import, the one important point on which I appear to differ from all other symbolists is the following. The null class o, which they define as containing no members, and which I, for convenience of symbolic operations, define as consisting of the null or unreal members o_1, o_2, o_3, etc., is understood by them to be *contained in every class*, real or unreal; whereas I consider it to be *excluded from every real class*. Their convention of universal inclusion leads to awkward and, I think, needless paradoxes, as, for example, that "Every round square is a triangle," because round squares form a *null* class, which, by them, is understood to be *contained in every class*. My convention leads, in this case, to the directly opposite conclusion, namely, that "No round square is a triangle," because I hold that every purely *unreal* class, such as the class of round squares, is necessarily excluded from every purely *real* class, such as the class of figures called triangles.

I may mention, as a fact not wholly irrelevant, that it was in the actual application of my symbolic system to concrete problems that I found it absolutely necessary to label realities and unrealities by special symbols e and o, and to break up the latter class into separate individuals, o_1, o_2, o_3, etc., just as I break up the former into separate individuals, e_1, e_2, e_3, etc. It is a vital principle in the evolution of any effective symbolic system that we should modify, and, whenever possible, simplify our notation, in order to adapt it to the varying needs of different classes of problems. It is this elastic adaptability to circumstances—this readiness to change the meaning of any symbol (not excepting *zero*), and even of any conventional arrangement of symbols, whenever it suits the purpose of the investigation—that enables my symbolic system to solve certain classes of problems (especially in mathematics) which lie entirely beyond the reach of any other symbolic system within my knowledge. In saying this I do not mean in any way to suggest that other symbolic systems may not have the advantage over mine in regard to other classes of problems which I have never studied. Mr. Russell's system in particular seems to be specially constructed to deal with problems which lie altogether out of the line of my researches. Different kinds of work require different kinds of instruments.

The following arrived too late to be added to the article on "Symbolic Reasoning":—

[My statement in § 3 (of "Symbolic Reasoning"), that "if any two classes X and Y are mutually exclusive, the complementary classes 'X and 'Y overlap," requires qualification. I should have said " if any two *non-complementary* classes, etc.". This I discovered symbolically as follows ; though, of course, it may be proved more briefly. Let ϕ denote the *unqualified* statement. We get

$$\phi = (xy)^\eta : (x'y')^{-\eta} = (xy)^\eta (x'y')^\eta : \eta$$
$$= (x : y') (y' : x) : \eta = (x = y') : \eta = (x = y')^\eta.$$

Thus, ϕ is equivalent to the statement that *the class* X *cannot be the complement of the class* Y, a statement which only holds good when X and Y are understood throughout to be *non-complementary*.]

HUGH MACCOLL.

[1905s]: The Existential Import of Propositions. Mind, Vol. 14, No. 56. pp. 578-580.

THE EXISTENTIAL IMPORT OF PROPOSITIONS.

My reply to Mr. Russell in the last number of MIND will also, on all essential points, serve as an answer to Mr. Shearman's note in the same number. Mr. Shearman, like most symbolists, maintains that "it is not self-contradictory to say $(0\!\prec\!A)$, whether A stands for 'existent' or for any other term". Now, consider the formula $AB\!\prec\!A$, which I believe all Boolians accept as valid for all values of A and B. It asserts that the class AB is always wholly contained in the class A. Let the classes A and B be both real but mutually exclusive. By our data, the class A, consisting of the individuals A_1, A_2, etc., really exists; so does the class B consisting of the individuals B_1, B_2, etc.; but the class AB, consisting of the individuals $(AB)_1$, $(AB)_2$, etc., supposed to be common to A and B, has no real existence; it is an unreal class, so that the unreal individuals composing it may be denoted by 0_1, 0_2, etc. Can we consistently assert, as the formula $AB\!\prec\!A$ (or its equivalent in this case $0\!\prec\!A$) asserts, that the *unreal* (and therefore non-existent) individuals 0_1, 0_2, etc., are contained in the class of *real* individuals A_1, A_2, etc.? It is hardly an answer to say that the symbol 0, as logicians usually define it, does not denote an unreal class made up of unreal members, as I define it, but a null or empty class containing no members; for is not a null class containing no members logically equivalent to an unreal class made up of unreal members? As I said in my reply to Mr. Russell, the crucial point which here separates me, I believe, from all other symbolists is that I regard the class 0, whether empty or made up of unrealities, as necessarily *excluded from every real class;* whereas they all regard it as *contained in every class whether real or not.*

Mr. Shearman says that in the note which he criticises I confuse the

[1] We cannot, of course, question such a symbolic statement unless we know the circumstances of the concrete case which it represents—here we do know them, and here the symbolic statement is misleading.

NOTES.

term 'existent' with the existence of things denoted by the term, and the term 0 with the non-existence of things denoted by it. A little further on, he says—and if he will read §§ 11, 12 of my article in the number of MIND containing his note, he will find that in this I quite agree with him—that "the presence of this term 'existent' does not imply the existence of things denoted by the term". Now, is it not curious that the confusion of ideas which Mr. Shearman imagines he finds in my note in regard to the words 'existent' and 'non-existent' is precisely the defect which I think I find in his note and in the reasoning of other symbolists? How is this? The explanation from my point of view is, that the confusion is solely on their side, and that it arises from the fact that they (like myself formerly) make no *symbolic* distinction between *realities* and *unrealities*, which I now respectively represent by the symbols e and 0. With them, 'existence' means simply existence in the Universe of Discourse, whether the individuals composing that universe be real or unreal; and the symbol 0, as they understand it, merely denotes *absence* from that universe. With me, the symbol e denotes *realities*, and 0 denotes *unrealities*, both of which may, or may not, co-exist in the Universe of Discourse or Symbolic Universe, S. Absence from the Universe of Discourse I hold to be illogical. Once anything (real or unreal) is spoken of, it must, from that fact alone, belong to the Symbolic Universe S, though not necessarily to the universe of realities e.

I cannot see the relevancy of Mr. Shearman's argument commencing with the statement that "with two terms 0 and 'existent' the universe of discourse is necessarily divided into four compartments". Even if the classes corresponding to these four compartments were all mutually exclusive, as he seems erroneously to assume, his criticism in this paragraph would not touch the principle of my note. The argument of my note, and my argument still, is that, since the statement $(XA = X)$, or its equivalent $(X \prec A)$, is understood by Boolian logicians to assert that "Every individual of the class X is also an individual of the class A," consistency requires that the statement $(0e = 0)$, or its equivalent $(0 \prec e)$, shall similarly assert that "Every individual of the class 0 is also an individual of the class e"; and that this being a self-contradiction, the formula $(0A = 0)$ or $(0 \prec A)$ fails when $A = e$ ('existent').

The following is a point of some importance. Let A and B be any two classes; let S be an individual taken at random from our universe of discourse; and let (AB), within brackets, denote the class of individuals common to A and B. So long as A and B are real and not mutually exclusive, we have

$$S^A S^B = S^{(AB)} = \theta.$$

But when A and B are real but *mutually exclusive*, the class (AB) is *unreal*, so that in this case we have

$$S^A S^B = \eta, \text{ but } S^{(AB)} = S^0 = \theta.$$

Thus, $S^A S^B$ and $S^{(AB)}$ are equivalent when A and B are real and not mutually exclusive; but they are *not* equivalent when A and B are real and mutually exclusive.

In my *general* logic, or logic of *statements*, the implication (AB : A)—which is *not* (as some have supposed) equivalent to the Boolian (AB$\prec$A) —is always true; for by definition we get

$$AB : A = (AB \cdot A')^\eta = (AA' \cdot B)^\eta = (\eta B)^\eta = \epsilon.$$

Similarly may be proved the truth of my formula $(\eta : A)$, which is *not* equivalent to the Boolian $(0 \prec A)$. A little consideration will show that though the implication $(\eta : A)$ is valid, the implication $(Q^\eta : Q^A)$ is not.

NOTES.

The latter fails both in the case $Q^\eta A^\epsilon$ and in the case $Q^\eta A^\theta$. In the first case, putting $Q = \eta$ and $A = \epsilon$, we get

$$Q^\eta : Q^A = \eta^\eta : \eta^\epsilon = \epsilon : \eta = (\epsilon\eta')^\eta = (\epsilon\epsilon)^\eta = \eta\,;$$

and in the second case, putting $Q = \eta$, $A = \theta$, we get

$$Q^\eta : Q^A = \eta^\eta : \eta^\theta = \epsilon : \eta = \eta, \text{ as before.}$$

The difference between my symbol of implication (:) and the usual symbol of class inclusion (—<) will appear from the fact that the statements $\eta : \epsilon$ and $\eta : \theta$ are both true, while the statements η—<ϵ and η—<θ are both false. For example, η—<ϵ asserts that every *impossibility* is a *certainty*, which is absurd; whereas $\eta : \epsilon$ only asserts $(\eta\epsilon')^\eta$, which is self-evident.

HUGH MACCOLL.

[1906k]: Symbolic Reasoning (VIII). *Mind,* Vol. 15, pp. 504-518.

IV.—SYMBOLIC REASONING (VIII.).[1]

By Hugh MacColl.

1. The main subject of this article will be paradoxes. We meet with them everywhere—in logic, in mathematics, and in science generally. They nearly always spring from the ambiguities and obscurities more or less inherent in all languages—the symbolic languages of logic and mathematics not excepted. The same words or symbols suggest different concepts to different minds, and even to the same mind at different times. Take the word *infinite* or *infinity*, which mathematicians usually represent by the symbol ∞. Works on modern geometry often speak of a series of straight lines meeting at "*the* point at infinity," when, as a matter of fact, their points of intersection at infinity, instead of being one, may be many, or may even be non-existent. And they also speak of a series of points being all in "the line at infinity," when there may be no real line containing all the said points, either at infinity or elsewhere. Similarly algebraists sometimes speak of the infinity ∞ as if it were one definite huge number or ratio which differed from a million or a billion in only one respect, that of being much larger; whereas there are numberless infinities, each of which differs from a million or a billion not only in being much larger, but also in another important quality of which I shall speak presently. The truth is that the real 'infinity' of mathematics denotes not a single individual ratio but a whole class, and that the symbol ∞, *when it represents a reality* (which it does not always), sometimes stands for an infinite ratio ∞_1, at another time or in another place for a different infinite ratio ∞_2, at another time or place for an infinite ratio ∞_3, and so on. Thus, when we meet such a statement as $\infty = 2\infty = \frac{1}{2}\infty$, which seems to assert the absurdity that infinity is equal to its double and also to its half, we must understand it to mean $\infty_1 = 2\infty_2 = \frac{1}{2}\infty_3$, a perfectly self-consistent statement which only asserts that the infinity ∞_1 is double the infinity ∞_2 and

---

[1] For VII. see Mind, July, 1905.

SYMBOLIC REASONING.

half the infinity ∞_3. Similarly an infinity ∞^m may be either infinite or infinitesimal in comparison with another infinity ∞^n.

2. But, it may be asked, what is it exactly that separates the infinite from the finite? Where is the exact line of demarcation? Let F denote the class of *finite* positive numbers or ratios, made up of the individual ratios F_1, F_2, F_3, etc.; and let H denote the class of positive *infinities*, H_1, H_2, H_3, etc. Logical consistency requires that these two classes shall be considered mutually exclusive; for it is clear that, speaking of any number or ratio A, the two statements A^F and A^H are mutually inconsistent. If any real ratio A is finite, it cannot be infinite; and if it is infinite it cannot be finite. Neither can it, however large, be on the borderland between the two classes. These statements may be expressed by a single formula, $(A^F A^H)\eta$.

3. For example, let M denote a *million*. The number M^M (which means $M \times M \times M \times \ldots$ etc., up to a million factors) is inconceivably large—so large that the volume of the earth (or even of the sun, or of a sphere enclosing our whole solar system) divided by that of the smallest drop of water, would be an exceedingly small number in comparison; yet the number M^M belongs to the finite class F and not to the infinite class H. So does the number M^{MM}, which is inconceivably large even in comparison with the inconceivably large number M^M; and so does M^{MMM}, which is inconceivably large even in comparison with M^{MM}. And we might carry the ascending comparison further, till the hand got weary of the repetition of exponents, without finding any number or ratio that belongs to the infinite class H, or that does not belong to the finite class F. Are the finite F and the infinite H then both indefinable? Is there no quality Q which we can assert of every F and deny of every H? There is, and it is this: *Every finite number or ratio F, however large, is expressible, either exactly or* (like π) *approximately, in the decimal or some other conventional notation, in terms of some finite number or ratio* (such as 10 or 100 or 1,000,000, etc.); *whereas no infinite number or ratio is, either exactly or approximately, expressible solely in terms of any finite number or numbers.* Thus, M^M, M^{MM}, M^{MMM}, etc., though inconceivably large, are all finite, because they are all expressible in terms of the finite and known number M; whereas $H_1, 2H_1, \frac{1}{10}H_1$, and generally FH (whatever be the finite number F and the infinite number H) are all infinities, because they are too large to be expressible, either exactly or approximately, solely in terms of any known numbers or ratios however large.

4. Similarly we may define the class of *infinitesimal* ratios, a class which we will here denote by h. Just as every infinite number or ratio H is too large to be expressible, either exactly or approximately, solely in terms of any finite number F, or finite numbers F_1, F_2, etc., so every infinitesimal ratio h is too small to be expressible, either exactly or approximately, solely in terms of any finite number or ratio F, or finite numbers F_1, F_2, etc. Thus, just as M^M, though inconceivably large, is still not infinite, so its reciprocal $1 \div M^M$, though inconceivably small, is still not infinitesimal.

5. Using the symbol A^x to assert that the number or ratio A belongs to the class x, where x may stand for F or H or h, these conventions or definitions give us several evident formulæ, of which the following are a few:—

(1) $(FH)^H$; (2) $(Fh)^h$; (3) $\left(\dfrac{F}{H}\right)^h$; (4) $\left(\dfrac{F}{h}\right)^H$; (5) $\left(\dfrac{H}{F}\right)^H$;

(6) $\left(\dfrac{h}{F}\right)^h$; (7) $(F \pm h)^F$; (8) $(H \pm F)^H$; (9) $H_1 + F = H_2$;

(10) $\dfrac{H_2}{F} = H_3$; (11) $\left(\dfrac{H}{h}\right)^H$; (12) $\left(\dfrac{h}{H}\right)^h$.

What leads to much confusion is the fact that mathematicians also use the word *infinity* and the symbol ∞ to denote such expressions as $\dfrac{1}{0}, \dfrac{2}{0}$, etc., *which represent no real ratios at all but pure non-existences*, such as in my two preceding articles I have denoted by the symbol 0. But as these pseudo-ratios $\dfrac{1}{0}, \dfrac{2}{0}$, etc., form a different class of non-existences from the pseudo-ratios $\dfrac{0}{1}, \dfrac{0}{2}$, etc., it would be convenient in *mathematical* reasoning to restrict the symbol 0 to the latter class, and the symbol ∞ to the former. Thus, the symbol 0 represents, as it were, the *death of a real infinitesimal ratio h* in passing from the positive to the negative state, or *vice versâ*; while the other non-existence symbol ∞ (which may be called *pseudo-infinity*) similarly represents, as it were, the *death of a real infinite ratio H*, in making the same transition. We shall then get the following self-evident formulæ, in which (to prevent ambiguity) the symbol $::$ will be used (instead of $=$) to assert equivalence of *propositions*, not equivalence of *ratios*.

(13) $\left(\dfrac{A}{B}\right)^H :: \left(\dfrac{B}{A}\right)^h$; (14) $\left(\dfrac{A}{B}\right)^0 :: \left(\dfrac{B}{A}\right)^\infty$; (15) $\left(\dfrac{\infty}{F}\right)^\infty$; (16) $\left(\dfrac{F}{\infty}\right)^0$;

(17) $(F\infty)^\infty$; (18) $(F0)^0$; (19) $(\infty \pm F)^\infty$; (20) $\left(\dfrac{F}{0}\right)^\infty$; (21) $\left(\dfrac{0}{F}\right)^0$.

SYMBOLIC REASONING.

If we denote negative infinity by K, negative infinitesimal by k, and use the symbol $\tan^x A$ as an abbreviation for the *statement* $(\tan A)^x$, and similarly for other trigonometrical ratios, we get

(22) $\tan^H\left(\dfrac{\pi}{2}-h\right)$; (23) $\tan^K\left(\dfrac{\pi}{2}+h\right)$; (24) $\tan^\infty\left(\dfrac{\pi}{2}\right)$;

(25) $\cot^0\left(\dfrac{\pi}{2}\right)$; (26) $\cot^h\left(\dfrac{\pi}{2}-h\right)$; (27) $\cot^k\left(\dfrac{\pi}{2}+h\right)$;

(28) $\sec^\infty\left(\dfrac{\pi}{2}\right)$; (29) $\cos^0\left(\dfrac{\pi}{2}\right)$; (30) $\sec^K\left(\dfrac{\pi}{2}+h\right)$;

with numberless others on the same principle of notation.

6. The following is a geometrical illustration bearing both on the ambiguity (as commonly employed) of the word *infinity*, and on the (to my mind inadmissible) paradox of the non-Euclidean geometry, that a point moving always in the same straight line and in the same direction may nevertheless finally find itself at the point of starting.[1] Let a

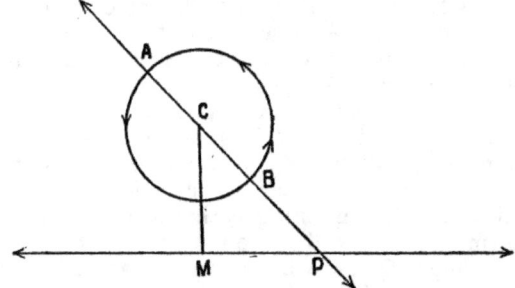

straight line of unlimited length, such as AB produced both ways indefinitely, revolve uniformly in the unscrewing direction round a fixed point C, and cut a *fixed* straight line, also of unlimited length, at the variable point P. Let MC, the perpendicular from C upon the fixed straight line, be our linear unit. As the moving line revolves uniformly round C, the point P moves farther and farther, and with fast increasing velocity, to the right of M, while the angle PCM (or BCM) increases continuously and uniformly. Just before this angle BCM becomes a right angle (the difference being infinitesimal) the straight line MP (which also represents tan PCM, since MC is our linear unit) passes through a

---

[1] The paradox is, of course, perfectly admissible in regard to a line that is *virtually* straight but not straight absolutely (see § 8). Such are some of the so-called "straight lines" of the non-Euclidean geometry.

numberless succession of increasing real positive infinite values, H_1, H_2, H_3, etc., that is to say, positive values all too large to be expressible solely in terms of any finite values (see § 3). When the angle BCM becomes exactly a right angle, the point P, the infinite straight line MP, and the infinite ratio $\frac{MP}{MC}$, which represents tan PCM, all vanish, and the revolving line becomes parallel to the fixed line. Then, the revolution still continuing, another and *different* point Q, *representing the intersection of the other branch (the branch CA produced) of the revolving line with the fixed line*, springs into existence at an infinite distance to the *left* of M, and the straight line QM, after passing through a numberless series of really infinite but diminishing values, eventually becomes finite and continues to diminish till it finally vanishes, or becomes zero, just as the variable point Q coincides with M.

7. The mistake made by non-Euclideans, when they appeal to geometrical examples like the preceding, is that they wrongly identify the variable point P which moves always *through contiguous positions* to the *right* of M, and farther and farther away from M till it finally vanishes, with the point Q which immediately after springs into existence at an infinite distance to the *left* of M, and then moves through contiguous positions nearer and nearer to M till it finally coincides with it. When this coincidence takes place, the infinite branch containing both the fixed point C and the revolving point B will be pointing perpendicularly upwards, with B above C, and will contain neither the point P, *which it lost when parallel to the fixed line and never recovered after*, nor the point Q, *which had never belonged to it but to the other infinite branch containing the revolving point* A. The fallacy of the non-Euclideans is analogous to that of the lawyers who assert that "the king never dies," because the instant the king P dies, his successor Q becomes, *ipso facto*, king in his place. To both we may return analogous answers: to the lawyers we reply that the dead king P is nevertheless *not* the living king Q; and to the non-Euclideans we reply that the vanished point P is nevertheless *not* the new point Q. When the angle BCM is a right angle there is no point P, and there is no point Q, so that in this position of the revolving line (a position parallel to the fixed line) the distances MP and QM are *pseudo-infinities* which have only symbolic existence. Suppose the revolving line to remain for a moment stationary in this parallel position. If it then revolves through an infinitesimal angle in the screwing direction, we get the

SYMBOLIC REASONING.

point P to the right of M, and a *real* infinite distance MP. If, on the contrary, it revolves from its parallel position through an infinitesimal angle in the unscrewing direction, we get the point Q to the left of M, and a *real* infinite negative distance MQ.

8. It would conduce to logical accuracy in dealing with these questions if we introduced the term *virtually* into our reasoning and defined it as follows. Two straight lines are said to be *virtually* parallel[1] when they meet at some real infinite distance, H or −H. Suppose, for example, we have three straight lines, A, B, C, and that A meets B at an infinite distance H_1, and meets C at another but still infinite distance H_2. Here we may accurately assert that A, B, C are *virtually* parallel, for, by our very definition of the words *infinite* and *infinitesimal*, it follows that the error of deviation from the parallel position, though theoretically real, must for ever remain too small for the most perfect instrument to detect, and for the most powerful notation accurately or approximately to express in finite terms. In like manner, if we suppose A and B to be two ratios, finite, infinite or infinitesimal, the statement that "A and B are *virtually*[2] equal" means $\frac{A}{B} = 1 \pm h$, in which h denotes any infinitesimal ratio, as defined in § 4. In the infinitesimal calculus (including the differential and integral) an equation (A = B) often asserts *virtual* and not real equality; but logical accuracy may be secured by the tacit convention that whenever we have the statement (A = B), it is to be understood as asserting that "A is *either absolutely or virtually* equal to B". Let A, B, C be three points on the surface of a sphere of radius R, forming the spherical triangle ABC. Whether R be finite, infinite, or infinitesimal, as defined in §§ 3, 4, if the ratios $\frac{AB}{R}, \frac{BC}{R}, \frac{CA}{R}$ be all three infinitesimal, the sum of the three angles A, B, C is *virtually* but not really equal to two right angles. If the three points A, B, C be on a *real plane surface*, then the sum of the three angles A, B, C is *really* equal to two right angles. Again, a finite section AB of a curve may be called *virtually* straight when at every

---

[1] Similarly two straight lines are *really* parallel when their (non-existent) point of intersection is at some pseudo-infinite distance, such as $\frac{1}{0}$, or $\frac{2}{0}$ or $\frac{\infty}{3}$.

[2] For example, the statement $\left(1+\frac{1}{2}+\frac{1}{2^2}+\frac{1}{2^3}+ \ldots +\frac{1}{2^H}=2\right)$ asserts *virtual* and not *absolute* equality.

point of the curve between A and B the radius of curvature is infinite, and the curvature consequently infinitesimal. Even an infinite section AB may be called *virtually* straight when $\frac{AB}{R}$ is infinitesimal, R being any of the radii of curvature;[1] for in this case the infinity AB (infinite in regard to any finite unit) is infinitesimal compared with the infinity R. The symbolic and linguistic conventions here proposed would, I think, greatly increase the logical accuracy of modern geometry without in the least impairing its great power as a practical instrument of discovery and research.

9. Metaphysicians sometimes ask whether space — the actual space of our perceptive experience—the space filled with the entity called *matter*, or with the entity called *ether*, or with both—is really infinite. Considering we can give no satisfactory definition either of 'matter' or of 'ether,' the question hardly admits of a clear and intelligible answer, or, at any rate, of any answer that logicians, metaphysicians, and physicists would be all likely to accept. The purely ideal space of the mathematician is far easier to understand, and this space must, I think, be pronounced infinite—that is to say, infinite in the sense of the word explained in § 3—for the simple reason that the opposite supposition plunges us at once into logical contradictions. We can all, without any conflict of opposing concepts, imagine a sphere whose radius is too large to be expressible (whatever be our conventional notation) in terms of finite ratios alone, and that sphere is, *by our very definition*, infinite. And we cannot stop there. By our definitions and conventions also, it follows that $H_1 + H_2 = H_3$; that the sum of the *real* infinities H_1 and H_2 make a third *real* infinity H_3 greater that either. Similarly we get $H_1 + H_2 + H_3 = H_4$, and so on for ever. But we cannot, without further data, assert $H_1 - H_2 = H_3$; for $H_1 - H_2$ may $= F_1$, or may $= 0$, since neither the supposition $H_1 = H_2 + F$, nor the supposition $H_1 = H_2$, involves any formal inconsistency.

10. It may be objected that the definitions which I here propose of the *finite*, the *infinite*, and the *infinitesimal* are quite arbitrary. To this I reply, firstly, that all definitions are more or less arbitrary; secondly, that however arbitrary my definitions may be, they are mutually consistent, and in

---

[1] By this is meant that $\frac{AB}{R_1}, \frac{AB}{R_2}, \frac{AB}{R_3}$, etc., are respectively infinitesimal for the separate points P_1, P_2, P_3, etc., between A and B. The infinities R_1, R_2, R_3, etc., are generally, though not necessarily, unequal.

SYMBOLIC REASONING. 511

no way clash with, but, on the contrary, render more precise, the rather vague significations usually attached to the words in ordinary language and even in mathematics; and, lastly, that they are very convenient and, if accepted, would materially increase the formal accuracy of our reasoning whether the questions discussed be metaphysical, logical, or mathematical. Clear working definitions of the words *finite*, *infinite* and *infinitesimal* are imperatively needed; and if those I here propose be not found suitable, others should be substituted. By "working definitions" I mean definitions which, by their formal precision, would prevent ambiguities and staggering paradoxes—paradoxes that may be true or false according to the meanings attached to the words in which they are expressed. In ordinary informal speech variability of meaning is, of course, permissible, as the context generally prevents all ambiguity. No one, for instance, would misunderstand the meaning of such a statement as "He took *infinite* pains, yet his gains were *infinitesimal*"; but in logic and mathematics the case is different. Here also, it is true, the context as a rule prevents ambiguity, but not always; and when it does not, the errors into which we fall are often serious. We should especially remember that such expressions as $\frac{1}{0}, \frac{2}{0}, \frac{\infty}{3}$, like their reciprocals $\frac{0}{1}, \frac{0}{2}, \frac{3}{\infty}$, etc., are not real ratios at all, but pure unrealities, though they have their utility as symbols.

11. Paradoxes also arise from the fact that our unit of reference is not always constant. A pound of tea is lighter at the equator than in London or Paris or Melbourne; yet a pound of tea always weighs a pound, neither more nor less, wherever we may weigh it, provided the scales we use be correct. And if we took our pound and scales and weights to the moon or to the planet Mars, the result, in spite of the greatly diminished attraction, would be the same; the pound of tea would be much lighter, but so would our unit of comparison, the metallic pound with which we weighed it. It is the same with all spatial dimensions. We can never be sure that our units of comparison remain constant; or, rather, we may be quite sure that they do not. The actual length of the standard yard or standard *mètre* varies with the temperature. True, so far as our experience goes, the variation owing to this cause is slight; but then our experience of the possibilities and actualities of nature is literally infinitesimal. For aught we know to the contrary, there may be other and far more powerful causes at work, causes which (unlike heat) act equally on all the perceptible substances of *our* universe,

though not equally, perhaps, or at all, on all the substances of other worlds beyond our ken; so that, in comparison with some constant standard unit lying hidden in some infinitely distant sphere, the dimensions of everything in *our* visible universe—of the sun, of the moon, of the earth, of its mountains, oceans, seas and rivers, of the houses upon its surface, and of the inhabitants who live in them, our own selves included, may be rapidly diminishing,[1] and in nearly the same relative proportions, so that the actual size of each one of us to-day may, in comparison with this constant unit, be only the hundredth or the millionth part of what he or she was yesterday. Personally I believe the actual variation to be much less serious; but this is an opinion for which, as for other cherished convictions, I can find no logical foundation.

12. Let us now examine the meanings of the words *finite*, *infinite* and *infinitesimal* in reference to time. Here too we have the actual and the ideal, and, as in the case of space, the ideal is easier to deal with than the actual. Actual time is measured by clocks and watches, and the correctness of these is tested by observations of the motions, or apparent motions, of the sun, moon and stars; but the respective motions of these, when mutually compared, are not uniform; so we take the apparent motion of one of them, the sun, and dividing its cycle into a certain number of equal parts, we take one of these as our constant unit of reference. For all practical purposes this answers all our needs; but what of the assumptions on which this theory of time is founded? How do we know that these solar cycles are even approximately equal? How do we know that, in comparison with some other unit of time, depending, say, upon the more uniform motion of some other heavenly body, far away in space beyond our power ever to discover, the motions of all the heavenly bodies, of our clocks, of our watches, of everything we know, including our very thoughts and sensations, may not be rapidly increasing in velocity, and in nearly the same relative proportions, so that, when measured by this standard unit, our years, days, hours, minutes and seconds, and, consequently, the duration of our lives may be infinitesimal in comparison with the years, days, hours, etc., and the duration of the lives of our fathers or grandfathers? The paradox arising from the possible variation of our standard and unit of space has thus its counterpart in the paradox

---

[1] Of course it is equally possible that the variation may be in the opposite direction.

SYMBOLIC REASONING.

arising from the possible variation of our standard unit of time. And so with all our units of comparison. For theoretical reasoning, as well as for the practical needs of daily life, we find it convenient to assume our units constant while other things vary; so we make this assumption the basis of all our reasoning without needlessly saddening our minds by a too logical analysis of its legitimacy.

13. Symbolic logic too has its paradoxes, that is to say, formulæ which appear paradoxical till they are explained, and then cease to be paradoxes. Such is the formula $\eta : \epsilon$, which asserts that "an impossibility implies a certainty". As soon as we define the implication $A : B$, by which we symbolise the statement that "A implies B," to mean simply $(AB')^\eta$, which asserts that the affirmation A coupled with the denial B' contradicts our data or definitions, the paradox vanishes. For then $\eta : \epsilon$ is seen simply to mean $(\eta \epsilon')^\eta$, which is a clear truism.[1]

14. Another paradox at first sight is the statement that the simple affirmative A, though *equivalent* to A^τ, which asserts that A is true, is not *synonymous* with A^τ; and that, in like manner, the denial A', though *equivalent* to A^ι, which asserts that A is false, is yet not *synonymous* with A^ι. Other symbolic systems, it is true, do not draw this distinction; but mine does, and so, I believe, do all civilised languages. The fact that they do is a *prima facie* presumption in favour of my opinion that the distinction is real and corresponds to a logical need. Surely it is more than a coincidence that every civilised language should have two separate expressions, one corresponding (taking an example at random) to the English statement "It rains," and another to the English statement "It is true that it rains"; and also two separate expressions, one corresponding to "It does not rain," and another to "It is false that it rains". The two statements A and A^τ are *equivalent* because neither can be true without the other being so also; but they are not *synonymous;* otherwise we could always substitute A^τ for A and *vice versa* in no matter what expression without altering its meaning. That this cannot always be done may be shown as follows. Let A denote a variable statement θ_r, that is to say, a statement, such as "It rains," which we assume to be true now or in the case considered, but which is not necessarily or always true. Then, on the one hand, we get

---

[1] Observe that though $\eta_x : \epsilon_y$, like $\epsilon_x : \epsilon_y$, is a *formal certainty*, yet $\eta_x \therefore \epsilon_y$, unlike $\epsilon_x \therefore \epsilon_y$, is a *formal impossibility*. Non-Euclideans seem to me to forget this when they say that their systems are as "self-consistent" as the Euclidean.

$$A^\theta = (\theta_\tau)^\theta = \epsilon\,;$$

for every variable, whether it be a variable of the class θ_τ (a statement true now but not always) or a variable of the class θ_ι (a statement false now but not always) remains still a variable, so that $(\theta_\tau)^\theta$, like $(\theta_\iota)^\theta$ is a formal certainty. On the other hand, we get

$$(A^\tau)^\theta = (\theta_\tau^\tau)^\theta = \epsilon^\theta = \eta\,;$$

for θ_τ^τ means $(\theta_\tau)^\tau$, which is a formal certainty, and a certainty cannot be a variable, since certainties and variables form two mutually exclusive classes by definition. This might also be proved in a slightly different way as follows. Let $A = \theta_1 = \tau_1$. We get

$$A^\theta = \theta_1^\theta = \epsilon\,;\text{ but }(A^\tau)^\theta = (\tau_1^\tau)^\theta = \epsilon^\theta = \eta.$$

Similarly we may show that A', though equivalent to A^ι, is not synonymous with A^ι. For let $A = \theta_\iota$. We have on the one hand

$$(A')^\theta = A^\theta = \theta_\iota^\theta = \epsilon,$$

and on the other

$$(A^\iota)^\theta = (\theta_\iota^\iota)^\theta = \epsilon^\theta = \eta.$$

The statements A and A' (like A^B and its denial A^{-B}) are of *the same degree*, whereas A^x (whether x stands for τ or ι or ϵ or η or θ, or any other class of statements) is *one degree higher*, and may therefore be called *the revision of the judgment* A. Two contradictory judgments, A and A', are placed before us, and we have to decide which is true. If we decide in favour of the affirmative A, we say that "A is true" and write A^τ; if we decide in favour of the negative A', we say that "A is false" and write A^ι. But the question to be decided may be not merely to decide whether A is true or false, but whether A follows necessarily from, or is inconsistent with, our definitions or admitted and unquestioned data. In that case we write A^ϵ when we decide that A *does* follow necessarily from our data; we write A^η when we decide that A is inconsistent with our data; and we write A^θ when we decide that A neither follows from nor is inconsistent with our data. Similarly, A^{xy}, or its synonym $(A^x)^y$, is a revision of the judgment A^x; and so on. It will be noticed that A^{xy} and A_x^y, which means $(A_x)^y$, are quite different statements, since A^x and A_x are different. The statement A^x *asserts* that A belongs to the class x; the statement A_x *takes this for granted*. For example, A_θ^ι, which means $(A_\theta)^\iota$, asserts that "the *variable* statement A is *false*"; whereas $A^{\theta\iota}$, which means $(A^\theta)^\iota$, asserts that "it is *false* that A is *variable*". The statements A and A_x are of the same

SYMBOLIC REASONING.

degree; the statement A^x is of a degree higher. The statement A^{xyz}, since it means $(A^{xy})^z$, is of the *first* degree as regards its subject A^{xy}; but since it also means $(A^x)^{yz}$, it is of the *second* degree as regards A^x, and of the *third* degree as regards the *root-statement* A. Now, suppose A stands for $Q^{\alpha\beta}$. In that case, though A is the *root-statement* as regards A^{xyz}, it is a statement of the second degree as regards the root-statement Q; and A^{xyz}, which is of the *third* degree as regards A, is of the *fifth degree* as regards Q; for, expressed in terms of Q, it means $Q^{\alpha\beta xyz}$.

15. Let me here say a few words in reply to the logicians who maintain that statements can only be classed as *true* and *false*, and that my introduction of such classes as *certainties, impossibilities,* and *variables,* and of any others that may concern our argument or researches, is wrong, or at any rate, outside the proper domain of logic, and especially of symbolic logic. This is very much as if one argued that since animals are only divisible into two classes, males and females, it is no business of true zoology to consider the respective characteristics of such creatures as lions, tigers, and leopards, to say nothing of others still more objectionable. All such attempts to surround symbolic logic by a Chinese wall of exclusion are futile.

16. Another argument against my system is that *variable* statements, statements which are sometimes true and sometimes false, have no real existence; that a statement if once true is true always, and if once false is false always. But surely this is a mere play upon words, and it does not seem to me very accurate even as that. A servant, in reply to an inquiry at the street door in the morning, says, and says truly, that "Mrs. Brown is not at home". The same servant, *in reply to the same inquiry* in the afternoon, says again, and this time, in obedience to instructions, says falsely, that "Mrs. Brown is not at home". She makes exactly the same statement as in the morning, because *she uses exactly the same form of words;* but this statement, this form of words, which was true in the morning, because in the morning *it conveyed true information*, is false in the afternoon, because in the afternoon *it conveyed false information*.

17. Let us look at the matter from another point of view. Suppose we have no data but our definitions or symbolic and linguistic conventions. Let A, B, C respectively denote the three statements "7 is greater than 5," "6 is greater than 9," "x^2 is greater than x". Is it not clear that with these meanings of the symbols we may truly and confidently make the three-factor compound statement $A^\varepsilon B^\eta C^\theta$? For,

by our very definitions of the words *certain*, *impossible*, *variable*, respectively represented by the symbols ϵ, η, θ, is not the first statement A *certain*, because *it follows necessarily from our data*, which are here limited to our definitions and linguistic conventions? Is not the second statement B *impossible*, because *it contradicts* (or is *inconsistent with*) *our data*? And is not the third statement C a *variable*, because, though perfectly intelligible, *it is neither certain nor impossible*? To say that C is *neither true nor false* would be incorrect; for it may be either. It is true when x is greater than 1; it is false when x is not greater than 1.

18. To take another case; suppose two men are playing dice, and that, just before a throw, three spectators make the three following statements, which we will denote by A, B, C: "The number that will turn up is less than 8" (A), "The number that will turn up is greater than 8" (B), "The number that will turn up is 5" (C). Since by our data, or tacit conventions, the only numbers possible are 1, 2, 3, 4, 5, 6, is it not clear that we must have $A^\epsilon B^\eta C^\theta$? Is not A *certain* because it follows necessarily from our data? Is not B *impossible* because it is inconsistent with our data? And is not C a *variable* because it neither follows from nor is inconsistent with our data? In the language of probability, the chance of A is 1, the chance of B is 0, and the chance of C is neither 1 nor 0 but a proper fraction. What that proper fraction is the statement C^θ does not say; but we know it to be $\frac{1}{6}$. Taking the three denials A', B', C', the chance of A' is 0, the chance of B' is 1, and the chance of C' is $\frac{5}{6}$; so that we have $(A')^\eta (B')^\epsilon (C')^\theta$. This shows that here, as always, the denial of any certainty A is an impossibility A', the denial of any impossibility B is a certainty B', and the denial of any variable C is also a variable C'.

19. Other paradoxes arise from the fact that each of the words *if* and *implies* is used in different senses. Putting A and B for two propositions, the statements "If A then B" and "A implies B," which, in my symbolic system, I find it convenient to treat as synonymous and as having the meaning which I represent symbolically by any of the three synonymous symbols $(A:B)$, $(AB')^\eta$, $(A'+B)^\epsilon$, are used by some logicians not only in the above sense, but also in the weaker sense which I attach to the mutually synonymous symbols $(AB')^\iota$ and $(A'+B)^\tau$; because these logicians erroneously consider my ϵ to be equivalent to my τ, and my η to my ι. But there is yet another sense in which we all sometimes use the word *implies;* for when we say "A implies B" we sometimes mean not only $(AB')^\eta$, that it is

impossible for A to be true without B being also true, or the equivalent statement that the affirmation of A coupled with the denial of B is inconsistent with our data, but also that A *contains* B, that is to say, that B *is a particular case of* A. In this sense, I think it would be better to use the word *contains* rather than the word *implies*. For example, we may say that the formula $(x^m x^n = x^{m+n})$ *contains* the formula $(x^3 x^2 = x^5)$ as a particular case, and similarly that $(x^3 x^2 = x^5)$ *contains* $(6^3 6^2 = 6^5)$. Representing the first and most general of these three statements by the functional symbol $\phi(x, m, n)$, it follows from our definition of a logical function that $\phi(x, 3, 2)$ must denote the second, and that $\phi(6, 3, 2)$ must denote the third. It also follows that $\phi(x, m, n)$ *contains* $\phi(x, 3, 2)$, and that $\phi(x, 3, 2)$ *contains* $\phi(6, 3, 2)$. Again let $\phi(A, B, C)$, or simply ϕ, denote the Barbara of *general* logic, namely,

$$(A : B)(B : C) : (A : C),$$

in which A, B, C may be any statements whatever—statements which may or may not have the same subject; and let $\phi_s(A, B, C)$, or simply ϕ_s, denote the Barbara of the *traditional* logic, namely,

$$(A_s : B_s)(B_s : C_s) : (A_s : C_s),$$

in which the statements A, B, C are understood to have the same subject S. We may then say that ϕ *contains* ϕ_s as a particular case. Also, since ϕ and ϕ_s are both *certainties*, we can assert not only $\phi : \phi_s$, that ϕ *implies* ϕ_s, but also $\phi_s : \phi$, that ϕ_s *implies* ϕ, since any certainty ϵ_x *implies* any other certainty ϵ_y. For, by definition, we have $\epsilon_x : \epsilon_y = (\epsilon_x \epsilon'_y)^\eta = (\epsilon_x \eta)^\eta = \epsilon$. We cannot however assert that ϕ_s *contains* ϕ, for it is ϕ_s that is a particular case of ϕ, and not ϕ that is a particular case of ϕ_s.

20. Misunderstandings and consequent paradoxes also arise from the fact that each of the words *because, therefore, prove,* and *infer* has more than one meaning; but a serious discussion of these would unduly lengthen the present article.

Post-scriptum. The preceding was written before I read Mr. Russell's kind and appreciative review of my *Symbolic Logic and Its Applications* (Longmans) in MIND, No. 58. The points of difference between Mr. Russell's views and mine are, as he says, small in comparison with the points of agreement, and the former would I feel sure be smaller still if we could discuss them orally face to face.

Mr. Shearman in his recently published book, *The Development of Symbolic Logic*, submits my symbolic system to much hostile criticism; but as he has evidently failed to grasp the simple elementary notions on which my formulæ and operations are founded, his whole reasoning is, from beginning to end, irrelevant. On page 153, for example, he mixes up logic with psychology and defines a statement as *impossible* when, and only when, nobody can believe it.[1] Now, the belief in witchcraft is not yet dead. It follows therefore from Mr. Shearman's definition that it is still possible for old women to ride through the air on broomsticks, and that so long as the belief lasts, the possibility will last also.

[1] Mr. Shearman's exact words are as follows: "As an instance of the way in which statements described by these three terms are to be dealt with, take the following: 'It is impossible that x is y'. This would appear in such a form as 'A thinker who can believe that x is y does not exist'." The "three terms" to which Mr. Shearman refers are the words *certain, impossible, variable*, respectively denoted by my symbols ϵ, η, θ (see §§ 15-18).

[1908b]: 'If' and 'Imply'. Mind, Vol. 17, pp. 151-152.

'IF' AND 'IMPLY'.

Mr. Shearman's review of Mr. Russell's *Principles of Mathematics*, in last April's MIND, contains a quotation from that work which merits the serious attention of logicians. Adopting the usual view among logicians, that the implication "A implies B" (or "If A then B") is always equivalent to the disjunctive "Either A is false or B true," Mr. Russell is quoted as saying that "It follows from the above equivalence that of any two propositions there must be one which implies the other". A very brief symbolic operation will show that (*assuming his premises*) Mr. Russell is quite right; but surely the paradoxical conclusion at which he

152 NOTES.

arrives should give logicians pause. For nearly thirty years I have been vainly trying to convince them that this assumed invariable equivalence between a conditional (or implication) and a disjunctive is an error,[1] and now Mr. Shearman's quotation supplies me with a welcome test case which ought, I think, to decide the question finally in my favour. Take the two statements "He is a doctor" and "He is red-haired," each of which, observe, is a *variable*, because it may be true or false. Is it really the fact that one of these statements implies the other? Speaking of any Englishman taken at random out of those now living, can we truly say of him "If he is a doctor he is red-haired, or if he is red-haired he is a doctor"? Is it really a certainty that "either all English doctors are red-haired, or else all red-haired Englishmen are doctors"? To throw symbolic light upon the question, let the two symbols $(A:B)_1$ and $(A:B)_2$ respectively denote "A implies B," first in the sense which logicians in general give to this expression, and next in the sense which I give to it. Putting D for "He is a *doctor*" and R for "He is *red-haired*," Mr. Russell's reasoning (I presume) is

$(D:R)_1 + (R:D)_2 = (D' + R) + (R' + D) = (D' + D) + (R' + R) = \epsilon + \epsilon = \epsilon.$

My reasoning is

$(D:R)_2 + (R:D)_2 = (D' + R)^\epsilon + (R' + D)^\epsilon = (DR')^\eta + (RD')^\eta = \eta + \eta = \eta.$

Thus, Mr. Russell, arguing correctly from the customary convention of logicians, arrives at the strange conclusion that (among Englishmen) we may conclude from a man's red hair that he is a doctor, or from his being a doctor that (whatever appearances may say to the contrary) his hair is red. My argument, founded on what seems to me a more natural convention, and one more in accordance with ordinary linguistic usage, arrives at the (to me) self-evident result, that in neither case does the conclusion follow from the premises—that an Englishman may be red-haired without being a doctor, and that he may also be a doctor without being red-haired.

Let us next take an example from mathematics. Suppose A, B, C to be random points on the circumference of a circle, forming the angular points of a triangle ABC. Take the two variable statements "AB is greater than AC" and "The angle A is an acute angle". Is it not the fact that neither statement implies the other? How then can it be true that "Either the first implies the second, or else the second the first"?

Is it too much to hope that this test case will at last open the eyes of logicians to the necessity of accepting my three-fold division of statements (ϵ, η, θ) with all its consequences? How far-reaching these consequences are can hardly be judged from my scattered papers in philosophical and scientific magazines. For a fair account of my system I must refer logicians to my recently published little volume on *Symbolic Logic and its Applications* (Longmans).

Hugh MacColl.

[1] The equivalence supposed holds good in the case of *constants* (*i.e.*, certainties and impossibilities), but not necessarily when the two statements are both *variables*.

[1908c]: 'If' and 'Imply'. *Mind,* Vol. 17, pp. 453-455.

IX.—NOTES.

'IF' AND 'IMPLY'.

WITH the Editor's kind permission I should like to say a few words in answer to Mr. Russell's and Prof. Mackenzie's critical notes in the April number of MIND.

If we look into the matter closely, it will, I think, be found that in mathematical reasoning the simple disjunctive $A' + B$ is always understood to really mean $(A' + B)^\epsilon$, which asserts that "it is *certain* (true in all cases) that either A is false or B true". This being understood, I admit that in *mathematics* (the theory of probability excepted) the disjunctive may be considered equivalent to the conditional or implication, so that the validity of Mr. Russell's valuable work on the *Principles of Mathematics* is in no way affected by our discussion. But in the wider field of general logic the assumption of equivalence is unsafe. Suppose, for example, we find it predicted in some astrological almanac that "This year a great *war* will take place in Europe," and also that "This year a disastrous *earthquake* will take place in Europe". Each of these assertions is a *proposition* both in Mr. Russell's sense of the word and in mine; and as neither of them can be called *certain* or *impossible* I regard them both as *variable*. Let W denote the first proposition and E the second. It is surely an awkward assumption (or convention) that leads here to the conclusion that "either W implies E or else E implies W". *War* in Europe does not necessarily imply a disastrous *earthquake* the same year in Europe; nor does a disastrous *earthquake* in Europe necessarily imply a great *war* the same year in Europe.

Mr. Russell, if I understand him aright, does not consider a propositional form of words, such as "He is a doctor," a real proposition till it is actually employed to give information. I should call it a proposition whether so employed or not. I see no reason why an instrument should not have the same name when lying idle as when employed for a purpose. But this mere question of name does not, so far as I can see, affect the real question at issue. The real cause of the misunderstanding is the fact that with Mr. Russell the proposition "A implies B" means $(AB')^\epsilon$, whereas with me it means $(AB')^\eta$. The symbol $(AB')^\epsilon$ asserts that AB' is *false*; the symbol $(AB')^\eta$ asserts that AB' is *impossible*. Our hypothetical predictions of the astrological almanac may, and probably will, turn out *false*; but as they contradict no certain or admitted *data*—no law of nature or of linguistic consistency—they can hardly be called *impossible*.

Prof. Mackenzie's reasoning may be symbolically illustrated as follows: Putting D for "He is a *doctor*," and R for "His hair is *red*," his four "essential steps" are:—

(1) $(D + D')(R + R')$, which $= \epsilon\epsilon = \epsilon,$
(2) $(D + D') + (R + R')$, which $= \epsilon + \epsilon = \epsilon,$
(3) $(D + R') + (D' + R)$, which $= (D + D') + (R' + R) = \epsilon + \epsilon = \epsilon,$
(4) $(R : D) + (D : R)$, which $= (RD')^\eta + (DR')^\eta = \theta\eta + \theta\eta = \eta + \eta = \eta.$

454 NOTES.

We might also write (4) thus:—
$$(R : D) + (D : R), \text{ which} = (R' + D)^\epsilon + (D' + R)^\epsilon = \theta^\epsilon + \theta^\epsilon = \eta + \eta = \eta.$$

The first three steps being certainties, each of them implies the other two, but none of them implies the fourth, as the fourth is an impossibility. Prof. Mackenzie says that "it may be doubted whether the transition from (2) to (3) is legitimate". But his doubt arises, I believe, from his thinking of (3) as meaning
$$(D + R')^\epsilon + (D' + R)^\epsilon,$$
which, as already shown, is an impossibility. Thus regarded, (2) would *not* imply (3), as a certainty cannot imply an impossibility. The statement $D + R'$, since (like its denial RD') it may be true or false, is a variable (θ), and therefore cannot be a certainty (ϵ). Hence, $(D + R')^\epsilon$ is a self-contradiction (η). Similarly, $(D' + R)^\epsilon$ is a self-contradiction.

Prof. Mackenzie says that our discussion on these points "leads us to see that the implications of a double disjunctive have not yet been sufficiently brought out by logicians". To this I would add that it is scarcely possible to treat the subject adequately without the aid of symbolic logic, and that no system of symbolic logic can do this effectively unless it takes symbolic account of the distinctions between certainties (ϵ), impossibilities (η), and variables (θ). Consider the standard syllogism *Barbara*. Let A denote its premisses and B its conclusion. It is usual to consider the syllogism as expressible by $A \therefore B$ ("A *therefore* B"). In this *therefore*-form the syllogism is not necessarily valid, for thus stated, *it asserts the truth of A, and is therefore false when A is false*. The proper symbolic form is $A : B$ ("*If* A *then* B"). In this form it is always true whether A be true or false, for in this implicational form it only asserts $(AB')^\eta$, that "it is never the case that the premisses are true and the conclusion false". Observe again that $A : B$ is equivalent to $B' : A'$, but that $A \therefore B$ is *not* equivalent to $B' \therefore A'$. For $A \therefore B$ *affirms* both A and B, while $B' \therefore A'$ *denies* both A and B. In the *therefore*-form ($A \therefore B$) the syllogism is open to the attacks made upon it by Mill, Bradley and others. In the "*If*" or *implicational* form ($A : B$) its validity is unassailable.

I am not one of those who would make a clean sweep of the traditional logic and all connected with it. On the contrary, I consider its subject matter, and especially its syllogistic problems, as exceedingly valuable in a first course of symbolic logic. But, as I have amply shown in my '*Symbolic Logic and its Applications*,' the whole subject of formal logic needs recasting.

As a simple syllogistic problem solved symbolically, take the following:—

Given the premiss that all the boys in a certain school who learn *Latin* also learn *French*, what is the weakest premiss that must be joined to this so that the combination may imply that some who learn *French* do not learn *Greek* ?

Speaking of a boy taken at random, let L, F, G respectively affirm that "he learns *Latin*," that "he learns *French*," that "he learns *Greek*". Let W denote the weakest premiss required. The syllogism will then be
$$(L : F) W : (F : G)'.$$
Transposing and changing signs, so as to get rid of the non-implicational $(F : G)'$, we get
$$(L : F)(F : G) : W'.$$
But (by *Barbara*, in its proper implicational form) we have
$$(L : F)(F : G) : (L : G).$$
Equating consequents, we get $W' = (L : G)$ and $W = (L : G)'$. The required permiss is therefore $(L : G)'$, which asserts that L does not

NOTES.

imply G, or (in other words) that "Some who learn *Latin* do not learn *Greek*". Supplying this missing premiss, we get a syllogism of the form *Bokardo*.

Observe that this method of transposition, by which all valid syllogisms can be reduced to the form $(A:B)(B:C):(A:C)$, would be logically inadmissible if (following the usual erroneous custom) we employ the sign $\therefore$ ("therefore") instead of the proper sign : ("implies") between the premisses and the conclusion. For though the conditional $AW:C'$ is equivalent to the conditional $AC:W'$, as each means $(AWC)\eta$, the categorical $AW \therefore C'$ (which affirms W and denies C) cannot be equivalent to the categorical $AC \therefore W'$ (which affirms C and denies W).

HUGH MACCOLL.

II.—LINGUISTIC MISUNDERSTANDINGS.[1]

By Hugh MacColl.

Part I.

I. Non-Euclidean Geometries.

It is a common saying among the distinguished mathematicians who have cultivated these fascinating studies that non-Euclidean geometries in general, and the Lobachevskian and Riemannian systems in particular, are no less " valid " than the common Euclidean with which we are all familiar. They do not assert, and since these three systems are, as they admit, mutually incompatible, they cannot very well assert, that all three are *true*. This seems to me somewhat perplexing. If the Euclidean, the Riemannian, and the Lobachevskian systems be founded on mutually incompatible principles, it follows that only one of them (if any) can be true : in what sense then can they be affirmed to be all three *valid ?*

It is sometimes said that an argument (whether syllogistic or other) may be perfectly valid quite independently of the truth or falsehood of its premisses. This is a dangerous doctrine from which I emphatically dissent. I have given my reasons elsewhere (see Mind, N.S., 43, 53, and my *Symbolic Logic*, pp. 47-49), and need not here repeat them. The premisses and the conclusion are, in my opinion, the most important factors of an argument, and if either of these be false—what-

---

[1] Some of Mr. MacColl's most important work has appeared in our pages. The above article reached us very shortly before his death (in his seventy-third year) on 27th December last, and thus appears without his revision. Mr. MacColl was a man of great mathematical and logical ability and of a real philosophic depth which the readers of Mind were among the readiest to recognise. Mr. MacColl's circumstances were not too favourable to the development of his powers, and he is to be congratulated on having done so much excellent work. He died at Boulogne, where he had resided for forty-four years. Here he had been engaged principally in the teaching of Mathematics. Mr. MacColl, who was a B.A. of London, began his studies at Glasgow, and had been engaged in teaching at Oxford.—Editor Mind.

LINGUISTIC MISUNDERSTANDINGS.

ever be the nature of the links connecting them—the argument should not be considered valid. Of course, this also may be considered a mere convention; but if so, I think it is one founded on common sense and practical convenience. Non-Euclideans also say that if the principles of the Riemannian or Lobachevskian geometries were unsound, they would lead to absurd and inadmissible conclusions. But to the simple unsophisticated intellect of the ordinary educated thinker that is precisely what has happened. For example, the principles of the Riemannian system lead necessarily to the conclusion that a point moving always in the same straight line, and never reversing its course, will at last arrive at its original position. Why should not this be regarded as a *reductio ad absurdum* of the Riemannian principle? The principle on which non-Euclideans secure validity, or apparent validity, for their reasoning seems to me of doubtful legitimacy. Without warrant or warning, they quite change the usual meanings of certain words and symbols, and especially that of the word 'straight'. With them this word invariably refers to some kind of *curve*, but of such a huge size that at every point the radius of curvature is *infinite* (in the sense of *inexpressibly large*), and the curvature consequently *infinitesimal*. The curvature is never quite *zero*, as then the lines which they call straight would also be straight in the ordinary acceptation of the word, in which case the non-Euclidean geometry would in all respects coincide with the ordinary Euclidean. But on this principle of arbitrarily changing the commonly understood meanings of words and symbols we might plausibly or paradoxically maintain that January has 37 days, February 34, and the whole year 555. We need only slyly change the base of our common arithmetical notation from ten to eight. Thus, 37 would mean $3(8)+7$, 34 would mean $3(8)+4$, and 555 would mean $5(8)^2+5(8)+5$.

M. Poincaré, in his *La Science et l'Hypothèse* (p. 67), says that the question whether the Euclidean geometry (or any other) is true is meaningless. "Autant demander," he remarks, "si le système métrique est vrai et les anciennes mesures fausses . . . une géometrie ne peut pas être plus vraie qu'une autre; elle peut seulement être plus commode."

But this is surely carrying liberty of conventions a trifle too far. In logic, as in practical politics, unlimited freedom is apt to degenerate into inconvenient licence, and ultimately into downright destructive anarchy. Every formula, even the most reliable, has its limits of validity, namely, the accepted conventional meanings of the words or symbols in which it is expressed. Otherwise, we might legitimately convert any false statement into a true, or *vice versa*, by simply agreeing to change the ordinarily accepted meanings of the words or other symbols in which it is expressed. I cannot go quite so far as some extreme 'pragmatists,' who, from their language, would appear to consider the *true* as almost, if not quite, synonymous with the *useful*; but I sympathise strongly with pragmatism in the emphasis which it lays on the latter word, the *useful*. Even in the pursuit of abstract truth the most important discoveries usually fall to those who always keep in view the possible practical applications of their abstract researches. The Euclidean geometry seems to me to be the only true one, not merely because it is admittedly the simplest and most convenient, but also, and chiefly, because it is the only system that frankly accepts the customary conventions of ordinary language.

There is a limit to the utility of definitions. We should explain the obscure or the complex in terms of the simple and comprehensible, not *vice versa*. The idea of straightness is one of those elementary notions which cannot well be conveyed by a formal definition. A simple illustration, such as a stretched string or a line drawn by the aid of a ruler, will convey it much better. Similarly, an illustration on paper of a circle, an ellipse, an hyperbola, etc., will immediately give the general idea of a curve, though here formal definitions are necessary to distinguish between the various classes. The gradual prolongation of an hyperbola away from its vertex will make clear even to a schoolboy how, when the radius of curvature increases without limit, the curvature gradually becomes infinitesimal, when, of course, the curve cannot by any possible measurement be distinguished from an absolutely straight line. Similarly, any one can grasp the fact that no possible measurement by the most delicate of instruments can ever detect the curvature of any finite arc AB when the radius and circumference of the circle to which it belongs are infinite in comparison. That is to say, by express definition of the *finite*, the arc AB is expressible, either exactly or approximately, in terms of some recognised unit (as a yard or a mile), while, by express definition of the *infinite*, the radius or circumference is too large to be so expressible. These definitions of the *finite*, the *infinite*, and the *infinitesimal* appear to me to be the only workable ones. I have seen no others that do not involve some self-contradiction.

Prof. Keyser, in the *Hibbert Journal*, January, 1909, defines a class or collection as infinite when, and only when,

LINGUISTIC MISUNDERSTANDINGS. 189

it " contains a part or sub-collection that is numerically equal to the whole". Now, when the symbol ∞ denotes some pseudo-infinity, such as $\frac{1}{0}$ or $\frac{2}{0}$ or the tangent of a right angle, our symbolic operations sometimes lead to such statements as $\left(\frac{1}{2} \times \infty = \infty\right)$, $\left(\frac{1}{3} \times \infty = \infty\right)$, etc., which seem to assert that infinity may be equal to its half, or its third, etc. But our symbolic operations also lead sometimes to such statements as $\left(\frac{1}{2} \times 0 = 0\right)$, $\left(\frac{1}{3} \times 0 = 0\right)$, etc.; so that a class or ratio whose part is equal to the whole may also be zero. The explanation of the seeming paradox is this:—

When x diminishes without limit for any positive finite value, say 1, till it becomes negative, the fraction $\frac{1}{x}$ passes through all possible positive infinite values, and the fraction $\frac{x}{1}$ through as many positive infinitesimal values, till both become negative when x becomes negative. When x vanishes into non-existence, as it passes from the positive to the negative state, the fractions $\frac{1}{x}$ and $\frac{x}{1}$ vanish into non-existence also, but with this difference, that the former is then represented by the symbol ∞, and the latter by the symbol 0. Thus, $\frac{1}{2} \times \infty$, $\frac{1}{3} \times \infty$, $\frac{\infty}{1}$, $\frac{\infty}{2}$, $\frac{1}{0}$, $\frac{2}{0}$, etc., represent one class of non-existences, the pseudo-infinities, while $\frac{1}{2} \times 0$, $\frac{1}{3} \times 0$, $\frac{0}{1}$, $\frac{0}{2}$, $\frac{1}{\infty}$, $\frac{2}{\infty}$, etc., represent another class of non-existences, the pseudo-infinitesimals. The secant of a right angle belongs to the first class; its inverse, the cosine of a right angle, belongs to the other.

Prof. S. Alexander, in the *Hibbert Journal*, October, 1909, says that the system of numbers 1, 2, 3, 4, etc., is infinite,

> "not merely because we can never get to the end of it, but for quite a different reason. Perform on each number of the system an operation, say, adding 1 to each number; you have 2, 3, 4, 5, 6, 7, etc., *which is a part of the original system*. Or double each number; *the resulting infinite series 2, 4, 6, 8, etc., is already contained in the original.*"

Now, the two statements which I have italicised in the above quotation seem to me somewhat wanting in clearness. What does the word *part* mean in the one, and the word *contained* in the other? Is it not usually understood that wherever there is a *part* there must also be a *whole*, and that this whole contains the part? It is true that if we continue the series 1, 2, 3, 4, etc., long enough, say, to an infinite number H_1 (infinite in the sense already given by definition); the series 2, 3, 4, 5, etc., to an infinite number H_2; and the infinite series 2, 4, 6, 8, etc., to an infinite number H_3; then the series 1, 2, 3, . . . H_1, contains the series 2, 3, 4, 5, . . . H_2, *provided H_2 does not exceed H_1*, and it contains the series 2, 4, 6, 8, . . . H_3, *provided H_3 does not exceed H_1*; but unless these relations hold between the infinities, the two statements in italics seem to me inadmissible. I cannot well conceive of a real whole class C (with members C_1, C_2, C_3, etc.) being destitute of some *last* member C_n; though I can quite conceive of n as infinite in the sense that it is far beyond the power of the decimal or any other arithmetical system of notation to express. Like the living population of a town, or of the earth, or of the real material universe, the number n may be conceived of as continually increasing, but at any given moment it exists. There can be no real *totality* without it.

II. Axiom, Inference, Implication.

From the statement that 'A implies B' it does not at all follow that B is a legitimate *inference* from A. As commonly understood, *inference* involves psychological considerations; *implication* does not. When we say that we *infer* B from A, we are understood to assert that we actually obtain our knowledge of B from our previous knowledge of A; but when we say that 'A *implies* B,' we usually mean, and in syllogistic implications we only mean, that the affirmation of A coupled with the denial of B constitutes an *impossibility;* that is to say, that this compound statement is either a linguistic inconsistency or else a statement incompatible with our admitted and unquestioned *data*. Just as a statement incompatible with our admitted *data* or linguistic conventions is called an *impossibility*, so a statement that forms a part of, or necessarily follows from, our admitted *data* or linguistic conventions is called a *certainty*. As used in formal logic, these two antithetical words do not of necessity involve any psychological considerations. It does not follow that a statement is a *certainty* because it is so considered. The statement that the earth is bigger than the sun was once universally but erroneously reckoned among the certainties; now it is universally and correctly reckoned among the impossibilities.

LINGUISTIC MISUNDERSTANDINGS. 191

That is why some of the operations of formal logic, like some mathematical operations, may be accurately performed mechanically, like sewing or knitting, by unconscious inanimate calculating machines. This does not at all imply that formal logic is absolutely independent of psychology—that they have nothing to do with each other. That would be as erroneous as to assert that a clock or a watch has nothing to do with mind, because, once arranged and wound up, it will automatically record the progress of time without our intervention. As a never absent pre-condition of a working logical or mathematical formula, just as of an automatic inanimate machine, we find the inventive human intellect. From the very meanings of the words, as well as from universal experience, *mechanism* always implies *mind*, though mind does not necessarily imply mechanism. Everywhere in the universe, mechanism without mind—mind of some kind, human or superhuman—is a contradiction in terms. None the less, it is convenient in scientific researches to consider the two as far as possible apart. Just as the workings of the forces of nature are most simply explained by considering them apart from all questions of theology, so the operations of machines and of logical or mathematical formulæ are most simply explained by considering them apart from the mentality of their inventors.

There are, however, perfectly intelligible statements which, though necessarily either true or false, are neither certainties nor impossibilities. That is to say, they do not necessarily follow from admitted and unquestioned *data*, nor do they contradict such *data*. Such statements I call *variables*. To illustrate these three mutually exclusive classes of statements we may give " Australia is larger than Ireland " and " six is larger than five" as examples of *certainties*, a class denoted by the symbol ϵ; "Ireland is larger than Australia " and "five is larger than six " as examples of *impossibilities*, denoted by the symbol η; while, *when we have no data except our linguistic conventions*, the statement that "my horse will win the race" and the statement that "the number that will turn up is less than nine" may be taken as examples of *variables*, denoted by the symbol θ. Thus, in my symbolic system, the complex symbol $A^\epsilon B^\eta C^\theta$ asserts that the statement A is a *certainty*, that B is an *impossibility*, and that C is a *variable*. We might, however, have special *data* which would force us to class the above or other variables as certainties or impossibilities. For example, if the number possible be restricted by our data to the numbers 1, 2, 3, 4, 5, 6 (as in dice-throwing), the statement that "the number that will turn up is less than nine" is a certainty, and the statement that "the number that will turn up is greater than nine" is an *impossibility;* while the statement that "the number that will turn up is greater than four" is a *variable*. Expressed in the technical language of probability, the chance of the truth of the first statement (or of the event which it affirms) is 1; the chance of the truth of the second statement (or of the event which it affirms) is 0; and the chance of the truth of the third statement (or of the event which it affirms) is two-sixths (or one-third). Regarded thus from the standpoint of probability, a *certainty* (ϵ) is a statement whose chance of being true is 1; an *impossibility* (η) is a statement whose chance of being true is 0; and a *variable* (θ) is a statement whose chance of being true is some fraction between 0 and 1.

Some logicians however maintain that it is incorrect to speak (as I do in the case of variables) of a statement or proposition as "sometimes true and sometimes false". I cannot see the incorrectness. It is purely a matter of convention, just as it is a matter of convention to speak of an event (such as the turning up of an ace in a game of cards) as sometimes happening and sometimes failing. Surely every time an event happens, a statement or proposition (whatever be the form of words) that affirms the occurrence is true; and every time it fails, this statement or proposition (*expressed in exactly the same form of words*) is false. The objectors to my view might similarly, and with greater plausibility, argue that no event ever happens more than once, since each fresh so-called recurrence is really a fresh and different event. I say "with greater plausibility," because, as a matter of fact, the events really *are* different, while the statement or proposition—that is to say, *the form of words*—may remain the same. The statement (or form of words) "an ace will turn up," pronounced before the event, or the statement "an ace has turned up," pronounced after the event, is surely true (or expresses a truth) whenever an ace does turn up, and false whenever it does not. Will it be objected that these are not real *propositions*, but mere "propositional forms"? Grammarians—and even objecting logicians when they are off their guard—often bring forward locutions like "The bird has flown," or "The boy has eaten his dinner," as examples of "propositions," though there may be no question of any real bird, boy, or dinner. Of course, we may agree to call such locutions "propositional forms" when they are not actually used to give real information, and only call them "propositions" when they are so used. But so we might agree to call a sword a 'slashing-weapon' when it is lying idle in its scabbard, and only call it a 'sword,' or 'weapon,' when it is actually wielded in a serious battle. All such linguistic conventions are, of course, logically permissible; but are they needed, or would they be useful? With regard to the particular question at issue, would it not be better to restrict the expression 'propositional forms' to such forms as 'All X is Y,' 'A implies B,' 'A is greater than B,' etc., in which the letters X, Y, A, B are always understood to represent mere *blanks* that may be replaced or filled up by any words which would convert these meaningless forms into real intelligible propositions?

What is an axiom? No clear line of demarcation can be drawn between an axiom and any other general proposition or formula that is known and admitted to be true. So far as its formulæ and operations are concerned, symbolic logic ignores the distinction altogether. Indeed it could not very well take notice of the distinction without introducing psychological considerations, which are in general foreign to its purpose. A proposition that may appear axiomatic to one person may appear doubtful to another, until he has obtained a satisfactory proof of it; after which he treats it as an axiom in all subsequent researches. Apart from psychological considerations what is meant by 'proof' or 'inference'? What is meant by such an assertion as that "B is an illegitimate inference from A" when A and B are known previously to be both true? To an omniscient mind would not all true propositions be equally axiomatic? Would it not be absurd to speak of such a mind as *inferring* B from A? This, of course, is an extreme case, but such cases are precisely those that most effectively test the validity of a principle. On the same principle, does it not seem absurd to speak of *inferring* B from A, whether "legitimately" or "illegitimately," when A and B are truths which have been arrived at independently, or when B is self-evident apart from all consideration of A? As a concrete example, take the proposition that 'any two sides of a triangle are together greater than the third,' of which Euclid gives a formal proof. Seeing that none of his so-called axioms is more self-evident, why did he consider a proof necessary? Strict Euclideans consider no proof valid, however convincing, if it takes anything for granted that is not founded on Euclid's twelve axioms, although as a matter of fact, Euclid himself, in several of his formal proofs, tacitly assumes axioms which are absent from his given list. But the question now before us is: What is really meant by *inferring* (or *deriving*) a proposition B from another proposition A (whether axiomatic or not) when B needs no proof, or is known to be true apart from all thought of A ? The answer is not far to seek. It involves another meaning of the word '*implies*'. The proposition B is said, in this sense, to be 'inferred from,' or 'derived from,' or 'implied in,' or 'contained in' A, when it is either a particular case of A, or when all the statements made in the so-called 'proof' of B are particular cases of the axioms or propositions which constitute A. For the sake of clearness, it would be better to express this kind of implication by the word 'contains' rather than the word 'implies'. Thus, when A 'contains' B, it follows that A also 'implies' B ; but the converse does not necessarily hold. For example, the formula $(x^2 - a^2) = (x - a)(x+a)$ both contains and implies the statement $(365^2 - 364^2) = (365 - 364)(365 + 364)$; and the converse also holds as regards the word 'implies,' since every certainty necessarily 'implies,' though it does not necessarily 'contain' every other certainty. But the converse does not hold as regards the word 'contains,' for though the general algebraic statement contains the particular arithmetical as a particular case, the arithmetical statement does not (in this sense) contain the algebraic.

A good illustration of these principles will be afforded by deducing the syllogism *Baroko* from *Barbara*. The syllogism *Barbara* (in its proper conditional or implicational form) is
$$(x : y)(y : z) : (x : z),$$
which we will denote by the functional symbol $\phi(x, y, z)$. *Baroko*, in its proper conditional or implicational form (beginning with the minor premiss), is
$$(z : y)(x : y)' : (x : z)'$$
which, by transposition, that is, by virtue of the formula $(AB' : C' = CA : B)$, is equivalent to
$$(x : z)(z : y) : (x : y),$$
which, by definition, is equivalent to $\phi(x, z, y)$. This shows that *Baroko* is equivalent to a syllogism which is a particular case of *Barbara*, as it is obtained from *Barbara* by interchanging y and z. Here the formula of transposition, $(AB' : C' = CA : B)$, as it holds for all values of A, B, C, evidently *contains* as well as *implies* the statement that the complex implication
$$(z : y)(x : y)' : (x : z)'$$
is equivalent to the complex implication
$$(x : z)(z : y) : (x : y);$$
for when we substitute $(z : y)$ for A, $(x : y)$ for B, and $(x : z)$ for C, we see at once that the complex equivalence expressed in terms of x, y, z is only a particular case of the simple equivalence expressed in terms of A, B, C. The implication $(AB' : C')$

LINGUISTIC MISUNDERSTANDINGS.

is equivalent to the implication (CA : B), because the former, by definition, means $(AB'C)^\eta$, and the latter, by definition, means $(CAB')^\eta$, which only differs from the former statement in the order of the factors in the brackets. Each asserts that the compound statement (CAB') is an *impossibility*. It is noticeable that the statement of equivalence, $(CA : B) = (CAB')^\eta$, is itself contained, and therefore implied, in the still simpler statement of equivalence, $a : \beta : (a\beta')^\eta$, from which it is obtained by changing a into CA, and β into B. It is also noticeable that though the implication $(a : \beta)$, which asserts that a implies β, is equivalent to $(\beta' : a')$, which asserts that the denial of β implies the denial of a, this equivalence does not hold when we substitute the sign of inference ($\therefore$) for the sign of implication (:). For it is clear that $(a \therefore \beta)$, or ("a *therefore* β"), which asserts both a and β, cannot be equivalent to $(\beta' \therefore a')$, which *denies* both a and β.

We may conveniently divest the word 'therefore' of all psychological meaning by agreeing to the convention that the symbol $(A \therefore B)$ shall simply mean $A (A : B)$, which both asserts A and that A implies B, in the sense already given to the word *implies*. On this convention, it of course necessarily follows that $(A \therefore B)$ is always true whenever A is true and B is a certainty; for, on this convention, $(A \therefore \epsilon)$ means $A (A : \epsilon)$, which $= A (A\epsilon')^\eta = A (A\eta)^\eta = A\eta^\eta = A\epsilon = A$; so that when B is a certainty (whether known to be so or not) the statement $(A \therefore B)$ simply asserts A, which is true by hypothesis.

Examples of inferences which finally lead to self-evident certainties are not uncommon in mathematics. Take the following. Suppose we have given us the statement of inequality

$$\frac{13\,x}{8} + \frac{1}{2} > \frac{3\,x}{4} - \frac{6-7\,x}{8} + 1,$$

in which, as usual, the symbol $>$ means "is greater than". Multiply each of these unequals by 8. We get
$$13\,x + 4 > 6\,x - (6 - 7\,x) + 8,$$
from the implicational formula (or axiom)
$$(m > n) : (Pm > Pn),$$
in which P is any positive number or ratio. That is, we get
$$13\,x + 4 > 6\,x - 6 + 7\,x + 8$$
Therefore $13\,x + 4 > 13\,x + 2$.
Subtracting $13\,x$ from each of these unequals, we get $(4 > 2)$, which is a self-evident certainty.

In this case we have deduced—we cannot, in the usual sense of the word, say "proved"—the obvious from the non-obvious, both being real certainties, though not both equally evident. By reversing the process, and suitably choosing our axioms, or fundamental formulæ of appeal, we might deduce the non-obvious certainty with which we began from the obvious certainty with which we concluded. Similarly, by a proper choice of axioms, or assumed formulæ of appeal, we might deduce any certainty from any other certainty.

III. Antinomies, Logical and Philosophical.

Symbolic logic, like the Kantian philosophy, has its antinomies; that is to say, apparently valid arguments that lead to contradictory conclusions. But there can be no such thing as a " reconciliation " of antinomies, Kantian or other. One at least of the arguments must contain an error somewhere, though it may be difficult to find out where. The following antinomy arrested me for a while in the development of my symbolic system.

The symbol A^θ, in my system, is short for $(A^\theta)^\theta$ and asserts that the statement A^θ is a variable.[1] The antinomy consists in the conflict of two arguments, of which the one professes to prove that the second-degree proposition $A^{\theta\theta}$ is an impossibility or self-contradiction; while the other professes to prove that it is not. The first argument is this:—

The statement A (assuming it to be intelligible) must be either a certainty, an impossibility, or a variable.

First, let A be a certainty, the certainty ϵ_1. Then A^θ means ϵ_1^θ and asserts that a certainty is a variable, which is impossible. Thus, when A is a certainty the statement A^θ is an impossibility. Call it η_1. Now, $A^{\theta\theta}$ means $(A^\theta)^\theta$, that is η_1^θ, and therefore asserts that the impossibility η_1 is a variable, which is a self-contradiction. Hence, when A denotes a certainty, $A^{\theta\theta}$ is an impossibility.

Next, let A be an impossibility, the impossibility η_2. Then A^θ will mean η_2^θ and asserts that the impossibility η_2 is a variable, an assertion which is an impossibility. Call it the impossibility η_3. Thus, $A^{\theta\theta}$, or its synonym $(A^\theta)^\theta$, means η_3^θ and asserts that the impossibility η_3 is a variable, which is a self-contradiction. Hence, when A is an impossibility, $A^{\theta\theta}$ also is an impossibility.

---

[1] A proposition of the form A^x is called a proposition of the *first degree*, because it has only one exponent, namely x. It asserts that the individual A belongs to the class x. That is to say, it asserts that the individual A represents one or other of the individuals x_1, x_2, x_3, etc. A proposition of the form A^{xy} is called a proposition of the *second degree*, because it has two exponents x and y. It means $(A^x)^y$. Similarly A^{xyz} means $(A^{xy})^z$ and is a proposition of the *third degree*. And so on.

LINGUISTIC MISUNDERSTANDINGS.

Lastly, let A be a variable, the variable θ_1. Then A^θ means θ_1^θ and asserts that the variable θ_1 is a variable, an assertion which is a self-evident certainty. Call it the certainty ϵ_2. Thus, in this case, $A^{\theta\theta}$ or its synonym $(A^\theta)^\theta$, means ϵ_2^θ and asserts that the certainty ϵ_2 is a variable, which is a self-contradiction. Hence, when A is a variable, $A^{\theta\theta}$ is an impossibility.

Thus we have apparently proved that whether A be a certainty, an impossibility, or a variable (and it must be one of the three) the second-degree proposition $A^{\theta\theta}$ is an impossibility.

The next argument, which professes to prove the opposite conclusion, namely, that $A^{\theta\theta}$ is *not* impossible, is as follows:—

Take any number of certainties ϵ_1, ϵ_2; any number of impossibilities η_1, η_2, η_3; and any number of variables $\theta_1, \theta_2, \theta_3, \theta_4$. Out of these nine statements take any statement at random, and call it A. If a certainty turns up we shall have A^ϵ, and the chance of this is 2/9. If an impossibility turns up we shall have A^η, and the chance of this is 3/9. If a variable turns up we shall have A^θ, and the chance of this is 4/9. Thus, the three statements $A^\epsilon, A^\eta, A^\theta$ are all variables, since they are neither certainties nor impossibilities, their respective chances being proper fractions between 0 and 1. Thus, on these perfectly admissible data, which may be put to the test of actual experiment, the statement $A^{\theta\theta}$, which means $(A^\theta)^\theta$, and only asserts that A^θ is a variable, is true.

Thus $A^{\theta\theta}$ involves no formal self-contradiction, and in certain conditions (such as those adduced) it is perfectly possible.

After some reflexion, I found that the second of these antinomies (namely, that $A^{\theta\theta}$ is *not* self-contradictory) is the true one. Where then is the error in the first argument? It consists in this, that it tacitly assumes that A *must* either be *permanently* a certainty, or *permanently* an impossibility, or *permanently* a variable—an assumption for which there is no warrant. On the second supposition, on the contrary—a supposition which is perfectly admissible—A *may change its class*. In the first trial, for example, A may turn out to represent a certainty, in the next a variable, and in the third an impossibility. When a certainty or an impossibility turns up, the statement A^θ is evidently false; when a variable turns up, A^θ is evidently true; and since (with the data taken) each of these events is possible, and indeed always happens in the long run, A^θ may be false or true, being sometimes the one and sometimes the other, and is therefore a variable. That is to say, on perfectly admissible assumptions, $A^{\theta\theta}$ is possible; it is not a *formal* impossibility.

But, *with other data*, A^θ may be either a certainty or an impossibility, in either of which cases $A^{\theta\theta}$ would be an impossibility. For example, if all the statements from which A is taken at random be exclusively variable, θ_1, θ_2, etc., then, evidently, we should have $A^{\theta\epsilon}$, and not $A^{\theta\theta}$. On the other hand, if our universe of statements consisted solely of certainties and impossibilities, with no variables, we should have $A^{\theta\eta}$, and not $A^{\theta\theta}$. Thus the statement $A^{\theta\theta}$ is *formally* possible; that is to say, it contradicts no definition or linguistic or symbolic convention; but whether or not it is *materially* possible depends upon our special or material data.

The Kantian and other antinomies of space and time may, I think, be similarly treated, if my definitions of the finite, the infinite, and the infinitesimal be accepted. Let us speak first of the abstract and purely conceptual spatial universe of the mathematician. This is a mere matter of convention and convenience. We may ascribe to it any shape and dimensions we please, provided they do not conflict with logical principles or human experience; but convenience and symmetry suggest that we should consider this conventional universe spherical with an infinite radius—infinite in the sense already defined. This will allow ample scope for all abstract speculation or theoretical reasoning, as well as for all the practical mathematical formulæ required by astronomers, present or future, in their stellar researches. For the numberless infinities, H_1, H_2, H_3, etc., which the imagination calls into existence, being each, by hypothesis, not only too large for any scientific instrument ever to measure, but also too large for any numerical notation ever to express even approximately, and having respectively also by hypothesis any ratios to each other we please to give them, finite, infinite, or infinitesimal, the imagination obtains unlimited range, while the sober reason is kept within the wholesome restraints of linguistic consistency. Of course, this definition of the word 'infinite' is not in strict accordance with its primary meaning; but if words were always restricted to their primary meanings no human language could ever have been developed, abstract ideas could never have been formed, and science and philosophy would never have come into existence. Words are mere symbols to which we may assign any convenient meaning that suits our argument, provided we make it perfectly clear, by definition or context, what that meaning is.

But this abstract and purely conceptual space is not, I think, the space which modern Kantians have in mind when they discuss Kant's antinomies. They refer to what may be roughly called the world of realities—the material world of

LINGUISTIC MISUNDERSTANDINGS.

phenomena that contains solids, liquids, gases, and the hypothetical ether, with the forces (conscious or unconscious) which we find acting on, through, and by means of those entities. The modern Kantians, adapting Kant's principles to the physical and psychical conditions revealed by modern research, maintain that, whether we start with the assumption that this world of realities is finite or with the assumption that it is infinite, we necessarily arrive at a conclusion which our reason rejects—a conclusion which, if not exactly a linguistic self-contradiction, is at any rate opposed to our *a priori* conceptions of reality. Now, if my definitions of the finite, the infinite, and the infinitesimal be accepted, the inevitable logical conclusion, as it seems to me, should be the exact opposite. Neither the assumption that the universe of realities is finite nor the assumption that it is infinite (as I understand these words) leads to any self-contradiction whatever; nor is either assumption opposed to our *a priori* conceptions of any reality. Why should there be any such opposition or contradiction, seeing that the spatial finite and the spatial infinite only differ in the fact that the latter is, and the former is not, utterly beyond our power of expression by comparison with any known unit, be it an inch, or a yard, or a mile, or a million million miles, or the circle or sphere of which any of these is the radius? This, of course, involves the conception of a bounded real and material ether-filled universe, finite or infinite as regards size or magnitude, with an absolutely blank, empty, etherless *nothingness* beyond —a purely conceptual abstract ultramundal space void of matter, void of ether, void of every kind of reality, sentient or non-sentient. This ultramundal vacuum is supposed by the Kantians to be an impossible conception. I do not find it so. On the contrary, what I find difficult to conceive is the non-existence of such a vacuum. I cannot picture to myself an absolutely boundless material or ether-filled universe existing everywhere with no absolutely empty etherless space anywhere. What is this pseudo-infinity but a resuscitation in another form of the meaningless old dictum that "Nature abhors a vacuum"? It is exactly paralleled by the pseudo-infinities $\frac{1}{0}$, $\frac{2}{0}$, etc., of mathematicians when they speak of the tangent or secant of a right angle—two trigonometrical ratios which do not exist, though the angle itself is a reality. Since, as it has been pretty well proved, nature has no particular abhorrence of an *airless* interplanetary and interstellar vacuum *within* the real universe, I see no valid reason why it should have any special abhorrence of an *etherless* vacuum *beyond* the universe.

III.—LINGUISTIC MISUNDERSTANDINGS.

By Hugh MacColl.

Part II.

IV. The Antinomies of Time.

Let us now consider the words *finite* and *infinite* in reference to time. Here the analysis is more difficult, because the primary conceptions are more complex. For we can have no clear idea of time till we are first in possession of clear ideas of space, number, change and motion. The commonest method of measuring time nowadays is by clocks and watches. But in this way what do we really measure? We measure the *spaces* traversed by the hour, minute or second hands. The astronomer measures time by measuring the angular changes produced by the motions of the heavenly bodies, taking a complete angular revolution of the sun among the apparently fixed stars as his final and most convenient unit of reference. We call this unit a *year*, and to suit our convenience we arbitrarily divide it into other units called *days, hours, minutes,* and *seconds*. But here again what we really measure is *space*, the apparent space described by the sun, which measures time for us by its apparent motion round a great circle of the sky, just as the hour or minute hand of a clock measures smaller portions of time for us by its real motion round the clock dial. If we perceived no change we could have no notion of time. Our case when we are lying in bed motionless, with shut eyes but awake, is no exception. For in this case we are conscious of our successive ideas or sensations, and from these mental changes we form a rough estimate of the progress of time. These examples sufficiently show how much more complex is the conception of time than the conception of space.

We roughly divide time into three divisions, present, past, future; but how far does each of these extend? By a commonly accepted convention, we agree that the past and the future should each be considered infinite; the latter, reckoned from the present, being conveniently called *positive;* the former *negative*. But what are the *limits* of the present? Here there is less agreement in linguistic usage. Some regard the present as a mere point in time, passing instantly into the past as soon as it is reached. From some points of view, and for some purposes, this convention has its utility; but they are exceptional cases. As a rule, it is more convenient to ascribe to the present an actual but limited duration, as when we speak of the present hour, the present year, or the present century. Thus viewed, the present presents no difficulty; but what about the infinite past, and the infinite future? These two infinities (direction apart) logically stand on the same footing, and thus ought to present no more difficulty than the infinity of space. Yet they do; at any rate, the former. A thing having once come into existence, be it an inanimate stone, or a sentient being, or the material universe, the conceptual supposition of its continuing to exist for ever in the future gives no shock to the reason. An endless as well as infinite future—infinite in the sense already defined—seems somehow less difficult to grasp, less of a self-contradiction, than a boundless as well as infinite spatial universe. But it is different with regard to a *beginningless* as well as infinite past. Many people, in thinking of an infinite deity, or of the finite or infinite universe, find this beginninglessness an impossible conception, and yet no less impossible the idea of anything, spiritual or material, springing suddenly and causelessly into existence out of an antecedent nothingness. These difficulties generally arise from reflexion and reasoning, for they do not seem to occur to very young children. These, I believe, as a rule, before they have heard of birth or death, take their own eternal existence, past and future, without beginning or end, simply and tacitly for granted. I knew a little girl who, when, at the age of four, she learnt for the first time that all must sooner or later die, her father, mother, brother, sister, and herself not excepted, gave way to a flood of tears, and for some days remained inconsolable. She could not resign herself to the bitter thought that for her, as for all near and dear to her, this happy life which she and they so thoroughly enjoyed must, after a limited but unknown length of time, come entirely to an end. What she thought of her past existence, before her parents had, as she supposed, "bought" her, I never heard, but I think it likely that she took for granted the beginninglessness of her past as she certainly did the endlessness of her future.

Why is an absolute void and nothingness in space an easier conception for us than an absolute void and nothingness in time? I think the reason is this, that in the measurement of

LINGUISTIC MISUNDERSTANDINGS.

time we have the principle of *repetition* after a complete revolution; in the measurement of space we have not. In measuring the distance from A to B we move from A to B in a straight line and never go twice over the same spot. In the measurement of areas and volumes we adopt the same principle; we never count a unit of area or volume twice over. But in the measurement of time (which we cannot accurately accomplish without also measuring space) we are obliged to adopt the principle of revolution and repetition. The minute or second hand of a watch performs a complete revolution on the circular dial in an hour or minute respectively, and then repeats the revolution again and again *over the same space*. Similarly, the sun appears to describe a complete revolution round a great circle of the heavens in a year, and then repeats the revolution year after year, *round the same apparent circle*. And, *a priori*, before we have studied the origin and evolution of the heavenly bodies, we can see no reason why this should not have gone on eternally in a beginningless past, and why it should not also go on eternally in an endless future. As an *a priori* conception, this beginningless and endless eternity of matter or spirit seems an easier conception than that of a deity or of a material universe starting suddenly into existence out of a preceding eternal nothingness at some infinitely remote point in the past. That the universe should, at some infinitely future date, suddenly explode into its previous hypothetical nothingness is a hardly less difficult supposition. Yet neither supposition involves any logical or linguistic inconsistency. Sudden and startling presentations of inexplicable phenomena, which as suddenly vanish, are not unknown to our experience. Shooting-stars, meteors, fire-balls, etc., may be cited as examples which perplexed our ancestors; and, coming to modern times, how many of the spectators of a cinematograph performance understand the cause or principle of the unexpected marvels which so completely deceive their eyes while they excite their imagination or tickle their sense of humour? Suppose one of these spectators, as might very well happen, were suddenly to drop off into unconsciousness in the middle of a scene, and afterwards, as suddenly, come to himself at the very same point in the middle of the same scene in a repetition of the performance. He would be wholly unaware of the flow of time during his unconsciousness; he would never suspect that he had been unconscious; the rest of the second performance would appear to him the simple and natural continuation of the first. To such a spectator his blank interval of unconsciousness would count as zero in his measurement of time.

"Imagine something analogous to happen to our whole universe, sentient and non-sentient.[1] Suddenly all the laws of nature are suspended. All motion ceases. Gravitation is no more. . . . The rising and falling waves stop as if suddenly sculptured on a sea of ice. . . . The preacher in his pulpit stops in the middle of the word *firstly;* the orator in parliament in the middle of the word *closure.* . . . The brain functions no longer. All thought, all feeling ceases. . . . The universe still exists, if existence it may be called, as dead matter, but its life has departed. Then, after a hundred years (as in the well-known fable), or a thousand years, or a million years, its life returns as suddenly as it left it. The earth resumes its revolution on its axis; the planets resume their course round the sun; motion rebegins everywhere. . . . The preacher continues his sermon; the parliamentary orator his angry protest against the closure. Everything goes on as if nothing had happened; nobody knows or suspects that anything *has* happened—that the life of the whole universe has been arrested for a million or more years.

"Now, from the strictly logical standpoint, how should this hypothetical suspension of all the laws of our universe, physical and psychical, be regarded? What about our scientific formulæ. As regards all formulæ bearing on the question of time in general, and age in particular, would they not be more simply workable, as well as more reliable in their application, if we considered the whole period of cosmic suspension, however long, as non-existent? . . . How many bankers would be willing, or would be able if willing, to pay the amount of interest that would have become due after more than a million years? . . . Confusion and perplexity would meet us everywhere. The only possible solution from the practical standpoint is the one that would be unconsciously adopted: everybody would regard the whole period of suspension as non-existent, as absolute zero."

Now, just as two atoms shot at random in the universe may conceivably collide, though the chance of the collision be infinitesimal, so the preceding hypothetical event may conceivably happen, since it involves no linguistic self-contradiction and is not incompatible with any known data. *It cannot be proved false.* Just as much and no more may be affirmed of many of the speculative hypotheses seriously advanced by serious scientists as serious explanations of the origin and evolution of the universe.

[1] This extract I quote (with some omissions) from my *Man's Origin, Destiny and Duty.*

But at present we are not discussing the possibilities of actual events, but the consistency of concepts, and the suitability of the terms in which we strive to express them. Can time exist—exist even as a clear concept—without the existence also of motion, or of the idea of motion? The illustration given will, I think, show that practical science at all event would, in its formulæ and calculations, have to ignore such existence.

V. Virtual Equality.

The so-called 'antinomies' of Kant appear to me to have sprung from a confusion between his clear apprehension of the primary or subjective meaning of the word *infinite* and his somewhat vague apprehension of the only meaning that can (in my opinion) be consistently attached to the word in exact mathematical researches. In practical mathematical researches it will be found that all valid formulæ containing reference to infinities or infinitesimals will retain their validity when these words convey the sense which I give them by express definition, namely, that the former is a number, magnitude, or ratio too large, and the latter a number, magnitude, or ratio too small, to be accurately or approximately expressed in the decimal or any other system of notation. Thus, every infinitesimal is the reciprocal of some infinity, and every infinity is the reciprocal of some infinitesimal; while every finite is the reciprocal of some other finite. On the other hand, the word *infinity*, in its primary or subjective sense, of *endlessness*, merely expresses the liberty claimed by the imagination of "beating the record," so to speak, whenever it chooses; that is to say, the liberty of surpassing any number, magnitude, or ratio, however large, by the conception of a number, magnitude, or ratio still larger. For example, if it be asked how many terms there are in the series A^1, A^2, A^3, A^4, etc., in which A denotes any number or ratio, we may legitimately reply that the number of terms is infinite. Here however the word *infinite* does not denote a real number at all, nor any property that can be attributed to any real number, nor any class to which any individual number belongs; and the statement that the number of terms is infinite is only another way of saying that though there is a definite *first* term there is no definite *last* term; or, in other words, that there is no fixed limit beyond which the imagination may not continue the series. When, on the other hand, we say that the sum of the series $A^1 + A^2 + A^3 +$ to infinity, assuming A to be a proper fraction between 0 and 1 (say, the fraction *one-half*) is $A/(1-A)$, what we really mean is that (denoting the class, or any individual of the class, of infinities, H_1, H_2, H_3, etc., by H, and the class, or any individual of the class, of infinitesimals, h_1, h_2, h_3, etc., by h) if we denote the sum

$$A^1 + A^2 + A^3 + \ldots + A^H$$

by S, then we shall have

$$\left(\frac{A}{1-A} - S\right)^h,$$

a symbolic statement which, in my notation, asserts that the difference between the greater ratio $A/(1-A)$ and the less ratio S is infinitesimal. Mathematicians commonly write

$$A + A^2 + A^3 + ad\ infinitum = A/(1-A);$$

but this equality never holds absolutely, for however far the series may be carried, the sum S is always less than $A/(1-A)$. Here the total number of terms in the series is the real though infinite number[1] H, and the last term of the series is A^H, which is necessarily an infinitesimal, since, by our data, A is a proper fraction less than 1. Hence also the statement $S = A/(1-A)$ asserts *virtual* and not actual equality; that is to say, there is a real difference between the two ratios asserted to be equal, but the difference is infinitesimal compared with (*i.e.*, divided by) either. (See my paper on *Symbolic Reasoning*, No. viii., p. 509, in MIND, October, 1906.) This example illustrates the sense in which mathematicians commonly use the word *infinite*, though the lack of an exact and satisfactory definition of the word in the generality of textbooks renders their language sometimes obscure and their statements apparently inconsistent. In this and similar cases we may write $S = A/(1-A)$, provided it be understood that *virtual* and not absolute equality is asserted. As defined in my paper in MIND, two quantities or ratios, finite, infinite, or infinitesimal, are said to be *virtually* equal when the difference between them is infinitesimal compared with (or divided by) either, and in the infinitesimal calculus it would be convenient if we adopted the convention that the symbol of equality ($=$) between two quantities or ratios in the statement ($A = B$) only asserts that the two are either virtually or absolutely equal. Thus the statement ($x = x + dx$) asserts *virtual* equality, whether x be finite, infinite, or infinitesimal, it being understood that dx/x is an infinitesimal ratio.

[1] The symbol A^H may either denote *the* Hth *power of* A (the exponent H being infinite by definition) or the *statement* that A *is infinite*. The context will always make clear in what sense it is employed in each case, for it is scarcely possible to mistake a statement for a ratio, or *vice versa*.

LINGUISTIC MISUNDERSTANDINGS.

Prof. Keyser, in the *Hibbert Journal* of October, 1909, page 188, says that—

> Mr. MacColl's conception of the *infinitesimal* is one that mathematicians have not been able to employ. As used by them, the term signifies, not a small quantity, but a variable that, under the conditions of the problem in which it occurs, may be *made* and kept small at will—a variable having zero for limit.

To this I reply that I believe the reason why mathematicians have not, so far, employed my conception of the *infinitesimal*—a conception which they all possess, however differently they may express it—is that my allied and complementary conception of "virtual equality" had never occurred to them. Restricting, as they do, the symbol of equality ($=$) to *absolute* equality, they could not consistently make the assertion ($a=a+x$) even when x is infinitesimal compared with a, so that, to preserve logical accuracy, they are obliged to be continually appealing to the round-about notion of a *limit*, an appeal which my proposed convention as to 'virtual equality' and the meaning of the symbol ($=.$) would render needless and irrelevant, without sacrificing one iota of logical accuracy.

In explaining the principle of a 'derivative' or 'differential coefficient,' modern writers on the infinitesimal calculus find it necessary to lay down various cautions to prevent beginners from misunderstanding the real meaning of their symbolic formulæ and operations.

Let me quote the following from an excellent work (*An Elementary Course of Infinitesimal Calculus*, by Horace Lamb, F.R.S.) which I have often recommended to pupils:—

> The symbol dy/dx is to be regarded as indecomposable, it is not a fraction, but the limiting value of a fraction. The fractional appearance is preserved merely in order to remind us of the manner in which the limiting value was approached.

Now, I agree that dx and dy should each separately be regarded as indecomposable, the letter d having no meaning apart from the letters x and y, which denote real quantities or ratios; but there is no necessity for so regarding the whole complex symbol dy/dx. I see no reason at all why we should not, like Leibnitz, regard dy/dx as a real fraction whose value depends upon the real values of its numerator and denominator dy and dx. The following simple example will, I feel sure, make this clear to every reader of MIND, whether he be acquainted with the infinitesimal calculus or not:—

Let $y=x^2$, and let dx be infinitesimal compared with x, so that dx/x is an infinitesimal ratio. Also let dy denote the increment received by y (that is, by x^2) in consequence of an infinitesimal increment dx received by x. Otherwise expressed, let $dy = (x+dx)^2 - x^2$. We get

$$dy = 2x\,dx + (dx)^2.$$

Hence, $dy/dx = 2x + dx$. So far, the sign of equality has denoted *absolute* equality. Now, since (by definition) dx is infinitesimal compared with x, it must also, *a fortiori*, be infinitesimal compared with $2x$, so that we get $2x + dx = 2x$. Here the sign of equality denotes *virtual* and not *absolute* equality. Thus, finally, we get $dy/dx = 2x$, an equality which is again *virtual* and not absolute. From this point of view there is no reference to a limit, as the conception of a limit is not needed. In this case, dx and dy are two really existing infinitesimals, and my assertion that the ratio of the latter to the former is *virtually* equal to $2x$ (as I define the word *virtually*) is really equivalent to what mathematicians mean when they say that $2x$ (technically called the "differential coefficient of y with regard to x") is the *limit* to which the fraction $\delta y/\delta x$ approaches as the increment δx (which they never speak of as an *infinitesimal*) approaches zero.

Let it be clearly understood that an assertion of virtual equality, such as $(x+dx=x)$, not merely asserts that dx is *negligible* compared with x, that practically it may be omitted because of its extreme smallness in comparison, but that it *must* be omitted in all possible calculations, because (by express definition) no arithmetical notation, and *a fortiori*, no instrument however delicate, can ever take account of its existence.

If a regular polygon of M^M sides (in which M denotes a million) be supposed inscribed in a circle, the difference by which its perimeter falls short of the circumference is certainly negligible, and more than negligible, compared with either as regards all practical calculations, but the ratio is not *infinitesimal*, because, though inconceivably small, it is still arithmetically expressible; that is to say, it can be expressed approximately by certain conventional collocations of the ordinary digits. In this case, therefore, we cannot consistently assert that the perimeter is *virtually* equal to the circumference. And if for M^M we substitute its millionth power M^{MM}, or any other huge but arithmetically expressible number, the result will be the same; the excess of the circumference over the inscribed perimeter, though utterly negligible, will, from our very definitions of the terms, be *finite* and not *infinitesimal*. But if we suppose a regular inscribed polygon of H sides, then the difference between the perimeter

LINGUISTIC MISUNDERSTANDINGS.

and the circumference would be infinitesimal compared with either (since H is, by definition, infinite); and for this reason, we can here assert that the perimeter is *virtually* equal to the circumference, and express this virtual equality in the form (P=C), in which P stands for *perimeter*, C for *circumference*, and the symbol (=) for an assertion of *virtual equality.*

Leibnitz founded his infinitesimal calculus on the notion of infinitesimals, which he merely regarded as extremely minute quantities, without clearly indicating in what respect an infinitesimal differs from a very small finite. Newton founded his calculus on the notion of the 'ultimate ratios' of vanishing quantities; that is, the ultimate ratio of the increment δy to the increment δx when the latter (and consequently, as a rule, the former), by continual decrease, reaches the limit zero; in which case $\delta y/\delta x$ takes the form 0/0. Both were right in their respective conceptions, but they expressed those conceptions awkwardly and in apparently self-contradictory language, which led the logicians, and many even of the mathematicians, of the day to question the legitimacy both of their reasoning and of their symbolic operations. Modern mathematicians have adopted Leibnitz's notation as more convenient than Newton's, but they have completely rejected the conception of negligible infinitesimals on which Leibnitz founded his notation, on the ground that it is logically inadmissible. And logically inadmissible the conception undoubtedly is so long as the symbol of equality (=) is restricted to *absolute* equality; for it is clear that A cannot be *absolutely* equal to A+h so long as h has any real value however small. But my convention, that the symbol of equality shall only denote an equality that may be either absolute or virtual (as I define the word *virtual*), entirely removes from Leibnitz's symbolic formulæ and operations the reproach of inconsistency. Modern mathematicians, following Newton's conception, have secured for it a certain measure of consistency, but at a heavy and needless sacrifice of brevity and simplicity. They have replaced Newton's conception of an 'ultimate ratio' of the form 0/0 by the more consistent idea of a *limiting* ratio, which they express in the form dy/dx. They regard dy and dx, however, not necessarily as infinitesimals or other small quantities, or indeed as necessarily quantities at all. They merely insist that the composite symbol dy/dx shall denote the exact limiting ratio which it is employed to represent. They thus studiously avoid Leibnitz's notion of *infinitesimals* by dispensing even with the word. Instead of speaking of infinitesimals, they nearly always speak of *limits.*

If x and a be real ratios (finite, infinite, or infinitesimal) and virtually equal, it generally follows that any function of x (if a real ratio) is virtually equal to the same function of a. When two infinities H_1 and H_2 are virtually equal, their difference $H_1 - H_2$, or $H_2 - H_1$, though necessarily infinitesimal compared with (or divided by) either, may be finite, infinite, or infinitesimal compared with any finite. All that the definition of "virtual equality" requires is that the fractions $(H_1 - H_2)/H_1$ and $(H_1 - H_2)/H_2$ shall each be either a positive or negative infinitesimal.

If x and a (whether finite, infinite, or infinitesimal) be virtually equal, the fraction $(x^n - a^n)/(x - a)$ is virtually equal to na^{n-1}. Mathematicians usually express this by saying that na^{n-1} is the *limit* to which the fraction $(x^n - a^n)/(x - a)$ indefinitely approaches as the variable x approaches the finite constant a. But by my proposed convention as to "virtual equality" and the meaning of the symbol of equality ($=$), the ratio a need not be finite. The proposition holds good universally, whether x, a, n (individually or collectively) be finite, infinite, or infinitesimal.

Let $A_1, A_2, A_3, \ldots, A_n$ be any ratios in ascending order of magnitude, and such that A_1 is virtually equal to A_2, A_2 to A_3, A_3 to A_4, and so on. Then, *if n be not infinite*, the smallest ratio A_1 is virtually equal to the largest ratio A_n, so that we can write $A_1 = A_2 = A_3 = \ldots = A_n$.

From this theorem it follows that in mathematical researches involving n affirmations of *virtual equality* (however large the finite number n may be), no error can possibly enter into the final result through the repeated omission of infinitesimals in successive affirmations of virtual equality. The theorem is a simple corollary from the easily proved formula $(Fh)^h$, which asserts that the product of a finite and an infinitesimal is an infinitesimal. The formula holds however large the finite F may be—even if it denote the millionth power of the millionth power of a million.

VI. Matter and Mind.

Among the antinomies discussed by metaphysicians are the arguments for and against the possibility of the real existence of space, time, and the material universe, apart from the existence of a human or superhuman mind to perceive them. To enter seriously and fully into such a discussion would be a formidable undertaking. No one can do so profitably without first making sure that he and his readers attach the same meanings to the words he employs. Otherwise he enters a labyrinth of ambiguities from which it is scarcely

LINGUISTIC MISUNDERSTANDINGS.

possible for him to find an exit. And the readers who venture to follow him commonly share his fate. The words *real* and *existence* especially need defining, and definitions of them are not easy. We all understand these words in various senses according to the context; and often also, even with the context to guide us, we wofully *mis*understand them. In common parlance we speak of real existence and of unreal (or imaginary) existence, and, logically enough, we regard these two classes of existence as mutually exclusive. Yet, on close inspection, it is not easy to find the exact line of demarcation. Should the abstractions *truth* and *error* be considered unrealities because they have no form, weight, or substance? Hardly, though their existence certainly depends upon that of the persons who understand or misunderstand them. Is an unreal or imaginary existence a contradiction in terms, a linguistic inconsistency? What is the difference between an unreal or imaginary existence, such as that of a fairy, and an absolute non-existence? May we reply that unrealities, like fairies and fictitious characters in novels, exist in the mind, and must therefore have at least a *subjective* existence? "In the mind?" What does the preposition *in* here mean? Is the mind (or soul) then a substance of some kind, material or immaterial, in which another substance, real or imaginary, can exist? Or are we merely talking figuratively and—rather vaguely—because the ideas which we strive to convey are too vague for exact expression? Is the mind the same as the soul? And if not, what is the difference? Can the distinction be clearly shown by an exact definition of each? Materialistic philosophers— or those who call themselves such, for these words also are ambiguous—sometimes speak of the soul as a "function of the brain," and sometimes as an "emanation from the brain". What do they mean? If they were pressed hard for definitions or explanations, and answered frankly, I think they would be forced to own that they did not know—that—to put it bluntly—they had been talking nonsense. In mathematics the word *function* has a clear and definite meaning. As a rule, when we can correctly say that y is a function of x, we can also say correctly that x is a function (though generally a different function) of y. If in this or some analogous sense, the soul can be said to be a function of the brain, can we, following the mathematical analogy, say that the brain is also a function of the soul? Idealists might plausibly maintain this view, but not materialists, as it is directly opposed to the latter's fundamental conceptions. Again, taking the other materialistic view, how can the soul be an "emanation," or *flowing*, from the brain? Can the soul, even as an analogy or metaphor, be likened to a liquid or gas flowing or escaping from a reservoir? The points of unlikeness are surely far more numerous than those of resemblance. And of the points of unlikeness the most striking is the fact that the mind or soul is *conscious of its own existence*, while the flowing gas or liquid is *not*.

In connexion with this question of the soul, I may be allowed to say a few words in reference to an objection raised by Prof. Taylor in his kind and appreciative criticism of my *Man's Origin, Destiny and Duty*, in MIND, July, 1909, pp. 451-453. In that book I define the soul as "*that which feels*," and argue that, wherever it may be situated, there is no proof that it is in the brain or nervous system, as these, judging from observation and experiments, appear (like the rest of the body) to be mere insensible channels through which some unknown force is transmitted (we know not whence) *to* the soul, causing sensations, and *from* the soul—generally by exertion of the *conscious will*—producing actions. In regard to this view, which he does not seem wholly to reject, Prof. Taylor writes:—

> But the view is still retained that this subject [the soul or real subject of consciousness] is extended and occupies a region in the physical space of ordinary perception. Thus Mr. MacColl's contention against the ordinary materialist takes the form of maintaining that the 'soul' must be *somewhere*, but the *somewhere* need not be "in the brain": it may be millions of miles away. Now, I should prefer to ask whether the question "Where is the soul?" has any meaning at all. Is it more reasonable to ask whether *e.g.* my belief that $2+3=5$ is in my brain or ten millions of miles away, than to ask what is the distance between Piccadilly Circus and the middle of next week? . . . The seat of consciousness is removed to a distant and possibly extra-stellar point, but the question still remains whether there would be any sense in saying that thought and sensation are "at" this point? Does a thought or feeling take up any extension at all? I think the author would see, on further reflexion, that the unity of our mental life, on his theory, would commit us definitely to the view that the soul literally is a mathematical point, and such a view is surely as unintelligible when that point is said to be millions of miles away as when it is said to be "in" the brain. The real absurdity surely lies in assigning presence "at" a point to the self at all.

Now, I admit at once that these are serious objections to my theory or hypothesis that the soul (whether material or etherial, or composed of some other substance entirely imperceptible to our present human senses) may possibly have, at any given moment, some definite though unknown form or size, and may occupy, like a planet, sun, or atom, some definite though unknown position in space. Great however as are the difficulties that lie in the way of this hypothesis,

LINGUISTIC MISUNDERSTANDINGS. 349

those that confront the opposite hypothesis, the hypothesis that the soul (the sentient entity by my express definition) has no spatial existence, seems to me more formidable still. But here again, perhaps, Prof. Taylor and I, like so many other sincere controversialists, do not always attach quite the same meanings to the same words. Even if my hypothesis led to the conclusion he supposes, that, as regards size, the soul corresponded to the conception of a mathematical point —a conclusion which can hardly follow from my premises, since these leave its size and form unknown and indefinite— the conclusion would involve no inconsistency. For, from my point of view—which I admit however to be different from that of mathematicians in general—a point may have any size whatever, provided the unit of reference be infinite in comparison. If any portion of matter, whether an atom, an electron, or something else still smaller, be infinitesimal compared with any nameable finite unit, be it the volume of a drop of water or that of the earth, then, and not otherwise, it may be regarded as a *point*. And this infinitesimal point may also consistently be conceived of as infinite in comparison with another point still smaller; and so on *ad infinitum*. It is all a matter of ratio or comparison, and depends entirely upon our arbitrary unit of reference. A ratio h_1/h_2 between two infinitesimals, like a ratio H_1/H_2 between two infinities, may be finite, infinite, or infinitesimal; but a ratio F_1/F_2 between two finites *must* be finite, from our very definition of the word. Assuming the soul to have a spatial existence, we have no data at present for determining its size, form, or position at any given moment, or whether these be fixed or variable. If any man chooses to assert that his soul (spatially considered) is at this moment finite, or that it is infinite, or that it is infinitesimal (these words being understood as I define them), I can neither verify nor disprove his assertion, though the first hypothesis—I cannot in the least explain why—seems to me the most likely. "But why consider the soul spatial at all?" Prof. Taylor would ask. My reply is that otherwise I must regard it as belonging to the class of entities which most grammarians lump together under the name of *abstractions*, such as *hunger, hardness, battle*, etc., and that such abstractions are but disguised predicates which cannot be separated from some non-abstract subject understood. There can be no hunger without a hungry person or animal; there can be no hardness without some hard substance; and there can be no battle without some sentient beings (human or non-human) who struggle for mastery. Similarly, I cannot conceive of a *thought* apart from a *thinker*, or of a *feeling* or sensation without a soul or *feeler*. The last word is here used in a somewhat novel sense, but the context explains it. And as, by express definition, I class thoughts and mental emotions in the category of sensations, it follows that the soul (including mind and spirit) is the thinker as well as the feeler. This extension of the meaning of the word *feeling* or *sensation* may not be in accordance with the usage of psychologists or physiologists, but it is, I think, in accordance with the usage of all of us in ordinary speech; for don't we all employ such expressions as "I feel sure," "I feel the force of the argument," etc.? Besides, words are after all mere symbols, like the mathematician's x, y, z, to which we may give any convenient meaning that suits our purpose, provided the context leaves no doubt as to what that meaning is. Can an idea, or emotion, or sensation, or occurrence be consistently spoken of as occupying any definite position in space? Yes, provided the speaker's or writer's meaning can be inferred clearly from the context. No one misunderstands the meaning of such a remark as "her thoughts are far away with her absent children," and nobody in this case is under the delusion that the thinker is in one place, and her thoughts far away in another. Don't we speak of the site of such and such a battle, though the abstract conception of a battle has in itself no form or position apart from those of the combatants? The conclusion arrived at by some modern psychologists, that "the thoughts themselves are the thinkers," seems to me as much a linguistic inconsistency as would be the statements that "the combats themselves are the combatants," that "the receipts themselves are the receivers," and that "the speeches themselves are the speakers". The statement (which I have quoted from memory) that "the thoughts themselves are the thinkers," is, if I am not mistaken, due to Prof. James. I do not suggest that it expresses Prof. Taylor's opinion, but it seems to me that his opinion that the soul cannot consistently be spoken of as occupying any spatial position necessarily leads to the conclusion expressed by the quoted statement.

VII. MATTER AND MIND.—*Continued*.

But then it may be asked, "If the thought itself is not the thinker, where *is* the thinker?" Superficially considered, the question sounds absurd. "*There* is the thinker," it may be replied—"that one-armed and one-legged man, sitting on that bench in the park, with that far-off look in his eyes." "Yes, but how much of him constitutes the thinker?" Since

LINGUISTIC MISUNDERSTANDINGS.

he can still think, and think as well as when he had two arms and two legs, the missing arm and leg formed no indispensable portions of the thinker, any more than his hair, or his nails, or his clothes. He is still the "thinker" without them. How far can we carry on this slicing away of non-indispensable portions of the material body and still leave the "thinker"? What is very curious is that that one-armed and one-legged thinker will tell you that when the weather is damp he still feels pains in his fingers and toes, not merely in those of the arm and leg which he still possesses, but also in those of the arm and leg which have been amputated and no longer, so far as he is concerned, exist. Point out to him the absurdity—at least from the linguistic standpoint—of this statement, and he will own it; but yet he will assure you that if the evidence of his eyes, aided by that of his sense of touch, did not convince him of the contrary, he would be under the illusion that he still possessed the missing members also; for that the sensation of pain, apparently felt in the no longer existing toes and fingers, is exactly similar now, after the amputation, to that which, in damp weather, he had previously felt before it. The hasty physiologist will say that, in spite of the apparent direct evidence of his senses, a man really feels pain *in his brain*, the seat of all pain and pleasure, mental or physical, as of all other kinds of sensation or consciousness. The evidence for this conclusion was never quite convincing, and modern experiments and observations tend more and more to discredit it. The brain, it is true, is an important medium through which the soul receives sensations, and generally the possibility of thought, sensation or consciousness, depends upon its physical condition—just as generally—not always—the sensation of warmth depends upon the physical condition of the atmosphere. In the same condition of the atmosphere one person may experience an uncomfortable sensation of heat, while another, standing near him, may shiver and suffer from the very different sensation of cold. Generally speaking, the sensation of seeing depends—in part at least—on the condition of the eyes, that of hearing on that of the ears, and that of smelling on that of the nose; but if recent experiments made by eminent doctors and physiologists can be trusted, the soul—which is the sentient entity by definition, and therefore the real person—can, in certain exceptional cases, see, hear, and smell through other channels than the organs through which those sensations are usually transmitted. If then the eyes, ears, and nose are not absolutely indispensable organs for the transmission of their special sensations to the soul, why should the brain be an absolutely indispensable organ for the transmission to the soul of sensations in general, thoughts and mental emotions included? So far, if I am not mistaken, Prof. Taylor and I hold the same views; we neither of us believe that the brain or any other material part of the body feels or thinks; but we differ as to the propriety of assigning spatial position anywhere, either within the body or without the body, to the mysterious entity that does.

I may remark that, in their essential principles, the arguments which led some philosophers to deny the spatial existence of the soul lead others to deny, not only the spatial existence of the soul, but the real existence, except as a concept, of space itself, and as a necessary corollary, the real existence of time, with which the existence of space seems to be inseparably connected. Thus, the whole objective and material world is resolved by them into Berkeley's subjective immaterial "ideas," or, as I find it more convenient (because more widely suggestive) to call them, *sensations*. I prefer the word sensations, because some philosophers deny 'ideas' to the lower animals, whereas sensations, and therefore *souls*, as I define the word, belong, as they admit, to both. Now, the question at issue is this: Does matter exist except as a mere sensation, or as a mere collection of sensations? The true answer seems to me to be this, that matter, in its common conditions of solids, liquids, and gases, is a very prominent *cause* of certain sensations, but should not be identified with the sensations themselves; which, whether elementary or compound, should be considered as its *effects*. It leads to self-contradiction to identify a cause (whether known or unknown) with its effects. Just as we infer a cause from its effect or effects, as, for example, the existence of a lion from its roaring or footprints, so we infer the existence of matter (whatever its ultimate constitution) from one or more of the sensations which it produces on the subjective or sentient portion of us which I call the *soul*.

To make my meaning plainer, let me give a homely illustration similar to one of Berkeley's, but somewhat differently presented and developed. I see before me something which, from its appearance, I infer to be an apple; but as the evidence afforded by the sensation called *seeing* is not always reliable, when unconfirmed by other sensations, I approach, touch, lift, and smell it. Let us suppose that the fresh sensations thus obtained strongly confirm my former conclusion, but do not convince me entirely, as, after all, the supposed apple might be a small turnip shaped and painted by a clever practical joker to resemble an apple, and rubbed by him with apple essence so as to deceive the sense of smell. As a final test

LINGUISTIC MISUNDERSTANDINGS.

therefore I cut a small slice and bite and chew it. If the new sensation of taste thus added to my former data confirm my former conclusion, may the conclusion be now regarded as absolutely certain? Not quite, for in spite of the evidence afforded by all those sensations, it might conceivably happen that a fresh sensation, namely, the discovery of a hard substance like a peach-stone at its centre, would prove to me that this was a new kind of fruit which, while it possessed many of the properties usually connoted by the word *apple*, possessed others which the word, as hitherto understood, did not connote. What in such a case should we do? Extend the meaning of the word *apple* so as to include the new fruit? or invent a fresh name, such as a *peach-apple*, or *apple-peach*, to distinguish it? Neither course would involve any logical inconsistency, but the latter would be more convenient. Now, in this case, should the word 'apple-peach,' or 'peach-apple,' denote the total collection of sensations? Should it not rather denote their cause?—or, to speak more accurately, the more salient portion of the infinite number of their causes? For, be it remarked, a cause is never single. The number of causes producing any sensation is really infinite. We could not, for example, have the sensation of seeing without the vibrating ether, nor the sensation of smell without the air to convey the effluvium, and so on for other causes, of which a few are more or less known, but of which the infinite majority must, to our very limited faculties, remain for ever unknown and unsuspected.

Now, the analysis which we have here applied to a word denoting a particular kind of matter may also be applied to the term *matter* in general. The first question to be settled is: What property or properties do we attribute to *every* kind of matter, apples included? Scientists in general agree that whatever entity comes under the designation of *matter* must possess at least one property, the property variously named *weight*, *ponderability*, or *attraction*. In other words, every particle of matter in the universe attracts and is attracted by every other particle of matter. We notice this property, this tendency to mutual approach, according to a certain observed law, in several substances, we infer it in others, and we agree that all the substances, known or unknown, which possess it shall be called *matter*. But what do we mean here by the words 'notice' and 'infer'? 'Noticing' and 'inferring,' like all other percepts and concepts, come (according to my definition) under the general name of *sensations*, so that, otherwise put, *matter* is, as in the former analysis, the name of the most prominent of the many causes of these sensations.

But there is one perplexing entity which this definition of

matter appears to exclude. I speak of the hypothetical ether. I say "hypothetical," not because I doubt the existence of some space-occupying substance whose vibrations and other properties cause the sensation of seeing, as well as other sensations which (or whose causes) we connect more or less with such vague words as *heat, electricity, magnetism*, etc., but because the various properties commonly attributed to the ether are difficult to conceive as coexisting in the same substance; and also because no two scientists quite agree as to the list of properties which would constitute a self-consistent and satisfactory definition of the substance which the word *ether* is supposed to represent. Prof. Haeckel, in his *Riddle of the Universe*, speaks of the ether as "imponderable matter," which (if ponderability is the one quality by which we distinguish matter from other entities) is a clear contradiction in terms. We may consistently call the ether an imponderable *substance*, or an imponderable *entity*, and conceive of it as occupying a portion (finite or infinite) of abstract space, but we cannot consistently speak of it as imponderable *matter*.

Pushed to their extreme logical limit, the arguments against the spatial existence of the soul—of the entity that feels—would be equally valid against the spatial existence of the ether, or even of matter itself. If matter or the ether be space-occupying *causes* or *transmitters* of sensations, why should not the soul be similarly a space-occupying *receiver* of sensations? Why also should not one soul similarly transmit (more or less modified) the sensations received, or similar sensations to another soul, human, superhuman, or infrahuman? As to the exact volume of space (fixed or variable) occupied at any given moment by any individual soul, from an infinitesimal to an infinite, nobody has any data for asserting; but then nobody has any absolutely sure data for making a similar assertion about the space or volume occupied at any given moment by any individual material body, or even by the whole material universe. Suppose the volume or space, compared with some imaginary fixed and constant unit of reference, occupied by the whole material universe at this moment were rapidly diminishing, but that all our units of distance, area, volume, time, and forces were also changing in corresponding proportions. We should for ever, generation after generation, remain ignorant of the appalling circumstance. When our mile had become an inch we should still call it a mile, and our new inch would have the same ratio to the new mile as our old inch had to the old mile. When our year had become a second, our new second would be reduced in proportion, and so would the

LINGUISTIC MISUNDERSTANDINGS.

duration of our thoughts, actions and lives. When our bodies had become microscopic, in proportion to their present size, we should still consider them of the same size, weight, and volume as we consider them now; and, relatively to the respective new units of size, weight, and volume, the same numbers and fractions would represent them. Yet, when our material universe had thus shrunk (as regards volume) to an atom or an electron, with our stars, our planets, and ourselves, all inside it as at present, and at the same relative distances, we should still be logically obliged to consider it as real as we consider it now, and, for exactly the same reasons. On the other hand, the other mathematical, abstract, empty, soulless spatial universe of nothingness beyond would, as now, have any dimensions we chose to assign to it, subject, as now, to the condition that (solely for the convenience of symbolic reasoning and calculation) we should assume those dimensions to be infinite.

But, it may be asked, if the whole material and ethereal universe, and *a fortiori* our own bodies, might thus eventually become infinitesimal in comparison with our present units without our ever suspecting it, does not this reasoning arrive at virtually the same conclusion as that of the idealists, who maintain that matter, space, and time are mere conceptions of the mind, and that, except as mental conceptions, they have no real existence? To this I can only reply that as regards the empty, abstract, airless, etherless space of the mathematician, the conclusion seems to me correct, and that, since time, even as a concept, cannot well exist apart from space, the conclusion may be correct as regards time also; but I cannot quite see how we can consistently speak of *matter* as a mere concept. The question of size is wholly irrelevant. An ant, or even a microbe, is every whit as real as a whale or an elephant. We must define our words so as, if possible, to avoid self-contradiction, and if matter be defined as a *cause* or *transmitter* of a certain defined class of sensations, this cause or transmitter must (like the soul or *receiver* of sensations) be as real as the sensations themselves. This follows from three fundamental assumptions with which neither science nor logic can well dispense without linguistic inconsistency. They are: first, that every effect must have a cause or combination of causes; secondly, that an effect can never be its own cause, nor a cause its own effect; and, thirdly, that if any effect be considered real, its cause or causes must be considered real also. The words cause and effect are here used in their customary scientific sense, without any implication, affirmative or negative, as to the real existence of one single *First Cause*.

www.ingramcontent.com/pod-product-compliance
Lightning Source LLC
Chambersburg PA
CBHW051333230426
43668CB00010B/1246